Microtechnology and MEMS

Series Editors

The series *Microtechnology and MEMS* comprises textbooks, monographs, and state-of-the-art reports in the very active field of microsystems and microtechnology. Written by leading physicists and engineers, the books describe the basic science, device design, and applications. They will appeal to researchers, engineers, and advanced students.

More information about this series at http://www.springer.com/series/4526

Gregory Barbillon · Alain Bosseboeuf ·
Kukjin Chun · Rosaria Ferrigno ·
Olivier Français
Editors

Engineering of Micro/Nano Biosystems

Fundamentals & Applications

Springer

Editors
Gregory Barbillon
Faculty of Engineering
EPF-Ecole d'Ingénieurs
Sceaux, France

Kukjin Chun
Seoul National University
Seoul, Korea (Republic of)

Olivier Français
ESYCOM Laboratory
ESIEE Paris
Noisy-Le-Grand, France

Alain Bosseboeuf
Centre de Nanosciences et de
Nanotechnologies (C2N) CNRS
University of Paris-Sud, University
of Paris-Saclay
Orsay, France

Rosaria Ferrigno
Institut des Nanotechnologies de Lyon
Université Claude Bernard Lyon 1
Villeurbanne, France

ISSN 1615-8326
Microtechnology and MEMS
ISBN 978-981-13-6548-5 ISBN 978-981-13-6549-2 (eBook)
https://doi.org/10.1007/978-981-13-6549-2

This Springer imprint is published by the registered company Springer Nature Singapore Pte Ltd.
The registered company address is: 152 Beach Road, #21-01/04 Gateway East, Singapore 189721, Singapore

Contents

Contributors

Mehdi Ammar Center for Nanoscience and Nanotechnology (C2N), CNRS, University Paris-Sud, University Paris Saclay, Palaiseau, France

Alain Bosseboeuf Center for Nanoscience and Nanotechnology, CNRS, University Paris-Sud, University Paris-Saclay, Palaiseau, France

Kukjin Chun Department of Electrical and Computer Engineering, Seoul National University, Seoul, South Korea

Annalisa De Pastina EPFL-STI-IGM-NEMS, Lausanne, Switzerland

Anne-Laure Deman Institut des Nanotechnologies de Lyon (INL), CNRS, UMR 5270, Université Claude Bernard Lyon 1, Université de Lyon, Villeurbanne, France

Thi Hong Nhung Dinh C2N/CNRS, Université Paris Sud, Palaiseau, France

Norbert Dumas CNRS, ICube (UMR 7357), University of Strasbourg, llkirch, France

Magalie Faivre Institut des Nanotechnologies de Lyon INL-UMR5270, CNRS, Université de Lyon, Villeurbanne, France

Olivier Français ESIEE Paris, Cité Descartes, Noisy Le Grand, France

Marie Frenea-Robin Laboratoire Ampère, UMR 5005, CNRS, Université de Lyon, Villeurbanne, France

Denis Funfschilling CNRS, ICube (UMR 7357), University of Strasbourg, Strasbourg, France

Pierre-Yves Joubert C2N/CNRS, Université Paris Sud, Palaiseau, France

Jinhyun Kim Seoul National University, Seoul, Republic of Korea

Sung Jae Kim Department of Electrical and Computer Engineering, Seoul National University, Seoul, Republic of Korea

Sunghoon Kwon Seoul National University, Seoul, Republic of Korea

Bruno Le Pioufle ENS Paris-Saclay, Cachan, France

Damien Le Roy Institut Lumière Matière, Campus LyonTech La Doua, Université Lyon 1, Villeurbanne, France

Amos Chungwon Lee Seoul National University, Seoul, Republic of Korea

Howon Lee Seoul National University, Seoul, Republic of Korea

Hyomin Lee Department of Chemical and Biological Engineering, Jeju National University, Jeju-si, Jeju, Republic of Korea

Morgan Madec CNRS, ICube (UMR 7357), University of Strasbourg, llkirch, France

Jinseong Noh Seoul National University, Seoul, Republic of Korea

Stephane Serfaty SATIE/CNRS, Université Cergy-Pontoise, Cergy-Pontoise, France

Claire Smadja University of Paris-Sud, Orsay, France

Wilfried Uhring CNRS, ICube (UMR 7357), University of Strasbourg, Strasbourg, France

Luis Guillermo Villanueva EPFL-STI-IGM-NEMS, Lausanne, Switzerland

Yeong Eun Yoo Department of Nano Manufacturing Technology, Korea Institute of Machinery and Materials, Daejeon, South Korea

Chapter 1
Introduction

Alain Bosseboeuf and Kukjin Chun

Micro/nano biosystems are miniaturized systems able to handle and detect or analyze biological samples for medical, environmental and biological applications. They are most often considered as a subset of MicroElectroMechanical Systems (MEMS) and called bioMEMS even when they do not include mechanical micro or nanotructures because they often share similar technological processes. They are usually designated as micro total analysis systems (μTAS) when they include all steps required to perform biological sample collection and its chemical or biochemical analysis as usually done in a laboratory. A lab-on-chip (LoC) is an highly integrated version on a single chip of a μTAS function or of a whole μTAS.

From the end of seventies up to the beginning of nineties integrated sensing systems were mainly based on microactuators and on microsensors of physical quantities (pressure, acceleration, magnetic field, temperature,…). They were typically fabricated by processes coming from microelectronics completed by dedicated bulk and surface micromachining of silicon or Silicon-On Insulator (SOI) wafers and eventual assembly with a glass wafer. Since about 1990, these integrated sensing systems were designated as MEMS and this denomination is now commonly used for all kind of miniaturized sensing and/or actuation systems. Later on, at the end of nineties, optical functions and/or components were added to build optical MEMS and Micro OptoElectroMechanical Systems (MOEMS). Finally at the beginnning of 21st century, nanostructures and nanomaterials began to be integrated, and NanoElectromechanical Systems (NEMS) were born.

A. Bosseboeuf (✉)
Center for Nanoscience and Nanotechnology, CNRS, University Paris-Sud University Paris-Saclay, 10 Boulevard Thomas Gobert, 91120 Palaiseau, France
e-mail: alain.bosseboeuf@c2n.upsaclay.fr

K. Chun
Department of Electrical and Computer Engineering, Seoul National University, 1 Gwanak-ro, Gwanak-gu, Seoul 08826, South Korea
e-mail: kchun@snu.ac.kr

G. Barbillon et al. (eds.), *Engineering of Micro/Nano Biosystems*,
Microtechnology and MEMS, https://doi.org/10.1007/978-981-13-6549-2_1

Nowadays, MNOEMS have reached a high level of sophistication and maturity and are fabricated by using a large variety of materials and processes and mobile communication devices such as cell phones enjoy MEMS technology for radio frequency signal reception and transmission. MNOEMS represent 80% of the industrial market of sensing and actuating miniaturized systems with an annual growth rate larger than 10% for 10 years.

Microfluidic devices in glass technology and labs-on-chip appeared later than physical microsensors but with a rapid growth since the beginning of nineties. Since, a very broad range of specific fabrication technologies largely based on polymers, and of various sensing principles were developped. BioMEMS have a wide variety of applications for biology, medicine and agriculture. In 2000, microfluidics systems and bioMEMS represented only a few % of the MEMS industrial market but it is now more than 20% and a compound annual growth rate of 25% is predicted up to 2023. Nevertheless, μTAS and labs-on-chips still represent a small fraction of this market and the challenge is to overcome the commercialization, ethic and legal barriers to benefit of the high potential of these bioMEMS.

The size of biological objects is ranging from a few nm to a few hundreds of μm (Fig. 1.1). So a basic requirement of bioMEMS is the ability to handle and detect objects at micro/nanoscale: manipulation methods to separate, concentrate, filtrate, and identify cells, proteins, viruses and DNA using electrophoresis, electroosmosis, and electrowetting. This is notably achieved by integrating biosensors as depicted in Fig. 1.2, A biosensor typically includes a microfluidic system with micro/nano channels, pumps, valves, filters and electrodes to transport, concentrate, separate, mix, localize,… the biological objects or micro/nano particles carrying them. It also usually includes biofunctionnalized surfaces or traps to capture targeted species with a high selectivity. Detection is often performed in research works by optical microscopy or by other external means but to benefit from the full potential of bioMEMS, miniaturized transducers or indicators should be integrated to detect captured species and to convert the result into an electrical or optical signal, or to a specific color. As detailed in the different chapters of this book, various electrical, optical, mechanical, magnetical and coupled phenomena can be used for fluids and bio species handling and detection at micro/nanoscale.

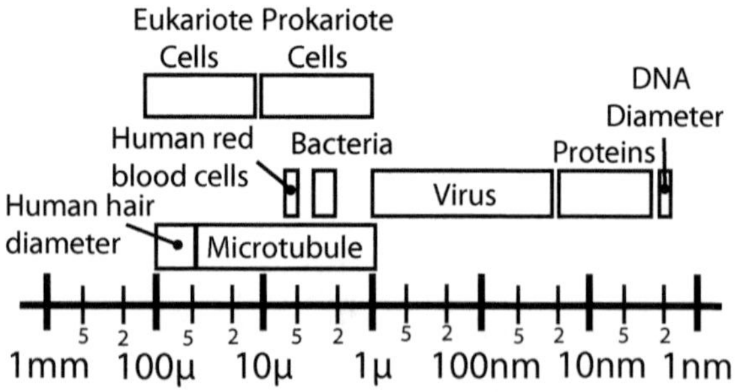

Fig. 1.1 Typical size range of biological species and comparison with hair diameter

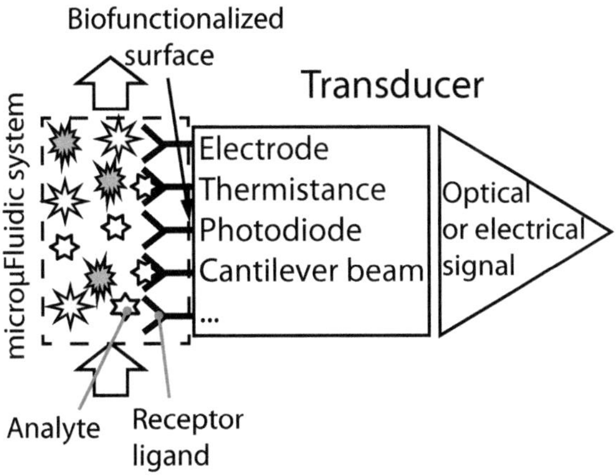

Fig. 1.2 Schematic drawing of a bio sensor

Owing to micro/nano fabrication technologies, bioMEMs integrating biosensors like in Fig. 1.2 can be highly miniaturized and fabricated at low cost by batch processing on various substrates (glass, silicon, polymers). This offers the possibility to get disposable systems that avoid maintenance operations like cleaning, purge, calibration, etc…and reduces the need of trained staff for their operation. BioMEMS have also many other proven or potential benefits when compared to conventional systems [1–25]:

- BioMEMS have the ability to handle small amounts of expensive chemical or pharmaceutical products and are sufficiently sensitive to detect small concentration of biological objects. This contributes to cost reduction and faster biochemical analyses but also reduces waste generation and minimize exposure to hazardous products. For human diagnostics, it reduces, for example, the amount of critical corporal liquids like blood that must be collected. For fundamental research, it also allowed, for the first time, operations at the single cell level.
- BioMEMS can allow faster measurements than with conventional systems because diffusion of chemicals, flow switching and diffusion of heat are faster at micro/nano scale. Eventually faster measurements can provide real-time process control and monitoring. As an example, integration of Polymerase Chain reaction (PCR) in a lab-on-chip allows a ten times faster DNA amplification.
- BioMEMS are able to integrate a large number of fluidics and biochemical functions on a single system and to perform parallel measurements. This largely reduces the cost of individual analysis and makes easier complex operations. For medical applications this allows a faster diagnosis of possible deseases and identification of the best suited antibiotic or antiviral.
- BioMEMs contributes to the reduction of human errors. Indeed, ease of use, reduced handling and automation of complex analyses should greatly reduce the risk of human errors.
- The small size and weight of bioMEMS favours a low consumption and make them more transportable.

All these abilities make BioMEMS suitable for point-of-care diagnostic that could revolutionize medicine by reducing the cost and time of bioanalyses as well as the training and infrastructure needs. This will make them usable by doctors in their offices, by nurses at hospitals and even by patients themselves. A rather unique successful example of the later case, is glucose monitoring for diabetics. For this application, portable systems based on lab-on-chip are already commercially available. Biosystems connected to smartphones begin to emerge.

This high potential of labs-on-chip and μTAS has been been largely recognized worlwide. As evidenced by the numerous review published in the last 15 years [1–25], a large amount of works, mainly focused for applications to human diagnostics and DNA analysis, or for the characterization of biological objects or bioprocesses, have been done. These works already demonstrated the large benefits of these miniaturized sensing systems. Some striking examples are the following one:

For DNA/RNA amplification and detection and for proteomics, labs-on-chip offer high gains in terms of speed. For example, LoC are able to sequence the human genome several order of magnitude faster than conventional systems and immunoassays with labs-on-chips can be done in a few tens seconds instead of a few tens minutes. Likewise, by integrating all steps, protein analyses can be reduced from hours to a few minutes. The small sise of fluidic microchannels in LoC is well suited for cell analysis and operations at the single cell level was demonstrated for the first time owing to labs-on-chip. Such a capability is demanded for physico-chemical modelling of biological processes and for the evaluation of parameters for cell changes identification. LoCs also provide high speed flow cytometry and cell sorting.

One holy grail of BioMEMs is to realize a low cost lab-on-chip able to perform a very large number of fully automated multiplexed bio chemical analysis of real complex biological samples like blood or saliva, A lot of work remains to be done to fully exploit the potential of bioMEMS and reach this goal. This tutorial book is intended for graduate students and researchers readers motivated by this challenge and having different backgrounds in physical science and electrical/mechanical engineering.

It provides the basic and advanced knowledge required to design, realize and test new LoCs or μTAS dedicated to the handling and sensing of biological samples. It is clear from Fig. 1.2 that building a state of the art fully integrated LoC or μTAS is demanding. Indeed, it is necessary to master micro/nanofluidics, biochemistry and bio analysis, sensing techniques, microfabrication processes and associated characterization and testing techniques. Practical and up-to-date information on all these topics is gathered in this book: First, Chap. 2 provides the basics of micro/nano fluidics and biology as well as information on the main ex-situ and in-situ characterization techniques. Then, fabrication and packaging technologies useful for BioMEMS are examined in Chap. 3. The four following chapters provide a detailed analysis of the main phenomena used in micro/nano fluidics systems and for bio-analysis. Finally the limitations of bioMEMS and current challenges will be presented in the concluding chapter.

References

1. H. Anderson, A. van der Berg, Microfluidic devices for cellomics: a review. Sens. Actuators **B92**, 315–325 (2003)
2. A.D. Sheehan, J. Quinn, S. Daly, P. Dillon, R. O'Kennedy, The development of novel miniaturized immuno-sensing devices: a review of a small technology with a large future. Analyt. Lett. **36**(3), 511–537 (2003)
3. R. Bashir, BioMEMS: state-of-the-art in detection, opportunities and prospects. Adv. Drug Deliv. Rev. **56**, 1565–1586 (2004)
4. A.C.R. Grayson, R.S. Shawgo, A.M. Johnson, N.T. Fkynn, Y. Li, M.J. Cima, R. Langer, A BioMEMS review MEMS technology for physiologically integrated devices. Proc. IEEE **92** (1), 6–21 (2004)
5. J. El-Ali, P.K. Sorger, K.F. Jensen, Cells on chips. Nature **442**, 403–411 (2006)
6. H. Craighead, Future lab-on-a-chip technologies for interrogating individual molecules. Nature **442**, 387–393 (2006)
7. D. Janasek, J. Franzke, A. Manz, Scaling and the design of miniaturized chemical-analysis systems. Nature **442**, 374–380 (2006)
8. P. Yager, T. Edwards, E. Fu, K. Helton, K. Nelson, M.R. Tam, B.H. Weigl, Microfluidic diagnostic technologies for global public health. Nature **442**, 413–418 (2006)
9. P. Abgrall, A.-M. Gué, Lab-on-chip technologies: making a microfluidic network and coupling it into a complete microsystem—a review. J. Micromech. Microeng. **17**, R15–R49 (2007)
10. J. Mairhofer, K. Roppert, P. Ertl, Microfluidic systems for pathogen sensing; a review. Sensors **9**, 4804–4823 (2009)
11. A. Arora, G. Simone, G.B. Slaieb-Beugelaar, J.T. Kim, A. Manz, Latest developments in micro total analysis systems. Anal. Chem. **82**, 4830–4847 (2010)
12. C.D. Chin, V. Linder, S.K. Sia, Commercialization of microfluidic point-of-care diagnostic devices. Lab Chip **12**, 2118–2134 (2012)
13. P. Crawand, W. Balachandran, Isothermal nucleic acid amplification technologies for point-of-care diagnostics: a critical review. Lab Chip **12**, 2469–2486 (2012)
14. J. Chen, J. Lin, Y. Sun, Microfluidic approaches for cancer cell detection, characterization, and separation. Lab Chip **12**, 1753–1767 (2012)
15. A. Foudeh, T.F. Didar, T. Veres, M. Tabrizian, Microfluidic designs and techniques using lab-on-a-chip devices for pathogen detection for point of care diagnostics. Lab Chip **12**, 3249–3266 (2012)
16. D. Mark, F. von Stteten, R. Zengerle, Microfluidics Apps for off-the-shelf instruments. Lab Chip **14**, 2464–2468 (2012)
17. P. Lisowski, P.K. Zazycki, Microfluidic paper-based analytical devices (μPADS) and micro total analysys systems (μTAS): development, applications, and future trends. Chromatographia **76**(19–20), 101–1214 (2013)
18. M.L. Kovarik, D.M. rnoff, A.T. Melvin, N.C. Dobes, Y. Wang, A.J. Dickinson, P.C. Gach, P. K. Shah, N.L. Allbritton, Micro total analysis systems: fundamental advances and applications in the laboratory, clinic and field. Anal. Chem. **85**(2), 451–472 (2013)
19. K. Manon, R.A. Joy, N. Sood, R.K. Mittal, The applications of BioMEMS in diagnosis cell biology, and therapy: a review. BioNanoscience **3**(4), 356–366 (2013)
20. A. Van Reenen, A.M. de Jong, J.M.J. den Toonder, M.W.J. Prins, Integrated lab-on-chip biosensing systems based on magnetic particle actuation: a comprehensive review. Lab Chip **12**, 1966–2986 (2014)
21. P. Ertl, D. Sticker, C. Kasper, V. Charwat, G. Lepperdinger, Lab-on-a-chip technologies for stem cell analysis. Trends Biotechnol. **22**(5), 245–253 (2014)
22. W. Jung, J. Han, J.-W. Choi, C.H. Ahn, Point-of-care testing (POVCT) diagnostic systems using microfluidic lab on-a-chip technologies. Microelectron Eng. **132**, 46–57 (2015)

23. F. Gorjikhah, S. Davaran, R. Saheli, M. Bakhtiari, A. Hasanzadeh, Y. Panahi, M. Emamverdy, A. Akbarzadeh, Improving «lab on a-chip» technique using biomedical technology: a review. Artif. Cells Nanomed. Biotechnol. **44**(7), 1609–1614 (2016)
24. E. Samiei, M. Trabizian, M. Hoorfar, A review of digital microfluidics as portable platforms for lab-on a-chip applications. Lab Chip **16**(13), 2376–2396 (2016)
25. D. Moschou, T. Tserepi, The lab on PCB approach: tacking the μTAS commercialization bottleneck. Lab Chip **17**, 1388–1405 (2017)

Chapter 2
Basics of Micro/Nano Fluidics and Biology

Olivier Français, Morgan Madec, Norbert Dumas, Denis Funfschilling and Wilfried Uhring

Abstract Mastering the basics of theoretical and experimental aspects of fluidics and biology is an obvious requirement for the design and evaluation of micro/nano biosystems. This chapter provides a review of the physics of micro/nano fluidics systems as well as basics of biochemistry and biological functions. It also gives detailed information on common experimental techniques used to characterize fluidic and biological systems as well as typical examples of application.

Keywords Microfluidics · Biological functions · Hydrodynamic flow · Capillary flow · Droplet generation · Biomolecular interactions · Enzymatic model · Micro particle image velocimetry · Fluorescence spectroscopy

2.1 Introduction

Micro-and nano-biosystems are, by nature, multidisciplinary systems involving both the common fields of engineering (such as electronics, mechanics, thermal physics, etc.) but also related domains such as biology and or fluidics. It is almost

O. Français (✉)
ESIEE Paris, Cité Descartes, 93160 Noisy Le Grand, France
e-mail: olivier.francais@esiee.fr

M. Madec · N. Dumas
University of Strasbourg/CNRS, ICube (UMR 7357), 300 bd Sébastien Brandt, 67412 Illkirch, France
e-mail: n.dumas@unistra.fr

D. Funfschilling
University of Strasbourg/CNRS, ICube (UMR 7357), 2 rue Boussingault, 67000 Strasbourg, France

W. Uhring
University of Strasbourg/CNRS, ICube (UMR 7357), 23 rue du Loess, 67037 Strasbourg Cedex, France

G. Barbillon et al. (eds.), *Engineering of Micro/Nano Biosystems*, Microtechnology and MEMS, https://doi.org/10.1007/978-981-13-6549-2_2

impossible to gather all the knowledge and know-how required for the development of such system within the same person or the same team. Thus, projects around biosystems often involve several teams from several area of expertise. In this context, the biggest challenge which come forward before technological locks, is the ability to communicate between the different actors of the project and understand the issues of each other.

For this purpose, this chapter aims at giving the reader the theoretical background in the domain of fluidics and biology and detailed information on some experimental techniques used for micro/nano biosystems assessment. It is composed of brief introductions on different concepts on which the reader can rely in order to understand the other chapters of this book. Extra bibliographic resources are explicitly suggested to the reader throughout the chapter for an in-depth learning of specific aspects that are only overviewed here. The chapter is divided in four parts.

The first section deals with the physical concepts related to fluidics phenomenon at micro and nanoscale. The impact of miniaturization on standard hydrodynamics laws (Navier-Stokes equation, Poiseuille flow, convection and diffusion in microfluidics channels) are discussed in the four main domains of microfluidics, i.e. the continuous flow (lab-on-chip), the centrifugal flow (lab-on-disc), the digital flow (droplet microfluidics) and the capillary flow (paper-based microfluidics).

The second section is an introduction to biology and biomolecular interactions. It starts with a brief reminder on biochemistry (atoms, molecules, functional groups, proteins). Then, the relationship between the structures of the proteins and the associated biological functions is demystified. In particular, the role of DNA, RNA, enzymes and antibodies are explained. Metabolic pathways and gene regulatory networks, which realize biological functions are also discussed. Finally, mathematical tools that are commonly used to model and simulate such systems are also presented.

The two last parts are focused on measurement techniques, and more specifically the micro-particle image velocimetry technique (MicroPIV) and some ex situ characterization techniques mostly based on optical properties. Standard PIV techniques as well as the consequences of the scale down to microPIV are discussed. First, the properties of fluids at microscopic level (density, viscosity, surface tension, interfacial tension, contact angle) are reminded. Then, experimental aspects and the main applications of microPIV (e.g. droplet formation, blood flow measurement, wall shear stress measurement, rheology, etc.) are described. The ex situ characterization techniques considered are colorimetry and fluorescence spectroscopy. For both, the physical principles, the experimental set-up, the characteristics of elements that compose the set-ups and their applications are described.

2.2 Microfluidics and Micro and Nanoscale (Olivier Français)

Microfluidic deals with the flow study at microscale and its implementation within micro-channel networks. The applications concern a large number of industrial domains such food sector, biotechnologies or chemical analysis with the advantages of small volumes and well-controlled manipulation. Microfluidic is at a crossroad between biology, physic, chemistry and engineering. It benefits from progress in microelectronics for its achievement and physical effects that occur at this scale with the size reduction. Nowadays, microfluidic can be seen as a toolbox for scientific research [1] but also for industrial engineering [2].

Concerning the domains associated, the microfluidic world can be divided into four principles (Fig. 2.1): (i) continuous flow [3] (case of classical microfluidic used in Lab On Chip where pressure driven handle the flow), (ii) centrifugal flow [4] (case of Lab On Disc devices based on spinning effect), (iii) digital flow [5] (case of two phase flow with droplet generation) and (iv) capillary flow [6] (associated to paper based microfluidic).

In this section, the Physics behind these four categories will be covered from a basic point of view, having in mind the impact of the miniaturization [7]. First of all, continuous hydrodynamics must be discussed taking into account the scaling factor of miniaturization. In fact, fluidic flow at micro-scale is mainly driven by pressure and viscous losses. As consequences, the Navier-Stokes equation governs the mass transport and the diffusion associated. For the diphasic flow (droplet), capillary phenomena dominate and enable stable droplet generation. With centrifugal flow, inertial effects are exploited to move liquid within microchannel embedded inside the disc. Finally, the surface tension drives the fluid front inside paper and thus induces liquid and mass transport towards analysis zone.

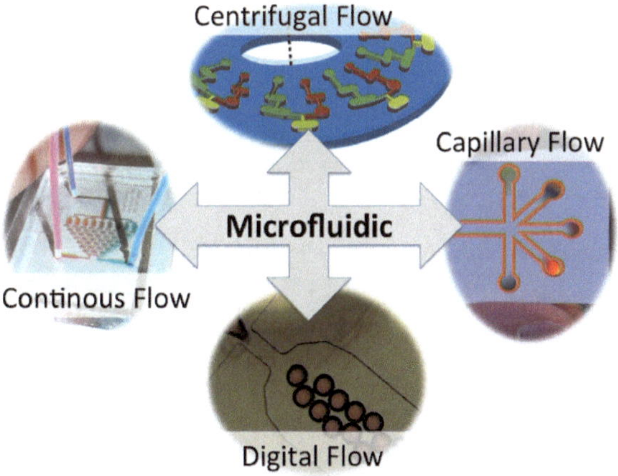

Fig. 2.1 Microfluidic devices: using physics to handle liquid at micro-scale

2.2.1 *Taking Benefits from Size Reduction Within Microfluidic Devices*

2.2.1.1 Scaling Laws

Flow behavior in miniaturized channels can be estimated using dimensionless parameters, which indicate the predominant physical phenomena, by a ratio study. These scaling laws rule the world, define life behavior and have many applications concerning miniaturization [8].

As a first approach, with size reduction, volumetric effects will decrease faster compare to surface and linear effects. The classical effects of fluid physics at human scale may be completely opposite at microscale. To illustrate the scaling laws, one example is the capability for insects to walk on water compared to human that sink into it. The physic is the same for the two scales, but the consequence not… It's a matter of balance between the gravitational force (F_g), which depends on mass of the object and thus depends on three power with size ($F_g \approx [s]^3$), and the capillary force (F_c) associated to the surface tension coefficient multiplied by the contact perimeter, with a dependence of one power with size ($F_c \approx [s]^1$).

Thus, at microscale the capillary force dominates compare to gravity force. The ratio between the capillary and gravity forces increases with the size reduction. It becomes greater than one at millimeter scale.

Based on this scaling approach, the predominant physics for liquid handling at microscale can be predicted and confirmed with the classical parameter values used in the field of microfluidic.

The volumes involve in microfluidic are in the order of the nano-liter. A microchannel with a hydraulic diameter D_H of 50 μm and a length of 1 cm corresponds to a volume of 20 nano-liters. To compare, a metric droplet corresponds to 50 μl, which fills 25 m of a channel with D_H = 50 μm.

2.2.1.2 Fluid Physics at Microscale: Dimensionless Parameters Analysis

At microscale, classical physical laws can be used but the predominance forces or effects must be carefully examined to well understand the principle of microfluidic device [9].

Reynolds number: R_e

The Reynolds number (Re) represents the ratio between fluid inertial forces and viscous forces (Eq. 2.1). Its value gives the regime of the flow. For a value below one, the regime is laminar and thus reversible with well-defined and stable flow path. Above three thousands, the regime is turbulent and gives complex pathway for the flow velocity line, which is totally instable [10].

Re is defined by the formula:

$$R_e = \rho \frac{VL}{\eta} \rightarrow [S^2] \tag{2.1}$$

As L represents the characteristic length of the channel and V is a distance over the time, the dependence with size for Re is a two power. It means that reducing the size by a factor 'k' will impact Re value by a factor k^2 and tends to greatly reduce its value.

From values used in microfluidic (see Table 2.1), Re varies in the range:

$$0.001 < \text{Re} < 0.2 \tag{2.2}$$

Consequently, it means that the flow within microfluidic channel can be well controlled and studied with a deterministic behavior (Fig. 2.2). The flow is laminar; the viscous effect dominates compared to inertial effect.

Peclet number: P_e

The Peclet number (P_e) characterizes the nature of the mass transport by a ratio between the time of diffusion (t_D) and the time of convection (t_c) (Eq. 2.3) [11]. The diffusion time, due to the Brownian's motion, is inversely related to the diffusion coefficient (D) of the molecule while the convection time is directly linked to the flow velocity (V). For a distance 'L' of transport, the expression of t_D and t_c are:

Table 2.1 Typical value parameters encountered in microfluidic

Channel hydraulic diameter	D_H	10 – 200 μm
Average flow velocity	V_o	0.1 – 10 mm.s^{-1}
Water dynamic viscosity (20 °C)	η	1e^{-3} Pa.s
Water density	ρ	1e^3 kg.m^{-3}
Surface tension (20 °C): – Water/Air – Water/Oil	σ	72.94 mN.m^{-1} 25 mN.m^{-1}
Diffusion coefficient in water: – Hemoglobin (Hb) – DiOxyde (O_2) – Sodium ion (Na^+) – Fluorescine dye – 10 nm nanoparticle	D	0.06e^{-9} m^2.s^{-1} 2.0e^{-9}m^2.s^{-1} 13.3e^{-10}m^2.s^{-1} 5.2e^{-10}m^2.s^{-1} 2.2e^{-11}m^2.s^{-1}

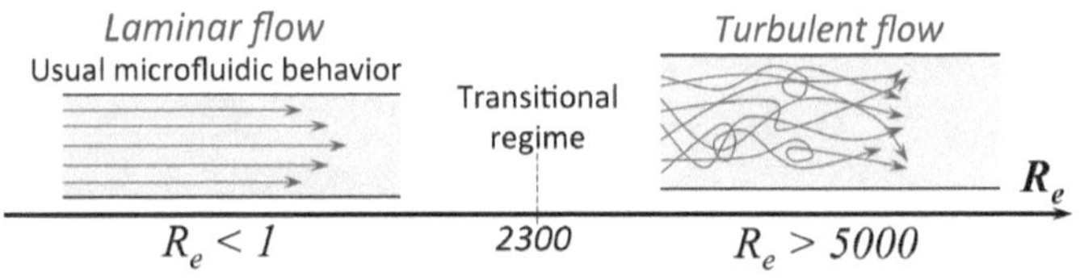

Fig. 2.2 Flow behavior in relation with the Reynolds number

$$t_D = \frac{L^2}{D} \text{ and } t_c = \frac{L}{V} \tag{2.3}$$

It gives the expression for P_e:

$$P_e = \frac{t_D}{t_c} = \frac{LV}{D} \tag{2.4}$$

For a value of P_e above 1, the flow is mainly convective and the mass transport is governed by the streamlines. Below 1, diffusion dominates and the molecules spread within the medium, the velocity having no influence on the transport, at the scale of a distance 'L'. As for Reynolds number, its size dependence is a two power, which tends to give low value for Pe, considering a mass transport over a distance 'L' (Eq. 2.4). Combined with the laminar flow behavior, it means a straight mass transport where mixing between two mediums will not be very efficient, as only diffusion occurs.

Advantages will be the design of chemical gradient generator taking into account the diffusion law within a laminar flow [12]. Applications are towards biology (chemotaxis analysis), particle handling in chemistry [13] or study of reaction in the case of molecule synthesis [14].

Capillary number: C_a

In case of a biphasic flow, the behavior between the two liquids (usually water and oil) can be estimated using the capillary number (C_a) [15]. C_a is the ratio between the viscous forces and the capillary forces (Eq. 2.5). It is defined with the expression:

$$C_a = \frac{\eta V}{\sigma} \tag{2.5}$$

The capillary force tends to reduce the interfacial area, in opposition the viscous force tries to extend it. At C_a number below one, droplet will be stable and reproducible (monodisperse) with a spherical shape. The flow behavior is similar to a dripping jet. With C_a number above one, droplets are stretched and unstable. The flow behaves with jetting shape between the two phases.

In microfluidic, as C_a is directly proportional to the scale of the object, size reduction favorises low value for C_a. Usual value for C_a are between 0.001 and 1.

Bond number:

The Bond number (Bo) compares the gravity force over the capillary force [16]. Its formula is given by:

$$Bo = \frac{\rho l^2 g}{\sigma} \tag{2.6}$$

Based on this relation, the capillary length l_c is defined when Bo = 1:

$$l_c = \sqrt{\frac{\sigma}{\rho g}} \sim 2.6\,\text{mm} \tag{2.7}$$

Thus, at microscale the capillary is dominant (Bo < 1) and can be used: (i) to travel liquid (Paper based microfluidics) without real impact of the gravity or (ii) to get round shape droplet with no deformation induce by their weight.

2.2.2 *Micro-Hydrodynamic Flow*

2.2.2.1 The Navier-Stokes Equation

Hydrodynamic of liquid flow remains fluid down to several nanometers. So in classical microfluidic, the fluid is assumed to be a continuous medium. Considering an incompressible Newtonian fluid, the Navier-Stokes equation (Eq. 2.8) can be used to describe the velocity field [17]:

$$\rho\left[\frac{\delta\vec{v}}{\delta t} + \left(\vec{v}.\vec{\nabla}\right)\vec{v}\right] = -\vec{\nabla}P + \eta\Delta\vec{v} + \vec{f}_{vol} \tag{2.8}$$

The two left terms of the equation deals with the inertial effect, which can be neglected within microfluidic channel at steady state. The right terms corresponds to the pressure force, the viscous force and the external volumetric forces where the gravity can be neglected.

2.2.2.2 Poiseuille Flow: Stokes Equation

At steady state, taking into account the Reynold's number low value (<<1) meaning that inertial term can be neglected, the Navier-Stokes equation can be reduced to its simple expression with no external force:

$$\vec{\nabla}P = \eta\Delta\vec{v} \tag{2.9}$$

Equation 2.9 corresponds to Poiseuille flow. It shows the balance between the pressure force and the viscous force.

Case of a cylindric channel

It can be easily solved in case of a circular channel, considering a no slipping boundary condition.

Supposing a stationary laminar flow in the z direction within a channel with a radius R_h, due to symmetric simplifications the velocity profile only depends on radius direction and the pressure drop is on the 'z' direction. In these cylindric coordinates, Eq. (2.9) becomes:

$$\frac{\delta P}{\delta z} = \eta \frac{1}{r}\left(\frac{\delta}{\delta r}\left(r\frac{\delta v_r(r)}{\delta r}\right)\right) \tag{2.10}$$

The pressure drops linearly along the channel. For a channel length L, with boundaries conditions such as p(z = 0) = P_o and P(z = L) = 0 given a pressure evolution:

$$p(z) = -\frac{P_o}{L}(z - L) \to \frac{\delta P}{\delta z} = -\frac{P_o}{L} \tag{2.11}$$

Supposing no slipping condition along the wall channel ($v(R) = 0$) and maximum velocity value at r $= 0\left(\frac{\delta v(0)}{\delta r} = 0\right)$, the velocity profile expression is then:

$$v_r(r) = \frac{-P_o}{2\eta L}\left(\frac{r^2 - R^2}{2}\right) = \frac{P_o R^2}{4\eta L}\left(1 - \frac{r^2}{R^2}\right) \tag{2.12}$$

The velocity profile has a parabolic shape within the microchannel (2.12). The maximum velocity v_{max}, the average velocity V and the flow rate Q are then:

$$v_{max} = \frac{P_o R^2}{4\eta L} \; and \; V = \frac{1}{R}\int_O^R v(r)dr = \frac{2}{3}v_{max} \tag{2.13}$$

$$Q = \int_O^R 2\pi r.v(r)dr = \left(\frac{\pi R^4}{8\eta L}\right)P_o \tag{2.14}$$

Case of squeeze film flow: Hele Shaw cell

In microfluidic, due to technical constraint, the channel section is rectangular. It can be defined by its width w and height h (Fig. 2.3).

The laminar flow is in the z direction and the velocity profile depends mainly on the y axis: assuming h < < w, edge effects are neglected in the x direction. Thus, following the same methodology of resolution used for circular channel, but here in Cartesian coordinates for thin rectangular channel, the solution of the stokes equation gives:

$$\frac{\delta P}{\delta z} = \eta\left(\left(\frac{\delta^2 v_z}{\delta^2 y}\right)\right) \Rightarrow v_z = -\frac{P_O}{2\eta L}\left(y^2 - \left(\frac{h}{2}\right)^2\right) et v_x = v_y = 0 \tag{2.15}$$

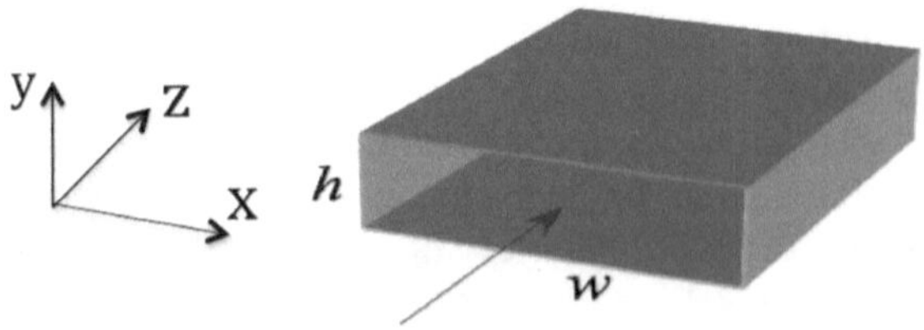

Fig. 2.3 Schematic view of a rectangular channel

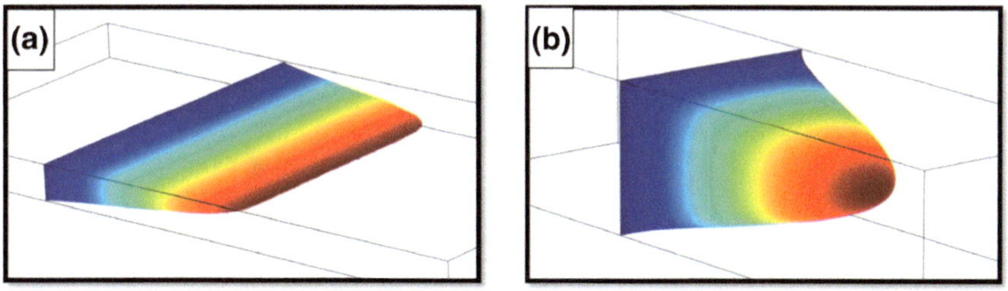

(c)	Circular	Planar	Square	Rectangular
profile		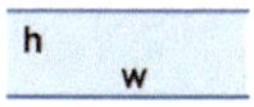	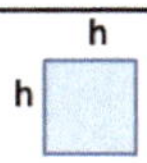	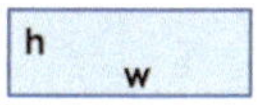
Parameters	Radius r	h<<w	h=w	0.2< h/w <1
Hydraulic resistance R_H	$\frac{8}{\pi}\eta L\frac{1}{r^4}$	$12\eta L\frac{1}{h^3 w}$	$28.4\eta L\frac{1}{h^4}$	$\frac{12\eta L}{1-0.63(h/w)}\frac{1}{h^3 w}$

Fig. 2.4 **a** Flow profile in a planar channel; **b** Flow profile in a square channel; **c** Hydraulic resistance expression with the channel shape

The profile obtained is parabolic (Eq. 2.15) with a maximum velocity at the middle of the channel (Fig. 2.4a). The relation between the pressure applied and the flow rate is then:

$$P_o = \frac{12\eta L}{wh^3}Q \tag{2.16}$$

In case of square channel, similar calculation can be done but simple analytical calculation can be achieved. The profile is still parabolic (Fig. 2.4b).

Hydraulic resistance

From previous calculation describing the laminar flow within circular or planar channel, the relation between the pressure drop ΔP and the flow rate Q (Eq. 2.14) is proportional (Eq. 2.16). This behavior is equivalent to a hydraulic resistance R_H defined as:

$$R_H = \frac{\Delta P}{Q} \tag{2.17}$$

The value of the hydraulic resistance R_H is defined by the channel profile (circular, square, rectangular…) [18]. In Fig. 2.4c, the R_H expression used in microfluidic is given for four cases.

The general formula for a rectangular channel can be obtained with a serial Fourier analysis and be implemented under mathematic software if accurate results are desired:

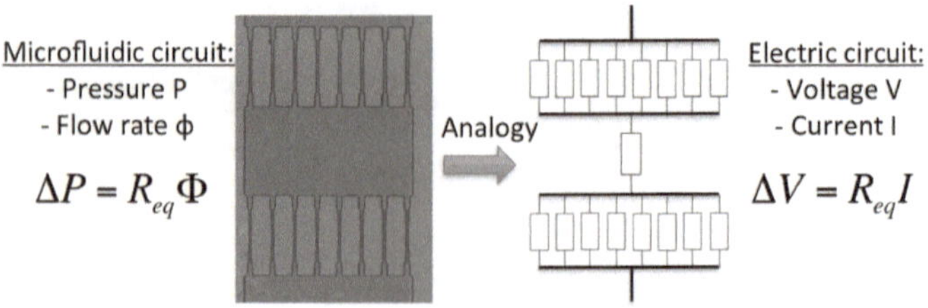

Fig. 2.5 Equivalent representation of a microfluidic network with the help of an electrical circuit

$$R_H = \frac{12\eta L}{wh^3}\left[1 - \frac{h}{w}\left(\frac{192}{\pi^5}\sum_{n=1,3,5,\ldots}^{\infty}\frac{1}{n^5}\tanh\frac{\pi n w}{2h}\right)\right]^{-1} \tag{2.18}$$

Analogy with an electrical network

Thus, a direct analogy of a microfluidic network can be done with an electric network. The electrical potential is equivalent to the pressure, the current corresponds to the flow rate. In consequence, this analogy can simplify the design of a microfluidic network and the study of its flow behavior. The Kirchoff's law can be applied (node and mesh laws) (Fig. 2.5) [19].

A microfluidic circuit can be analysed with an equivalent electrical schematic based on electric resistance for all pressure and flow rate distributions.

2.2.3 Mass Transport Within a Microfluidic Channel

2.2.3.1 Convection and Diffusion

As depicted in part 2.2.1, the flow in microfluidic is generally laminar. As consequence, two miscible liquids in contact will induce a bi-dimensional flow, the absence of turbulence restrains the mixing only by diffusion (Fig. 2.6) [20].

Taking advantage of this behavior is a way to design gradient generator of chemical species within microfluidic channel for chemotaxy analysis for example [21, 22]. It's also a way to study the chemical reaction kinetic between two species. In fact along the two liquids, the contact position becomes the time of reaction, which is a crucial parameter to clearly understand the evolution of a chemical reaction.

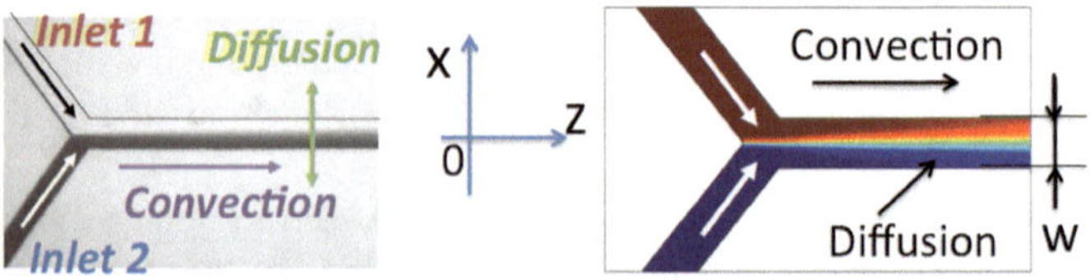

Fig. 2.6 Mass transport between two miscible liquids within a microfluidic network : convection with the flow velocity and diffusion by Brownian motion

2.2.3.2 Study of Concentration Profile

For microfluidic, in a straight channel under flow, the mass transport will be a competition between the natural diffusion of the particle in a transverse direction of the flow (here x direction) and its convection due to the velocity profile following the flow (here z direction). The fick's law and the mass conservation combination give the equation governing this transport:

$$\frac{\delta c(t)}{\delta t} = D\Delta c(t) - \vec{v}\vec{\nabla}c(t) \tag{2.19}$$

Solving this equation needs to take into account the velocity profile to determine the concentration evolution with time and position within the microchannel. It's a complex process to find the real solution of the mixing evolution, but with some specific approximations this 3D problem can be reduced to an approximate 1D problem:

- the Peclet's number associate to the flow is assumed to be high (Pe > 1) compare to the size of the channel under study.
- Having Pe > 1, the diffusion and convection process can be isolated and studied independently.
- Diffusion occurs in the x direction (transverse of the flow) and behaves like and 1D domain.
- Convection occurs in the z direction with a uniform average velocity. The time of diffusion will be connected with the position in the channel with the relation: $z = Vt$.

2.2.3.3 Mass Transport in the Context of Microfluidics

Thus the mass transport for the diffusion part can be reduced to a 1D problem (Eq. 2.20), in the transverse direction compared to the flow direction:

$$\frac{\delta c(x,t)}{\delta t} = D\frac{\delta^2 c(x,t)}{\delta^2 x} \tag{2.20}$$

Starting with a concentration profile at t = 0 (corresponding to z = 0):
For $x<0\, c(x,t) = C_o$ and $x > 0\, c(x,t) = 0$
The profile evolution of the concentration is depicted by:

$$c(x,t) = \frac{1}{2}C_o\left(1 - erf\left(\frac{x}{2\sqrt{Dt}}\right)\right) \tag{2.21}$$

where erf is the Gauss error function:

$$\mathrm{erf}(k) = \frac{2}{\sqrt{\pi}}\int_o^k exp(-u^2)du \tag{2.22}$$

It is common to use as diffusion length scale the parameter l_D, which gives a concentration close to half the final solution:

$$l_D = \sqrt{Dt} \tag{2.23}$$

Here starting from a centered profile [Co, 0] in x direction, the final solution will be a homogenous concentration of Co/2. To get the condition of half the final solution across the width (w) of the channel, the time of diffusion can be put in relation with the length of the channel L with the average velocity field V (Eq. 2.24).

$$\frac{w}{2} = \sqrt{Dt}\,\text{with}\,t = \frac{V}{L} \tag{2.24}$$

At steady state, the concentration profile with the microchannel is then:

$$c(x,z) = \frac{1}{2}C_o\left(1 - erf\left(\frac{x\sqrt{V}}{2\sqrt{Dz}}\right)\right) \tag{2.25}$$

This profile evolution is presented in Fig. 2.7. It shows the progressive diffusion of species along the microchannel.

The formula (Eq. 2.25) can be used to estimate the passive mixing between two medium (case of the T-channel), to study diffusion of nanoparticles [23], for estimation of D, or concerning sorting molecules with there size in relation with their coefficient of diffusion (case of H filter) [24].

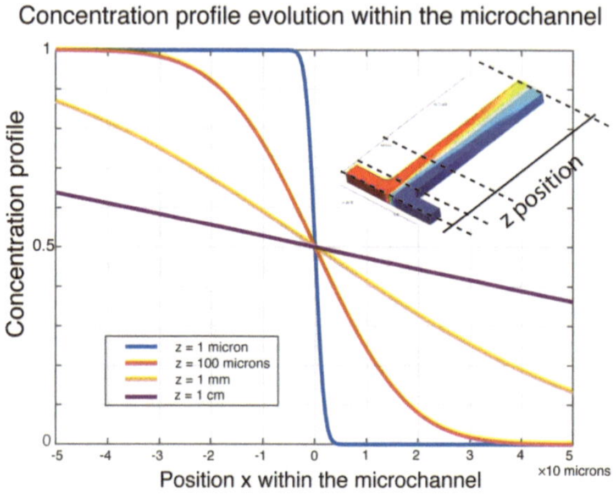

Fig. 2.7 Concentration profile evolution of a specie: time of diffusion is linked to convection travel in the channel

2.2.4 *Two Phase Flow: Droplet Generation Within Microfluidic Devices*

The use of droplet within microfluidic chip permits the development of device with high numbers of unit system on which biological or chemical species can be isolated [25]. Each droplet becomes a micro-reactor with its own properties. It reduces drastically the contamination or cross-pollution during the experiments. To do so, basic operations such as generation, fusion, breakup or storage of the droplets have to be mastered [26]. The capillary phenomenon is used to fabricate and control droplet within a microchannel. Considering immiscible fluids, such as water and oil, the surface tension between the water/oil interface will drive the capability of generating droplet and keep it stable.

2.2.4.1 Surface Tension: Basics

The origin of the surface tension is due to the molecular force discontinuity, which appears at the interface between materials (Fig. 2.8). The interactions between the molecules in the interfacial zone induce a surface energy to keep the interface stable. It's expression (γ or σ) is in [mN/m] or [mJ/m^2] corresponding to an energy per unit surface (δW) or a force per unit length (δF). It corresponds to the energy needed to increase of 1 m^2 the surface (Eq. 2.26).

As consequence, pressure discontinuity is the first effect of surface tension. In fact, surface tension corresponds to the pressure jump ΔP that is induced between the two faces of an interface between two liquids in relation with its curvature. Its expression comes from Laplace's law applied to a droplet.

$$\delta W = \sigma dS \, or \, \delta F = \sigma dl \text{ and } \Delta P = P^{+} - P^{-} = \frac{2\sigma}{R} \tag{2.26}$$

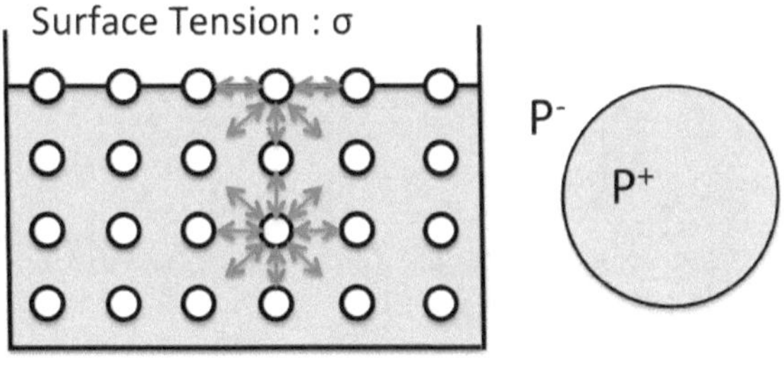

Fig. 2.8 Surface tension at the interface between two medium

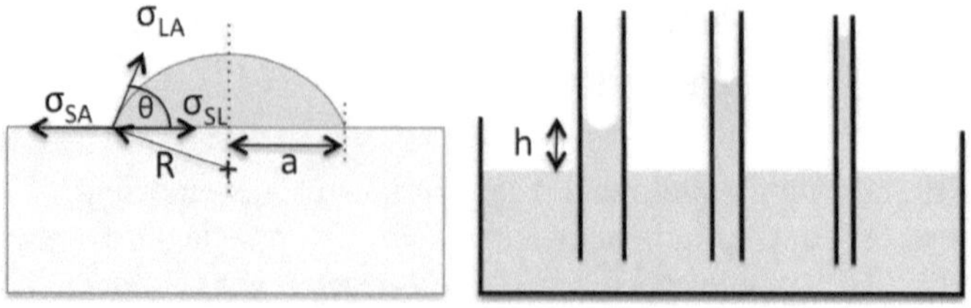

Fig. 2.9 Capillary effect induces by surface tension

2.2.4.2 Droplet on a Surface: Wetting Effect

The curvature of the interface gives the pressure discontinuity. This effect occurs also with capillary phenomena when a triple interface is achieved, corresponding to a Solid/Liquid/Air interface. This is the case of a liquid droplet on a surface. At the equilibrium, from a local point of view, the droplet contact angle θ (Fig. 2.9) depends on the three surface tension parameters with the relation:

$$\sigma_{SL} + \sigma_{LA} \cos \theta = \sigma_{SA} \tag{2.27}$$

The pressure within the droplet can be calculated taking into account the interface curvature. The pressure drop is:

$$\Delta P = \frac{2\sigma_{LA}}{R} \text{ with } R = \frac{a}{cos\theta} \tag{2.28}$$

The solid surface wetting property influences the curvature of the interface. Depending on contact angle value, the surface is Hydrophilic for $\theta < 90°$ and Hydrophobic for $\theta > 90°$.

A well-known effect is the Jurin's law describing the rising (or falling) of a fluid column within a hydrophilic (or hydrophobic) tube (with a radius a). The capillary pressure at the interface induces an equivalent height (h) of water column by hydrostatic equilibrium (Fig. 2.9). The height of liquid is given by:

$$h = \frac{2\sigma \cos \theta}{\rho g a} \tag{2.29}$$

2.2.4.3 Application in Microfluidics: In Situ Droplet Generation in Two Phase Flow

Taking advantage of miniaturisation (Re < <1: laminar flow, Bo < <1 and Ca <<1: capillary dominates and stable interface), combining two immiscible fluids such as water and oil is a way to generate highly reproducible droplets within microchannel (Fig. 2.10).

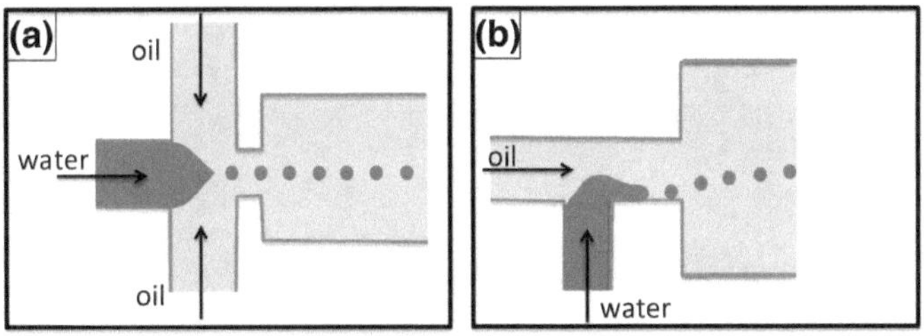

Fig. 2.10 Droplet generation by **a** flow focusing or **b** T-channel configuration

To modify the interface between the two fluids and generate a segmented flow composed of droplets, the shear stresses (viscous forces) must be able to break the surface tension (capillary forces) at the interface. It had been demonstrated by Thorsen et al. (2001) using a T Junction channel [27] (Fig. 2.10b). A column of fluid (water) is pinched and breaks by another fluid (oil) coming perpendicularly. The flow focusing design (Fig. 2.10a), based on the injection of the segmented phase coaxially into the continuous phase, had been proposed by Anna et. al in [28]. Droplets are emitted periodically at the nozzle of the flow focusing.

These two topologies are commonly used in droplet based microfluidic technology due to their simplicity of fabrication and use [29]. Highly monodisperse microdroplets are produced at kHz rates and generally made of aqueous solution dispersed in a continuous non-miscible inert carrier oil phase [15].

In order to stabilize the droplets and avoid coalescence between them, a surfactant is added to the continuous phase. Surfactants are molecules composed of two parts: one hydrophobic (tail) and one hydrophilic (polar head). It creates a layer at the interface that prevent coalescence between contacting droplets [5]. Of course, it affects the surface tension value by decreasing it.

In order to estimate the droplet radius formation, a balance between the surface tension and the viscous stresses can be calculated (Eq. 2.30). The first tends to reduce the interfacial area in order to minimize the energy, the second tries to extend the interface following the stream. At the limit of the droplet formation, the equilibrium can be express as the equality between the two stresses at the interface associated to the pressure effect:

$$\Delta P = \frac{2\sigma}{R} \leftrightarrow \tau_w = -\eta \frac{\delta v}{\delta y} = \eta \frac{2v_{max}}{h} \tag{2.30}$$

We can extract an estimate droplet radius expression:

$$R = \frac{\sigma}{\eta v_{max}} h \cong \frac{h}{C_a} \tag{2.31}$$

From that expression (Eq. 2.31), the capillary number is introduced to show the dependence of the droplet radius with C_a which is in relation with the balance between interfacial pressure (ΔP) and viscous stress τ_w.

2.2.5 Centrifugal Device: Lab on Disc

Compare to pressure driven microfluidic, where the flow is induced by differential pressure or by using syringe pump, spinning device used the centrifugal forces to drive liquids within microchannel (Fig. 2.11). Usually named as Lab On Disc microfluidic [30], it's similar to a Compact Disc with microchannel array embedded within its material.

2.2.5.1 Lab on Disc Principle

Inertial forces are exploited to induce the liquid displacement [31]. Considering a punctual mass (M) within the LOD submitted to the angular velocity Ω, three forces act on M:

- Centrifugal force: $\vec{F}_{cent} = M.r\Omega^2\overrightarrow{u_r}$, which is a radial force similar to the gravity $(g_c \sim r.\Omega^2)$ acting on the mass, oriented towards the external of the LOD.
- Coriolis force: $\vec{F}_{cor} = 2M\frac{\overrightarrow{dr}}{dt} \otimes \vec{\Omega}$, oriented perpendicular to both the angular velocity and liquid velocity vector.
- Euler force: $\vec{F}_E = Mr.\overrightarrow{u_r} \otimes \frac{\overrightarrow{d\Omega}}{dt}$

These three forces are handled in order to manipulate liquid within microchannel. In the context of microfluidics, the force density (f) is used and associated to the negative gradient of pressure with the relation: $\vec{f} = -\overrightarrow{grad}P$.

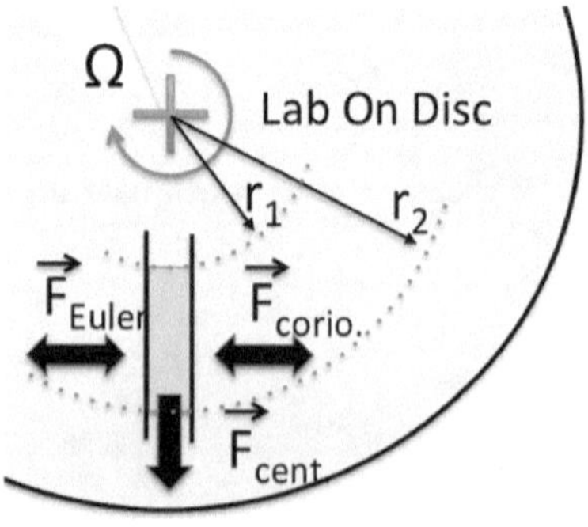

Fig. 2.11 Use of a Compact Disc format for centrifugal driven flow

2.2.5.2 Centrifugal Flow Within Microfluidic Channel

Considering a rotating channel filled with a liquid column from distance r1 and r2 to the center, the pressure density induces by the centrifugal force can be written as:

$$\rho r\Omega^2 = -\frac{\delta P}{\delta r} \quad (2.32)$$

The pumping pressure generated is thus:

$$\Delta P_{2-1} = -\frac{\rho\Omega^2}{2}\left(r_2^2 - r_1^2\right) \quad (2.33)$$

This differential pressure (Eq. 2.32), defined as the centrifugal pressure applied on the liquid, drives the liquid radially outward and can be associated to a laminar flow (microfluidic context) as studied in previous section. Supposing a circular channel (diameter D_H), the mean velocity expression obtained is:

$$V = \frac{\rho D_H^2 \Omega^2}{32\eta}\left(\frac{r_1 + r_2}{2}\right) \quad (2.34)$$

Based on a micro-channel with a diameter of 100 µm, with a filling of water from $r_1 = 2$ cm, $r_2 = 4$ cm and a spinning of 10 round/sec, the centrifugal effect gives:

$$\Delta P_{1-2} = 2.4\,kPa - V = 3.7\,\text{cm.s}^{-1}$$

2.2.5.3 Valve in the Flow: Capillary Effect

In order to control the fluid injection, the use of capillary effect or local hydrophobic surface enables to stop the flow under a limited pressure (Fig. 2.12), which above the flow can pass and continue [32].

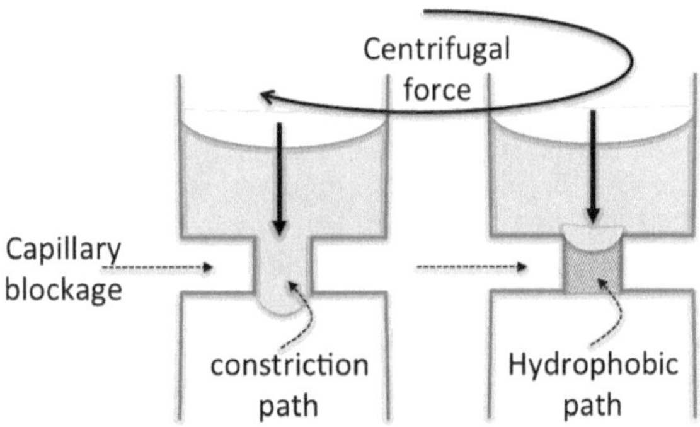

Fig. 2.12 Capillary effects to control the flow path with centrifugal force

Fig. 2.13 Use a chamber for metering sample volume

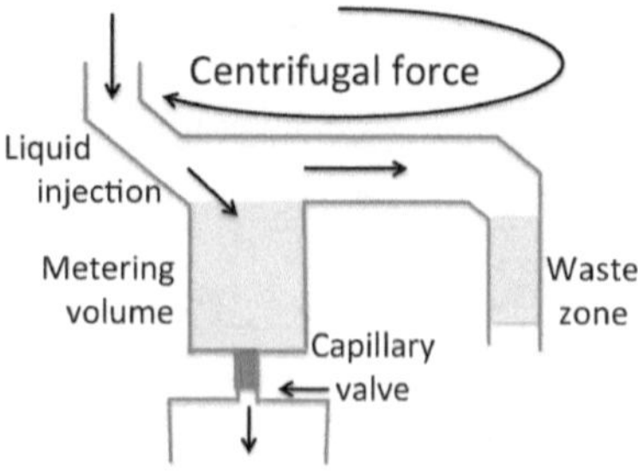

Using previous demonstration, supposing a circular channel, the valve blocks the liquid until (Eq. 2.35):

$$\Delta P_{2-1} = -\frac{\rho\Omega^2}{2}\left(r_2^2 - r_1^2\right) \leftrightarrow \Delta P_\sigma = \frac{4\sigma cos\theta}{d} \tag{2.35}$$

It give a burst angular velocity above which the liquid can pass (Eq. 2.36):

$$\Omega_b = \sqrt{\frac{8\sigma cos\theta}{d\rho\left(r_2^2 - r_1^2\right)}} \tag{2.36}$$

This principle is used for example when a quantified liquid need to be delivered (Fig. 2.13). Controlled volume [33] is stored in a chamber and is delivered by using an angular velocity above Ω_b.

In microfluidic, due to the rectangular section of the channel, the principle is still the same but the relation is a little more complex in order to take into account the cross section of the channel. For mixing application, Euler force is used by angular velocity oscillation with time $\left(\frac{d\Omega}{dt}\right)$. The rapid change in spinning velocity permits advective flow that induces the mixing [4].

2.2.6 *Paper Based Microfluidics: Lateral Flow*

2.2.6.1 Capillary Driven Flow Within Paper

In 2005, Pr. Whitesides had proposed the principle of developing microfluidic within paper [34]. Aim is to get low cost medical devices for point of care diagnostic. The idea is to use the capillary effects to induce diffusion of liquid inside paper and to shape the flow with "wax" integration. The paper is composed of pressed cellulosic fibres which behave like a hydrophylic porous material [35]. Here, the diffusion is guided by hydrophobic wall integrated within the paper by printing technic associated to thermal inclusion (Fig. 2.14) [6]. By this principle the fluid can be driven towards specific reaction zone in order to achieve chemical analysis.

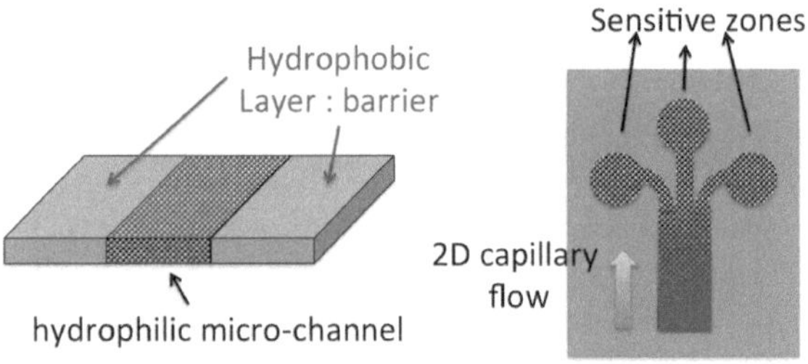

Fig. 2.14 Paper based microfluidic: lateral flow using capillary effect within paper drives the liquid

2.2.6.2 Simple Model Associated: Lucas-Washburn Law

The Lucas-Washburn law describes at a first approximation the fluid displacement within a paper. It gives the distance L traveled be the fluid front for a time t with R_p the average pore radius (Fig. 2.15) [36]. In this description, evaporation and gravity effects are not taking into account. The driving force is the capillary effects with the surface tension σ, the viscous force defines the speed rate with equivalent hydraulic resistance $\left(R_{Hp}\right)$ of the pore and the flow rate (Q).

Supposing an equivalent channel of radius R_p, filled of liquid on a distance L, the flow rate generated by the capillary forces is:

$$\Delta P_c = \frac{2\sigma\cos\theta}{R_p} = R_{Hp} * Q = \left(\frac{8}{\pi}\eta\frac{L}{R_p^4}\right) * \left(\pi R_p^2\frac{dL}{dt}\right) \tag{2.37}$$

The evolution of the fluid front to a distance L is thus:

$$L\frac{dL}{dt} = \frac{\sigma\cos\theta R_p}{4\eta} \Rightarrow L(t) = R_p\sqrt{\frac{t}{\tau_{adv}}} with\, \tau_{adv} = \frac{2\eta R_p}{\sigma cos\theta} \tag{2.38}$$

The dynamics of the fluid front is square root dependence with time (Eq. 2.38). The characteristic time of liquid advection within the paper depends on the porous radius and viscosity to surface tension ratio. This average 1D model for fluid transport with scaffold of cellulose is a first approach to control the interaction time between reagents in an assay (Eq. 2.36). More complex models need to include shape of the channel, the evaporation rate and gravity effect.

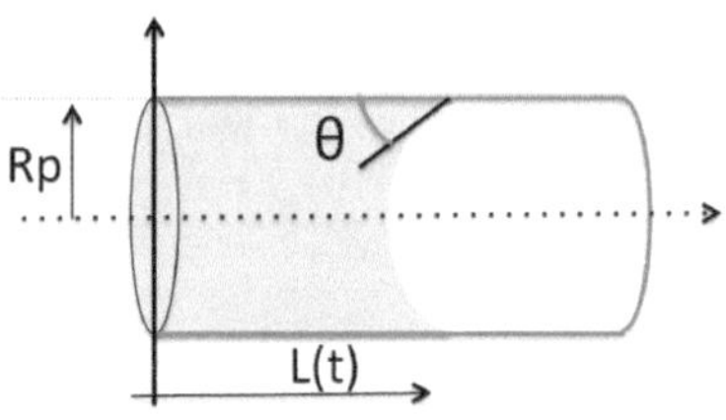

Fig. 2.15 Fluid front representation in a circular channel

2.3 Biomolecular Interactions (Morgan Madec)

This chapter aims at giving the reader a background in biochemistry required to understand the structure of the main chemical compounds of living matter (carbohydrates, proteins, DNA…), their role in biological functions as well as their interactions. The following questions are discussed: What a protein is made of? How does DNA give a protein? How in turn proteins act on living matter to perform biological functions? What is the role of other molecules such as carbohydrates or lipids? What is an antibody? An enzyme? How do they all interact? Finally, the last part of this chapter deals with mathematical modeling associated with those biological mechanisms.

2.3.1 From Atoms to Proteins

This section condenses the main principles of biochemistry. Only the concepts required for the understanding of biological mechanisms described along this book are discussed. Some of them have been simplified in order to the facilitate reader's understanding. For a deeper dive, we suggest the book Biochemistry written by Berg et al. [37].

2.3.1.1 Atoms and Molecules

Mendeleyev periodic table of chemical elements is composed of 118 entries. In this chapter, focus is put on six of them which are the most abundant in living matter. There are often designated by the acronym CHONPS for carbon (C), hydrogen (H), oxygen (O), nitrogen (N), phosphorus (P) and sulfur (S). They are summarized in Table 2.2. The residual is composed of ions and other atoms, the most relevant being the calcium which is as abundant as sulfur in some organisms.

Except for some elements, namely the noble gases, a standalone atom is not stable because of an incomplete external electron shell. To reach stability, they

Table 2.2 List of the most common atoms in living matter with associated chemical properties

	Symbol	Atomic number	Valence	Atomic mass	Electro-negativity	Abundance rate (%)
Hydrogen	H	1	1	1.008	2.30	63
Oxygen	O	8	2	15.999	3.44	24
Carbon	C	6	4	12.001	2.55	10
Nitrogen	N	7	3	14.007	3.04	1.4
Phosphorus	P	15	3 or 5	30.973	2.19	0.2
Sulfur	S	16	2, 4 or 6	32.060	2.58	0.1

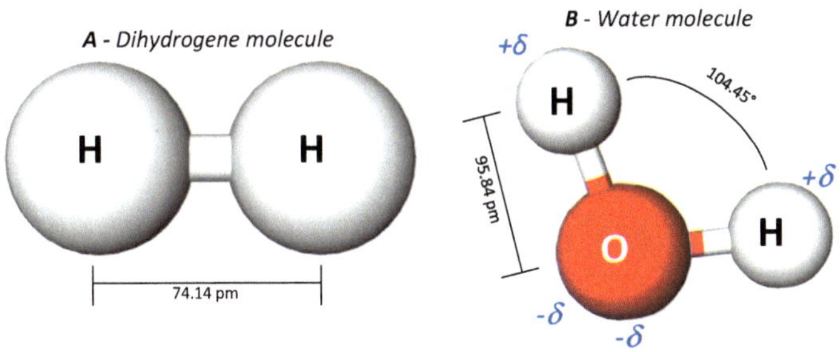

Fig. 2.16 Representation of Dihydrogen and water molecules

share electrons with other atoms in order to form molecules. A shared pair of electrons is a covalent bond and the number of covalent bond an atom should have to be stable corresponds to its valence (Table 2.2). The simplest example is the dihydrogen (H_2) molecule, composed with two H atoms sharing one covalent bond (Fig. 2.16a).

Electron sharing is not always equitable because some atoms can be more avid of electron than others. In this case, the electron pair is pulled by the most avid atom and becomes a polarized molecule. The "pulling power" of atoms is quantified by their electronegativity (Table 2.2). In the example of dihydrogen, both atoms have the same electronegativity and the covalent bond is equitably shared between atoms. The molecule is nonpolar. In the case of water (H_2O, Fig. 2.16b), as the electronegativity of oxygen is much higher than that of hydrogen atom, the two electron pairs are pulled by the oxygen atoms, and the molecule becomes polar. It is commonly admitted that a molecule is considered to be polar as soon as the electronegativity gap between the atoms involved in covalent bonds is over 0.4. As a consequence, H–N, C–O and C–N bonds leads to polarized molecules whereas H–C and all bonds with two similar atoms are considered as nonpolar.

Moreover, the arrangement of atoms in a molecule always corresponds to the configuration that minimizes its free energy induced by electrostatic forces at the microscopic level. For instance, a water molecule is V-shaped and both 0.95Å-long O–H bonds form a 104.45° angle. For small molecules, these configurations can be deduced from theoretical consideration. For larger molecule, computer software such as Avogadro [38] or MolView [39] should be used.

2.3.1.2 Functional Groups

There are plenty of molecules in living matter corresponding to different combinations of atoms. To analyze and predict the properties of a given molecule, it is often easier to split the molecule into functional groups. There are dozens of functional groups. A non-exhaustive list of those most commonly met in living matter is given in the following (Fig. 2.17).

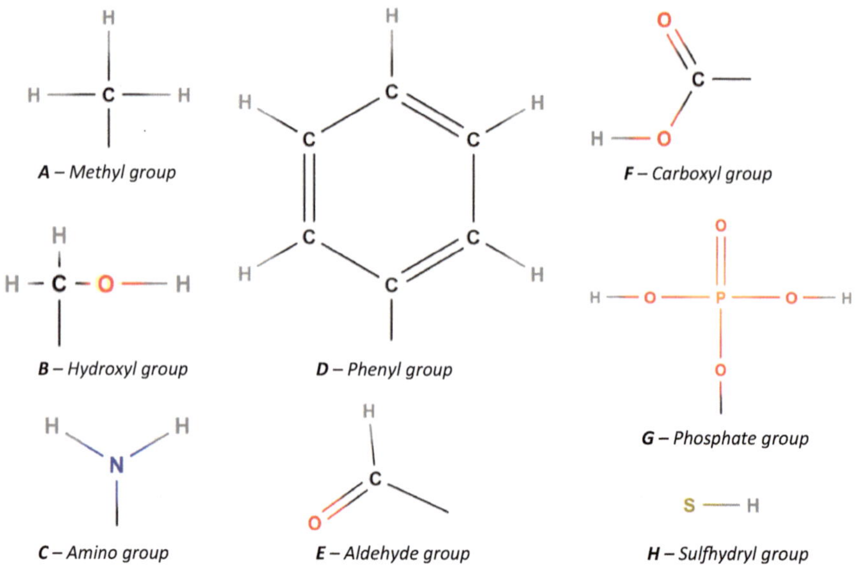

Fig. 2.17 Representation of the most common functional groups in living matter

i. *Methyl group* is only composed of a C–H bond (the number of H being adjusted to meet the valence of C). Methyl groups can be connected together to form a carbon chain with single, double or triple C–C covalent bonds. Molecules made of methyl groups are hydrocarbons.
ii. *Phenyl group* is a 6-C cycle with alternating single and double C-C bonds. Molecules constituted of such a group are benzene derivatives.
iii. *Hydroxyl group* is an O–H bond that replaces an H in a carbon chain. A carbon chain with at least one hydroxyl group is an alcohol (e.g. ethanol). The O–H bond renders a molecule polarized.
iv. *Amino group* is composed with N–H bond. An amino group can be inserted in a carbon chain, forming an amine (e.g. methylamine).
v. *Aldehyde group* is a methyl group in which a double C = O bond replaces a C–H bond. The carbon chain with at least one aldehyde group is an aldehyde (e.g. formaldehyde). C = O bond make them polarized.
vi. *Carboxyl group* combines on the same carbon an aldehyde and a hydroxyl group. Carboxyl groups can tail a carbon chain to form a carboxylic acid (e.g. acetic acid), which is also a polarized molecule.
vii. *Phosphate group* is composed of a phosphorus atom with two P–OH bonds, single (P–O) and a double (P = O) bond. One of the oxygen atom of the P–O bond can be connected to carbon chains, other phosphate groups or more complex molecules. Phosphate groups can also lose one or several H of P–OH bonds to become phosphate ions.
viii. The sulfhydryl group is an equivalent of the hydroxyl group where sulfur replaces oxygen.

2.3.1.3 Weak Bonds

Beside covalent bonds, other weaker interaction may occur between molecules or between atoms from the same molecule. These interactions are classified into four main categories:

i. *Ionic bonds* are the most straightforward. Ions are charged molecules for which one or several atoms have a number of covalent bond that are either above or below the expected one (valence). When a negatively charged ion meets a positively charged ion, they attract each other and form an ionic bond. For example, positive sodium ion bind to negative chloride ion in order to form NaCl (salt).
ii. *Hydrogen bonds* rely on the same principle except that they involve polar molecules instead of ions: the positive pole of a polar molecule attracts the negative pole of other molecules. The term "hydrogen bond" comes from the fact that because of its weak electronegativity, hydrogen is always the positive pole of a polar molecule. Hydrogen bonds occur for instance between two water molecules: one of the hydrogen atoms of a molecule attracts the oxygen atom of another one. These bonds are responsible for the crystalline structure of liquid water and ice.
iii. *Hydrophobic bonds* occur between hydrophobic molecules. As water is polar, nonpolar molecules cannot bind to water. As a consequence, nonpolar molecules are called hydrophobic. If several hydrophobic molecules are put in contact with water, they tend to clump up together by forming hydrophobic bonds in order to reduce their contact surface with water. Hydrophobic bonds occur for instance between oil molecules in water and lead to emulsions.
iv. *Van der Walls bonds* are induced by Van der Walls forces at atomic level and may happened between atoms or molecules, even if there are nonpolar.

2.3.1.4 Four Main Classes of Organic Macromolecules

In this fourth subsection, the four most common classes of macromolecules encountered in living matter are described.

Carbohydrates

Carbohydrates are macromolecules composed of carbon, hydrogen and oxygen. There are also called sugars and the most common one is the glucose ($C_6H_{12}O_6$). In living matter, glucose (as well as many other carbohydrates) is involved in a complex set of reactions (glycolysis metabolic pathway) in order to be converted into other carbohydrates, amino acids (see below), water and carbon dioxide. Glucose is also considered to be a source of energy for life as the glycolysis release a large amount of energy.

Lipids

Lipids are often assimilated to fats. Most of them are long hydrocarbon chains headed or tailed by specific functional group. The long hydrocarbon chain makes them hydrophobic or at the least amphiphilic (composed of a hydrophilic head and a long hydrophobic tail). Lipids have two main function in living matter.

Firstly as for the glucose, the degradation of lipids (especially the triglycerides) by specific metabolic pathways releases a great amount of energy. Secondly, lipids can make hydrophobic bonds with one other in order to form long damp-proof macrostructures called micelle, liposome or bilayer sheets. For instance the cell membrane consists of a bilayer of amphiphilic phospholipids whose hydrophobic tails are bond together and hydrophilic heads exposed to the inside and outside of the cell (cf. Sect. 2.4.2.3).

Amino acids and proteins

An amino acid is a molecule composed of a core with two carbons, one with a carboxyl group and the other with an amino group and a side chain (radical) which differs from an amino acids to another (Fig. 2.18a, b). There are only 20 different amino acids. They are listed in Table 2.3.

Amino acids form peptide bonds together to create long polypeptidic chains called proteins. A peptide bond is produced by the reaction between the N of the amine group of a given amino acid and the C of the carboxyl group of the other one. Thus, it releases a water molecule (Fig. 2.18c). The number of amino acids in a protein chain may vary from ten to thousands. Due to the electrical properties of each amino acid, the shape of the chain is very complex and can be declined at four levels:

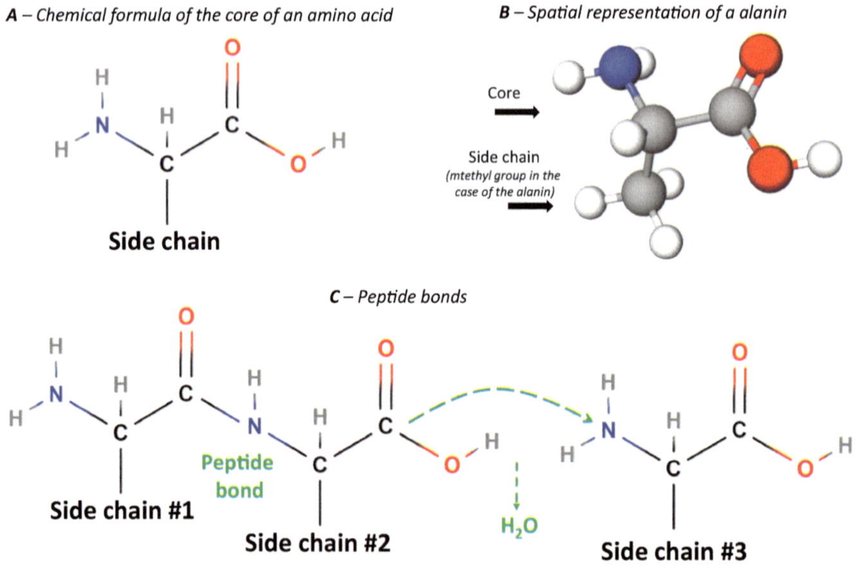

Fig. 2.18 General formula of an amino acid (**a**), 3D representation of the alanine (**b**) and the peptide bond reaction between successive amino acids to form protein (**c**)

Table 2.3 List of the 20 amino acids

Name	3-letter symbol	1-letter symbol	Side Chain formula	Property
Alanine	Ala	A	$-CH_3$	Hydrophobic
Arginine	Arg	R	$-(CH_2)_3-NH-CNH_2(NH)$	Basic (pos. ion)
Asparagine	Asn	N	$-CH_2-CONH_2$	Polar uncharged
Aspartic Acid	Asp	D	$-CH_2-COOH$	Acidic (neg. ion)
Cysteine	Cys	C	$-CH_2-SH$	Disulfide bridges
Glutamic Acid	Glu	E	$-(CH_2)_2-COOH$	Acidic (neg. ion)
Glutamine	Gln	Q	$-(CH_2)_2-CONH_2$	Polar uncharged
Glycine	Gly	G	$-H$	Structural flexibility
Histidine	His	H	$-CH_2-C_3H_2NNH$	Basic (pos. ion)
Isoleucine	Ile	I	$-CH-C_2H_5(CH_3)$	Hydrophobic
Leucine	Leu	L	$-CH_2-CH-CH_3(CH_3)$	Hydrophobic
Lysine	Lys	K	$-(C_2H_4)_4-NH_2$	Basic (pos. ion)
Methionine	Met	M	$(CH_2)_2-S-CH_3$	Hydrophobic
Phenylalanine	Phe	F	$-CH_2-C_6H_5$	Hydrophobic
Proline	Pro	P	$-C_3H_6$ looped	Structural rigidity
Serine	Ser	S	$-CH_2OH$	Polar uncharged
Threonine	Thr	T	$-CH(OH)-CH_3$	Polar uncharged
Tyrosine	Tyr	T	$-CH_2-C_6H_4OH$	Hydrophobic
Tryptophan	Trp	W	$CH_2-C_4NH-C_4H_4$	Hydrophobic
Valine	Val	V	$-CH-CH_3(CH_3)$	Hydrophobic

i. The *primary structure* corresponds to the sequence of amino acids.
ii. The *secondary structure* describes the way the side chains are organized. The organization is mainly driven by hydrogen bonds occurring between successive side chains. The two most common structures are the alpha-helix and the beta-sheet. The secondary structure is mainly influenced by the position of glycines and prolines in the chain.
iii. The *tertiary structure* is the way the protein is folded at the molecule scale. Folding is driven by a set of weak bonds described in 2.3.1.3. Protein folding can be predicted with the help of complex computing algorithms such as PyMOL [40].
iv. A protein is sometimes an assembly of several polypeptide chains called sub-units. The *quaternary structure* describe this assembly. Hemoglobin, a protein used to fix and transport oxygen in blood, is an example of tetramer composed of two couples of sub-units, the α-chain and the β-chains.

Nucleotides and nucleic acids

A nucleotide is molecule composed of a five-carbon sugar, up to three phosphate groups and a nitrogenous base. Nucleotides are the building blocks of nucleic acids.

The most famous nucleic acid is probably the deoxyribonucleic acid (DNA), which encodes the genome that can be considered as the blueprint of living matter. In this case, the five-carbon sugar is a deoxyribose 5-phosphate and the nitrogenous base is an adenine (A), a thymine (T), a guanine (G) or a cytosine (C) (Fig. 2.19a).

The sugar of a nucleotide can bind to the phosphate group of the next nucleotide in the chain (phosphodiester bond) in order to form a DNA strand (Fig. 2.19b). The structure of DNA is a double-helix composed of two complementary DNA strands, as predicted by Watson and Crick in 1953. In this structure, A always faces T and C always faces G. This complementarity is due to the fact that A and T exhibit 2 potential hydrogen bonds while G and C exhibit three (Fig. 2.19c, f). To identify the carbon is the deoxyribose, they are numbered from 1′ to 5′, as depicted in Fig. 2.19A. This notation is also used to identify the first (5’) and the last (3′) nucleotide of a DNA chain.

Binding and separation of DNA strands are respectively called hybridization and denaturation. The ratio of hybridized form over denatured form of DNA molecule is an empirical low which mostly depends on the temperature T, the length of DNA strand N, the ratio η of G-C pairs (as the triple hydrogen bond of G-C is stronger than the double for A-T) and the concentration of positive ions such as calcium $[Na^{+}]$ (which operate as a screen preventing the two DNA strands to repulse each other due to negatively charges backbone). The temperature at which 50% of the DNA strands are hybridized is given by [41]:

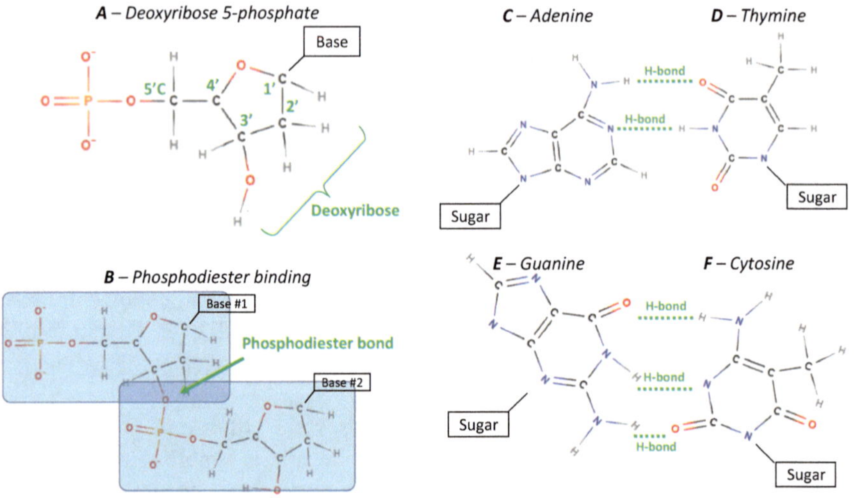

Fig. 2.19 Representation of a Deoxyribonucleotide (**a**), a phosphodiester bond between two deoxyribonucleotide (**b**) and the four nitrogenous bases of DNA (**c** to **f**). Hydrogen bonds between complementary bases are also represented in a green dotted line

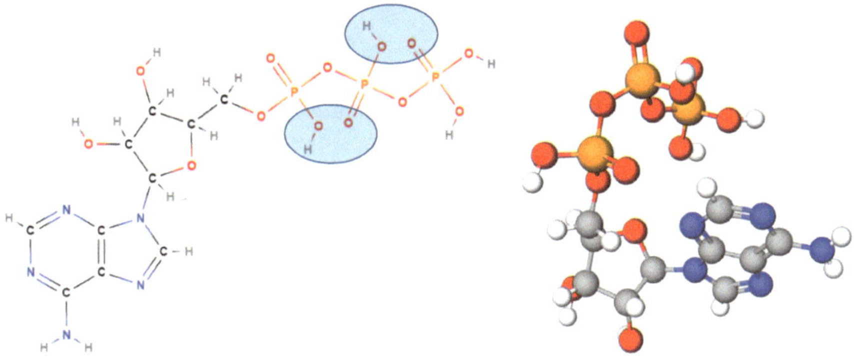

A – ATP Chemical Formula *B – ATP 3D structure*

Fig. 2.20 Adenosine Triphosphate Chemical formula and 3D representation

$$T_{50\%} = 81.5 + 16.6 \cdot \log_{10}([\mathrm{Na}^+]) - \frac{600}{N} + 0.41 \cdot \eta \tag{2.39}$$

The second most common nucleic acid is the ribonucleic acid (RNA) composed of a ribose (which resembles a deoxyribose except that a OH replace the H on carbon 2', see Fig. 2.34a), a phosphate group and a nitrogenous base (same as for DNA except that the uracil replaces the thymine). Relationships between DNA and RNA from genetic point of view will be described in the next subsection.

The last couple of nucleotides that deserve to be discussed are the adenosine triphosphate (ATP) and the adenosine diphosphate (ADP). For ATP, the sugar is a ribose, the nitrogenous base is an adenine and it is tailed by three phosphate groups (Fig. 2.20). Structures of ADP and ATP are the same except that ADP has only two phosphate groups. ATP's internal energy is very high due to the proximity of polarized P-O covalent bonds. ATP to ADP hydrolysis releases a large amount of energy which is often used to force reaction to occur (see Sect. 2.3.3.4). Conversely, ADP to ATP phosphorylation may also be coupled to reactions with very favorable energy balances in order to harvest released energy. In other words, the role of ATD and ADP in metabolism can be compared to the role of a battery in a hybrid vehicle (with the fuel and thermal motor corresponding respectively to glucose and glycolysis). ATP and ADP are used in every metabolic pathway: it is estimated that the mean mass of ATP in human body is about 250 g but total mass of ATP consumed and produced is about 50 kg a day.

2.3.2 *From DNA to Proteins*

The last part of this section complete the loop by describing how DNA and proteins are related.

2.3.2.1 A Brief History of Genetic

Genetic was born in the middle of the 19th century and is still an important topic of investigation topic [42]. Making a long story short, modern genetics started in 1863 with George Mendel who studied seven characteristics of peas. He pointed out that something (a gene) in living matter should be responsible of the inheritance of several characteristics from a generation to another. Moreover, from statistical analysis, he also made the assumption that each individual has two genes, one from every parent, and that a dominance relationship exists between the different versions of these genes. It was confirmed later by Thomas Morgan's experiments on the fruit fly. Morgan introduced two new notions: the fact that some characteristics depend of the sex of the individual and the fact that several gene are grouped in chromosomes.

From 1928 to 1952, many experiment (Frederick Griffith, Oswald Avery and Alfred Hershey) have been carried out in order to find the nature of the "something" that transmit characteristics from an individual to another. They conduct experiments with mice and mice killer bacteria. They first pointed out that the "killing" characteristics can be acquired by a harmless bacteria in contact with molecules extracted from killer bacteria. With this in mind, they isolated all these molecules and reintroduced them one by one in a harmless bacterium. They observed that the bacteria become a killer after introduction of DNA.

The last part of the mystery was to understand how DNA created the biological function. The notion of mRNA, tRNA, codon, genetic code … which will be described in the next session have been established during the 60s by several scientists such as Ochoa, Khorana, Holley and Nirenberg. Since the 60s, advances in biotechnology and computer science led to major breakthrough in the understanding of genetic mechanisms, the deciphering of the genetic code, the study of relationships between genes and some diseases, the actions of several proteins on metabolism and even the complete decoding of the about 25,000 genes in the human body (Human Genome Project kicked off in 1990 and achieved in 2003).

2.3.2.2 Transcription and Translation of DNA

The conversion from DNA to an active protein is a two (sometimes three) stage process. Firstly, one of the two DNA strands (namely the coding strand) is transcribed into a messenger RNA (mRNA) strand. Cells make a copies of the blueprint before manufacturing it instead of using its single original blueprint. The major actor of transcription is a specific protein, the mRNA polymerase, which reads one of the strand of DNA molecule (previously separated from the other strand by another molecule) and permits mRNA nucleotides to assemble together in front of the DNA coding strand. Of course, a cell does not transcript all its genes continuously. Most of the time, transcription is triggered by specific events and the transcription rate is regulated by the concentration of other molecules, the so-called transcription factors.

Secondly, mRNA is translated into an assembly of amino acid. Basically, three consecutive nucleotides in the mRNA sequence form a codon and each codon encodes for a given amino acid according to the genetic code (Fig. 2.21). With 64 (4^3) combinations for 20 amino acids, the genetic code is redundant. Translation consists in reading the sequence of mRNA three by three and assembling the corresponding amino acid. This operation requires a ribosome and transfer RNAs (tRNA). tRNAs are molecules that exhibit at one side an amino acid and at the other side an anticodon (i.e. the sequence of three nucleotides which are complementary to the codon which codes for this amino acid). As a consequence, tRNA's anticodon recognizes its cognate codon on mRNA strand and bind to it. Ribosomes are complex molecules made of dozens of proteins. The ribosome is composed of two subunits: the smallest is in charge of codon recognition on mRNA and the biggest one is in charge of making peptide bonds between the amino acids of each tRNA. The ribosome binds at the beginning of mRNA strand and slides along it up to the stop codon.

The third stage, which is optional, consist in adding post-translational modifications on the translated proteins to make them functional.

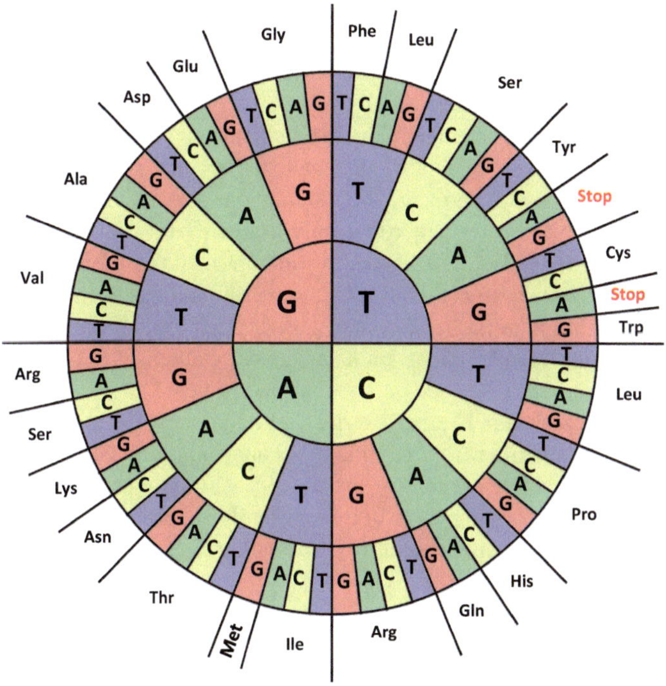

Fig. 2.21 The Genetic Code

2.3.3 *From Proteins to Biological Functions*

In this section, focus is put on proteins which are the main actors of biological functions. Basically, a biological function is induced by the binding of a protein with other molecules. This mechanism is described in the first section. The second and the third parts describe antibodies and enzymes which are specific proteins. Finally, the two last sections deal with regulations that may occur in metabolic pathways (regulation of the rates of chemical reaction) and/or at the transcriptional level (regulation of the expression of genes).

2.3.3.1 Protein-Ligand Binding

The shape of a protein is very important in biochemistry because it mostly determines the protein's biological function. From a macroscopic perspective, the shape of proteins can be highly diversified, but most of them exhibit at their surface specific amino acids sequences that form binding sites. Those binding sites have a shape and local electrostatic properties (e.g. positive or negative charges, propensity for hydrogen bonds ...) in order to bind small molecules called ligands. The analogy that is commonly used to represent these binding sites is that of the lock and the key. A given protein may have several binding site, each of them corresponding to the same or different ligands.

Protein-ligand binding is a natural and instantaneous phenomenon that occurs as soon as the protein and the ligand are put together in a solution. However, the reaction is not always total and there are different ways to quantify its efficiency. At the macroscopic level, we consider that only a given ratio θ of sites are occupied and that this ratio depends on the ligand concentration and an affinity parameter K_A. From a microscopic perspective, binding and unbinding occurs stochastically during time with a given probability and get balanced around the ratio θ.

Another crucial notion about protein-ligand binding is the conformational change (allostery). We have to keep in mind that proteins are flexible and reach a shape that minimizes their internal energy. When a binding site is occupied by a ligand, the distribution of energy may locally change and the shape of the rest of the molecule may also change in turn. In particular, other binding sites may be affected, thus increasing or decreasing their affinity with the ligand. Allostery of proteins give a natural way to control the affinity of binding sites for one molecule by the concentration of other molecules.

The mechanism is illustrated of Fig. 2.22 on the protein that exhibit 4 binding sites, two for the ligand L_1, one for a ligand L_2 and one for a ligand L_3. Positive cooperativity occurs between both L_1 binding site. When a binding site for L_1 is occupied, the shape of the other binding site for L_1 change, increasing its affinity (Fig. 2.22b). Negative cooperativity occurs between L_1 and L_2 sites. When a binding site for L_1 is occupied, the binding site for L_2 also change, decreasing its affinity (Fig. 2.22d). When the affinity decreases to almost zero, binding sites becomes competitive.

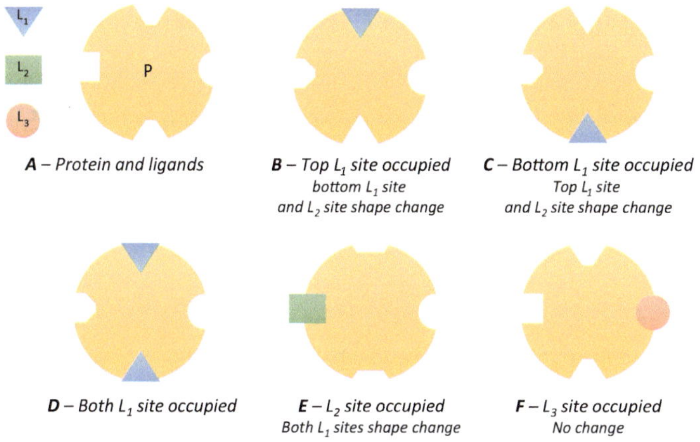

Fig. 2.22 Different conformations of the protein P depending on the binding of three ligands on protein binding sites

Finally, L_3 is not affected by the fact that L_1 is occupied. The binding site are independent (Fig. 2.22f). Conformational changes are symmetric and reciprocal. In our example, the increase in affinity of the top L_1 site when the bottom L_1 site is occupied is the same as that of the bottom L_1 site when the top L_1 site is occupied (Fig. 2.22c). Similarly, the decrease in affinity of the L_1 binding site when L_2 is occupied is the same as that of the L_2 binding site when L_1 is occupied (Fig. 2.22e).

2.3.3.2 Antibodies and Antigens

Antibodies are proteins used for the recognition of specific small amino acid sequences called antigens. The antibodies are Y-shaped (Fig. 2.23a) and are composed of four polypeptic chains: two heavy chains (HC) and two light chains (LC) interconnected by disulfide bridges. Each chain is composed of different

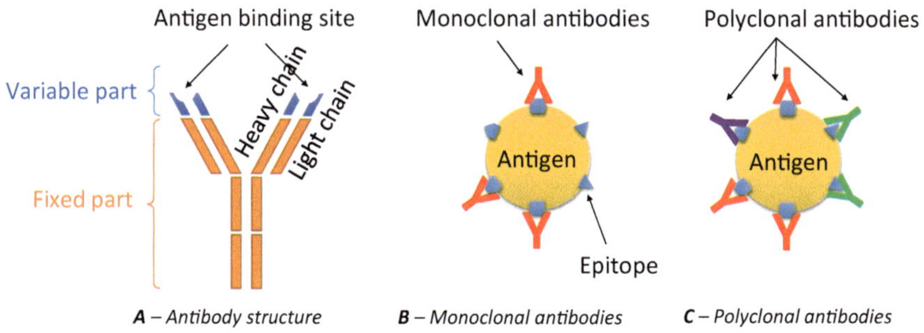

Fig. 2.23 Structure of an antibody (**a**) and distinction between monoclonal and polyclonal antibodies (**b** and **c**)

fragments: 3 or 4 constants and 1 variable for the heavy chains, 1 constant and 1 variable for light chains. The antigen that the antibody can recognize is determined by the two variable fragments. The binding between an antibody and an antigen leads to an immune complex mainly composed of hydrogen bonds. In practice, an antigen may have multiple binding sites (epitopes) for multiple antibodies.

There are several types of antibodies. The first distinction that is made is between the monoclonal antibodies and the polyclonal antibodies. Monoclonal antibodies detects only one single epitope of an antigen whereas polyclonal antibodies is a mixture of antibodies able to recognize multiple epitopes on the same antigen (Fig. 2.23b, c). The name of antibodies often ends by the suffix—mab or—pab. A complete nomenclature of antibodies has been established in 2008 in order to assign a unique and generic name to a given antibody as a function of its target (bacterium, fungus, immune system …) and its source (human, mouse, primate …).

Antibodies are used by living cells as key molecules for the immune response according to different mechanisms [43]. For example, different types of antibodies are present near or at the surface of immune cells in order to detect and fix toxins from pathogenic bacteria, preventing them from harming. Similarly, free antibodies can recognize specific antigens at the surface of pathogenic bacteria, bind to them with their variable side and bind in turn to macrophages (cells responsible for the digestion of foreign bacteria) with their fixed side.

Antibodies are also widely used in biotechnology. Their modular structure provide an easy way to detect and fix various proteins with a high level of selectivity. For instance, they are used in biosensors to fix proteins on a surface for further analysis (e.g. in piezoelectric biosensors, antibodies are spotted at the sensor surface and fix protein from the reaction chamber which lead to weight changes measured by the piezoelectric plate) [44]. Other application of antibodies are flow cytometry to differentiate cells [45] or in several experimental protocols to sort proteins of interest from a solution containing many different molecules [46].

2.3.3.3 Enzymes

Enzymes are another class of proteins that catalyzes chemical reactions. The name of enzymes always ends by the suffix—ase and the radix is often related to the substrate or the product of the reaction they catalyze. To understand the action of an enzyme in a chemical reaction, let us first consider a reaction converting a substrate S into a product P. Let G_P and G_S be the free energy of the product and the substrate. The reaction occurs only if the energetic balance $\Delta G^o = G_S - G_P$ is positive. But consider now that the conversion from S to P involves an intermediate states with free energy $G_I > G_S$. The difference $G_I - G_S$ is the activation energy E_A, i.e. the amount of energy we need to invest in the reaction to go over the energy barrier (Fig. 2.23). If E_A is high, the reaction will be very slow. It can be compared to a rock on a mountain. The rock would like to go down the mountain in order to decrease its potential energy. But if somewhere it is held by a rim, it will never reach the valley, except if someone gives it enough energy to go over the rim by pushing it.

The role of the enzyme is to provide another intermediate states with a lower and activation energy E'_A. As $E'_A < E_A$, the reaction is faster. Continuing the analogy with the rock, the enzyme creates another pathway from the top of the mountain to the valley on which rims are lower.

From a chemical point of view, enzyme act on the substrate like a nutcracker on a nut. Substrate and enzyme binds together to form a complex in which the transformation from the substrate to the product requires less activation energy in comparison to the direct transformation without complex.

Side chemical species may also be involved in the mechanism. Some of them are part of the reaction by providing or harvesting energy (ATP, see Sect. 2.3.1.4), atoms or functional groups. Some other are not consumed nor produced by the reaction but they improve the efficiency of the enzyme. These chemical compounds are called cofactors or coenzymes [47]. As a conclusion, it should be reminded that enzyme only speed up the reaction. It affects kinetics but not the energy balance of a given reaction.

2.3.3.4 Metabolic Pathways and Regulation

A metabolic pathways is a set of chemical reactions achieving a specific biological function. Such pathway includes substrates and products (starting and ending points of the pathway), intermediate species, enzymes and regulations. One of the most studied metabolic pathways is the glycolysis which transforms glucose into pyruvate (Fig. 2.24). Glycolysis involves 9 intermediate molecules, requires 10

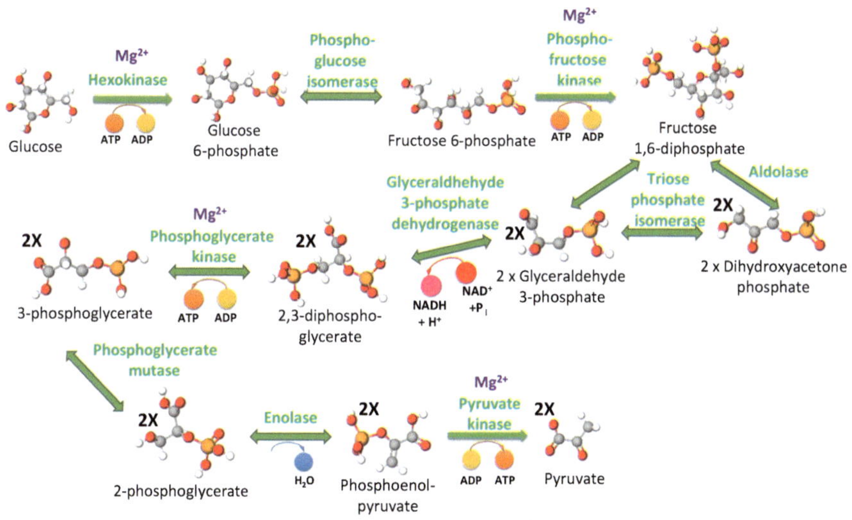

Fig. 2.24 The glycolysis pathway

enzymes and 2 cofactors and produces 2 ATPs from 2 ADPs (2 ATPs are consumed in the fists steps but 4 are harvested in the last ones).

Let us use this metabolic pathway to illustrate the role of the ATP and the effect enzymes. Firstly, focus is put on the first reaction of the pathway: the conversion of the glucose into glucose 6-phosphate (G6P). The equilibrium is reached when the concentration of the substrate and the product reach the following equation:

$$\Delta G^0 = -R \cdot T \cdot ln\left(\frac{[P]}{[S]}\right) \tag{2.3.2}$$

where R is the gas constant ($R = 8.31\,\mathrm{J} \cdot \mathrm{K}^{-1} \cdot \mathrm{mol}^{-1}$), T is the temperature and ΔG^0 is the standard free energy of the reaction, which is $14.3\,\mathrm{kJ} \cdot \mathrm{mol}^{-1}$ for the glucose + $H_3PO_4 \rightarrow$ G6P. As a consequence, at ambient temperature, at equilibrium the glucose concentration is about 300 time higher than G6P concentration. To alter this equilibrium the reaction, it is coupled to an ATP hydrolysis ($ATP + H_2O \rightarrow ADP + H^+ + H_3PO_4$) which ΔG^0 is $-30.5\,\mathrm{kJ} \cdot \mathrm{mol}^{-1}$. The overall ΔG^0 of the coupled reaction is now $-16.7\,\mathrm{kJ} \cdot \mathrm{mol}^{-1}$ which lead to glucose concentration at equilibrium 1000 times lower than the concentration of G6P.

Second, focus is put on another reaction of this pathway: the isomerization (same chemical formula but different atomic configuration) of glyceraldehyde 3-phosphatye (GAP) into dihydroxyacetone phosphate (DHAP). The reaction consists in two hydrogens transfer (Fig. 2.25) with the cis-enediol as an intermediate state. The reaction balance is $\Delta G^0 = -7.8\,\mathrm{kJ/mol}$ but the free energy of cis-enediol is very high ($E_A = 100\,\mathrm{kJ/mol}$). As a consequence, the reaction may occur but will be very slow. GAP isomerization can be catalyzed by an enzyme, the triose phosphate isomerase (TIM), which speed up the reaction by a factor 10^{10}. In practice, the GAP binding site of TIM exhibits an H^+ arm on the right and a negatively charged arm one on the left. When GAP binds on TPI, the two hydrogens that need to be transferred connect to these arms. In terms of energy, we are at the point C on the curve in Fig. 2.25e, which is obviously a non-stable equilibrium state. When the molecule is release, DHAP is preferred rather than GAP because its free energy is lower. According to Eq. (2), at equilibrium, there are 22 times more DHAP than GAP.

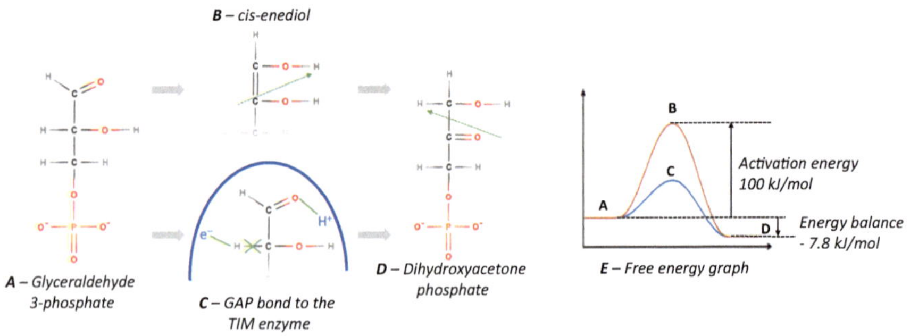

Fig. 2.25 Illustration of the enzymatic activity on the reaction glyceraldehyde 3-phosphate → Dihydroxyacetone phosphate

Metabolic pathways are not always active and cells have different ways to regulate them. First, it must be remembered that a reaction is never complete and the concentration ratio between substrate and products is fixed by energetic considerations (Eq. 2.39). Thus, a metabolic pathway can be blocked as soon as the concentration of a product reached the equilibrium. This natural mechanism prevents from an excessive accumulation of a products. Another way to increase/decrease enzyme's activity is to use cofactors. For instance, the concentration of Mg^{2+} increase the activity of several enzymes in glycolysis pathway (e.g. hexokinase, see Fig. 2.23). Moreover, most of the metabolic pathways are self-regulated with feedback loops. For instance, G6P is an inhibitor for hexokinase. It induces a conformational change that reduce the affinity of the glucose binding site of hexokinase.

The number of possible regulation and, as a consequence, the complexity of the pathway increase dramatically with the number of steps and the number of molecules involved in the pathway.

2.3.3.5 Transcriptional Regulation

Another way to regulate metabolic pathways is to control directly the transcription of genes that encode for the enzymes involved in the pathway. The principle of DNA-level regulation is the same as for metabolic regulation except that the regulation molecule controls the binding rate of the RNA polymerase to DNA instead of an enzyme activity.

In practice, there is a specific DNA sequence called the promoter upstream of the gene. RNA polymerase may bind on this promoter to initiate the transcription. Several mechanisms are implemented to regulate this transcription. First, some molecules have the appropriate conformation in order to bind to the DNA on or just after the promoter, preventing the binding of RNA polymerase and the transcription. Then regulatory DNA sequences can also be found before the promoter. Those sequences form binding site for regulating molecules (transcription factors) which modify the promoter structure and increase/decrease its affinity for RNA polymerase. Sometimes, the regulation mechanism is more complex. For instance, the human insulin gene can be regulated by up to 14 transcription factors [48].

To illustrate the concept, the example of the lactose operon is described hereafter. Lactose operon provides bacteria with a mechanism to digest and metabolize the lactose (Fig. 2.26). The operon has been first described in 1961 by Jacob and Monod, which is considered as the first breakthrough in the understanding of gene regulation mechanisms [49]. The Lac operon is composed of two parts: first, a constitutive promoter (always active) and a gene coding for LacI and second, a regulated promoter and three genes coding for LacZ (coding for β-galactosidase), LacY (coding for β-galactosidase permease) and LacA (coding for galactoside O-acetyltransferase) on the right. The idea of the regulation is the following: if there is no lactose around the bacteria, it is a waste to produce the enzymes letting lactose enter into the cell and metabolizing it. As a consequence, the gene expression should be repressed. The repression mechanism is performed by the LacI protein

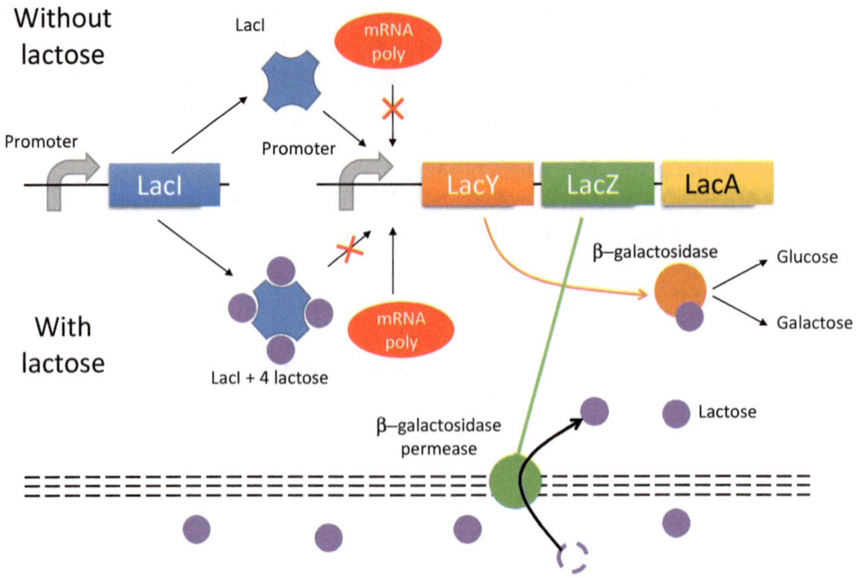

Fig. 2.26 The LacI operon

which binds to the Lac promoter, preventing the RNA polymerase from transcribing the genes. Conversely, if there is a large amount of lactose molecules around the cell, some of them penetrate into the bacteria and bind LacI. This induces a conformational change that prevents LacI from binding DNA. As a consequence, genes are expressed: LacY opens channels allowing more lactose to invade the cell and LacZ breaks lactose into glucose and galactose. When there is no more lactose to consume, the last molecules of lactose unbind from LacI which is able to repress gene transcription again.

Moreover, glucose remains the preferred energy source for bacteria. As a consequence, in presence of glucose, lactose metabolism is not required. Lactose operon is thus also repressed by glucose (not represented in Fig. 2.26).

Gene regulation can be cascaded (operons are constituted of genes that codes for transcription factors regulating other operons) in order to achieve more complex biological functions. In this case, we speak about gene regulatory networks. Moreover, regulatory network also exists at translational level: regulatory molecules modify the mRNA translation rates.

2.3.4 Mathematical Modeling

The last part of this sub-chapter aims at giving the background on mathematical modeling of the biological mechanisms described here above.

2.3.4.1 Generic Framework for a Biological System

A biological system is a set of chemical compound linked by reactions. To describe it in a mathematical way, four elements are requires. Firstly, the molecules involved in the system are listed. Let $\mathbf{X}(t)$ be an N-element vector composed of the concentration of the N involved molecules. Secondly, the reactions are also listed. Let $\mathbf{V}(t)$ be an M-element vector composed of the rates of the M reaction. Because of regulations, the elements of $\mathbf{V}$ are not constant but depends on some elements of $\mathbf{X}$. Thirdly, a stoichiometry matrix is build. It gives the number of molecules produced or consumed by each reaction. Let $\mathbf{S}$ be this $N \times M$ matrix. Finally, most of the molecules of living matter are not stable over time and degrade with a given ratio. Thus, we also define the decay vector $\mathbf{d}$, an N-element vector composed of the decay constant for each involved molecule.

A first way to simulate the system is to compute the temporal evolution of the concentration of each species. At each time step, the variation of concentration for each species is equal to the sum of the production, consumption and degradation rate. From the quantities define here above, we can write the set of ordinary differential equation (ODE) to solve in a vector form:

$$\frac{d\mathbf{X}}{dt} = \mathbf{S} \times \mathbf{V}(\mathbf{X}) - d \cdot \mathbf{X} \tag{2.39}$$

Such equation rate can be solved with standard ODE solver (such as Matlab, SciLab, Python), with dedicated tools such as COPASI [50] but also with SPICE-like simulators [51]. An alternative way to analyze this system is to find the steady states which correspond, for each species, to a balance between synthesis rate, consumption rate and decay rate. Steady states can be obtained by solving the following equation set:

$$\mathbf{S} \times \mathbf{V}(\mathbf{X}) - d \cdot \mathbf{X} = 0 \tag{2.40}$$

The number species and mechanisms to take into account in an actual biological system can be very large, which often leads to large and strongly coupled ODE sets. Simulations require long computation time and lead to convergence issues. There are some tricks to simplify the model of large systems. The most usual one consists in representing the system as an interaction graph and use graph theory to identify poles that corresponds to subsystems for which high-level macro-models can be used. The approach is exactly the same as for multi-level modeling in microelectronics (e.g. in electronic circuit, a behavioral description for operational amplifier instead of a transistor-level model). For instance, let us consider a complex system involving glycolysis. If enzymes and intermediate species involved in glycolysis do not interact with the rest of the system, the full pathway can be modeled by a single equation giving the production rate of pyruvate as a function of the concentration of glucose and the quantity of enzyme (supposed to be constant). This macro-model allow to get rid of 9 species and 10 reaction in comparison with the complete model.

Another way to simplify the equation set is to use the quasi steady state approximation (QSSA). QSSA can be applied as soon as several mechanisms are very fast in comparison to other ones. In this case, the fastest reaction are considered to be always "almost at equilibrium", i.e. between two time steps of the slowest reaction, the fastest one has reached the steady state. QSSA will be illustrated in Sect. 2.3.3.4.

2.3.4.2 First Order Reactions

In the following four subsections, the most common models for reaction rates are described. For a standard chemical reaction, such as $A+B \leftrightharpoons C+D$ the most common model is the first-order model. The rate of the reaction is given by:

$$v = k_{on} \cdot [A] \cdot [B] - k_{off} \cdot [C] \cdot [D] \tag{2.41}$$

where k_{on} and k_{off} are the forward and reverse rate constant. Equilibrium is reached when the reaction rate is zero. In this case, association constant K_A and dissociation constant K_D can be defined as following:

$$K_A = \frac{1}{K_D} = \frac{k_{on}}{k_{off}} = \frac{[C] \cdot [D]}{[A] \cdot [B]} \tag{2.42}$$

2.3.4.3 Hill's Equation

In this subsection, we consider a reaction involving a macromolecule P that may bind up to N ligands L. The system can be described by a set of first-order reaction $P+L \rightarrow PL$, $PL+L \rightarrow PL_2$, $PL_2+L \rightarrow PL_3$, … Let K_{A_1}, … K_{A_N} be the macroscopic association constant for each of these reactions. The total quantity of P is distributed among the different potential configurations:

$$P_{TOT} = [P] + [PL] + \ldots = [P] \cdot \left(1 + K_{A1} \cdot [L] + K_{A1} \cdot K_{A2} \cdot [L]^2 \ldots\right) \tag{2.43}$$

This expression is named the binding polynomial. We then associate a signal σ_k to each configuration of P. This signal can be either a reaction rate, a transcription rate for a promoter, the average number of occupied binding site, a fluorescence signal, etc. The signal polynomial is defined as following:

$$S = [P] \cdot \left(\sigma_0 + \sigma_1 \cdot K_{A1} \cdot [L] + \sigma_2 \cdot K_{A1} \cdot K_{A2} \cdot [L]^2 \ldots\right) \tag{2.44}$$

The total signal associated to the protein P is given by the rational fraction:

$$S_{TOT} = \sum_{i=0}^{N} \sigma_i \cdot \frac{[PL_i]}{P_{TOT}} = \frac{S}{P_{TOT}} \tag{2.45}$$

This approach can be extended to macromolecules with multiple binding sites for multiple ligand and the associated signal can always be written as a rational fraction [52].

Let us discuss about some simple cases. First, consider that P has a single binding site with the macroscopic dissociation constant K_D. The signal of interest is the average number of occupied sites θ. From Eq. (2.39):

$$\theta = \frac{K_{A1} \cdot [L]}{1 + K_{A1} \cdot [L]} = \frac{1}{1 + \frac{K_D}{[L]}} \tag{2.46}$$

Now consider that P has N equivalent binding sites. The average number of occupied sites is now given by:

$$\theta = \frac{\sum_{i=1}^{N} i \cdot \prod_{j=0}^{i} K_{Aj} \cdot [L]^j}{1 + \sum_{i=1}^{N} \prod_{j=0}^{i} K_{Aj} \cdot [L]^j} \tag{2.47}$$

where K_{Aj} is the affinity constant of binding site when $j-1$ other sites are occupied. If binding sites are equivalent and independent (i.e. $K_{Aj} = k_A^j$), it has been demonstrated that [52]:

$$\theta = N \cdot \frac{k_A \cdot [L]}{1 + k_A \cdot [L]} = \frac{1}{1 + \frac{k_D}{[L]}} \tag{2.48}$$

where k_A and k_D are the microscopic affinity for one each binding site. Comparison between Eqs. (2.46) and (2.48) highlights the fact that multiple binding independent binding sites behaves as one single binding site.

Now, let us consider that binding sites are cooperative, i.e. the affinity of binding site increases with the number of occupied sites. From a mathematical point of view, strong cooperativity leads to the following relationship: $K_{Ai} \gg K_{Ai+1} \cdot [L]$. In other words, binding and signal polynomial are dominated by the zero-order term and the highest-order term. Thus, the average number of occupied sites is given by:

$$\theta = N \cdot \frac{\left(\prod_{i=0}^{N} K_{Aj}\right) \cdot [L]^N}{1 + \left(\prod_{i=0}^{N} K_{Aj}\right) \cdot [L]^N} = \frac{1}{1 + \left(\frac{K'_D}{[L]}\right)^N} \tag{2.49}$$

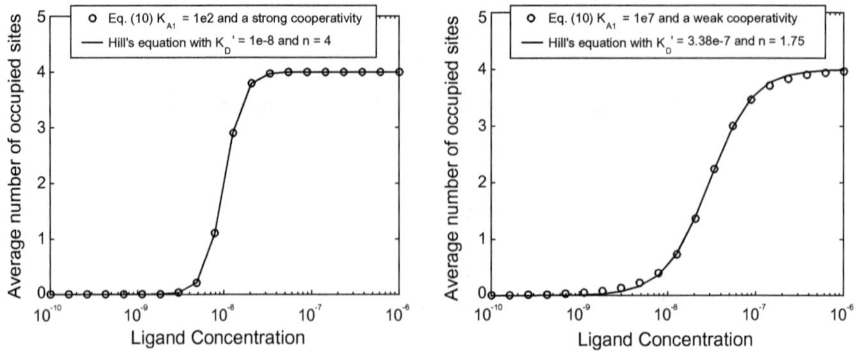

Fig. 2.27 Modeling multi-site cooperativity in protein-ligand binding with Hill's equation

where K'_D is the apparent dissociation constant. Equation 2.49 is the Hill equation. In practice, cooperativity is not absolute and the assumption used to deduce Eq. (2.49) is not completely met. But most of the time, a binding mechanism with cooperative sites exhibits a sigmoid response and can be modeled by a Hill equation with empirical Hill constant K'_D (concentration of $[L]$ for which half of the sites are occupied) and Hill number (between 1 and the number of sites) [53]. To illustrate the purpose, Fig. 2.27 shows examples of $\theta([L])$ curves for 4 binding sites with weak and strong cooperativity.

Hill's equation is a universal model that can be applied for most of the signal of interest that are proportional to the site occupancy rate.

2.3.4.4 Michaelis and Menten's Enzymatic Model

As described in Sect. 2.3.2.3 an enzymatic reaction can be decomposed into two stages: the binding of the substrate on the enzyme $(E+S \rightarrow ES)$ and the transformation of the enzyme-substrate complex into the product and the recycled enzyme $(ES \rightarrow E+P)$. Each reaction can be modelled by a first-order reaction rates. Let k_1 and k_{-1} be respectively the forward and reverse reaction constant for the first reaction and k_2 and k_{-2} for the second. Let us model the complete enzymatic reaction according to the formalism described in Sect. 2.3.3.3. The species vector, reaction rate vector and stoichiometry matrix are the following (decay is neglected and to simplify writing, square brackets around concentration have been omitted):

$$\mathbf{X} = \begin{bmatrix} E \\ S \\ ES \\ P \end{bmatrix} \quad \mathbf{V} = \begin{bmatrix} k_1 \cdot E \cdot S - k_{-1} \cdot ES \\ k_2 \cdot ES - k_{-2} \cdot E \cdot P \end{bmatrix} \quad \mathbf{S} = \begin{bmatrix} -1 & 1 \\ -1 & 0 \\ 1 & -1 \\ 0 & 1 \end{bmatrix} \tag{2.50}$$

The Michaelis-Menten model relies on two assumptions: (i) the second reaction is not reversible ($k_{-2} = 0$) and (ii) the second reaction is much faster than the first one. As a consequence, QSSA assumption can be applied on E and ES. Each time an E is consumed and an ES is produced by reaction 1, an ES is consumed and an E is produced by reaction 2. As a consequence, the time derivative of the concentration of E and ES are equal to zero. From (2.50), one can write:

$$\frac{d[ES]}{dt} = -\frac{d[E]}{dt} = k_1 \cdot [E] \cdot [S] - (k_{-1} + k_2) \cdot [ES] \tag{2.51}$$

Thus,

$$k_1 \cdot [E] \cdot [S] = (k_{-1} + k_2) \cdot [ES] \tag{2.52}$$

Let E_{TOT} be the total amount of enzyme ($E_{TOT} = [E] + [ES]$). It can be deduced from (2.51) to (2.52) that the production rate of P is equal to the consumption rate of S which is also equal to:

$$\frac{d[P]}{dt} = -\frac{d[S]}{dt} = \frac{v_{max} \cdot S}{K_m + S} \tag{2.53}$$

where $v_{max} = k_2 \cdot E_{TOT}$ is the maximal reaction rate (when S is high) and $K_M = \frac{k_{-1} + k_2}{k_1}$ is the Michaelis constant. Figure 2.28 shows simulation results for enzymatic reaction with complete model and with Michaelis-Menten model for different cases: one in which QSSA condition is met (simulation results fit perfectly), one in which it is almost met and one in which it is not met (error appear between the two curves).

Again, if an enzymatic reaction involved in a larger pathway pas be described by a unique reaction $S \rightarrow P$ with a Michaelis-Menten model as soon a neither E nor ES, interact with the other molecules involved in the pathway.

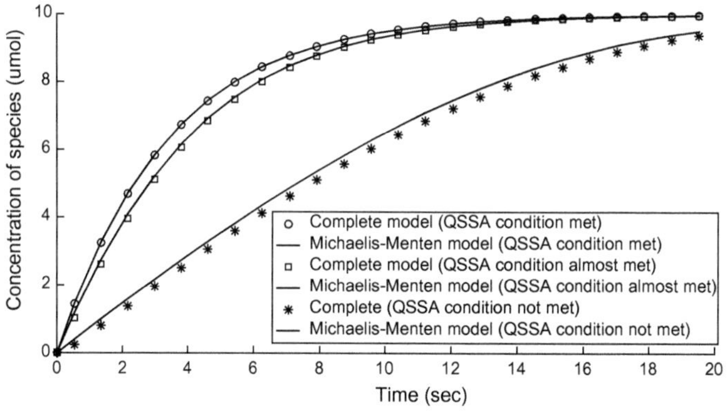

Fig. 2.28 Comparison of simulation results between complete model and michaelis-menten model

2.3.4.5 Transcription and Translation

Without regulation, the transcription rate of DNA into mRNA mostly depends on the efficiency of RNA polymerase. Most of the time in model, the concentration of RNA polymerase is considered to be high enough and does not impact the transcription rate k_{tr} which is a constant depending only on the length of the sequence. In practice, this assumption remains questionable. By the same way, the translation rate depends on the efficiency of the ribosome which is also considered to be in excess. The translation rate of one mRNA into a protein is given by a constant k_{tl} which also depends on the sequence length. As several copy of the same mRNA may exist and may be translated in parallel, the overall translation rate is given by $k_{tl} \cdot [mRNA]$.

DNA transcription and mRNA translation are very slow mechanism in comparison to molecule binding (timescale for the binding of transcription factor to DNA site is about 1 s whereas timescale for transcription and translation is more than 1 min [54]. As a consequence, regulation can be introduced in transcription/translation models under the QSSA approximation. Let us first consider a promoter with a regulating sequence with four cooperative binding site for a transcription factor A. We assume that the transcription rate increases monotonically with the number of occupied binding site so that the Hill's equation can be applied. QSSA condition allows to modulate directly the transcription rate with a term depending on the concentration of A:

$$k_{tr} = k_{tr,max} \cdot \left(\alpha + \frac{1-\alpha}{1+\left(\frac{K_A}{[A]}\right)^{n_A}} \right) \tag{2.54}$$

where α ($\ll 1$) is the promoter leakiness (even if the promoter is repressed, a basal transcription rate exists), $k_{tr,max}$ is the maximal transcription rate, K_A and n_A are Hill's parameters and number. K_A is sometimes called activator strength because it represents the concentration of activator required to reach about 50% of the maximal transcription rate. In the case of a repressor, transcription rate decreases with the increasing site occupancy ratio. As a consequence, the expression becomes:

$$k_{tr} = k_{tr,max} \cdot \left(\alpha + \frac{1-\alpha}{1+\left(\frac{[R]}{K_R}\right)^{n_R}} \right) \tag{2.55}$$

where K_R and n_R play the same role as K_A and n_A for the activator. Expression (2.54) and (2.55) can be merged for promoter with more than one transcription factor. Same principles can be applied to the translation rate for RNA-based regulation system.

2.4 MicroPIV Measurement Technique (Denis Funfschilling, Norbert Dumas)

2.4.1 Introduction

Among the different measurement techniques existing in microfluidics, two techniques have emerged: imaging (direct imaging, fluorescence imaging, shadow imaging and the following image analysis) and μPIV (Micro Particle Image Velocimetry) technique for flow field measurements. They both are valuable source of information for experimentalist. This chapter is focused on this latter technique. The origin of the μPIV dates back to the pioneer work of Santiago et al. [55] and Wereley et al. [56]. μPIV is considered as the most widespread method for measuring velocities in fluids [57].

The μPIV (Micro Particle Image Velocimetry) is an adaptation of already existing PIV technique to the microscale. The spatial resolution of a μPIV can be better than 1 μm [58]. It has the enormous advantage to measure the flow field in a complete plane, not like other techniques like LDV (Laser Doppler Velocimetry), or Ultrasonic Doppler Velocimetry which operate only point wise or line wise. In short, the PIV technique consists of (1) seeding the fluid with particles, (2) illuminating the fluid with a laser sheet (most of the time, a double cavity Yag Laser is used), (3) taking two images separated by a known interval of time, more precisely, each image is taken during the burst of the laser that lasts a few nanoseconds, (4) both images are divided into interrogational areas, and each of these interrogational areas is cross-correlated with the corresponding interrogational areas of the second image, (5) the mean velocity on each interrogational area is deduced by dividing the mean displacement of the particles of this area by the time between the two bursts. But unlike the PIV where the velocity field is measured in a plane created by a laser sheet obtained by expanding the laser beam in 2D by a cylindrical lens, the μPIV uses the depth of field of the imaging optics to image a "slice" of the flowing seeded fluid [59].

Several companies such as Dantec Dynamics are proposing complete μPIV set-up. (www.dantecdynamics.com), LaVision (www.lavision.de) or TSI (www.tsi.com). μPIV is particularly interesting for biology research in microfluidics because it gives a mine of information about the flow, the stresses, the movement of cells, bacteria…

In some research fields, it has been the major source of development in the last decade, for example:

- Droplet and bubble manipulation is a flagship for microfluidics applications. Droplets have been used as independent chemical and biological reactors for cell culture, protein crystallization, nanoparticles synthesis, enzyme kinetics [60] or biological assays for living cells [61] among others. A deeper understanding of the mechanisms driving the formation and motion of droplets and bubbles is necessary for the effective design of multiphase microfluidic devices [60]. μPIV is the most appropriate measurement technique to achieve this goal.

- Growth and evolution of cells like endothelial cell under flow can be measured by μPIV in microsystems [62].
- In the rheology domain, Rodd et al. [63] think that μPIV in micro channels is the most effective technique for reliably and quantitatively characterize the kinematics of complex flow over a wide range of flow regimes.

2.4.2 Preliminary Characterization of Fluids and Microsystems

Before beginning experiments in microsystems, experimentalists must collect the physical properties of the fluids used, like viscosity, density, surface tension, interfacial tension as well as the surface properties of the microchannels' walls. When pure fluids are used, these properties are most of the time well documented in books like CRC Handbook of Chemistry and Physics, Taylor & Francis Group [64]. Otherwise, when mixtures of fluids are used, in lab measurements may have to be undertaken.

2.4.2.1 Density

The density of a fluid can be measured by a pycnometer (a glass beaker of a known volume filled with the fluid and weighted on a high precision scale) or by a hydrometer. A hydrometer is represented schematically on Fig. 2.29. The measurement is based on the Archimedes principle. A calibrated glass body dips into the sample. The part of the graduated glass emerging from the fluid depends on the density of the fluid to be measured. Companies like VWR or Fisher Scientific are selling these inexpensive chemistry lab materials.

More elaborate density measurement apparatus are also available like (i) hollow glass tube. The tube is filled with the fluid. The frequency of vibration of the tube is related to the density of the fluid, (ii) radiometric density measurements. The radiometric measurements are non-invasive, non-intrusive and based on gamma-ray or X-ray transmission principle: i.e. the attenuation of the radiation is proportional to the density. Such measurements devices are manufacture for example by Berthold Technologies GmbH (Germany) or Ronan Engineering Company (UK).

2.4.2.2 Viscosity

The first rheological question arising is whether the viscosity of the fluid is *Newtonian* or *non-Newtonian*.

When the fluid is *Newtonian*, its viscosity is constant, i.e. its viscosity is independent of shear rate and time. As represented in Fig. 2.30, the shear stress is then a linear function of the shear rate, i.e.

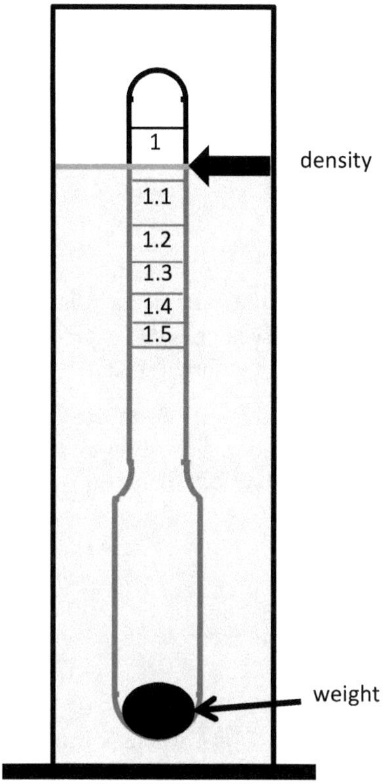

Fig. 2.29 schematic representation of a hydrometer

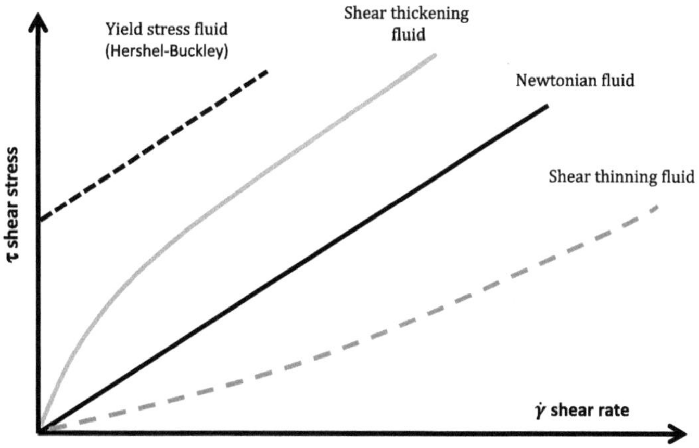

Fig. 2.30 Behaviour of Newtonian and non-Newtonian fluids (shear thickening, shear thinning and yield stress non-Newtonian fluids are represented)

$$\tau = \mu \acute{\gamma}$$

where τ [Pa] is the shear stress, μ [Pa.s] the viscosity, and $\acute{\gamma} = \frac{dU}{dx}$ [s^{-1}] the shear rate.

Most of the pure simple fluids around us are Newtonian (water, alcohol, glycerol, silicone oil…). In this case, the viscosity can be measured in a viscometer. Several affordable viscometers are available like capillary, falling ball or rotational viscometers.

- **Capillary viscometer** (see Fig. 2.31). This type of viscometer also named Ostwald viscometer is quite common in laboratories and is appreciated for its simplicity, precision and low cost. Its physical principle is based on the measurement of flow resistance (i.e. pressure loss) of the fluid flowing through a capillary.

The fluid is usually flowing under the earth gravitational force. The time necessary for a certain amount of fluid to flow through the capillary is measured. The viscosity is then calculated with the formula:

$$v = K_c \times t_f$$

where [m^2/s] is the kinematic viscosity, K_c [m^2/s^2] is the capillary factor and t_f [s] the time measured corresponding to the flow of the volume of fluid located between the start and stop marks (see Fig. 2.31).

The previous formula is only valid in the *laminar flow regime* and is based on the Hagen-Poiseuille Law i.e. $\frac{V}{t} = \frac{\pi R^4 \Delta P}{8L\mu}$, where R [m] is the radius of the capillary, L [m] the length of the capillary, and ΔP [Pa] the pressure drop through the capillary. Several capillary viscometers with different capillary diameters may be necessary to cover larger ranges of viscosities.

- **Falling ball viscometer**. The measurement is based on Newton's law of motion. The force balance refers to a falling ball which has reached its terminal velocity. Three forces are present: the buoyancy force, the weight force and the drag force. Measurement have to be done in the Stokes' flow regime (i.e. Reynolds number < 1). In this case, the drag force on the ball is:

$$F_D = 3\pi\mu u_t d$$

where d [m] is the diameter of the ball and u_t $[m/s]$ the terminal velocity.

The viscosity is then:

$$\mu = \frac{g(\rho_s - \rho_f)d^2}{18u_t}$$

where ρ_s and ρ_f [kg/m^3] are the densities of the ball and of the fluid respectively, and g [m^2/s] the acceleration of gravity.

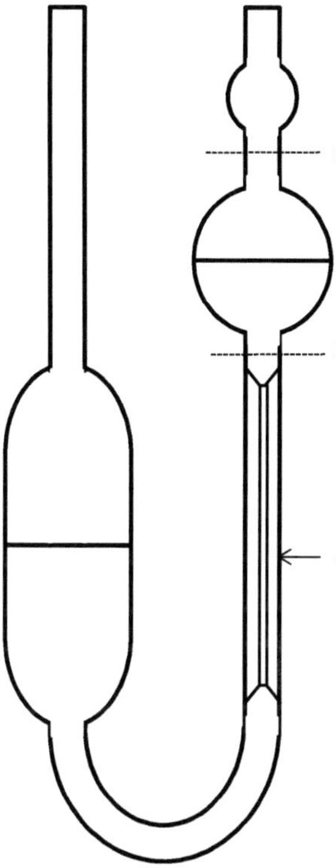

Fig. 2.31 Schematic representation of a capillary viscometer

- **Rotational viscometer**. In the case of rotational viscometers, the angular velocity and the torque of a rotating geometry (a disk, a bob, a spindle, a needle, a cylinder or a cup) are measured simultaneously. Knowing the exact dimensions of the geometry, and assuming a laminar flow, it is possible to calculate the viscosity. These kinds of viscosity measurements are well appreciated in industry, because the rotating geometry is directly immersed into the fluid and the reading is immediate. Rotational viscometers are not appropriate for non-Newtonian fluids viscosity measurements because of their limited accuracy and limited range of shear rates available for viscosity measurements. In this case, rheometers are necessary for a more precise and complete investigation of the viscosity of the fluid.

  When the viscosity is non-Newtonian (the fluid can be shear thinning, shear thickening, thixotropic (time dependent), have a yield stress….), measurements must be done in Rheometers.

Blood and most polymers and polymer solutions are non-Newtonian. The simplest non-Newtonian fluids are fluids whose viscosities can be expressed by a power law (see Fig. 2.32) :

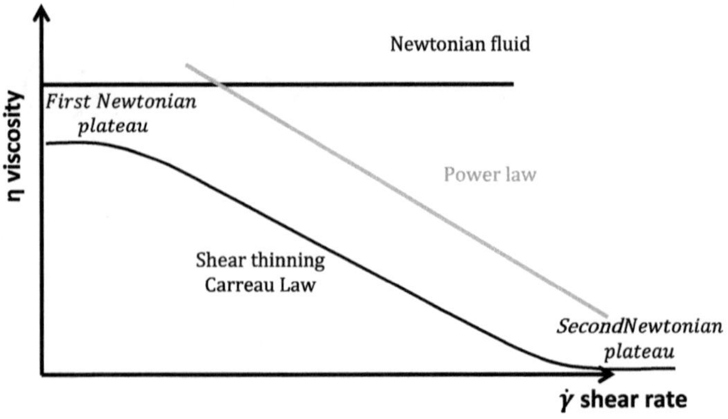

Fig. 2.32 Viscosity of Non-Newtonian fluids following a power law or a Carreau-Yasuda law

$$\eta = k\acute{\gamma}^{n-1}$$

η [Pa.s] is the viscosity, k [Pa.s^n] is the consistency index, and n [-] the flow behavior index.

Another simple model for non-Newtonian fluids is the Carreau-Yasuda model:

$$\frac{\eta(\dot{\gamma}) - \eta_\infty}{\eta_0 - \eta_\infty} = \left[1 + (\lambda\acute{\gamma})^a\right]^{\frac{n-1}{a}}$$

η_0 [Pa.s] is the zero shear viscosity corresponding to the first Newtonian plateau, η_∞ [Pa.s] is the infinite shear viscosity corresponding to the second Newtonian plateau, λ [s] is a time constant, n [-] the flow behavior index and a [-] represents the width of the transition region between η_0 and the power-law region.

Rheometers are necessary for rheology measurements of non-Newtonian fluids. They are expensive and sophisticated devices often used in research and able to measure viscosities in function of time shear rate, dynamic oscillations, or after a step in shear rate…. The main manufacturers are TA Instruments (http://www.tainstruments.com), Anton-Paar (http://www.anton-paar.com), Brookfield http://www.brookfieldengineering.com). They are based on the same principle as the rotational viscometers. They simultaneously measure the angular velocity and the torque. The most common geometries are cone and plate, plate and plate, Couette or double Couette cylinders.

It is difficult to present the rheology in more details in this chapter, but we invite the interested reader to consult the literature on rheology (see for example [65–67]).

2.4.2.3 Surface Tension/ Interfacial Tension/ Contact Angle

Most of the fluids can be separated in two groups: (i) hydrophilic fluids (fluids having a strong affinity for water. Typical hydrophilic fluids are water solutions, glycerol, alcohol…) and (ii) hydrophobic fluids (fluids which have little or no affinity for water, like oils and most of the polymers). Hydrophilic and hydrophobic liquids in general do not mix together and can form emulsions which are thermodynamically unstable. The hydrophilicity and hydrophobicity of the walls of the microsystems and the interfacial tension of the two fluids are important to know because they will determine if direct emulsions (oil in water) or inverse emulsions (water in oil) are created. It will also influence the stability of the droplets and emulsions formed.

Surface tension

The physical origin of the liquid/gas surface tension is the attraction existing between neighboring molecules in the liquid state. Molecules located at the liquid/ air interface have lost half of their cohesive interaction due to their immediate proximity to the gas interface. Molecules located at the interface are frustrated and therefore will tend to minimize interfacial area.

The surface tension γ can be seen as the energy of cohesion of a molecule U divided by its surface a^2 [68]:

$$\gamma = \frac{U}{2a^2}\,[\mathrm{J/m^2}]$$

U $[J]$ is the energy of cohesion, a $[m]$ is the molecule dimension.

Surface tension can also be seen as a capillary force per unit length

$$\delta W = \gamma dA$$

In the case of a squared liquid film with one of the side of length L $[m]$ the work of the surface tension force is expressed as

$$\delta W = Fdx = 2\gamma L dx$$

δW $[N \times m]$ is the work of the surface tension force, dA $[m^2]$ is the variation of surface, dx $[m]$ is the infinitesimal displacement and F $[N]$ the capillary force.

The surface tension can be measured by different techniques: the spinning drop technique, the Du Noüy ring, the Wilhelmy plate, pendant drop / sessile drop, maximum bubble pressure method to cite the most popular. We will present briefly three of them (i) the Wilhelmy plate method (ii) the rising/ pendant drop, and, (iii) the maximum pressure drop.

Wilhelmy plate method

The Wilhelmy method is based on the measurement of the capillary force applied on an object. The object is a high surface energy plate, typically in platinum, which is taken to a flame before each measurement to avoid contamination. The plate is slowly pulled out of the bath (see Fig. 2.33a). The force goes through a maximum when the capillary forces are aligned vertically. In that case the force is $F = l\gamma\, cos(\theta)$ with an angle $\theta = 0$, what becomes $F = l\gamma$.

l [m] is the length of the thin plate, F [N] is the measured force and γ [N/m] the surface tension. This method has a precision of about 1%. The Wilhelmy method is very popular due to its simplicity and low cost. The Du Noüy ring method is very close to the Wilhelmy method, but differs essentially by measuring the force on a ring instead of a plate.

Pendant drop and rising drop method

A bubble or a drop is formed at the tip of a small diameter capillary (see Fig. 2.33b). The equilibrium between the hydrostatic pressure and the Laplace pressure is [68]:

$$\gamma C = \rho g z$$

where ρ [kg/m^3] is the density, g [m/s^2] the acceleration of gravity, and z [m] the vertical coordinate, C is the curvature assuming an axisymmetric droplet,

$$C = \frac{-\frac{d^2 r}{dz^2}}{\left(1 + \left(\frac{dr}{dz}\right)^2\right)^{\frac{3}{2}}} + \frac{1}{r\left(1 + \left(\frac{dr}{dz}\right)^2\right)^{1/2}}$$

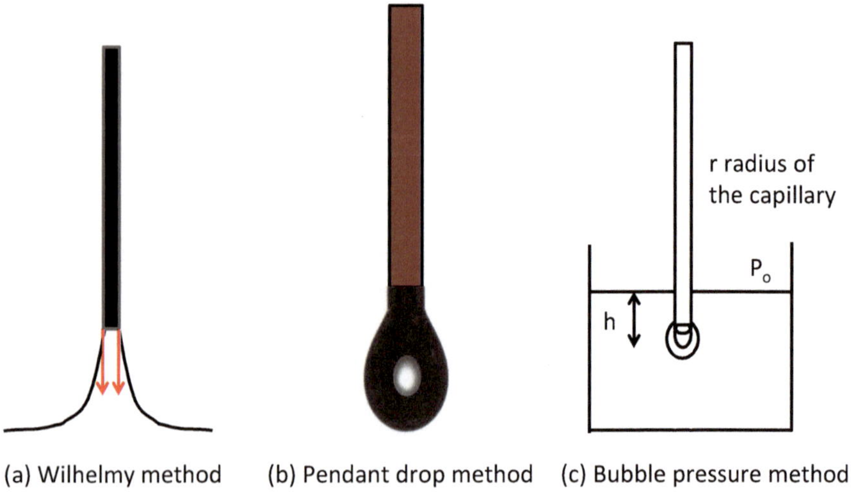

Fig. 2.33 Schematic representation of surface tension measurement apparatus

This method has an accuracy of the order of 1%, and is fully automated by manufacturer like *Trackers*^{TM} form Teclis Scientific. This method has the advantage to be able to measure surface tension (gas/ liquid interface) as well as interfacial tension (liquid/ liquid interface).

The maximum bubble pressure method

The pressure of a bubble forming at a capillary is measured during the growth and ejection. The measured pressure corresponds to (see Fig. 2.33c):

$$P(R) = P_0 + \rho g h + \frac{2\gamma}{R}$$

where P_0 [Pa] is the atmospheric pressure, h [m] the height of the water, $P(R)$ [Pa] the pressure measured in the capillary, and R the diameter of the bubble. This pressure goes through a maximum that corresponds to the bubble of the same diameter as the capillary. From that, the previous relation makes it possible to calculate the surface tension.

This method is interesting because it is able to measure the surface tension at short times and its evolution with the time, what is important for the study of surfactant migration or the presence of contamination.

Interfacial tension

Interfacial tension is the surface tension between two fluids, and it can be measured by pendant/rising drop method presented previously. In this case, the density difference between the two fluids has to be taken into account so that:

$$\gamma C = (\rho_d - \rho_c) g z$$

where $C = \frac{-\frac{d^2 r}{dz^2}}{\left(1 + \left(\frac{dr}{dz}\right)^2\right)^{\frac{3}{2}}} + \frac{1}{r\left(1 + \left(\frac{dr}{dz}\right)^2\right)^{1/2}}$, is the curvature and ρ_c, ρ_d [kg/m^3] are the densities of the dispersed and continuous fluids respectively.

In the case of two liquid phase microfluidics, the interfacial tension is of primary importance.

Contact angle

The surface of a material is named hydrophilic if the contact angle of a drop of water on its surface is less than 90°, otherwise it is named hydrophobic (see Fig. 2.34). It is important to know whether the channel walls are hydrophilic of hydrophilic, since it will determine if droplets of oil in water or droplets of water in oil are formed at the T or flow-focusing junction. In case of hydrophilic walls, oil droplets in water are formed, and inversely in the case of hydrophobic walls, water droplets in oil are formed. As a rule, the stable continuous phase is always the one that wets the surfaces.

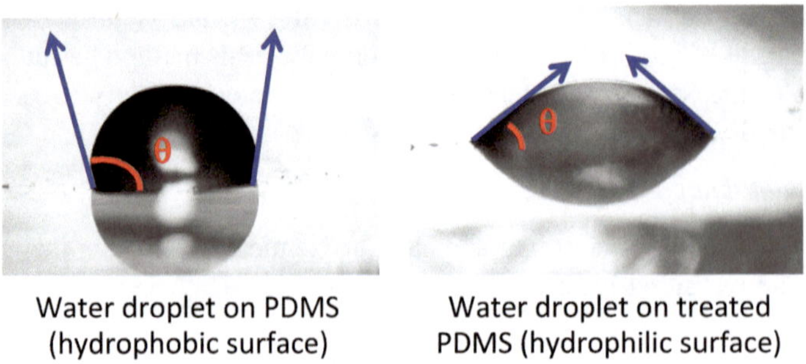

Fig. 2.34 Definition of the contact angle between a drop and a surface

The wetting is named "total" if the droplet spreads completely. It is the case of water on very clean glass or on metal surfaces which are high energy surfaces. If the droplet stays as a drop, the wetting is partial. The spreading parameter is defined as [68]:

$$S = E_{dry}^{substrate} - E_{wet}^{substrate}$$
$$S = \gamma_{SO} - (\gamma_{SL} + \gamma)$$

where $E_{dry}^{substrate}$ is the surface energy of the dry substrate, $E_{wet}^{substrate}$ is the surface energy of the wet substrate, γ_{so} is the surface tension of the solid/air interface, γ_{SL} is the solid/liquid surface tension, and γ is the liquid/air surface tension.

- If S > 0, total wetting, the contact angle $\theta_E = 0$
- If S < 0, partial wetting, and it is considered as rather wetting if $\theta_E \leq \pi/2$ and rather non-wetting if $\theta_E > \pi/2$
- The Young relation is obtained either by a balance of the capillary forces acting on the contact line (Fig. 2.35) and projecting them on the solid plan or by calculating the work of the forces for an infinitesimal displacement:

$$\gamma cos\theta_E = \gamma_{SO} - \gamma_{SL}$$

Among the different technique existing, we will present three of them which are particularly simple.

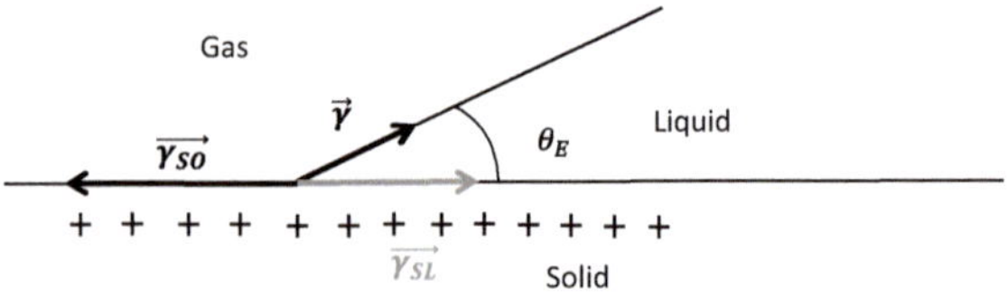

Fig. 2.35 Contact angle and surface tension

- The contact angle θ_E can be measured by shadowgraphy as represented in Fig. 2.34, either directly on droplets of a few microliters (typically about 5 μl) or for more precision by using a software fitting the profile of the droplet and calculating the contact angle.
- For contact angle $\theta_E < 40°$, the mirror droplet technique based on the optical reflectometry is to be favored [68]. The droplet is illuminated by a large laser beam perpendicularly to the substrate, and the reflection cone measured on a screen allows the determination of the mean contact angle with a precision of 0.1°.
- The last technique presented here is the measurement of the contact angle by Reflection Interference Contrast Microscopy. By counting the interference fringes and the number of them, it is possible to calculate the profile of the droplet, and by that, the contact angle with a precision of about 0.1°.

In the case of real, not perfectly clean surfaces, an advancing and receding contact angle can be measured.

Commercial surface tension and contact angle measurement apparatuses are available by manufacturers like Krüss Gmbh, Kibron Inc, Dyne Testing Ltd, USA KINO Industry CO. Ltd, Biolin Scientific or Data Physics Instruments, GmbH, Germany, among others.

Surfactant

Surfactants are usually added to the oil/ water mixture in order ease the emulsification process and to stabilize the emulsion. Surfactant are molecules that own a hydrophilic group (an anionic, cationic or non-ionic group named "head"), and a hydrophobic group (usually a hydrocarbon chain named "tail"). Because of their hydrophilic head and hydrophobic tail, they are prone to locate themselves at the interfaces. They are used at concentrations going from a fraction of a percent to a few percent. They modify the surface tension and stabilize the droplet formed. The Critical Micelle Concentration (CMC) is the concentration above which micelles i.e. aggregates of surfactant molecules are formed (see Fig. 2.36). It is important to be beyond the CMC to avoid interfacial tension variations by depletion, what may occur during the droplet formation process. The CMC can be determined

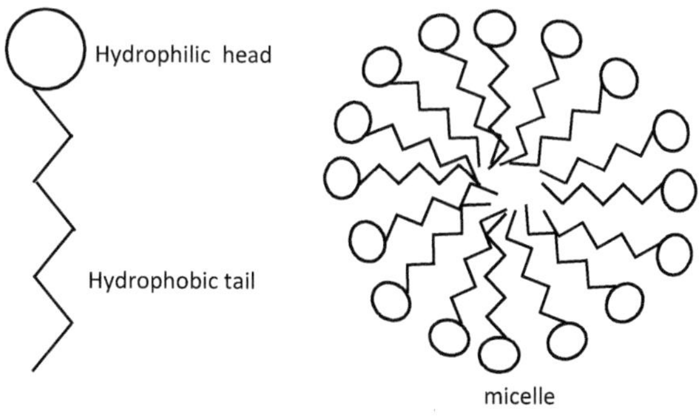

Fig. 2.36 Schematic representation of a surfactant molecule and micelle formed by molecules of surfactants

experimentally in measuring the surface tension for different concentrations of surfactant by one of the methods presented earlier. Below the CMC, the surface tension changes strongly with the concentration of surfactant while it remains almost constant above the CMC.

The previously presented physical properties are useful to calculate the Capillary number, i.e. the relative effect of the viscous force to the interfacial forces:

$$Ca = \frac{\frac{\mu U}{r}}{\frac{\sigma}{r}} = \frac{\mu U}{\sigma}$$

where μ [Pa.s] is the viscosity, U [m/s] the velocity, and σ [N/m] the interfacial tension.

The capillary number is the driving parameter of most of the two phase microfluidics systems, and the most suitable non-dimensional number characterizing regimes of formation of droplets or bubbles.

2.4.3 *Experimental Aspect of Microfluidics*

2.4.3.1 µPIV Experimental Set-Up

µPIV has been developed on inverted microscopes [69]. Numerous companies are building these kinds of microscopes: Zeiss, Leica, Olympus, Nikon…. The easy optical access of the microfluidic microsystems (usually in transparent materials like Glass, PDMS, Plexiglass or other transparent plastics), has been used to visualize the flow field on a plan. A typical set-up is represented schematically in Fig. 2.37.

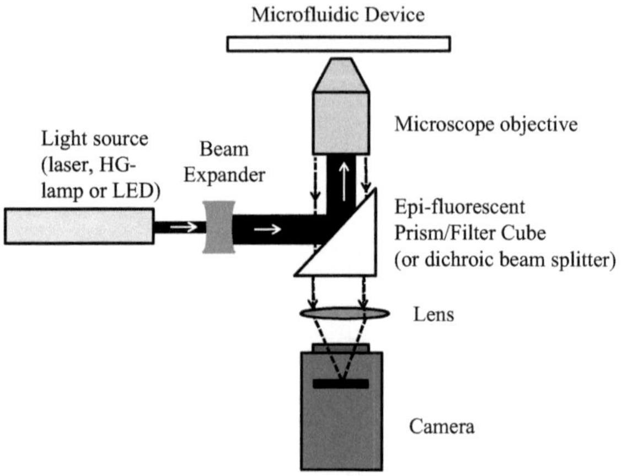

Fig. 2.37 Schematic representation of a µPIV set-up (see [57, 69, 70])

A strong light, usually a double-pulsed laser light (typically a Yag Laser), is used to illuminate fluorescent particles. But other light sources can be used like mercury lamp [55] or strong LED [72]. In the latter case, shadow imaging of non-fluorescent particles are used instead of fluorescent particles. In case of a laser, an epi-fluorescent prism transmits the laser light to the microfluidic microchannel through the objective lens of the microscope. The fluorescent particles absorb the laser light and re-emit light in a different wavelength. This fluorescent light crosses the objective lens and the epi-fluorescent prism to be received on the camera. The epi-fluorescent prism includes a filter that cuts the light at the wavelength of the laser and lets the fluorescent light go through in order to get rid of incoming laser light, of the scattered light, of the multiple reflections, of the background illumination inhomogeneity [69] and of the speckle patterns. Images are commonly taken on 12 bits CCD or CMOS camera. Double image cameras are chosen when double-pulsed laser are used.

Camera

Microfluidic experimentalists often rely on high speed cameras for imaging and high temporal resolution μPIV. The main brands of high speed cameras are Phantom, Photron and Optronis. They are useful for two reasons: (i) Displacements on images are large because of extensive zooming on microscopes. Therefore, short exposure times and high frequency imaging are necessary to capture temporal evolution of the velocity field or the movement or deformation of droplets or bubbles, (ii) The physical limit of light illumination on the microchannel must be compensated by a high sensitivity of the high speed cameras in order to freeze images (very low exposure times are necessary to avoid blurred images).

Exposure times as low as 1 μs and frequencies as high as tens of thousands of images per second are used routinely in microfluidics. Since microsystems are transparent, imaging of droplets by back illumination and image analysis on softwares like ImagJ or Matlab have been a major tool to estimate the size of droplets, mixing or droplet formation to cite a few.

Depth of field

Unlike the traditional PIV where the measurement plane and its thickness is defined by the thickness of the illuminating laser sheet, in the case of μPIV the illumination is a volume illumination, and the thickness of the measurement plan is defined by the depth of field [69] or depth of focus [71] of the optical system. The question on how to define the thickness of depth of field form the optical system has appeared very early in the development of μPIV systems [69–71].

The estimated depth of field δ_{Zm} of the microscope objective has been estimated by [69] to be :

$$\delta_{Zm} = \frac{3n\lambda_0}{NA^2} + \frac{2.16d_p}{tan\theta} + d_p$$

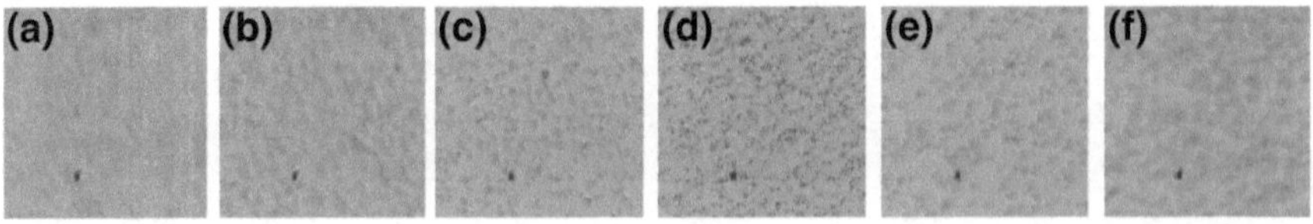

Fig. 2.38 Image of d_p = 1.09 μm diameter particle measured at different distance from the focal plane **a** 40 μm below the focal plane, **b** 20 μm below, **c** 10 μm below, **d** at the focal plan, **e** 10 μm above, **f** 20 μm above, the objective lens is × 20, the numerical aperture is 0.40 (image extracted from Carrier et al. 2015 [72])

where the different parameters are : NA, numerical aperture of the microscope objective, d_p, the particle diameter, λ_0 wavelength of the light, n index of refraction and θ the half angle of the objective ($NA = n\sin\theta$).

To compare the previous formula with experimental results (see Fig. 2.38), 1.09 μm particles have been placed between two glass plates and images have been taken at different objective-microsystem distances (see Fig. 2.38). The experimental measurements are in good agreement with the δ_{Zm} = 16.5 μm depth of field of the microscope calculated by the above analytical expression [69].

Olsen and Adrian [71] defined a *depth of correlation* as the depth over which particles significantly contribute to the correlation function. They give the analytical formula

$$Z_{corr} = \left[\frac{(1-\sqrt{\varepsilon})}{\sqrt{\varepsilon}}\left(f^2 d_p^2 + \frac{5.95(M+1)^2\lambda^2 f^4}{M^2}\right)\right]^{1/2}$$

where d_p is the particle diameter, M the magnification, λ the wavelength of the light emitted by the particle, f is the focal number of the lens, $f \approx n_0/(2NA)$, NA being the numerical aperture and n_0 the index of refraction of the immersion medium. ε is the ratio above which the weighting function contribution of a particle located beyond Z_{corr} is negligible. In practice, ε is usually taken as 0.01 when assuming that the out-of-focus particles have an intensity of 1% of the in focus particles [57]. Bourdon et al. 2004 [70] have validated experimentally and computationally the accuracy of this analytical formula.

Particles

Particles used as seeding particles in μPIV must follow the flow field, therefore (i) they must be well dispersed in the fluid (particles may be treated by surfactant grafting or by adding small amount of surfactant in case of polymer particles in a hydrophilic fluid), (ii) they must have a density close to the density of the fluid to avoid sedimentation or creaming effects, (iii) they must be small enough to follow the flow, large enough to avoid large Brownian motion that would add scattering to the velocity measurement, and large enough to be visible on several pixels of the camera. Numerous companies are providing calibrated particle, among them www.microparticles.de, Bangs Laboratory (www.bangslabs.com), www.sigmaaldrich.com, Merck-Millipore, Harvard Office of Technology Development, Duke Scientific (www.dukescientific.com), Invitrogen (www.thermofisher.com), Molecular Probes™ and the companies selling μPIV equipment. A general tradeoff is to use seeding particles of the order of 1 μm. In case of fluorescent seeding, particles can be smaller because not the particles, but the emitted fluorescent light is imaged.

Cross correlation

The seeded flow is illuminated by a laser pulse of a few picoseconds while the first image is taken. The second image is taken during the burst of the second pulse (see Fig. 2.39). Each image is divided into interrogation areas of 16 × 16 or 32 × 32 or 64 × 64 pixels or combination thereof. It is recommended to have around 10 particles per interrogational areas. The most common technique for calculating the flow field is to use a pixel by pixel cross correlation between corresponding interrogational areas.

$$C(s) = \iint_{Area} I_1(X) I_2(X - s) dX$$

The correlation produces a peak identifying the most common particle displacement dX (see Fig. 2.39). A velocity map is obtained by repeating the cross correlation over each interrogational area.

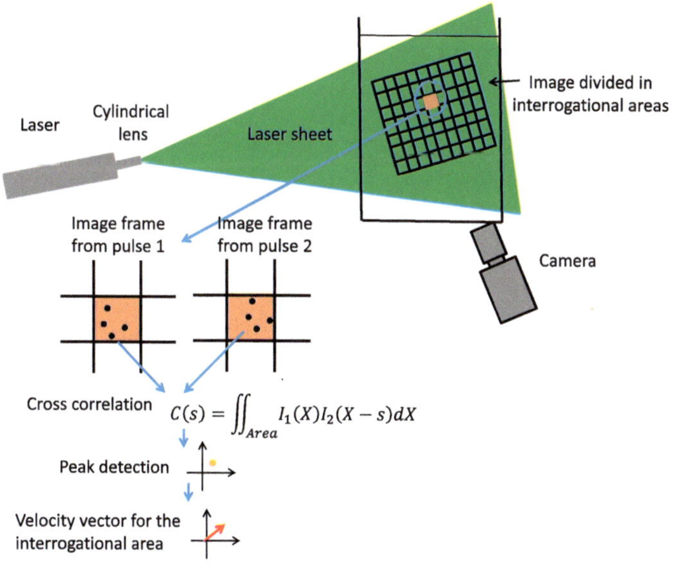

Fig. 2.39 Particle Image Velocimetry set-up and principle of the flow field measurement, i.e. Cross correlation between corresponding interrogational areas

Flow field can also been obtained by Particle Tracking, but in this case the working conditions are more severe like having less seeding particles and a good contrast so that particle can be identified clearly.

Brownian motion

At small scale, Brownian motion may play a major role. Calculation of the impact of Brownian motion on the velocity has been first studied by Santiago et al. [55]. For a time interval of Δt corresponding to the time interval between the two images used for the PIV, the mean square distance of diffusion is $\langle s \rangle^2 = 2D\Delta t$, where

$$D = \frac{kT}{3\pi\mu d_p}$$

where d_p is the particle diameter, k the Boltzmann's constant, T the temperature and μ the dynamic viscosity of the fluid. The displacement of a particle following a steady flow at a velocity u is $\Delta x = u\Delta t$ over a time interval Δt.

Therefore, the relative error due to Brownian motion is estimated by:

$$\varepsilon_B = \frac{\langle s^2 \rangle^{1/2}}{\Delta x} = \frac{1}{u}\sqrt{\frac{2D}{\Delta t}}$$

ε_B has to remain low.

Devasenathipathy et al. 2003 [73] notice that the diffusive uncertainty decreases as $1/\sqrt{N}$, where N is the number of particles per interrogational area.

Most of the μPIV are 2D, but 3D μPIV and tomographic have recently been developed. The reader interested should consult the recent reviews of Cierpa and Kähler [57] on these 3D μPIV.

2.4.4 *Experimental Field Where μPIV Has Been a Major Measurement Technique*

2.4.4.1 Droplet Formation

Droplet manipulation is one of the most successful applications of microfluidics (see Fig. 2.40). The reproducible formation of droplet serves various applications such as high throughput screening [20]. Our understanding of the droplet formation in microsystems has been significantly improved using the μPIV measurement technique (see [72, 74, 75]).

μPIV allows the measurement of the stress on an interface. Knowing the viscosity, the local velocity gradient $\dot{\gamma}$ can be calculated from the flow field, and the local shear stress on the droplet interface (or cell membrane in the case of a cell) can be calculated by $\tau = \mu\dot{\gamma}$ (see Fig. 2.41).

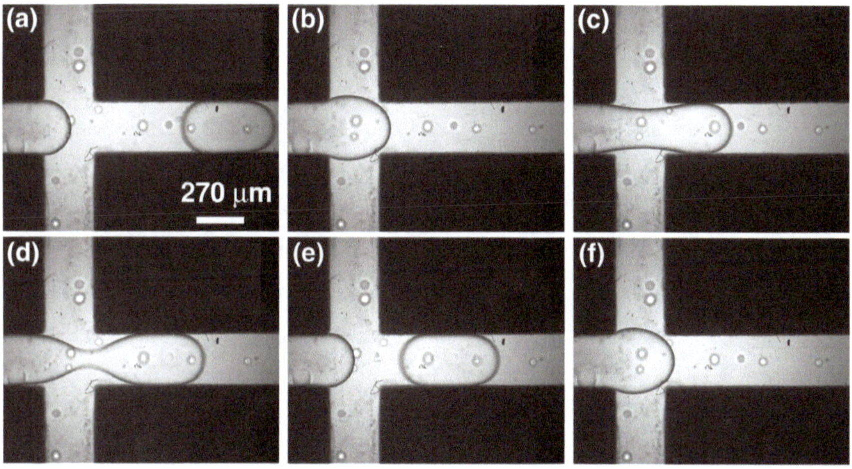

Fig. 2.40 Droplet formation in microsystems (image extracted from Funfschilling et al. 2009 [75])

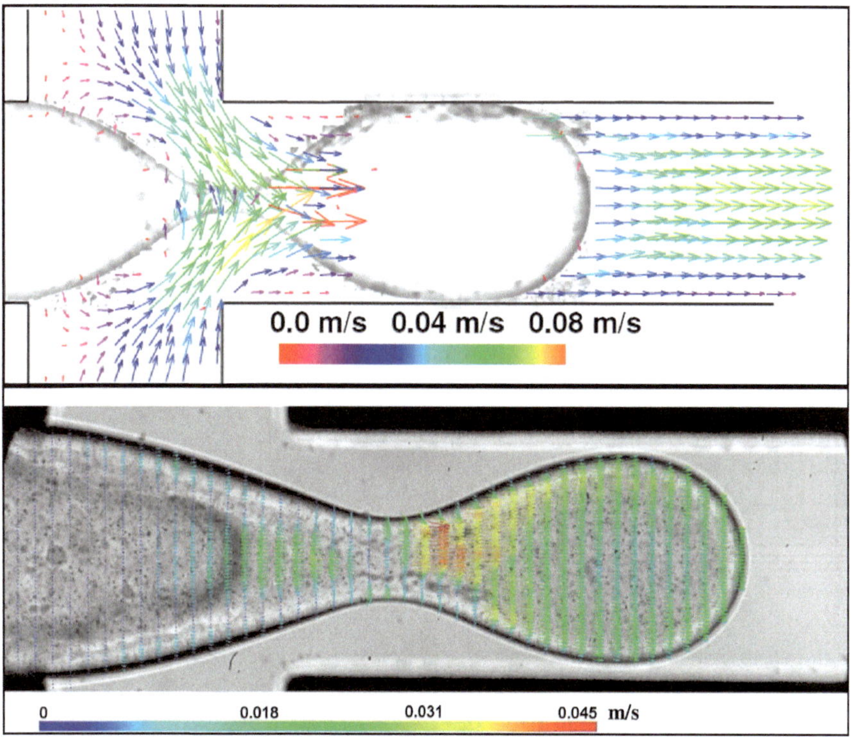

Fig. 2.41 μPIV measured flow field inside and outside a droplet in formation. The step of the neck thinning on the way to rupture is represented. The physical mechanism of the formation of droplet is well depicted by the flow field measurement (image extracted from Funfschilling et al. [75] and Carrier and al. [72])

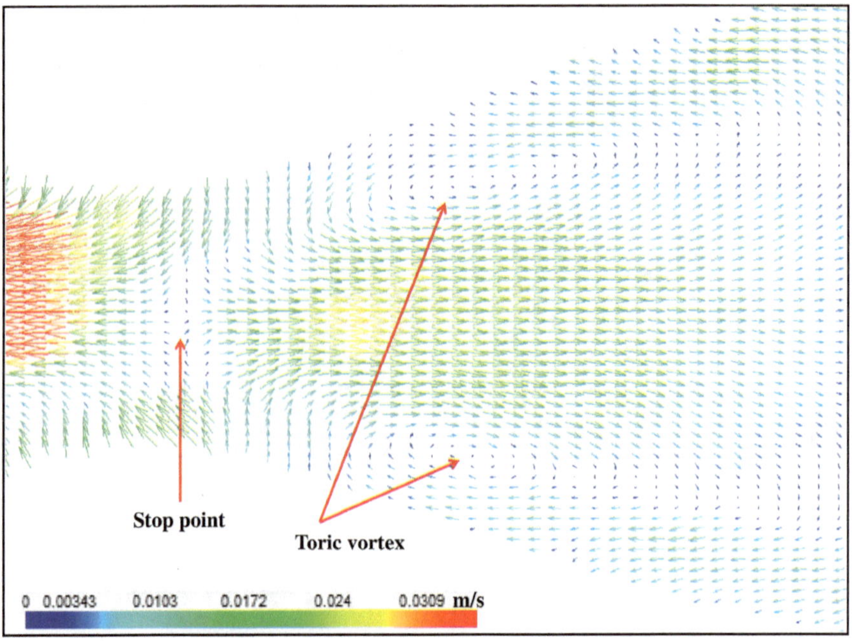

Fig. 2.42 Flow field in a droplet in formation. A stop point is clearly visible between two toroidal vortices respectively in the droplet to be detached and in the finger on the way to retreat (image extracted from Carrier et al. [72])

Elongational flows are also measured by this technique (see Fig. 2.42). It is important information that enables the measurement of mechanical properties of membranes of cells, or the elasticity of DNA [76].

2.4.4.2 Stress Effect of Blood Flow on Growing Endothelial Cell on the Arterial Wall

Abdominal Aortic Aneurism and Brain Aneurism are still difficult to prevent and to cure. Recent research pointed out the role of the endothelial cells and the wall shear stress to which they are exposed on the development of Aneurism [77]. Furthermore, shear-induced shear stress modulates the gene expression of endothelial cells [78]. μPIV in microchannel is the way (and probably the only reliable way) to study the behavior of endothelial cells under the periodic cardiac blood flow that induces a periodic shear stress on the arterial wall covered with endothelial cells. Micropumps reproducing the cardiac cycle can be used to control the flow in a Hele-Shaw cell (a thin cell with wide X and Y dimensions, so that the flow can be considered as nearly 2D), where the walls are covered with endothelial cells [62]. As shown in Fig. 2.43 in the case of a droplet in formation, the wall shear stress can be extracted from the flow field measured by μPIV by multiplying the near wall velocity gradient by the viscosity of the fluid $\tau = \mu\acute{\gamma}$.

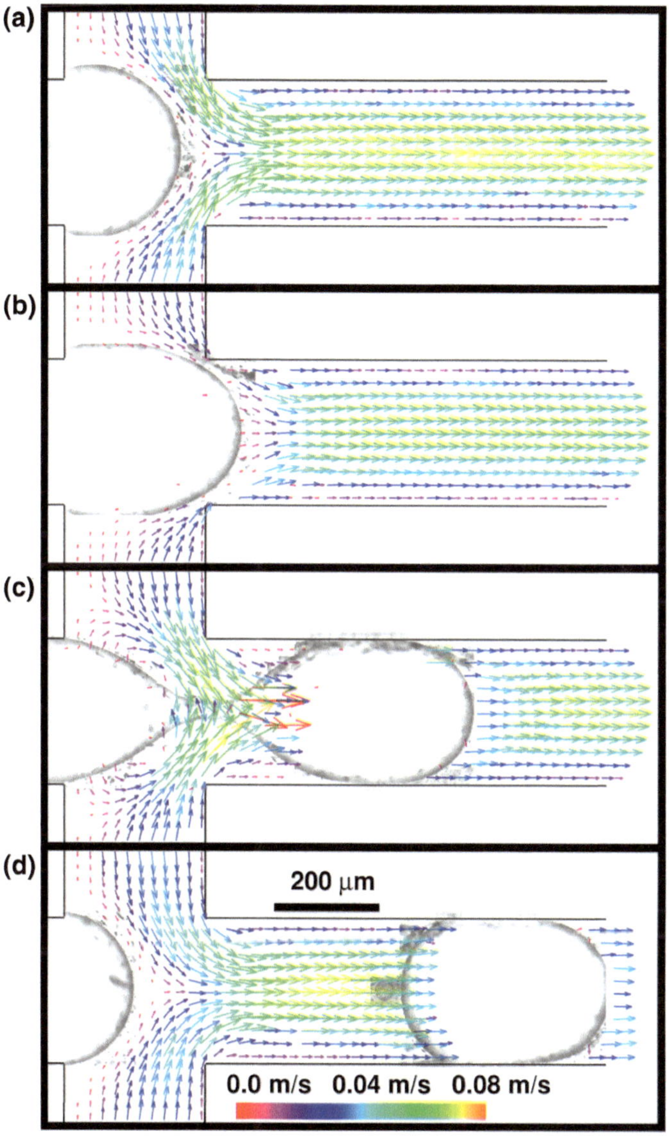

Fig. 2.43 Flow field around a droplet in formation. The shear stress applied locally on the droplet surface $\tau = \mu\acute{\gamma}$ can be measured knowing the viscosity of the fluid and the shear rate $\acute{\gamma}$ calculated from the flow field (image extracted from Funfschilling et al. [75])

Rossi et al. [62] have determined the topography and shear stress distribution over individual cells with subcellular resolution by the μPIV technique. In vitro experiments on endothelial cells in microsystems have the advantage of better established and more reproducible flow conditions and are less subject to biological variations.

The study of Rossi et al. [62] reveals that the shear responsive gene *KLF2* (Krüppel-like factor-2) increases its activity as the shear stress level increases.

2.4.4.3 Wall Shear Stress Measurements in the Arterial System of an Embryonic Chicken

Wall shear stress is strongly involved in the development of the circulatory system. In the embryonic stage, blood flow affects the development of the primitive heart and blood vessels [79]. The alteration of the flow in the heart leads to malformations in the cardiovascular system [79]. The biological response to the haemodynamic forces can be short term for the control of vascular tone or long term through gene expression [25]. It plays also a role in the etiology of a number of diseases, for example regions with low or oscillating wall shear stress have been linked to atherosclerosis [80]. The blood flow and wall shear stress have been measured in vivo in the outflow of an embryonic chicken heart [25, 26] by the μPIV technic with large temporal resolution. μPIV is a very reliable non-intrusive method for measuring the velocity profile with a sufficient spatial resolution to calculate the wall shear stress from the closed wall velocity gradient and therefore enables the study of the interaction between hemodynamics and the surrounding tissue [80].

2.4.4.4 Rheology

According to [63], μPIV in micro channels offers the possibility to develop rheometers suitable for probing rheological properties of weakly elastic fluids that appear as Newtonian under the conditions attained in a conventional rheometer. Sugii et al. [81] have measured the rheology of blood flow in microcirculation *in vitro* by measuring both red blood cell velocity and plasma velocity with a high-speed μPIV.

2.5 Main Ex-Situ Characterization Techniques for Bio-devices (Norbert Dumas, Wilfried Uhring)

2.5.1 Introduction

Microfluidics devices are preferably made of see-through material enabling the characterization of the biochemical reaction inside the chip, besides the possibility of visual inspections and flow characterization. This chapter focusses on main optical techniques that do not require integrated part inside the chip. There are other physical principles that would be classified as not requiring integrated part but they are not treated here (radioactive marker, NMR…) are less common. Some optical setups are presented in details so that one can make it from detached optical element and an optical table. This goal is also to explain role of each element. Alternatively, inverted microscope with some elements already assembled can be bought directly from a manufacturer. Microscopes are convenient in a biological lab environment because it offers the binocular visualization. Optical readers for microfluidics are also commercially available.

2.5.2 Colorimetry

2.5.2.1 Measurement Principle

Colorimetry is a well-established method for biological assay. The colored product can be the result of a reaction. For example, the Berthelot reactant produces a blue-green solution in a quantity depending on the ammonium ion concentration. It can also be used identify a sample solution it from its color. This is named barcoding.

The setup is based on the measurement of the absorbance of the sample, which is considered to be in a droplet. The basic Beer-Lambert law is used to characterize the absorption for a monochromatic light source propagating in a homogeneous and isotropic medium and can be expressed as:

$$I = I_0 . 10^{-\varepsilon_\lambda . l . C}$$

Or

$$A_\lambda = -\log_{10} \frac{I}{I_0} = \varepsilon_\lambda . l . C$$

where I/I_0 is the transmittance of the medium, A is the absorbance, also called the optical density, that are both unitless, ε_λ is molar extinction generally expressed in $L.mol^{-1}.cm^{-1}$, l is the length of the optical path in the medium and C in the molar concentration generally express in $mol.L^{-1}$.

The molar extinction ε_λ is related to the chemical species, the wavelength and the temperature. Considering a microfluidic canal of thickness l, if ε_λ is known, the concentration of the chemical species can be assessed by measuring the absorbance.

In presence of several chemical compounds the absorbance of each species are added so that the total absorbance of a droplet is given by:

$$A_\lambda = \sum_{i=1}^{n} A(\varepsilon_{\lambda i} . l . C_i)$$

As a consequence, n different concentrations of chemical species can be discriminated by using n different wavelengths (or bands) and by solving the

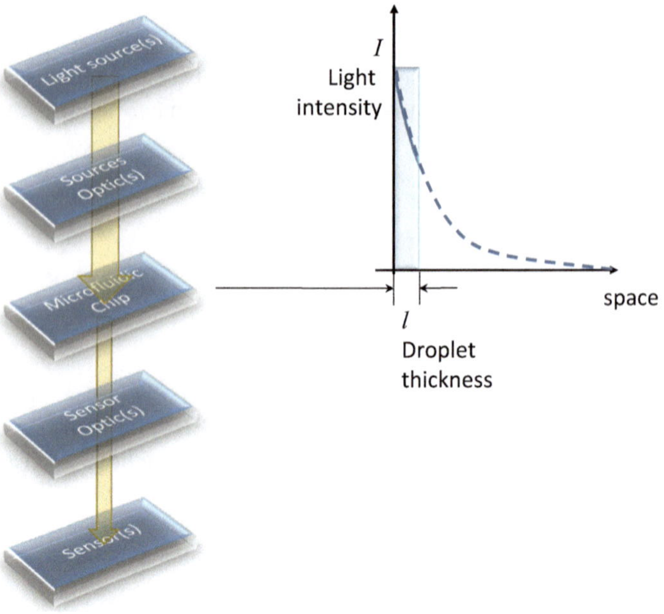

Fig. 2.44 General principle of a colorimetry optical setup

n resulting equations. Another method consists in varying the species concentrations within the sample while the wavelength is constant. For these measurements, the reference intensity I_0 has to be measured. The simplest method is to ensure that the targeted compound concentration is almost null in a specific sample. If the carrier fluid, the other species concentration and the reflective index do not change, I_0 is the intensity measured with this reference droplet(s). For paper based microfluidics, it can be difficult because a dry paper has a different reflective index than wet paper [82].

The scheme of the general principle optical setup is given in Fig. 2.44. Reflective based methods are possible but less sensitive. This is a very straight forward measurement approach and many parts of this setup can be very simple and cheap. The main challenges of this setup are:

(1) Ensuring that most of the light is passing through the sample, e.g. a droplet, in order to obtain the maximal sensitivity and to be able to carry out a quantitative assessment of the measured concentration.
(2) Ensuring a low noise measurement in order to obtain a sufficient signal to noise ratio. Indeed, in regards with the very low canal thickness l, the variation of the light intensity can be very weak, thus the signal to noise ratio must be kept high in order to be able to discriminate the lowest concentrations.

2.5.2.2 Implementation

Quantitative colorimetry requires a monochromatic measurement, thus monochromatic LED or laser diode are very suitable as a light source for this application [83]. Indeed, the commercially available LEDs are covering a large spectrum from UV to visible and Near infrared. Alternatively, polychromatic light sources such as white LED, incandescence/halogen lamp [84] and supercontinuum can also be used in addition with optical bandpass filter. For instance, thanks to its additional Bayer filter that acts as a bandpass filter, a simple mobile phone camera can be a cost effective devices for colorimetric device [85, 86]. Because it is cost effective, colorimetry is often used with very low cost paper-based microfluidics. For example, the use of a switched tri-color LED (Red, Green Blue) and a single photodetector allows transmittance colorimetry measurement with very low cost [82] ($\sim$50$).

However for a complete characterization, a wavelength sweep is required. It can be implemented with a white light source followed by a prism or a monochromator. A moving hole determines the wavelength. For example, it has been used to characterize more precisely a low cost colorimetric system [87]. Improvement of the technique includes modulating the light source and using a lock-in amplifier for readout. Placing the light source and the sensor in the canal axis is better, as the thickness is increased, but it requires chip modification. Sometimes the reference sample with no product is measured in separate chamber to that the analysis chamber is used only once.

2.5.3 *Fluorescence Spectroscopy*

2.5.3.1 *Principle of Fluorescence Spectroscopy*

Fluorescence spectroscopy is based on the absorption of light by a fluorescent substance which will re-emit a photon at higher wavelength [88]. This Stockes shift is exploited to separate the fluorescence light from the source (see Fig. 2.45). The fluorescent substance is called fluorophore or fluorochrome. It can either be the biomolecule we want to measure directly (intrinsic), or a fluorophore attached to a target molecule (extrinsic). A fluorophore can also be used as an indicator with fluorescence properties which depends on the presence of the target molecule. The progress in biochemistry provides many ways (from simple and direct to complex and indirect) to extract the information of interest for the biologist from a measurement of fluorescence. Once the biochemistry is set, three main techniques are applicable for Lab on Chip to read the fluorescence information: Intensity, Polarization and Lifetime. It is important to maximize the intensity of light that is measured for the three techniques because of noise and other parasitic signals. It is based on the principles described in this section that are also applicable to phosphorescence, which is another type of luminescence.

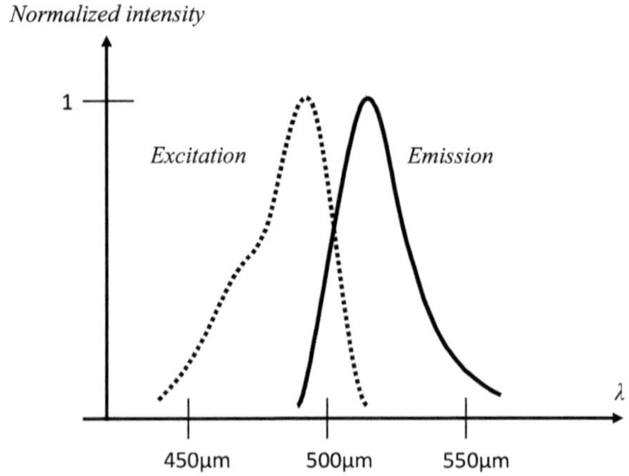

Fig. 2.45 Image of a fluorescence excitation and emission spectrum (fluorescein)

Fluorescence measurement offers a better contrast than colorimetry. For example let's consider that one wants to make the difference between no product and a product at low concentration. Colorimetry measures a small decrease of a large light intensity caused by absorption. As a consequence, the relative variation is weak. Although fluorescence depends also on absorption of photons, fluorescence is detected only if there is product because what is detected is only what is reemitted. Therefore the relative variation is much larger. This is important because, even for high throughput application, concentrations are often lower than 1 μM. The detection limit of fluorescence spectroscopy can be lower than 1 nM. In this respect, colorimetry is very limited as the detection limit is in the order of 1 μM.

Nonetheless it should be noted that, because this technique still depends on absorption, the fluorescent signal intensity increases with the effective number of fluorophore molecules that has actually absorbed light. In order to not waste the incident light, the light intensity can be concentrated using a collimated laser beam as a light source. However the relative shallowness of the microfluidics canal is a problem because for a typical concentration of product the laser beam is not totally absorbed and passes through the canal. In this respect, techniques depending only on the detection of product at surface level (i.e. CHEMFET, SPR) do not suffer from the thickness reduction. However these techniques require the integration of a sensing element inside the canal contrary to the fluorescence spectroscopy.

The product absorbs the following light intensity I_a according to the beer lambert law:

$$I_a = I_0\left(1 - 10^{-\varepsilon_\lambda \cdot l \cdot C}\right) \sim I_0 \cdot \varepsilon_\lambda \cdot l \cdot C \cdot \ln(10)$$

where I_0 is the incident light intensity from the source. If the laser spot is smaller than the canal and we neglect the loss on the light path from the laser to the inside of the canal, I_0 is the laser intensity. ε_λ is the molar extinction coefficient of the product and is dependent of the wavelength of excitation. l is the thickness for a light coming orthogonally to the plane of the microfluidics chip. C is the concentration of the product. For low concentration and thickness, the absorbed light is proportional all these four parameters (I_0, ε_λ, l, C).

The fluorescence emission depends on I_a but also on the quantum yield (Q.Y.) because not all absorbed photon are reemitted. For the fluorescein fluorophore this quantum yield is 95% in a 0.1 solution of NaOH at 22 °C but for Tryptophan, which is one of the amino acid responsible for the fluorescence of many proteins, the quantum yield is only 14% in water at 20 °C. This quantum yield depends on the non-radiative decay, quenching by other species (e.g. dissolved oxygen) and transfer of an exited state such as RET (resonance energy transfer). The latter can induce the emission of a photon but not from the considered fluorophore. When excited with a constant light source, the average reemitted intensity is $I_e = k\ Q\ I_a$, where k is a coefficient taking into account the Stockes shift on the entire emission spectrum. It is smaller than 1 because the emitted photons are less energetic than the absorbed ones. As this light is reemitted in all directions, only a fraction of it can be collected through a microscope objective.

The absorption of a photon is quasi instantaneous as soon as the photon arrives but it is not reemitted immediately. The average time to reemit the photon is called the lifetime (τ). The probability that the photon is emitted after a time t is:

$$p(t) = \frac{1}{\tau} e^{-\frac{t}{\tau}}$$

To observe this phenomenon, a very short pulse of light should excite the fluorophores and a high bandwidth light sensor should measure the temporal response that is an exponential decay. When the incident light is constant the emitted light is also constant because of averaging effects. Zooming in time and looking at the emission of photons one by one show that it is a Poisson process. This results in quantum noise that can affect the measure for low intensity [89].

If the light is polarized, only the fluorophores with a transition moment aligned with the electrical field of the incident light are excited. The polarization of fluorescence will depend on the orientation of the transition moment at the emission of the photon. Due to agitation the fluorophores move and rotate and one can define a rotational correlation times that depends strongly on the size of the fluorophore. It is typically in the order of ns or tens of ns for a macromolecule. If the lifetime is much higher than this time, the polarization of the light is lost. As the lifetime of phosphorescent molecules is in the order of ms, anisotropy cannot be applied in this case. Fortunately the lifetime of many fluorophores are typically in the order of ns or tens of ns. Therefore one can expect a change in the polarization of light if the fluorophore attaches to a macromolecule bigger than itself.

2.5.3.2 Considerations on the Setup for Lab on Chips

Research and development in biology make intensive use of spectrofluorometers, whether the sample is placed in a chamber or in microplate array for parallelization and achieving high throughput analysis. However, although the same principle is used and the optical setup is similar for analysis, most of the time it has to be adapted to read a microfluidics chips and the optical setup is built around the chip. Sometimes the chip is placed on an inverted epifluorescence microscope, offering great flexibility for visualization. Obviously the format of the microfluidics chips differs but also the way of achieving high throughput and small volumes is a completely different concept. The microplate readers use parallelization of the sensors and robotic arms while the samples are static in a well. Using microfluidics, the sample is preferred to be contained in a moving microdroplet although sensors can be easily parallelized due to small size. With a rate of 1000 droplets/s the same rate of analysis is achieved with a single sensor. This is much faster than using robotic arms while simplifying the setup.

Here, we will consider mainly that samples are contained in a flow of droplets passing in a canal, unless specified otherwise. Static analysis is possible with the same apparatus by moving the chip (with either a manual or a motorized system). One can also analyze static droplets in stop-flow or EWOD (electrowetting on dielectric) or also Dielectrowetting [90]. Differential measurement with moving droplets is easy because, when the oil separating two droplets is passing by, a reference signal can be read if the oil is non fluorescent. Ideally the spectrofluorometer can excite fluorescence and read it at any wavelength. It uses a white and stable light source (e.g. a Xenon lamp) that is either filtered or passed through a monochromator in order to select a wavelength. While it can still be used, it is more convenient to use a collimated laser source. Whereas the wavelength is fixed, the size of the beam offers an easy way to focus the intensity for more efficiency. If several wavelength of excitation are required, lasers beam are combined as well as the detectors for each spectrum of emission. Laser can also be coupled to an optical fiber to excite the fluorophore. The emission can also be collected with the optical fiber so that the probe location can be easily displaced.

A limitation of the laser beam intensity is the alteration of the microfluidic chips or the samples. At high intensity photobleaching of the fluorophore can be problematic for long measurement. Material should be chosen to be transparent to the wavelengths of interest, which are not necessary in the visible spectrum. For example, common borosilicate glass absorbs UV below 300 nm. The fluorescence of the material should be low and is problematic for weak intensity measurement. The fluorescence of the glass depends strongly on impurities inside. In [90], the experiments at 4 wavelengths show that a commercial borosilicate glass has a slightly lower autofluorescence than plastic materials (PDMS, PC, COC and PMMA). It also depends on the excitation wavelength. Photobleaching of the material is possible but it does partially recover [91].

A confocal setup also helps to remove the fluorescence of the material and other interfering lights by reducing the depth of focus. This way, only the light coming from the focal plane is collected. As only what happen in the thickness of the canal is of interest, the optical requirements are relatively low, i.e. much less than a confocal microscope with 500 nm or less axial resolution. In [92], authors use a relatively big pinhole size of 500 μm and a 20 × 0.75 NA objective. It results in a 14 μm probe depth, still smaller than the droplet thickness. Background parasitic signal can also be removed by two-photon absorption. In this case the measured fluorescence is concentrated at the focal point. The probe volume is very small, typically less than 1 μm side [93].

Small probe volume can be required for Single Molecule Detection (SMD). It is often achieved with Spatial Two-Photon Fluorescence Cross-Correlation Spectroscopy [94]. However microfluidics offers the advantage to perform SMD through dilution without using an ultra-small probe volume. It benefits from the high repeatability on a large number of droplets in a small time that makes possible good statistical analysis [95]. Integrated fiber optics (or optical waveguides) may also improve the spatial resolution but it is limited by focusing issue at the blunt end of the fiber if there is no special ending [96–98]. Integrating spherical lenses to benefit from photonic nanojet is effective [99, 100]. However these integrations require major changes in the process of fabrication of the chip.

Unless it is required, having a too small probe volume can be problematic. If the droplet content is inhomogeneous there is a risk that the fluorescent target is missed or in a region where the incident light is different (Fig. 2.46). This is a strong source of discrepancy for cell analysis in droplets for example. To avoid this, the light can be spread on a fine line across the canal width [101].

Other parameters to consider for fluorescence analysis in microfluidics are the pH (fluorophore are very sensitive to pH and a buffer is required), temperature (less sensitive), the formation of "excimer" with some fluorophores at high concentration. The latter causes a change of the fluorescence spectrum because of a change in the fluorophore nature. The biochemistry should take into account the environment

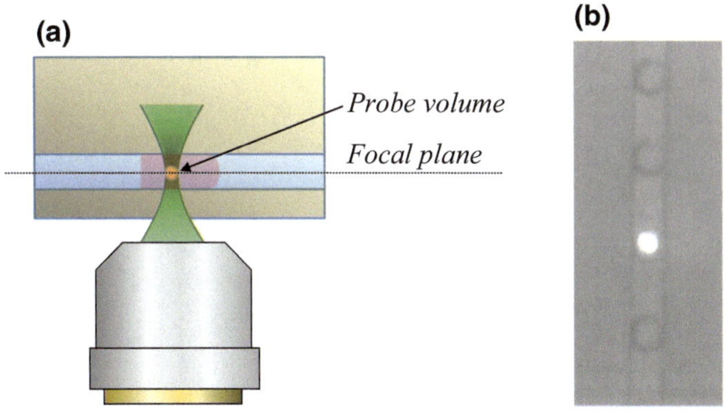

Fig. 2.46 **a** Schematics of focused green laser spot in a droplet emitting orange fluorescence, **b** top view image of a spot

of the fluorophore including the surfactant if there is any in the droplet. It is preferable to use a surfactant dissolved in the oil unless the surfactant is part of the biochemistry. For example, the surfactant is functionalized in [102]. With microfluidics, canal edges can also be problematic because the reflection of the laser beam on it can cause cross coupling in multiple points measurement for example. Another source of coupling is the diffusion of the fluorophores through the oil [103]. It is fortunately typically slow and depends strongly on the fluorophore, the liquids and the surfactant.

Sometimes, the biochemistry method needs a functionalized surface. Without modifying the process to functionalize the microfluidics chips it is possible to use functionalized magnetic bead. They are commercially available.

2.5.3.3 Setup for Fluorescence Spectroscopy

The Fig. 2.47 shows the optical setup for fluorescence spectroscopy with two static excitation monochromatic light sources that can be used simultaneously: blue (λ_0) and green (λ_1). Collimated laser sources suits very well for the application. Preferably they should be stable (power regulated if using laser diodes) because the measurement is sensitive to intensity. This setup also offers the possibility to read the fluorescence signal on two bands of interest: yellow with sensor S_0 and orange with sensor S_1. Each sensor measures the intensity over a certain time. This time can be long for a static observation in a microfluidics chamber. As the kinetic of the reactions are relatively slow, it is possible to average to obtain a good signal to noise ratio. For high throughput droplet analysis higher sensor bandwidth is required while the period of observation is shorter. Photomultiplier Tubes

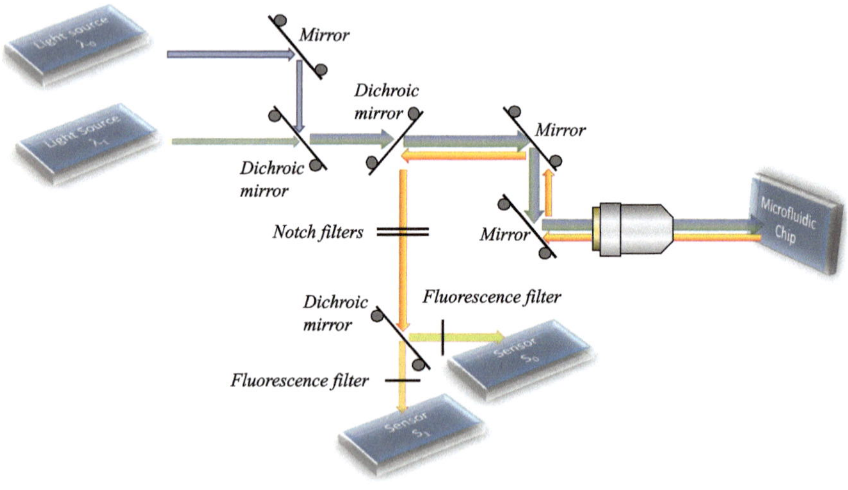

Fig. 2.47 Fluorescence spectroscopy with two wavelength

(PMT) are well suited in terms of performances (sensitivity, resolution, bandwidth, quantum yield). Furthermore the controlled amplification is adjusted to the signal level on a large dynamic range. The scintillator of the PMT should exhibit a good quantum yield at the considered wavelength. Alternatively Avalanche Photodiodes (APD) may also offer good performances [104].

The two light sources are combined in a single beam with a dichroic mirror which transmits the green light (λ_1) without deflection and reflects the blue one in a specular way. The transition from transmission to reflectivity of this kind of mirror is quite sharp because it is based on interference with thin film layer. The mirror on the blue (λ_0) source of light makes a 90° turn to the beam but its main purpose is to offer a convenient way to adjust the blue beam spot on the dichroic mirror onto the outcoming spot of the transmitted green beam (λ_1). It is preferable to have two fine knobs (tip and tilt, grey circle in Fig. 2.47) to adjust the orientation of the mirror plane and therefore the orientation of the laser beam and where it will hit the dichroic mirror. It should be noted that two degrees of freedom are enough to fully define the orientation of a beam. However, having two superposed spots on the dichroic mirror is not enough because the two beams can separate again. The two fine adjustment screws of the dichroic mirror will ensure that the beams are superposed and will hit the microfluidic chip at the same location. In theory, having two non-superposed laser beams hitting the chips at the same location is possible, if they are not too far from each other, but in practice superposition is recommended.

After passing through another dichroic mirror, the superposed beam reaches a telescopic system of two mirrors to align this beam along the axis of the objective. Without these 4 degrees of adjustment the position and orientation of the green laser would be necessary and it is much less convenient. The telescopic system enables also to change the direction of the beam if this setup is to be adapted on an inverted epifluorescence microscope for example. The objective can be the one of an epifluorescence microscope whether it is implemented in a microscope or not. If not mounted of a microscope, the objective or the chip should be placed on a translator to finely place the focal plane on the sample location in the chip. 2D translation of the chip in the focal place is also necessary to precisely hit the sample at its location. The size of the beam spot on the microfluidics chip is proportional to the wavelength, to the focal distance and inversely proportional to the initial beam diameter. These relations come from the limit of diffraction. For example, if the spot size is 200 μm with a focal distance of 12 mm and it will be 20 μm for 1.2 mm. A good numerical aperture will collect the maximum light but it requires a short focal distance for a given objective lens diameter. The microfluidics canal wall thickness will limit the focal distance.

The collected light is collimated when coming out of the objective. When reaching the first dichroic mirror on its path, it is reflected. The cutoff wavelength of this dichroic mirror should be between the excitation and the emission. However a small but still significant part of the laser beam, which comes from a reflection on the microfluidics chip, is also reflected by the dichroic mirror. It should be noted that dichroic mirror have a typical transmission coefficient of 90%. It means that 10% is reflected even for a wavelength inside the transmission band (it is normally a non-dissipative element). Two sharp notch filters of high attenuation order (typically 5 or more) are used to remove efficiently the remaining excitation signals. Another dichroic mirror is used to separate the two bands of wavelength (yellow and orange on the example of the Fig. 2.47). A fluorescence filter is used to further filter what is outside the band considered (e.g. environmental light of the room). These filters are also based on interference in order to be very selective. In case that it is not sufficient and that differential measurement cannot be used (no droplet), it is possible to use a classical scheme with modulated light source and a lock-in amplifier to read out the signals.

Based on the same principle, the presented setup can be adapted for a single or more excitation light source and a single or more optical sensor. In practice, it is more convenient to measure the fluorescence while observing the microfluidics chip. For example it is possible to see where the laser spot hits the chip while making sure that droplets are properly formed in a canal. A white source of light (e.g. halogen lamp) is filtered and lightens the part of the chip to be observed (see Fig. 2.48). The filter is either high pass, low pass or band pass depending on where is a window of wavelength available. This light should not excite the fluorophores but also should not be measured as a fluorescence signal by the sensors. A dichroic mirror is used to separate the image (toward the camera and its optics) from the fluorescence signal. In an alternative configuration to the Fig. 2.48 setup, the image can go straight and the rest is reflected. A part of the fluorescence signal can also go toward the camera to have an image of the spot superposed with the image of the canal. Inverted microscope can be featured with wheels to hold and select the filter or the dichroic mirror so that the setup is adaptable. A safety filter should be added before the binocular to cutoff the excitation wavelength because reflection in the eyes is a risk.

2.5.3.4 Setup for Fluorescence Polarization

To measure polarization some elements must be added to the setup of presented in Fig. 2.47. First, the light sources should be polarized. Lasers are usually polarized but a polarization filter can be added to keep the desired component of polarization. On the sensing side a polarization cube separates the two components (see Fig. 2.49). A polarization filter with the proper orientation is added before each sensor to further remove the unwanted component of the light. Finally the measurement is performed by calculating the polarization (P) or the anisotropy (r) from the intensity ($I_{//}$ and $I_{\perp}$) measured by the two sensors:

$$P = \frac{I_{//} - I_{\perp}}{I_{//} + I_{\perp}} \tag{2.5.6}$$

$$r = \frac{I_{//} - I_{\perp}}{I_{//} + 2 \cdot I_{\perp}} \tag{2.5.7}$$

Both values are null if the two components are equal. This is the case for non-polarized light. Because it is ratiometric, these two values do not depend on the global intensity level. It influences the signal to noise ratio but not the average value. For this reason polarization results are reproducible with different equipment whereas intensity measurement is used only to monitor change unless a reference sample is used. It is less sensitive to the light source stability and to all parameters of Eq. 1 (concentration, thickness…).

2.5.3.5 Setup for Time Resolved Fluorescence

Implementing time-resolved fluorescence does not modify the optical setup of Fig. 2.47. The light of source must be pulsed (see Fig. 2.50) at a rate higher than the droplet rate but smaller than at least 5 times the lifetime to measure. Pulse duration should be much smaller than the lifetime of the fluorophore (typically some ns). The pulse duration of laser diodes can be in the order of hundreds of ps [105]. It makes it suitable for many fluorophores. Many lasers offer better performance in terms of pulse duration or power but are more expensive and sometimes very long to stabilize.

For the implementation presented on Fig. 2.50, the sensor must have a high bandwidth. The lifetime τ is extracted from the exponential decay contained in the sensor signal. Another way to obtain the exponential decay is called Time Correlated Single Photon Counting (TCSPC). It consists in measuring the arrival time of each single photon after a pulse with a Time to Digital Converter (TDC). The resolution of the TDC is in the order of the lifetime of smaller. The histogram of these arrival times is the exponential decay from which the lifetime is extracted. Single photon counting offers very good sensitivity. However, it requires that no more than one photon is received after a pulse. Otherwise there is a risk that two photons arrive too close. The second would fall into the dead time of the photon detector and would not be detected. As the photon emission is a Poisson process the probability is never null but for 1 photon over 10 pulses the lifetime estimation is quite accurate (3%, [106]). This is reserved for low intensity. It has been shown to be applicable, using Single Photon Avalanche Diodes (SPAD) as detectors, for

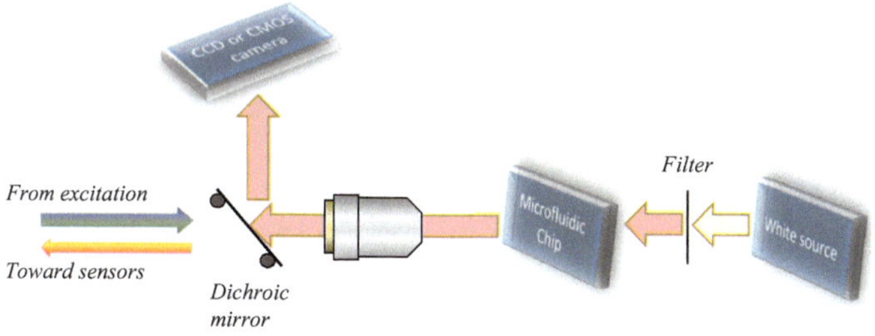

Fig. 2.48 Additional camera setup

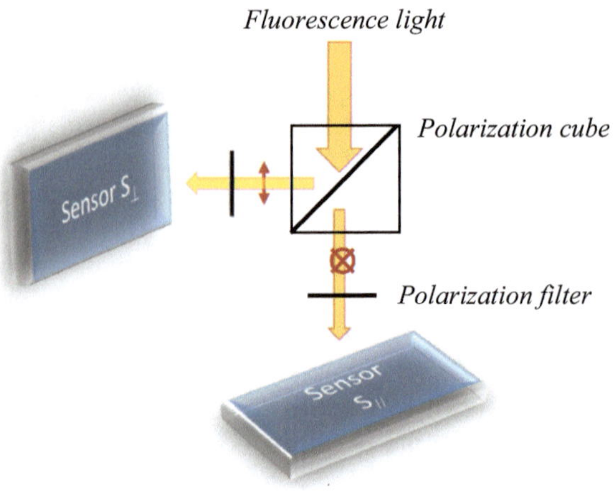

Fig. 2.49 Additional setup for polarization

Fig. 2.50 Implementation for time resolved fluorescence spectroscopy

droplet analysis [107] at about 1000 droplets per second. The standard deviation on the lifetime measurement is limited by the quantic noise and is $1/\sqrt{N}$, N being the total number of received photon. Similarly to polarization, although it offers different possibilities, time resolved spectroscopy does not depend on the global intensity level. Intensity only affects signal to noise ratio. Therefore it is less sensitive to parameters such as light source stability and drift, concentration or thickness of the canal.

2.5.3.6 2D Imaging

The setups for a single point measurement can be reused for 2D imaging. The sensor is replaced by a camera. There is a tradeoff of performance and cost for the choice of the camera: CCD, ICCD, EMCCD, CMOS, which can be cooled to enhance signal to electronic noise. The light source is less focused and often needs to be homogeneous on the area of observation. Although the performance are lower in terms of sensitivity, resolution, bandwidth on a single point (excitation light is not focused, each element of the sensing matrix is not optimized) it can be used to replace single point measurement and give supplementary information (e.g. the size and shape of a cell). It is possible to monitor cells labelled with fluorophore through time lapse microscopy in microfluidic culture [108]. To further improve the possibilities or the performances, confocal and two photons techniques are used. Dark field illumination also offers better performances because when the light comes from the sides there is less background signal [109]. On the sensor side, the intensity is not the only technique for imaging; time resolved fluorescence can be implemented with camera. This is called Fluorescence Lifetime Imaging (FLIM) [110].

2.5.3.7 Wavelength-Resolved Spectroscopy and Raman Scattering

Using three color cameras without fluorescence filter offers some capabilities for wavelength resolved imagery. It can be used for barcoding droplets. This is useful to identify the droplet that is analyzed, particularly in high throughput application [111]. However the spectral resolution is too poor for many applications. Using several sensors with different fluorescence filters as in Fig. 2.47 is better but the number of sensors is limited. A prism or a monochromator will achieve better spectral resolution. Due to dispersion a spot of light is transformed in a 1D line of light. Each sensors put on this 1D line measured the intensity at a different wavelength (Fig. 2.51). If the fluorescence comes from a 1D line, the result is a 2D surface to be measured with a camera. One dimension is the spectral resolution and the other is the spatial resolution. An alternative is to use one dimension for spectral resolution and the second dimension for time resolution with a streak camera [112].

Raman scattering spectroscopy demands a very good resolution for distinguishing the emission peaks that characterize the sample. The Raman scattering is the absorption of a photon by a material and reemission at a lower wavelength (anti-stockes) because the material has lost energy. The spectrum of the emitted signal contains peaks that are a fingerprint of its chemical nature and its surrounding. It can be measured with a setup similar to the fluorescence spectroscopy. However the Raman diffusion is too low and mixed with other phenomenon

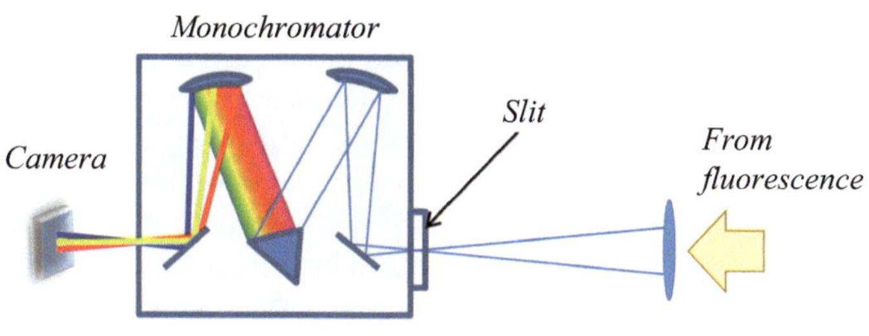

Fig. 2.51 Principle of spectrum decomposition using a monochromator

(e.g. Mie and Rayleigh diffusion). Surface Enhanced Raman Spectroscopy (SERS) provides enough intensity for analysis in a microfluidic chip [113]. This technique requires a metal surface close to the sample. Metal deposition in a chamber works but the chip process has to be modified. Alternatively, using metal nanoparticle does not necessitate metal deposition. The nanotubes are suspended in a colloid solution and should be mixed very efficiently with the sample. This mixing can be done in the microfluidic chip. The main difficulty is to preserve the high SERS activity of the nanoparticle. To solve this problem other chemical substances have to be mixed.

2.5.3.8 High Throughput Applications and Cytometry

High throughput screening is used for evaluating quickly the potential of very large number of molecules to have an interaction with a target (e.g. a cell). The information is mostly binary. If there is an interaction the molecule may be potentially a drug and is further studied with other techniques. Robotic equipment working with microplate array are intensively used to accelerate the test and reduced the quantity of chemicals per test. However, volumes that are smaller than microliter are very difficult to manipulate and the throughput is limited by the mechanical speed of the robot and parallelization. Microfluidics overcomes the difficulty of manipulating small volumes as a 100 pL or smaller droplet is a routine with this technology. The droplet rate offers much higher throughput than robot capabilities. On one hand, there is still a difficulty for injecting the large number of products in a sequential way in the chip. On the other hand, it is easy to dilute an injected product to test it at different concentrations and obtain dose response at high speed. It been implemented with fluorescence spectroscopy in [114]. As some products react only for a certain range of concentration, this prevents false negative test.

Flow cytometry consists in counting the number of cells in a solution flow and to characterize it. Based on the measured characteristics the cells are sometimes sorted individually. The solution is injected in a capillary and the cells are passing in the probe volume one by one because of the diameter of the capillary forces the cell to be aligned in a single line. If a light illuminates one side of the capillary, there is an intensity drop on the other side when a cell is passing by. A small microfluidic canal can replace the capillary. Sometimes, a sheath liquid is injected with a hydrodynamic focusing to run laminarly on the side of the sample solution and further reduce the width and forcing cell alignment. Fluorescence spectroscopy offers better sensitivity and more information. The laser spot size is important to distinguish between two close cells when they are in a continuous flow. If they are encapsulated in a droplet, special care should be taken to not encapsulate two cells in the same droplet. If a sorter is added the techniques is called FACS (Fluorescence Activated Cell Sorting). The same principle can be used for droplet sorting and is not limited to cell sorting [115].

References

1. L. Gervais, N. de Rooij, E. Delamarche, Microfluidic chips for point-of-care immunodiagnostics. Adv. Mater. **23**(24), H151 (2011)
2. C.D. Chin, V. Linder, S.K. Sia, Commercialization of microfluidic point-of-care diagnostic devices. Lab Chip **12**, 2118 (2012)
3. B.H. Weigl, R.L. Bardell, C.R. Cabrera, Lab-on-a-chip for drug development. Adv. Drug Deliv. Rev. **55**(3), 349 (2003)
4. J. Ducrée, S. Haeberle, S. Lutz, S. Pausch, F. Von Stetten, Zengerle, The centrifugal microfluidic Bio-Disk platform. J. Micromech. Microeng. **17**(7), S103 (2007)
5. S.-Y. Teh, R. Lin, L.-H. Hung, A.P. Lee, Droplet microfluidics. Lab Chip **8**(2), 198 (2008)
6. A.W. Martinez, S.T. Phillips, G.M. Whitesides, E. Carrilho, Diagnostics for the developing world: microfluidic paper-based analytical devices. Anal. Chem. **82**(1), 3 (2010)
7. T.M. Squires, S.R. Quake, Microfluidics: Fluid physics at the nanoliter scale. Rev. Mod. Phys. **77**(3), 977 (2005)
8. A. Ghosh, Scaling laws, in mechanics over micro and nano scales, ed. by S. Chakraborty Springer (New York, 2011) p. 61
9. D.J. Beebe, G.A. Mensing, G.M. Walker, Physics and applications of microfluidics in biology. Annu. Rev. Biomed. Eng. **4**, 261 (2002)
10. F.M. White, I. Corfield, *Viscous fluid flow 3* (McGraw-Hill, New York, 2006)
11. J.S. Kuo, D.T. Chiu, Controlling mass transport in microfluidic devices. Annu. Rev. Anal. Chem. **4**(1), 275 (2011)
12. N.L. Jeon, S.K.W. Dertinger, D.T. Chiu, I.S. Choi, A.D. Stroock, G.M. Whitesides, Generation of solution and surface gradients using microfluidic systems. Langmuir **16**(22), 8311 (2000)
13. M.H.V. Werts, V. Raimbault, R. Texier-Picard, R. Poizat, O. Français, L. Griscom, J.R.G. Navarro, Quantitative full-colour transmitted light microscopy and dyes for concentration mapping and measurement of diffusion coefficients in microfluidic architectures. Lab Chip **12**(4), 808 (2012)
14. M.H.V. Werts, V. Raimbault, M. Loumaigne, L. Griscom, O. Français, J.R.G. Navarro, A. Débarre, B. Le Pioufle, Optical microscopy and spectroscopy of analyte-sensitive functionalized gold nanoparticles in microfluidic systems, in Colloidal Nanocrystals for biomedical applications VIII. Proc. SPIE **8595**, 85950W (2013)
15. T.P. Lagus, J.F. Edd, A review of the theory, methods and recent applications of high-throughput single-cell droplet microfluidics. J. Phys. D Appl. Phys. **46**(11), 114005 (2013)
16. H. Gu, M.H.G. Duits, F. Mugele, Droplets formation and merging in two-phase flow microfluidics. Int. J. Mol. Sci. **12**(4), 2572 (2011)
17. K.W. Oh, K. Lee, B. Ahn, E.P. Furlani, Design of pressure-driven microfluidic networks using electric circuit analogy. Lab Chip **12**(3), 515 (2012)
18. N.A. Mortensen, F. Okkels, H. Bruus, Reexamination of Hagen-Poiseuille flow: shape dependence of the hydraulic resistance in microchannels. Phys. Rev. E **71**(5), 57301 (2005)
19. T. Bourouina, A. Bosseboeuf, J.-P. Grandchamp, Design and simulation of an electrostatic micropump for drugdelivery applications. J. Micromech. Microeng. **7**(3), 186 (1997)
20. M.A. Holden, S. Kumar, E.T. Castellana, A. Beskok, P.S. Cremer, Generating fixed concentration arrays in a microfluidic device. Sens Actuators, B Chem. **92**(1–2), 199 (2003)
21. Y. Wang, T. Mukherjee, Q. Lin, Systematic modeling of microfluidic concentration gradient generators. J. Micromech. Microeng. **16**(10), 2128 (2006)
22. N. Li Jeon, H. Baskaran, S.K.W. Dertinger, G.M. Whitesides, L. Van de Water, M. Toner, Neutrophil chemotaxis in linear and complex gradients of interleukin8 formed in a microfabricated device, Nat. Biotechnol. **20**(8), 826 (2002)
23. J.P. Brody, P. Yager, Diffusion-based extraction in a microfabricated device. Sens Actuators A Phys. **58**(1), 13 (1997)

24. P. Yager, T. Edwards, E. Fu, K. Helton, K. Nelson, M.R. Tam, B.H. Weigl, Microfluidic diagnostic technologies for global public health. Nature **442**(7101), 412 (2006)
25. T. Schneider, J. Kreutz, D.T. Chiu, The potential impact of droplet microfluidics in biology. Anal. Chem. **85**(7), 3476 (2013)
26. X. Casadevall i Solvas, A. deMello, Droplet microfluidics: recent developments and future applications. Chem. Commun. **47** (7), 1936 (2011)
27. T. Thorsen, R.W. Roberts, F.H. Arnold, S.R. Quake, Dynamic pattern formation in a vesicle-generating microfluidic device. Phys. Rev. Lett. **86**(18), 4163 (2001)
28. S.L. Anna, N. Bontoux, H.A. Stone, Formation of dispersions using 'flow focusing' in microchannels. Appl. Phys. Lett. **82**(3), 364 (2003)
29. R. Seemann, M. Brinkmann, T. Pfohl, S. Herminghaus, Droplet based microfluidics. Reports Prog. Phys. **75**(1), 16601 (2012)
30. R. Gorkin, J.-M.J. Park, J. Siegrist, M. Amasia, B.S. Lee, J.-M.J. Park, J. Kim, H. Kim, M. Madou, Y.-K. Cho, Centrifugal microfluidics for biomedical applications. Lab Chip **10**(14), 1758 (2010)
31. J. Ducrée, S. Haeberle, T. Brenner, T. Glatzel, R. Zengerle, Patterning of flow and mixing in rotating radial microchannels. Microfluid. Nanofluidics **2**(2), 97 (2006)
32. M. Madou, J. Zoval, G. Jia, H. Kido, J. Kim, N. Kim, Lab on a CD. Annu. Rev. Biomed. Eng. **8**(1), 601 (2006)
33. O. Strohmeier, M. Keller, F. Schwemmer, S. Zehnle, D. Mark, F. von Stetten, R. Zengerle, N. Paust, Centrifugal microfluidic platforms: advanced unit operations and applications. Chem. Soc. Rev. **44**(17), 6187 (2015)
34. A.W. Martinez, S.T. Phillips, M.J. Butte, G.M. Whitesides, Patterned paper as a platform for inexpensive, low-volume, portable bioassays. Angew. Chemie Int. Ed. **46**(8), 1318 (2007)
35. Y. Xia, J. Si, Z. Li, Fabrication techniques for microfluidic paper-based analytical devices and their applications for biological testing: a review. Biosens. Bioelectron. **77**, 774 (2016)
36. A.K. Yetisen, M.S. Akram, C.R. Lowe, Paper-based microfluidic point-of-care diagnostic devices. Lab Chip **13**(12), 2210 (2013)
37. J.M. Berg, J.L. Tymoczoko, G.J. Gatto, L; Stryer, "Biochemistry", 8th edn. (W.H. Freeman and Co Ltd, New York, 2015)
38. M.D. Hanwell, D.E. Curtis, D.C. Lonie, T. Vandermeersch, E. Zurek, G. Hutchinson, Avogadro: an advanced semantic chemical editor, visualization, and analysis platform. J. Cheminformatics **4**, 17 (2012)
39. T.J. Smith, MOLView: a program for analyzing and displaying atomic structures on the Macintosh personal computer, J. Molecular Graphics **13**(2), (1995). http://molview.org/. Accessed 3 April 2018
40. [The PyMOL Molecular Graphics System, Version 1.8 Schrödinger, LLC. Website: http://pymol.org. Accessed 3 April 2018
41. E.T. Bolton, J. McCarthy, A General method for the Isolation of RNA complementary to DNA. Proc Natl. Acad. Sci. USA **48**(8), 1390 (1962)
42. A.H. Sturtevant, *A History of Genetics*, (Cold Spring Harbor Laboratory Press, 2001)
43. A. Abbas, A.H. Lichtman, S. Pillai, *Cellular and molecular immunology*, 9th edn. (Elsevier, Philadelphia, 2018)
44. X.L. Su, Y. Li, A self-assembled monolayer-based piezoelectric immunosensor for rapid detection of Escherichia coli O157:H7. Biosensors and Biolectronics **19**(6), 563 (2004)
45. C. Prussin, D.D. Metcalfe, Detection of intracytoplasmic cytokine using flow cytometry and directly conjugated anti-cytokine antibodies. J. Immunol. Methods **188**(1), 117 (1995)
46. B.J. Takács, Protein purification: theoretical and methodological considerations, in *Encyclopedia of Analytical Chemistry: Applications, Theory and Instrumentation* (Wiley & Sons Ltd, 2006)
47. J. Gao, S. Ma, D.T. Major, K. Nam, J. Pu, D.G. Truhlar, Mechanisms and free energies of enzymatic reactions. Chem. Rev. **106**(8), 3188 (2006)
48. D. Melloul, S. Marshak, E. Cerasi, Regulation of insulin gene transcription. Diabetologia **45** (3), 309 (2012)

49. F. Jacob, J. Monod, Genetic regulatory mechanisms in the synthesis of proteins. J. Mol Biology **3**, 318 (1961)
50. S. Hoops, S. Sahle, R. Cauges, C. Lee, J. Pahle, N. Simus, M. Singhal, L. Xu, P. Mendes, U. Kummer, COPASI—a complex pathway simulator. Bioinformatics **22**(24), 3067 (2006)
51. Y. Gendrault, M. Madec, C. Lallement, J. Haiech, Modeling biology with HDL languages: a first step toward a genetic design automation tool inspired from microelectronics. IEEE Trans. Biomed. Eng. **61**(4), 1231 (2014)
52. J. Haiech, Y. Gendrault, M.C. Kilhoffer, R. Randjeva, M. Madec, C. Lallement, A general framework improving teaching ligand binding to a macromolecule. Biochim. Biophys. Acta **1843**(10), 2348 (2014)
53. J.N. Weiss, The Hill equation revisited: uses and misuses. FASEB J. **11**(11), 835 (1997)
54. U. Alon, An introduction to systems biology: Design principles of biological circuits, (Chapman & Hall/CRC, 2006)
55. J.H. Santiago, S.T. Wereley, C.D. Meinhart, D.J. Beebe, R.J. Adrian, A particle image velocimetry system for microfluidics. Exp. Fluids **25**(4), 316 (1998)
56. S.T. Wereley, J.G. Santiago, R. Chiu, C.D. Meinhart, RJ. Adrian, Micro-resolution particle image velocimetry, in *Proc Micro- and Nanofabricated Structures and Devices for Biomedical Environment Applications*. Proceedings. SPIE vol. 3258, p. 122 (1998)
57. C. Cierpa, C.J. Kähler, Particle imaging techniques for volumetric three-component (3D3C) velocity measurements in microfluidics. J. Vis. **15**, 1 (2012)
58. S.T. Wereley, CD. Meinhart recent advances in micro-particle image velocimetry. Annu. Rev. Fluid Mech. **42**, 557 (2010)
59. H.F. Li, M. Yoda, Multilayer nano-particle image velocimetry (MnPIV) in microscal Poiseille flows. Meas. Sci. Technol. **19**, 075402 (2008)
60. S.L. Anna, Droplets and bubbles in microfluidic devices. Ann. Rev. Fluid Mech **48**, 285 (2016)
61. R. Seemann, M. Brinkmann, T. Pfohl, S. Herminghaus, Droplet based microfluidics. Rep. Prog. Phys. **75**, 016601 (2012)
62. M. Rossi, R. Lindken, B.P. Hierck, J. Westerweel, Tapered microfluidic chip for the study of biochemical response at subcellular level of endothelial cell to shear flow. Labon a Chip **9**, 1403 (2009)
63. L.E. Rodd, J.J. Cooper-White, D.V. Boger, G.H. McKinley, Role of the elasticity number in the entry flow of dilute polymer solutions in micro-fabricate contraction geometries. J. Non-Newtonian Fluid Mech. **143**, 170–191 (2007)
64. J. Rumble (ed.),*CRC Handbook of Chemistry and Physics*, 98th edn. (Taylor & Francis Group, 2017)
65. C.W. Macosko. *Rheology: Principle, Measurements, and Application*, (WILEY-VCH, 1994)
66. G. Astarita, G. Marrucci, L. Nicolais (Eds.), *Rheology, Volume 1: Principles* (Springer,1980)
67. H.A. Barnes, J.F. Hutton, K. Walters, *Vol.3: An Introduction to Rheology*, 3rd edn. (Elsevier, Amsterdam-London-New York-Tokyo, 1993)
68. P.-G. De Gennes, F. Brochard-Wyart, D. Guere, *Capillary and wetting phenomena -drops, bubbles, pearls* (Waves, Springer, New York, 2004)
69. C.D. Meinhart, S.T. Wereley, M.H.B. Gray, Volume illumination for two-dimensional particle image velocimetry. Meas. Sci. Technol. **11**, 809 (2000)
70. C.J. Bourdon, M.G. Olsen, A.D. Gorby, Validation of an analytical solution for depth of correlation in microscopic particle image velocimetry. Meas. Sci. Technol. **15**, 318 (2004)
71. M.G. Olsen, R.J. Adrian, Out-of-focus effects on particle image visibility and correlation in microscopic particle image velocimetry. Exp. Fluids **29**, S166 (2000)
72. O. Carrier, F.G. Ergin, H.Z. Li, B.B. Watz, D. Funfschilling, Time-resolved mixing and flow-field measurements during droplet formation in a flow-focusing junction. J. Micromech. Microeng. **25**, 084014 (2015)
73. S. Devasenathipathy, J.G. Santiago, S.T. Wereley, C.D. Meinhart, K. Takehara, Particle imaging techniques for microfabricated fluidic systems. Exp. Fluids **34**, 504 (2003)

74. V. Van Steijn, M.T. Kreutzer, C.R. Kleijn, μ-PIV study of the formation of segmented flow in microfluidic T-junctions, Chem. Eng. Science **62**, 7505 (2007)
75. D. Funfschilling, H. Debas, H.Z. Li, T.G. Mason, Flow-field dynamics during droplet formation by dripping in hydrodynamic-focusing microfluidics. Phys. Rev. E **80**, 015301 (2009)
76. D.E. Smith, S. Chu, Response of flexible polymers to sudden elongational flow. Science **281**, 1335 (1998)
77. L. Boussel, V. Rayz, C. McCulloch, A. Martin, G. Acevedo-Bolton, M. Lawton, R. Higashida, W.S. Smith, W.L. Young, D. Sloner, Aneurysm growth occurs at region of low wall shear stress. Stroke **39**, 2997 (2008)
78. J.N. Tropper, M.A. Gimbrone, Blood flow and vascular gene expression: fluid shear stress as a modulator of endothelial phenotype. Mol Med Today **5**, 40 (1999)
79. C. Poelma, K. Van der Heiden, B.P. Hierck, R.E. Poelmann, J. Wersterweel, Measurements of the wall shear stress distribution in the outflow track of an embryonic chicken heart. J. R. Soc. Interface **7**, 91 (2010)
80. C. Poelma, P. Vennemann, R. Lindken, J. Westerweel, In vivo blood flow and wall shear stress measurements in vitelline network. Exp. Fluids **45**, 703 (2008)
81. Y. Sugii, R. Okuda, K. Okamoto, M. Takeda, Velocity measurement of both red blood cells and plasma of in vitro blood flow using high-speed micro PIV technique. Meas. Sci. Technol. **16**, 1126 (2005)
82. K.E. Audrey, T.P. Scott, C.S. Adam, A.M. Katherine, W.M. Andres, S. Pierre, J. Nina, P. Mara, M.W. George, Quantifying colorimetric assays in paper-based microfluidic devices by measuring the transmission of light through paper. Analyt. Chem. **81**(120), 8447 (2009)
83. M. O'Toole, D. Diamond, Absorbance based light emitting diode optical sensors and, sensing devices. Sensors **8**, 2453 (2008)
84. A. Hsiao, M.R. Gartia, T.-W. Chang, X. Wang, P. Khumwan, G.L. Liu, Colorimetric plasmon resonance microfluidics on nanohole array sensors. Sens. Bio-Sens. Res. **5**, 24 (2015)
85. M.T. Koesdjojo, S. Pengpumkiat, Y. Wu, A. Boonloed, D. Huynh, T.P. Remcho, A.V.T. Remcho, Cost effective paper-based colorimetric microfluidic devices and mobile phone camera readers for the classroom. J. Chem. Educ. **92**(4), 737 (2015)
86. Y. Jung, J. Kim, O. Awofeso, H. Kim, F. Regnier, E. Bae, Smartphone-based colorimetric analysis for detection of saliva alcohol concentration. Appl. Opt. **54**(131), 9183 (2015)
87. G. Minas, J.C. Ribeiro, R.F. Wolffenbuttel, J.H. Correia, On-Chip Integrated CMOS optical detection microsystem for spectrophotometric analyses in biological microfluidic systems, in *IEEE International Symposium on Industrial Electronics*, Dubrovnik, Croatia, June 2005
88. J.R. Lakowicz, *Principles of Fluorescence Spectroscopy*, 3rd Edn. (Springer, 2006)
89. M. Kollner, J. Wolfrum, How many photon are necessary for fluorescence-lifetime measurements? Chem. Phys. Lett. **200**(12), 199 (1992)
90. H. Geng, J. Feng, L.M. Stabryla, S.K. Cho, Dielectrowetting manipulation for digital microfluidics: creating, transporting, splitting, and merging of droplets. Lab Chip **17**(6), 1060 (2017)
91. A. Piruska, I. Nikcevic, S.H. Lee, C. Ahn, W.R. Heineman, P.A. Limbacha, C.J. Seliskar, The autofluorescence of plastic materials and chips measured under laser irradiation. Lab Chip **12**, 1348 (2005)
92. G.D.M. Jeffries, R.M. Lorenz, D.T. Chiu, Ultrasensitive and high-throughput fluorescence analysis of droplet contents with orthogonal line confocal excitation. Anal. Chem. **82**(23), 9948 (2010)
93. Y. Zeng, L. Jiang, W. Zheng, D. Li, S. Yao, J.Y. Qu, Quantitative imaging of mixing dynamics in microfluidic droplets using two-photon fluorescence lifetime imaging. Opt. Lett. **36**, 2236 (2011)
94. P.S. Dittrich, P. Schwille, Spatial two-photon fluorescence crosscorrelation spectroscopy for controlling molecular transport in microfluidic structures. Anal. Chem. **74**, 4472 (2002)
95. A.M. Streets, Y. Huang, Microfluidics for biological measurements with single-molecule resolution. Curr. Opin. Biotechnol. **25**(69), 2014 (2014)

96. X. Su, S.E. Kirkwood, M. Gupta, L. Marquez-Curtis, Y. Qiu, A. Janowska-Wieczorek, W. Rozmus, Y.Y. Tsui, Microscope-based label-free microfluidic cytometry, Optics Express **19**(1), 387 (2011)
97. Y.-C. Tung, M. Zhang, C.-T. Lin, K. Kurabayashi, S.J. Skerlos, PDMS-based opto-fluidic micro flow cytometer with two-color, multi-angle fluorescence detection capability using PIN photodiodes. Sens. Actuators B: Chem. **98**(2–3), 356 (2004)
98. M.L. Chabinyc, D.T. Chiu, J.C. McDonald, A.D. Stroock, J.F. Christian, A.M. Karger, G.M. Whitesides, An integrated fluorescence detection system in poly(dimethylsiloxane) for microfluidic applications. Anal. Chem. **73**, 4481 (2001)
99. P. Ghenuche, J. de Torres, P. Ferrand, J. Wenger, Multi-focus parallel detection of fluorescent molecules at picomolar concentration with photonic nanojets arrays. Appl. Phys. Lett. **105**, 131102 (2014)
100. Y.J. Fan, Y.C. Wu, Y. Chen, Y.C. Kung, T.H. Wu et al., Three dimensional microfluidics with embedded microball lenses for parallel and high throughput multicolor fluorescence detection. Biomicrofluidics **7**, 044121 (2013)
101. E. Schonbrun, P.E. Steinvurzel, K.B. Crozier, A microfluidic fluorescence measurement system using an astigmatic diffractive microlens array. Opt. Express **19**(2), 1385 (2011)
102. S. Ursuegui, M. Mosser, A. Wagner, Copper-free click chemistry for microdroplet's W/O interface engineering. RSC Ad. **6**(97), 94942 (2016)
103. P. Gruner, B. Riechers, B. Semin, J. Lim, A. Johnston, K. Short, J.-C. Baret, Controlling molecular transport in minimal emulsions. Nature Commun. **7**, 10392 (2016)
104. M.S.-Art, A.J. deMello, J.B. Edel, High-Throughput DNA droplet assays using picoliter reactor volumes. Anal. Chem. **79**(17), 6682 (2007)
105. W. Uhring, V. Zint, J. Bartringer, A low-cost high-repetition-rate picosecond laser diode pulse generator, in *Semiconductor lasers and laser dynamics*, Strasbourg,Sept.2004, Proc. SPIE 5452, 583 (2004)
106. P.B. Coates, The correction for photon 'pile-up' in the measurement of radiative lifetime. J. Physics E: Sci. Instrum. **1**(8), 878 (1968)
107. J. Leonard, N. Dumas, J.P. Caussé, S. Maillot, N. Giannakopoulou, S. Barre, W. Uhring, High-throughput time-correlated single photon counting. Lab-on-Chip **14**(22), 4297 (2014)
108. I. Santi, N. Dhar, D. Bousbaine, Y. Wakamoto, J.D. McKinney, Single-cell dynamics of the chromosome replication and cell division cycles in mycobacteria. Nat. commun. **4**, 2470 (2013)
109. Y. Zhao, D. Chen, H. Yue, M.M. Spiering, C. Zhao, S.J. Benkovic, T. Jun Huang, Dark-Field Illumination on Zero-Mode Waveguide/Microfluidic Hybrid Chip Reveals T4 Replisomal Protein Interactions. Nano Lett. **14**(4), 1952 (2014)
110. G. Giraud, H. Schulze, D.-U. Li, T.T. Bachmann, J. Crain, D. Tyndall, J. Richardson, R. Walker, D. Stoppa, E. Charbon, R. Henderson, J. Arlt, Fluorescence lifetime biosensing with DNA microarrays and a CMOS-SPAD imager. Biomed. Opt. Express **1**, 1302–1308 (2010)
111. Y. Ding, S. Stavrakis, X. Casadevalli Solvas, A.J. deMello, A high throughput droplet-based microfluidic barcode generator, in *17th International Conference on Miniaturized Systems for Chemistry and Life Sciences,* 27–31 October 2013, Freiburg, Germany
112. M. Zlatanski, W. Uhring, J.-P. Le Normand, Sub-500-ps temporal resolution streak-mode optical sensor. IEEE Sens. J. **15**(11), 6570 (2015)
113. Q. Zhoua, T. Kima, Review of microfluidic approaches for surface-enhanced Raman scattering. Sens. Actuators B: Chem. **227**, 504 (2016)
114. O.J. Miller, E. Harrak, T. Mangeat, J.-C. Baret, L. Frenz, B. El Debs, E. Mayot, M.L. Samuels, E.K. Rooney, P. Dieu, M. Galvan, D. R. Link, A.D. Griffiths, High-resolution dose–response screening using droplet-based microfluidics. Proc. Natl. Acad. Sci. U S A **109**(2), 378–383 (2002)
115. J.-C. Baret, O.J. Miller, V. Taly, M. Ryckelynck, A. El-Harrak, L. Frenz, C. Rick, M.L. Samuels, J.B. Hutchison, J.J. Agresti, D.R. Link, D.A. Weitz, A.D. Griffiths, Fluorescence-activated droplet sorting (FADS): efficient microfluidic cell sorting based on enzymatic activity. Lab Chip **9**(13), 1850–1858 (2009)

Chapter 3
Micro/Nano Fabrication and Packaging Technologies for Bio Systems

Yeong Eun Yoo, Claire Smadja and Mehdi Ammar

Abstract The investigation and development of micro/nano biosytem requires a sealed fluidic platform to separate, mix or control the flow of liquids and bio samples as well as a biochemical surface processing to selectively capture or repel biospecies. The first part of this chapter reviews the main techniques used for the fabrication of microchannels, reservoirs, pillars,… in various substrate materials. This includes direct machining techniques such as mechanical cutting, lithography and electroforming, as well as various replication techniques such as PDMS or UV curable resin casting, hot embossing and overall injection molding that is compatible with mass production. The second part describes the recent advances in the development of functionalized surfaces and their applications in biochips. First a focus is put on bioreceptors immobilization and a brief presentation of bioreceptors (antibodies and aptamers) is included. Next the polymers employed against plasmatic proteins fouling are reviewed and finally the surface chemistry preventing bacteria attachment is presented. The two approaches leading to bacteria repelling or killing, depending on the polymers employed, is discussed. The last chapter part is devoted to a critical analysis of bonding and welding techniques proposed to seal fluidic platforms.

Keywords Microfabrication · Machining processes · Microfluidics platforms · Bacteria · Proteins · Antifouling · Self-assembled monolayers

Y. E. Yoo
Department of Nano Manufacturing Technology, Korea Institute of Machinery and Materials, 156 Gajeongbuk-ro Yusung-gu, Daejeon, South Korea
e-mail: yeyoo@kimm.re.kr

C. Smadja
University of Paris-Sud, Orsay, France
e-mail: claire.smadja@u-psud.fr

M. Ammar (✉)
Center for Nanoscience and Nanotechnology (C2N), CNRS, University Paris-Sud, University Paris Saclay, Palaiseau, France
e-mail: mehdi.ammar@u-psud.fr

G. Barbillon et al. (eds.), *Engineering of Micro/Nano Biosystems*,
Microtechnology and MEMS, https://doi.org/10.1007/978-981-13-6549-2_3

3.1 Introduction

Basically, the full fabrication of micro/nano biosystems requires the mastering of three main steps which are covered in this chapter: manufacturing of the micro/nano fluidics system with its internal elements, surface biofunctionnalization to selectively capture or prevent attachment of biospecies, and packaging to seal and protect the fluidic system. These three steps are not necessarily consecutive and additional processing steps are often required for fluidic connections and, in fully integrated systems, for fluid injection with micropumps, for fluid control with microvalves and for sensors integration. The different types of microvalves and micropumps will be described in the first part of Chap. 4 and sensing/actuation techniques by mechanical, optical, electrical and magnetical ways are detailed in the following chapters of this book.

The choice of fabrication and surface functionalization technologies depends on many factors such as the substrate material used (polymer, glass, silicon, paper,…), the targeted application and its environment (ex vivo, in vivo, in vitro), the level of integration and of autonomy (portable systems), the technology readiness level (going from basic principles to real application product), the targeted cost, the need or not to get a disposable, re-usable, biocompatible and/or biodegradable system, etc.… Consequently, a very broad range of techniques have been developed and will be developed in the future for the micro/nanofabrication and biofunctionnalization of micro/nano biosystems. An exhaustive description is impractical but this chapter provides a critical analysis and a description of recent advances for a large selection of them.

Concerning microfabrication, this chapter will describe the main direct machining and replication techniques, their issues and some solutions. The subchapter on biofunctionnalization will first present bioreceptors and their immobilization on a surface, will then review polymers for plasmatic proteins fouling and will finally present surface chemistry that can be used to repel or kill bacteria. The last part of this chapter will discuss the different bonding or welding techniques that can be applied to the sealing of fluidic devices.

3.2 Micro/Nano Fabrication for Fluidic Devices (Yeong-Eun Yoo)

Nano or micro-fluidic platform for bio-chip or lab-on-a-chip consists of various fluidic elements such as channels, pillars, reservoirs, etc. in nano or micro-scale for separation, mixing, or controlling the flow of the samples. The various functions and performance of these micro-fluidic devices are determined significantly by the characteristics of the fluidic elements. Therefore, it is important to fabricate the various micro/nano structures on the fluidic platform for bio-chip or lab-on-a-chip.

As shown in Fig. 3.1, a basic fluidic platform can simply be thought of as having two parts: a plate with fluidic elements on the surface (a fluidic plate) and a plate to cover the fluidic plate (a cover plate). It may not often be clear to distinguish the fluidic plate from the cover plate, because both can have fluidic elements depending on the design of the device, or it may consist of several elements (Fig. 3.2).

The fabrication processes for micro/nano fluidic elements can be generally grouped into either direct machining of the structures on a workpiece or replicating the structures from a previously fabricated template.

The direct machining process includes mechanical cutting or forming, laser machining and chemical etching along with UV lithography for materials such as metal, silicon, glass, plastic or ceramic depending on the process. Replication processes include the following: (1) casting process using thermoset resins like PDMS or UV curable resins like photo-resist, (2) hot embossing process on thermoplastic substrate, (3) injection molding using thermoplastic materials such as

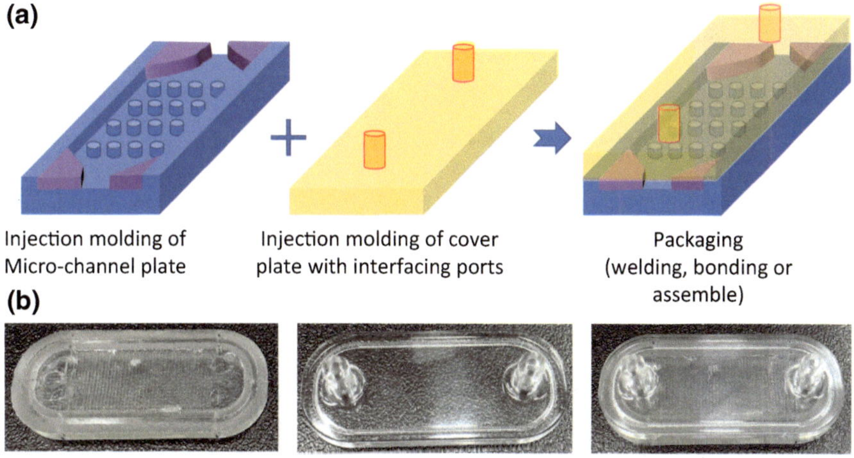

Fig. 3.1 Schematics and images for micro/nano fluidic platform consists of channel plate and cover plate

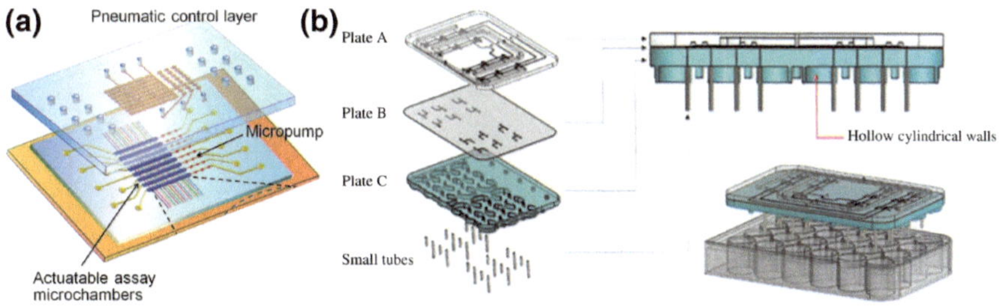

Fig. 3.2 Examples for **a** micro-fluidic channels or chambers on both plates [1] **b** micro-fluidic platform with multi-components for cell culturing [2] (with permission of Springer)

COC (cyclic olefin copolymer), PMMA (polymethylmethacrylate) or PC (polycarbonate). All these replication processes can fabricate the same structures or devices repeatedly.

3.2.1 *Direct Machining Process for Nano/Micro Fluidic Device*

The direct machining process may be efficient to see the feasibility of the fluidic system at the early stage of development from the view point of the cost or lead time etc. because there is no need on replication process which requires additional tooling, casting, embossing or molding process. Depending on the shape and size of the structure, the area for process, the material of the workpiece, etc., various direct machining processes are currently available for micro/nano structures and one of them can be applied. In this chapter, several direct machining processes will be reviewed briefly for fabrication of micro/nano structures.

3.2.1.1 Mechanical Cutting Process

Micro-shaping and micro-milling are representative mechanical cutting processes for machining micro-structures. Micro-shaping is efficient to machine micro-groove type of structures straight or circularly by reciprocation or turning of cutting tool or workpiece (Fig. 3.3). The groove can be machined to be order of 1 μm both in width and depth while the ridge between grooves can be machined in order of 1 μm in width, but tens of microns in depth depending on the shape of the groove. Although the micro-shaping is very efficient process for micro-groove, it is generally not available for the fluidic channel-like structures of irregular shape or path as shown in Fig. 3.4.

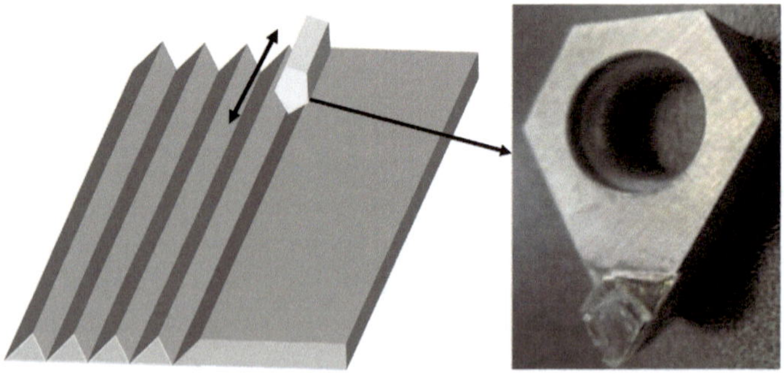

Fig. 3.3 A schematic for shaping of micro-groove with cutting tool (left) and a photo of micro diamond cutting tool [3] (with permission of Springer)

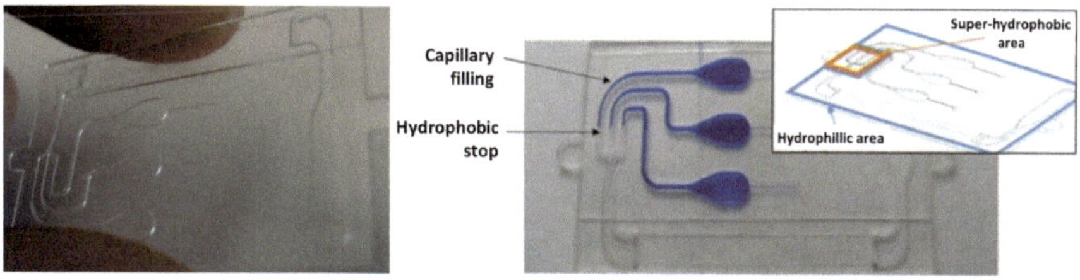

Fig. 3.4 Images of fluidic channel in irregular shape or path [4] (with permission of Elsevier)

The fluidic channel of irregular shape or path not straight as in Fig. 3.4 can be machined through micro-milling process using a rotating cutting tool (Fig. 3.5), which is available to cut the channel-type structures as small as several tens of micrometer or embossed structures even in order of few micrometers [5, 6].

Micro-milling depends on cutting speed or spindle speed, feed rate, depth of cut etc. as process parameters. The quality of the surface and surface structures micro-milled is also affected by tool design (Fig. 3.6) such as cutting-edge radius, rake angle, helix angle of the flute, number of flute, etc. The optimal conditions on process or tool may be quite different for the micro-milling process due to reduced mechanical stiffness of the micro-milling tool, comparable size of the cutting-edge radius or grain of the workpiece material to the depth-of-cut, etc. Extensive researches on micro-milling have been reported [7–14]. In terms of the dimension of structure to be machined, the mechanical machining may be limited by the size of the cutting tool. Some commercial micro-milling cutting tools are available currently around 25 μm, therefore the fluidic channel less than 25 μm approximately is hard to machine directly on the device. The limit of the dimension of surface structure, however, may become smaller for embossed or positive surface structures if the cutting area is larger than the cutting tool (Fig. 3.7). The positive surface structures on the workpiece as in Fig. 3.7b may be damaged by the

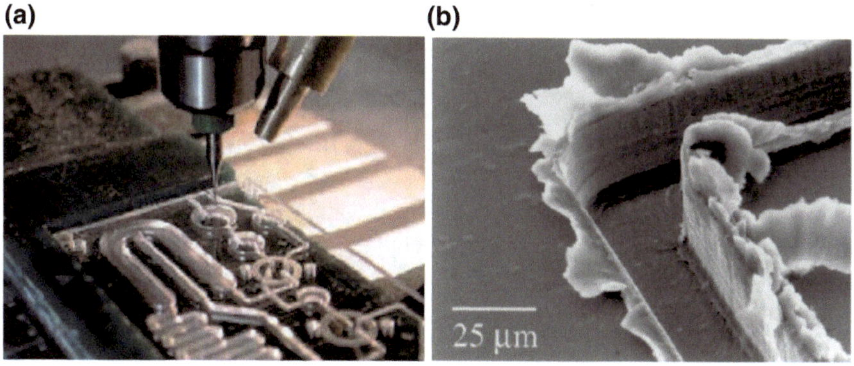

Fig. 3.5 Photos for **a** micro-milling process **b** a micro-channel machined by micro-milling [7] (with permisiion of Creative Commons)

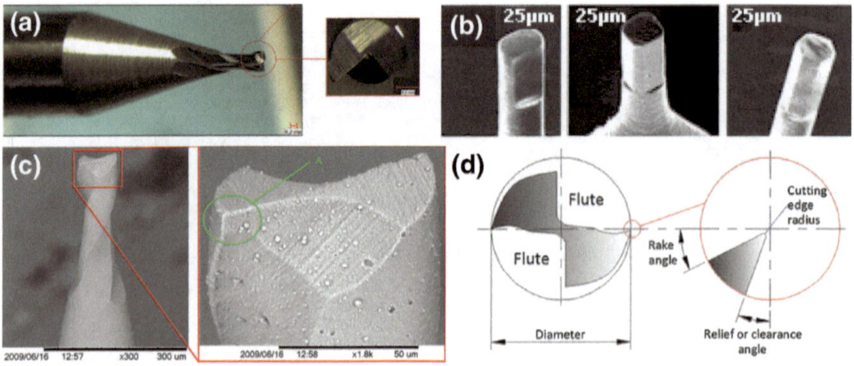

Fig. 3.6 Photos of various micro-milling tools in different size: **a** 300 μm in diameter [8], **b** 25 μm in diameter [9], **c** less than 100 μm in diameter [10] and **d** design on cutting edge and flute [8] (With permission of Springer)

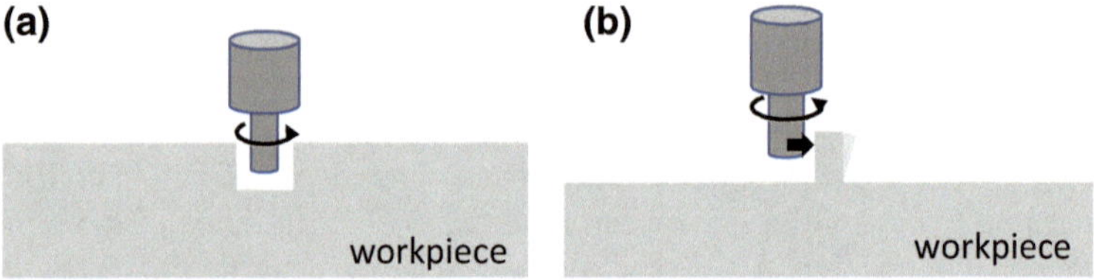

Fig. 3.7 Schematics for micro-milling of **a** negative structures like groove or channel **b** positive (embossed) structures

interaction with the cutting tool [5, 6]. This may limit the dimension of the embossed surface structures to be machined. Fluidic channels smaller than the machining limit can be fabricated by replicating the surface structures using polymeric resin based on the embossed surface structures machined on a workpiece.

Mechanical machining process can be used for various materials such as brass, copper, carbon steel or plastic. One of the advantages of the mechanical machining process is that it can machine various surface structures directly on durable materials like metal, which is crucial to mass production of polymeric device by replication process such as injection molding. Also using the mechanical cutting process, the cross-section of the structures can be designed to have more diverse shapes or wider range of slope for the vertical wall. Moreover, the vertical wall of structure machined mechanically has generally positive slope, which is an advantage for easy releasing of the replica during the replication process (Fig. 3.8).

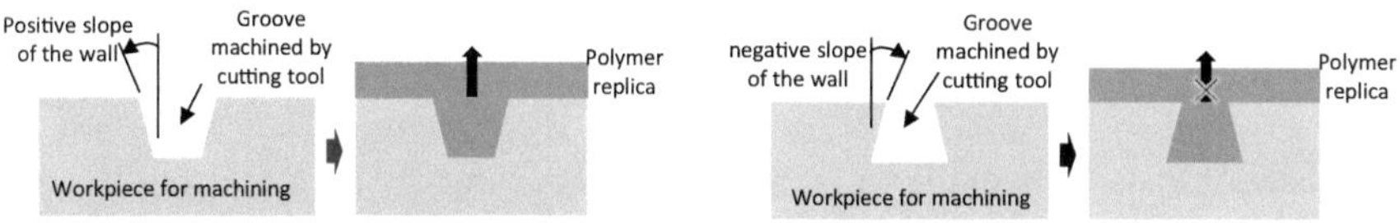

Fig. 3.8 An effect of the wall slope of the micro-groove machined on replication process

3.2.1.2 UV Lithography

Based on the reaction of the photo-resist(PR) when exposed to the UV light or E-beam, the PR layer coated on silicon or glass substrate can be patterned intrinsically by UV light exposure through a pattern mask or direct writing of E-beam. The smooth PR layer turns into a structured layer with channels, pillars or chambers etc. by a developing process through which the PR is removed as patterned. Depending on the PR type, negative or positive, the PR is removed from different region (Fig. 3.9). For negative PR, the exposed region remains through the developing process because the negative PR becomes more stable by cross-linking reaction when being exposed to UV light or e-beam. For the positive PR, the exposed region is removed after developing due to the degradation during exposure. A transition region exists between the exposed and unexposed region due to chemical diffusion, optical diffraction or reflection, etc. The minimum dimension of the structure fabricated depends on the size of the transition region, which is determined by contrast of the photo-resist. The surface structures obtained after developing process can be used as a barrier layer to etch surface structures selectively on silicon or glass substrate, etc., a platform for fluidic device or a master to fabricate Ni-stamper used for injection molding.

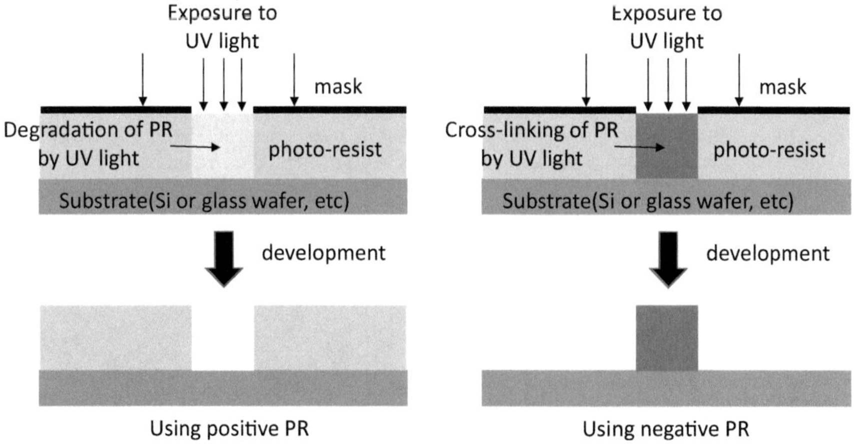

Fig. 3.9 Patterning by UV lithography based on positive or negative photo-resist

Dry or wet etching process can be applied to fabricate the structures on the hard materials such as silicon, glass or metal using the PR pattern as a barrier. The cross-sectional shape of the structure is determined by the anisotropy of the process and material. As shown in Fig. 3.10, through wet etching of Si-wafer using a barrier layer with a square opening in several micrometers, the Si-wafer is etched to have a negative pyramid in order of micrometers with 70° of slope approximately rather than being etched into hemispherical structure due to the anisotropic crystal direction. The dry etching is an anisotropic process because a reactive gas impinges generally in normal direction to the substrate, so a highly anisotropic structure like deep channel or hole with straight vertical wall can be fabricated (Fig. 3.11).

Although the surface structures may be fabricated directly on a device itself via the processes mentioned above, they can also be fabricated with additional processes on a master or template to replicate the structures repeatedly using polymeric materials. The Figs. 3.12 and 3.13 show one of processes for fabricating Ni stamper which is a typical durable nano/micro pattern template for injection molding. A PR master is fabricated first via UV lithography process using either of positive PR or negative PR and then Ni stamp is fabricated using nickel electroforming process explained in next section. From the viewpoint of the injection molding, the slop of side wall of the surface structure like channel is important because it affects greatly the productivity of the injection molding process or it even determines whether the structure can be injection molded or not. The channel of the PR master may have negative or positive slope of vertical wall depending on the PR type and the

Fig. 3.10 A negative micro-pyramid array fabricated by wet etching on Si-wafer (1 0 0) **a** mask for wet etching **b** top view for the array **c** cross-section for the micro-pyramids

Fig. 3.11 An image of cross-section for micro-channel array dry etched on Si-wafer

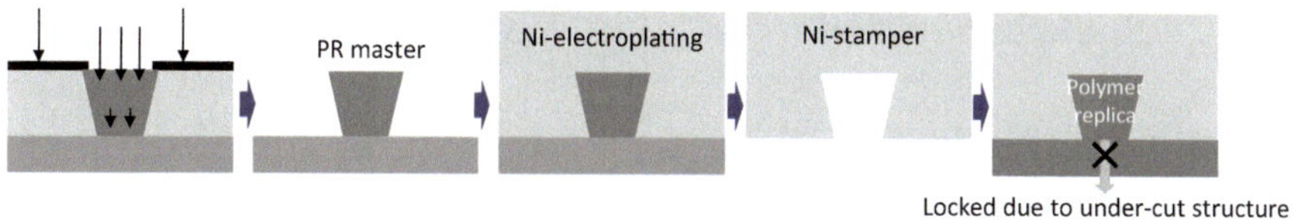

Fig. 3.12 Schematic of a fabrication process for micro-structure on a substrate using negative PR and an issue on replication due to the slope of the side wall

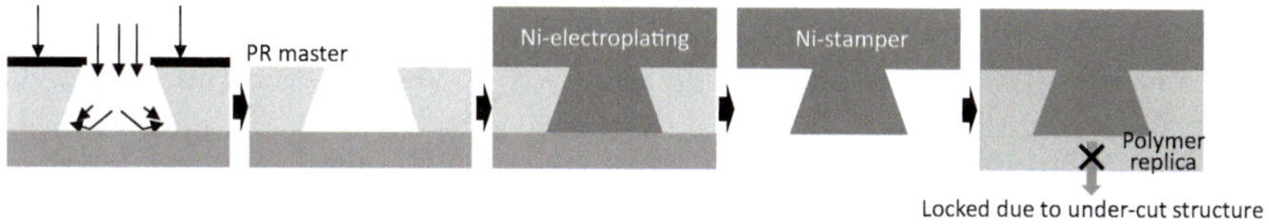

Fig. 3.13 Schematic of a fabrication process for micro-structure on a substrate using positive PR and an issue on replication due to the slope of the side wall

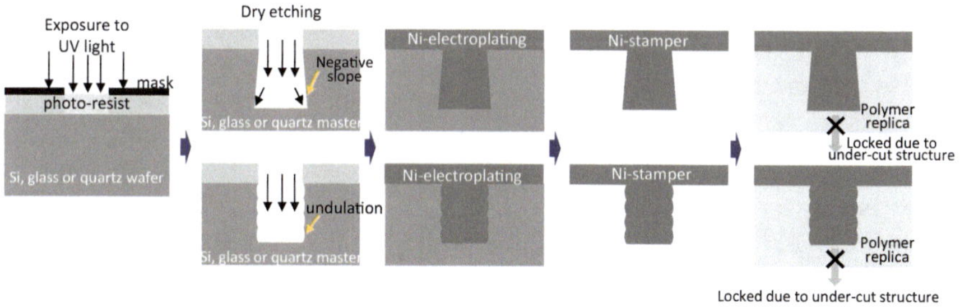

Fig. 3.14 Schematic of a fabrication process for micro-structure on a substrate by dry etching and issues on replication due to the slope of the side wall and undulation on the side wall

conditions for exposure to UV light because the PR is affected differently along the thickness direction due to the decrease of the UV light intensity, the diffusion of cross-link initiator, light scattering in the PR layer or reflection of the light on the substrate [15]. For example, the wider exposure along the depth due to reflection or scattering of the UV light at the bottom of the PR layer may cause negative slope of side wall with wider channel at the bottom. The negative slope of side wall may cause little issue for device itself. For master or stamp for replication, however, the negative slope results in severe failure during the fabrication due to locking on the stamper of the replica. The channel on the silicon or the glass substrate fabricated by dry etching may also have same negative slope of the side wall (Fig. 3.14). In addition to this, the dry-etched side wall has undulation in the depth direction and this undulation can also cause releasing issue of replica from the master or stamper (Fig. 3.14).

3.2.1.3 Ni Electroforming Process

The pattern master fabricated on silicon or glass wafer cannot be directly used for injection molding process due to the high pressure (higher than tens of MPa) and high temperature (higher than 200 °C) during the injection molding process. Nickel electroplating using a silicon, glass or PR pattern master is a process to fabricate a durable master (called Ni-stamper) for nano/micro surface structures which generally cannot machined directly on a metal workpiece. Ni-electroforming is a process depositing nickel layer on the pattern master by redox reaction. As in Fig. 3.15, the pattern master, as a cathode, coated with conductive seed layer typically over 100 nm of thickness and a nickel anode are dipped in electrolyte. Then Ni layer starts to grow on the pattern master by redox reaction and replicate pattern from the master. Ni layer can grow until it reaches the target thickness generally between 0.3 and 1.0 mm for sufficient stiffness during the injection molding process.

3.2.2 Fabrication the Fluidic Device Using Replication Process

For mass production or multiple fabrication of the nano or micro fluidic device, a replication process like casting or molding using polymeric material are required. Three types of polymeric materials including thermoplastic, thermoset and UV curable resin can be used for fabrication of the fluidic device based on replication process. Depending on these materials, several replication processes can be used

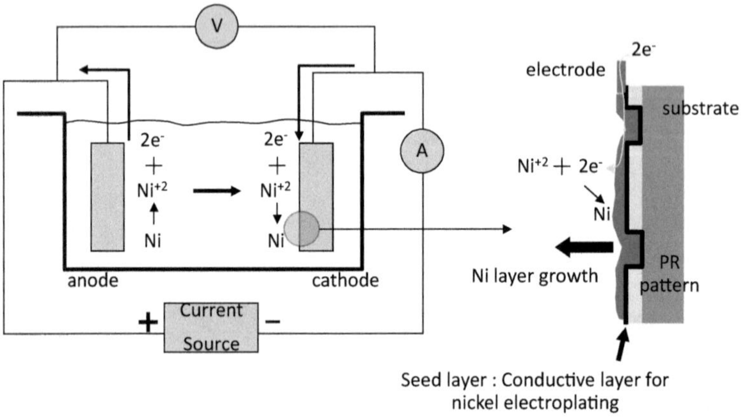

Fig. 3.15 Schematic for nickel electroplating process for fabrication of Ni-stamp

such as PDMS casting, UV curable resin casting, hot embossing or injection molding using thermoplastic resin. In this chapter, these casting, embossing and injection molding processes are reviewed

3.2.2.1 PDMS(Polydimethylsiloxane) Casting

PDMS is in liquid phase with low viscosity at room temperature, but it starts to solidify when being mixed with curing agent and the temperature is raised. PDMS is a material most widely used to fabricate a micro-fluidic device due to its low viscosity, low surface energy, good elasticity and very simple casting process. As shown in Fig. 3.16, the casting process with PDMS consists of mixing PDMS and curing agents, eliminating bubble, coating, curing and releasing the PDMS replica from the pattern master [16, 17]. The casting process can be done with ease in a laboratory without any complicate apparatus or process expert. Due to its low viscosity in liquid state and good elasticity in solid state, PDMS can fill the micro-structures well in the coating process and be released easily after being cured from the template. Moreover, the PDMS is excellent for sealing the fluidic channel of the device due to its good adhesion to many materials such as silicon, glass or plastic after oxygen plasma treatment which results in increasing of free radicals on the surface [18]. Also, it is convenient to apply interconnection port or tube for sample delivery or interfacing with other device because it is easy to punch a hole out or insert a port into the PDMS substrate [19–22].

As mentioned in this section, the PDMS casting process has many merits for micro-fluidic device fabrication at an early stage of development in laboratory. This casting process using PDMS, however, is only for fabrication of small number of device due to its poor productivity, durability or reproducibility, etc. The PDMS casting process requires quite long cycle time, typically longer than tens of minute. In addition, The PDMS replica tends to deform easily due to its low mechanical modulus and has less dimensional accuracy due to its relatively large shrinkage rate during the curing process.

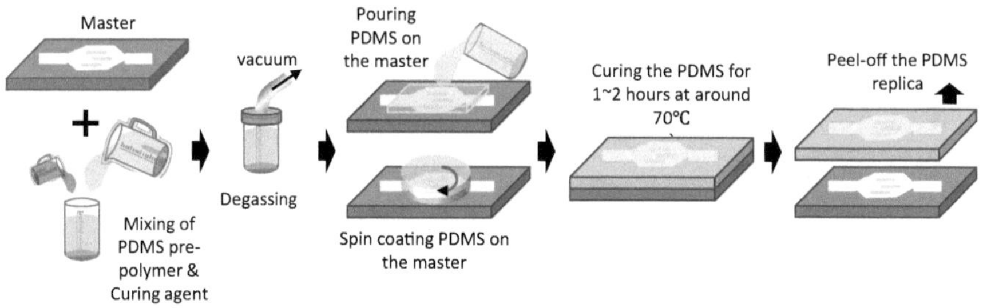

Fig. 3.16 PDMS casting process for micro-channel plate

3.2.2.2 UV Curable Resin Casting

UV curable resin can also be used for replicating the nano or micro-structures with supporting substrate. In terms of replication of nano or micro-structures and productivity, UV curable resin may have a good potential because it fills the structures easily due to its low viscosity and is cured very quickly when being exposed to UV light. A casting process may consist of resin coating on a transparent plastic film or substrate, embossing the structures using pattern master, curing by UV exposure and releasing from the master (Fig. 3.17). As an alternative process, the UV curable resin can be coated on the pattern master first and a transparent film or plate is laminated on the coated layer followed by curing and releasing. For replication of micro/nano structures, at least, one of the structures master or the substrate for pattern layer of the replica should be transparent to the UV light for curing the resin.

3.2.2.3 Hot Embossing

The hot embossing is a process that replicates nano or micro-structures from the master to thermoplastic substrate by heating the substrate over glass transition temperature (T_g), pressing, cooling and releasing (Fig. 3.18). The hot embossing process has a good potential for replicating high-aspect-ratio of nano or micro-structures precisely on a plastic substrate [23, 24]. The drawback of the hot embossing process is its low productivity due to the long cycle time over tens of

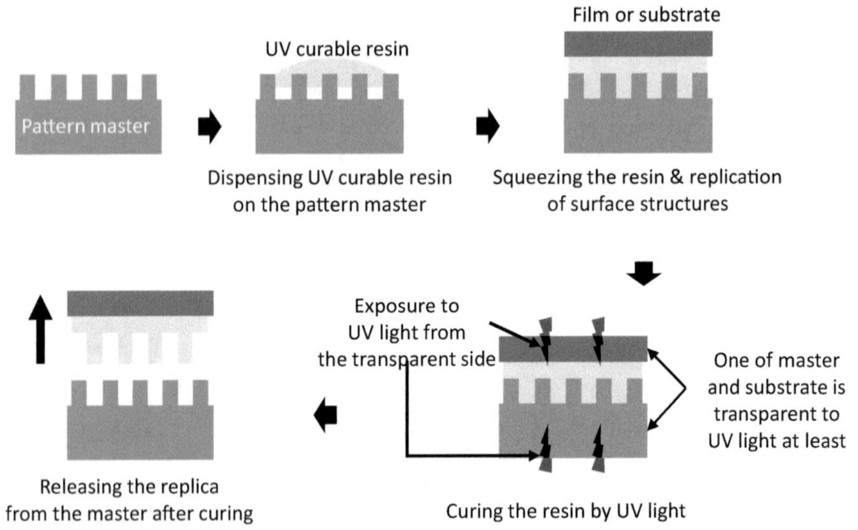

Fig. 3.17 A replication process for micro/nano structures using UV curable resin

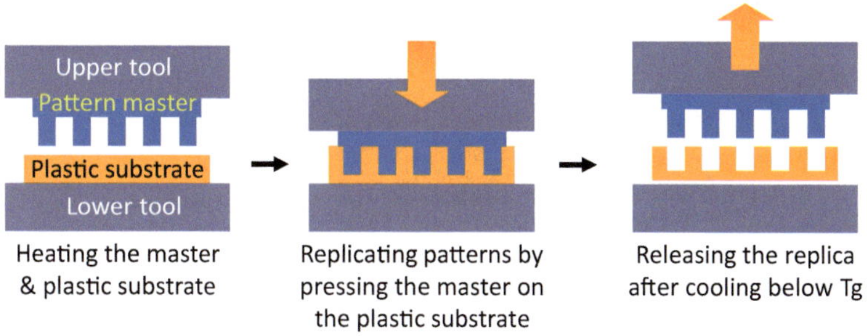

Fig. 3.18 A hot embossing process for replicating micro structures on a thermoplastic substrate

minutes for heating and cooling mostly. The poor productivity of the hot embossing process refrains this process from being used for the device requires a large scale of production.

3.2.2.4 Injection Molding

As the demand for mass production of the devices employing micro/nano structure increases, those devices are needed to be manufactured by injection molding process. The injection molding is a representative mass production process for plastics due to its excellent productivity with very short cycle time typically less than a minute or even several seconds for some commercial products. Injection molding process consists of four steps: filling, packing, cooling and releasing (Fig. 3.19). In the filling step, the polymer melt is injected into the mold cavity, where a stamp or a master is installed on the surface, and fills the intaglio structure on the stamp. In the releasing step, the injection-molded plate with micro/nano surface structures is mechanically released from the stamp after being solidified through cooling step. Because the polymer melt is hot (usually at least 200 °C) and the surface of the mold or stamp is relatively cold (usually less than 100 °C), the hot polymer melt starts to be cooled and solidified rapidly from the skin layer after it is injected into the mold cavity to contact the mold surface (Fig. 3.20). The rapid cooling and

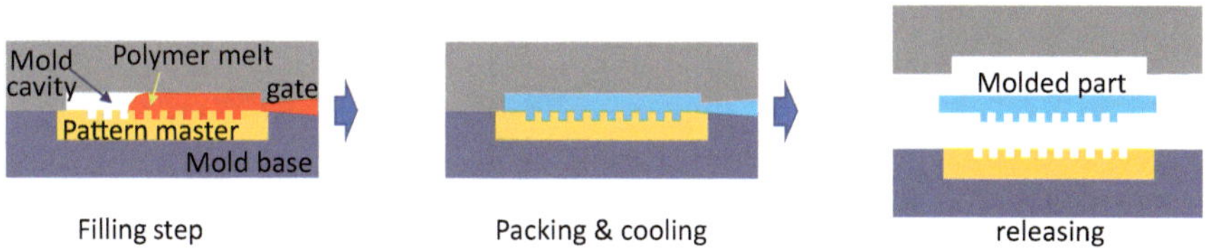

Fig. 3.19 Schematics of injection molding process

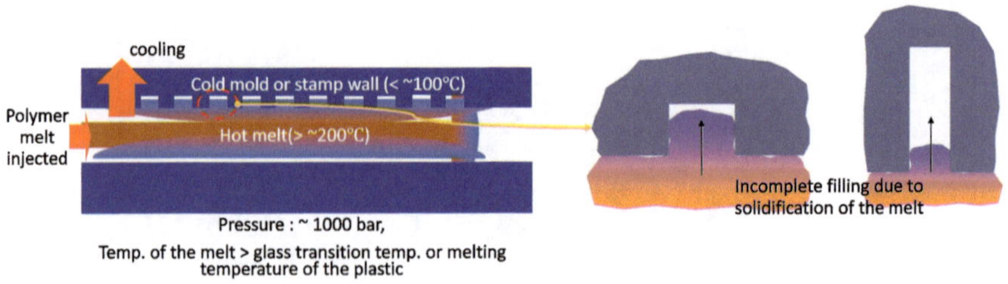

Fig. 3.20 Incomplete filling of the hot melt for injection molding of micro/nano structures

solidification of the melt from the skin layer makes it very difficult to replicate the surface micro/nano structures, especially in the case of a high aspect ratio of structure (Fig. 3.20).

In addition to this incomplete filling of the hot melt, deformation or fracture of the micro/nano structures during the releasing step is another critical issue for injection molding of a device employing micro/nano structures. This deformation issue at the releasing step is caused by an interference between the stamp and plastic structures (Fig. 3.21). These two issues, incomplete filling and deformation of the structures, are most concerns in applying the injection molding for mass production of the device with micro/nano structures and have been studied by many research groups.

For incomplete filling problem, many researches have been continued to optimize the process or mold design to fill the micro/nano structures on mold surface completely with melt before the melt solidifies. Considering the key parameters for melt filling process such as melt temperature, mold temperature, injection speed and packing pressure, replication of the micro/nano structure is generally known to enhance under conditions of higher melt temperature, higher mold temperature, faster injection speed and higher packing pressure. Higher temperature of melt or mold delays the solidification of the polymer melt, which allows the melt to fill the tiny structures on the mold. Faster injection makes the melt to fill the surface structure more quickly before being solidified too. Higher packing pressure can make the melt to flow into the surface structures even after being solidified slightly. For smaller or higher aspect ratio of structures, higher temperature, higher pressure or faster injection speed are generally required to replicate the structures better.

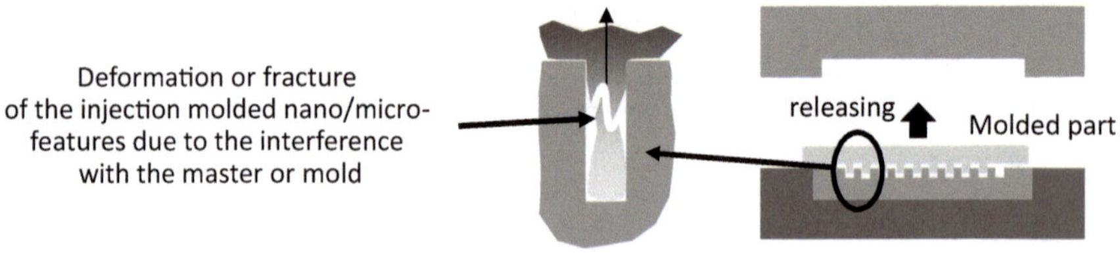

Fig. 3.21 Deformation of the injection molded micro/nano structure during the releasing step

These conditions, however, result in some negative effects such as longer cycle time or warpage of the substrate. For some micro/nano structures, it may be almost impossible to replicate the structure to the edge or corner precisely by optimizing conventional injection molding process using a mold set to a constant temperature during the process (Fig. 3.22). To overcome this issue, vario-themal injection molding processes, called rapid heat-and-cool injection molding process [25–27], are adapted for some devices with micro/nano structures [24, 28, 29]. In the vario-thermal injection molding process, the mold is heated up to the glass transition temperature of the molding material or higher temperature for filling step and cooled down below conventional mold temperature, for example 70 °C in case of PMMA, for cooling and releasing step. The key technology of the vario-thermal injection molding process is to heat up and cool down the mold very quickly to maintain the productivity of conventional injection molding process. Since the thermal mass of the mold is quite large, it takes too long time to heat up and cool down the entire mold system. So many molds for vario-thermal process employ a mold design isolating the surface with structures of the cavity thermally by separating the thin layer of the surface from the mold base [24, 25, 29], applying insulation layer beneath the surface or flow channels for heating and cooling close to the surface, etc. In Figs. 3.22 and 3.23, some injection molding results for micro/nano structures are shown using vario-thermal injection molding process.

Another advantage of injection molding is that various commercial materials are available to meet diverse requirements for the device in terms of mechanical, chemical or optical performance. PMMA, PC, PS and COC are used mostly for injection molding of a plastic nano or micro-fluidic device, but other thermoplastic resins for injection molding and their variations are also available.

Based on injection molding, the device can be designed and fabricated to have many functional elements such as port for tubing, load-bearing structures like rib or body, structures for assemble like welding bid, hole for bolting, snap-fit, etc. in addition to nano or micro-structures (Fig. 3.24).

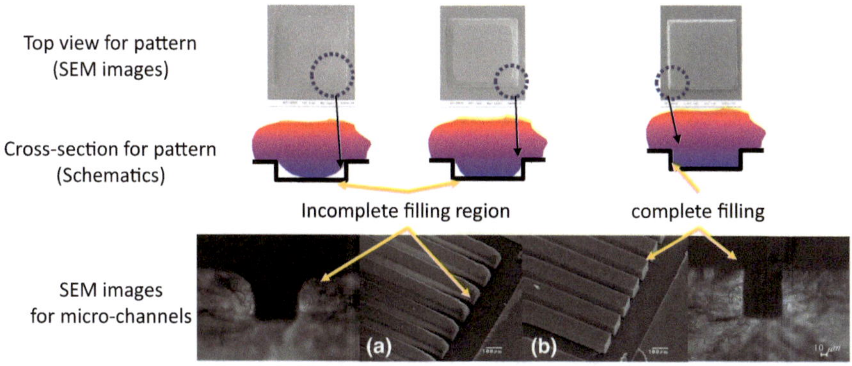

Fig. 3.22 Some images for micro-structures injection molded at different mold temperature

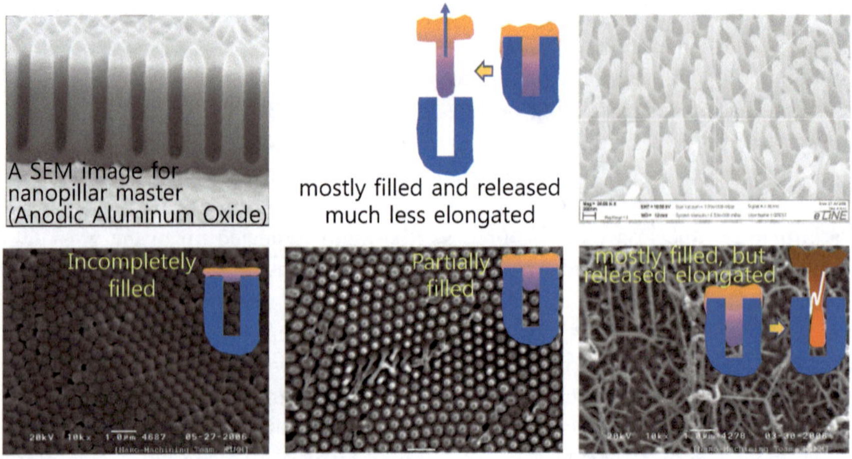

Fig. 3.23 Injection molded high aspect ratio of nanostructures using vario-thermal injection molding process [24] (Copyright (2009) The Japan Society of Applied Physics)

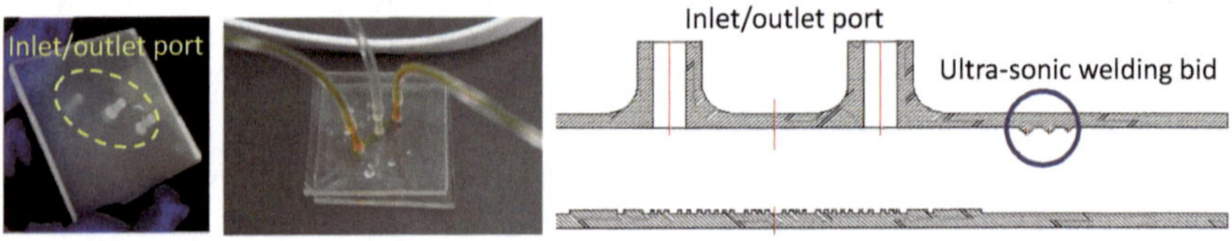

Fig. 3.24 Injection molded parts for fluidic channel device with functional elements

Considering these advantages along with excellent productivity and cost competitiveness, the injection molding process is most promising manufacturing process for commercialization of nano or micro-fluidic devices.

3.3 Surface Biofunctionalization Engineering for Bioreceptors and Antifouling Coatings (Claire Smadja and Mehdi Ammar)

3.3.1 Introduction

Biomolecules immobilization on solid supports is of paramount importance for biochip applications. Immobilization can be achieved by covalent immobilization, physical adsorption, and entrapment within polymer matrices or non-covalent attachment via intermediate biomolecular species. However, when analyzing complexes mixture such as blood, plasma or serum, unwanted interferences related to nonspecific adsorption of analytes occurred. To avoid this effect, coating allowing bioreceptors immobilization while reducing dramatically fouling phenomenon could be proposed [30].

The ultrasensitive biosensing methods are also combined to a high specificity provided by antibodies grafting. So far, the issue of self-deposited functionalization has been seldom addressed because of the absence of generic techniques allowing molecular diagnosis at the nanoscale. Therefore, a controlled and adaptive chemical preparation of the sensing surfaces is important for the bio-functionalization step. Modification of solid substrates by deposition of organic monolayers with controlled architecture [31, 32], such as thiols on gold, carboxylic acids on alumina or alkanes on silicon and silica are a subject of growing interest.

Therefore, several groups proposed a variety of functionalized surfaces coated to avoid fouling from complex media. Several techniques based on self-assembled monolayers (SAMs) are widely used as coating for proteins grafting but also to obtain anti-fouling properties towards proteins or bacteria. Poly[ethylene glycol] and its derivatives have been widely employed [30, 33]. In addition, advances in polymers synthesis lead to a novel class of non-fouling materials, including brushes of hydrophilic electro-neutral polymers [30, 33–35].

This chapter focuses on recent advances in the development of functionalizable surfaces and their applications in biochips. Several types of coating materials will be described. In the first part of this review, we will focus on bioreceptors immobilization. This section also includes a brief presentation of the bioreceptors (antibodies and apatmers). The next section reviews the coating anti proteins fouling. In this section, we also described results concerning fouling resistance against models plasmatic proteins and blood plasma or serum. The last section relied on functionalized antifouling coatings towards bacteria. This section includes several works on applications in this field. Different coating allowing antifouling against bacteria such as Pseudomonas aeruginosa, Staphylococcus aureus, Enterococci and Escherichia coli will be presented.

3.3.2 *Immobilization of Bioreceptors*

3.3.2.1 Biorecognition Elements

Antibodies. Immunoglobulines are the most widely used biorecognition element. They are produced by B cells for defensive immune response when exposed to foreign organisms [36]. Antibodies are Y-shaped immunoglobulins weighing approximately 150 kDa (Fig. 3.25). Each arm of an antibody contains a pair of antibody heavy ($\sim$55 kDa) and light ($\sim$25 kDa) chains that are connected by interchain disulfide bonds. An antibody can be divided into two distinct regions, variable (located at the top of the Y) and constant (located at the bottom of the Y) regions. An antigen binds to the paratope, variable region, of an antibody that is located at the top of the Y in a region [36].

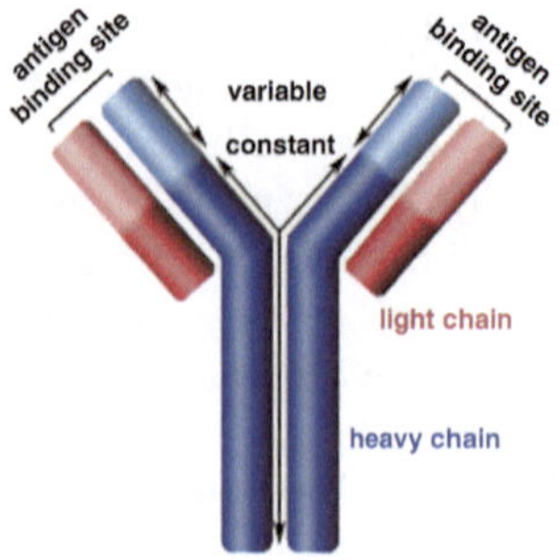

Fig. 3.25 Antibody structure (http://www.biology.arizona.edu/immunology/tutorials/antibody/structure.html)

Nucleic acids. DNA and RNA aptamers can be also used for the detection of proteins analytes [37]. The aptamer corresponds to a single strands nucleic acid that forms a 3-dimensional structure interacting specifically with a target via intricated mechanisms i.e. van der Waals, hydrogen bonding and electrostatic interactions.

3.3.2.2 Self-assembled Monolayers [SAMs] for Biochemistry

Self-assembled monolayers are ordered molecular assemblies of different organic materials (Fig. 3.26). SAMs are performed by simply immersing a substrate into a solution of the surface-active material. The driving force for the spontaneous formation of the 2D assembly includes chemical bond formation of molecules with the surface and intermolecular interactions. SAMs provide the needed design flexibility, both at the individual molecular and at the material levels, and offer a vehicle for investigation of specific interactions at interfaces, and of the effect of increasing molecular complexity on the structure and stability of two-dimensional assemblies.

In contrast to ultrathin films made by, for example, molecular beam epitaxy (MBE), and chemical vapor deposition (CVD), SAMs are highly ordered and oriented and can incorporate a wide range of groups both in the alkyl chain and at the chain terminal. For example, Self-Assembled Mono-layers (SAM) of organosilicon are currently used for coating glass slides, enhancing the biocompatibility of the surface and protecting proteins from denaturation during the immobilization step. In fact, maintaining protein conformations not only determine their tri-dimensional structure but also bioactivity and ability to interact with antigens, antibodies or substrates. The SAM technique is applied to interdisciplinary areas, and is widely engaged in research at the interface of chemistry with biorecognition.

3.3.2.3 Bioreceptors Attachment

Many biological and medicals applications, such as biochip, labchip and biosensors involved proteins—proteins or proteins ligand interactions. Attachment of

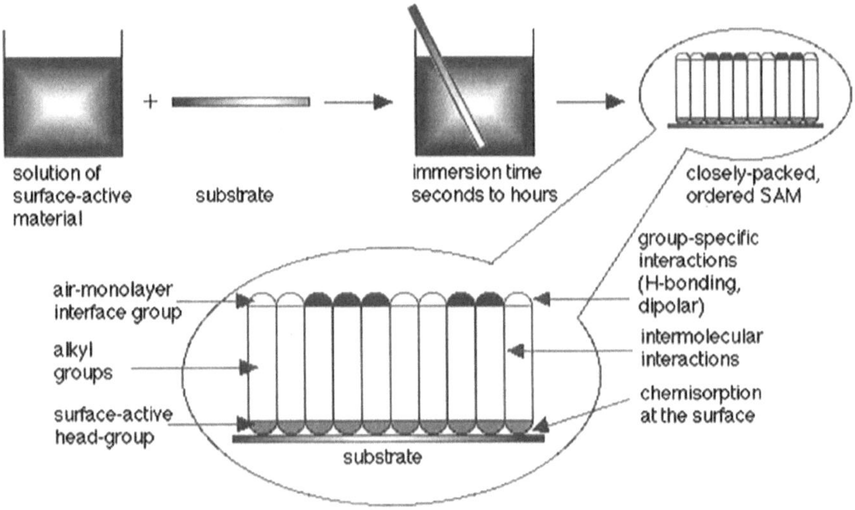

Fig. 3.26 Self Assembled Monolayers universal protocol. Reprinted with permission from [38]

biomolecules on different surfaces is therefore of paramount importance for micro-device development. Biomolecules could be immobilized on a large range of substrates; from carbone nanotubes, silicon, semiconducting polymers, magnetic nanoparticles [39]. In these surfaces, the biological properties of the proteins must be maintained. In the context of lab on chip or biosensors development, the aim is to detect target molecules (e.g. biomarkers) at low level in highly complex media such as human serum, plasma or urine. Therefore, for biosensing and lab-chip applications the proteins immobilized (e.g. antibodies) on surfaces, should meet the following criteria; (i) accessibility of the active site, (ii) an homogeneous surface coverage, (iii) stability of the grafting. In addition, proteins should be attached onto surfaces without affecting conformation and function to maintain their biological activity [32]. A large variety of immobilization techniques are proposed nowadays which are mainly based on three mechanisms: physical, covalent, and bioaffinity immobilization.

3.3.2.4 Physical Adsorption

Physical adsorption is one of the simplest methods [40, 41]. Proteins can adsorb on surfaces via weak interactions, mainly ionic interactions, hydrogen bonds, hydrophobic and polar interactions. Although physical adsorption provides several advantages such as method simplicity, the risk is to obtained heterogeneous layer with proteins randomly oriented, considering that each molecule can form many

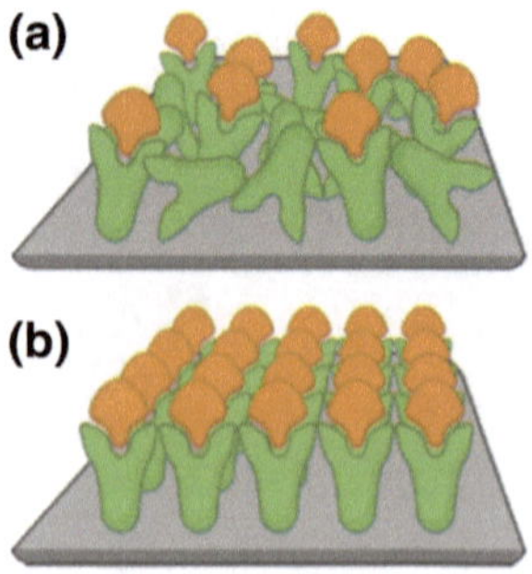

Fig. 3.27 Differences between random **a** and oriented **b** proteins. Reprinted with permission from [39]

contacts in different orientations for minimizing repulsive interactions with the substrate and previously adsorbed proteins [32]. In addition, conformational changes, reducing thereby proteins biological activity could also occur [39] (Fig. 3.27).

3.3.2.5 Non Site Specific Covalent Immobilization

In this approach, bioreceptors are covalently bound to the immobilization support through accessible functional groups of exposed amino acids. Covalent bonds are mostly formed between side-chain exposing functional groups of proteins with chemical modified supports [e.g. Self-assembled monolayers (SAM)], resulting in an irreversible binding and producing a high surface coverage [32]. In addition, the immobilization process should be performed under physiological conditions to avoid proteins denaturation during their coupling on the chemically modified surfaces [39]. Chemical binding via side chains of amino acids is often random, since it is based upon residues present at the proteins surfaces. The most widely used chemical groups are amines and thiols that are both good nucleophiles. Carboxylic groups can be activated to renders them reactive towards nucleophiles.

Amine chemistry. Lysine residues are the amino acids the most commonly involved for covalent grafting since they are frequently located on the proteins surfaces [39, 42]. N-Hydroxysuccinimide [NHS] is the most frequently used agent for the coupling between protein and surfaces, forming stable amide bonds (Fig. 3.28). This method has been successfully employed to graft antibody directed against Alzheimer diseases biomarkers, amyloid peptides, on silicon surfaces silanized by octenyl trichlorosilanes (OTS) (Fig. 3.28). A step of oxidation permit to emerge carboxylic groups. The optimized protocol rends possible to perform an immunoassay allowing thereby a sensitive detection of amyloid peptides on these surfaces. In this work, the carboxyl functionalized surface has been activated by

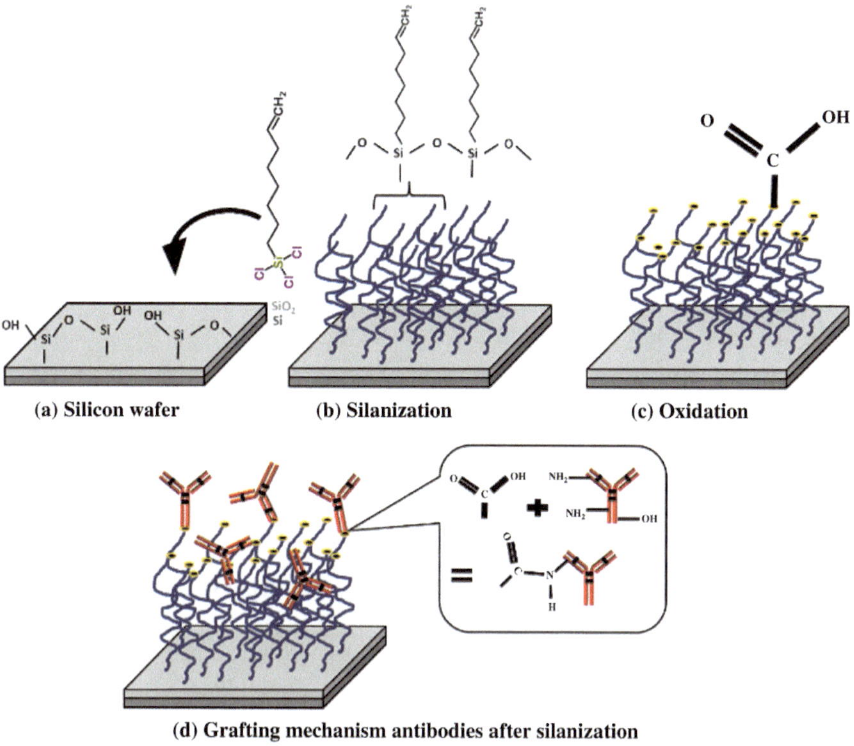

Fig. 3.28 Schematic representation of the 7-octenyltrichlorosilane grafting reaction on silicon and immobilization of proteins on emerged COOH groups. Reprinted with permission from [42]

1-ethyl-3-(3-dimethylpropyl) carbodiimide (EDC)/sulfo-NHS for coupling with the monoclonal antibodies directed against Aβ 1–42 [43].

Thiol chemistry. This methods required cysteine which is the unique aminoacids containing a thiol group. Since cysteines are not as abundant as lysines, proteins are less prone to random orientation rendering cysteine interesting for immobilization on sensor surfaces for instance [32].

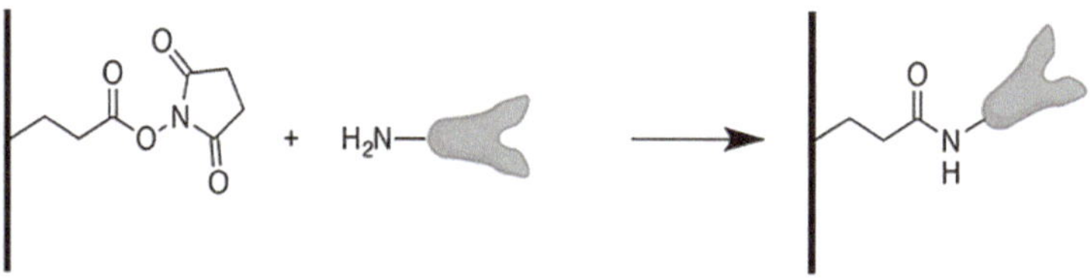

Fig. 3.29 Immobilisation anbtibodies via the amine groupe on N-hydroxysuccinimide (NHS) modified surfaces. Reprinted with permission from [39]

Carboxyl chemistry. Immobilization via carboxyl groups is also interesting opportunity considering the fact that aspartic and glutamic acid represent the major fraction of surface exposed groups on proteins. The coupling reaction could also involve EDC/Sulfo-NHS chemistry (Fig. 3.29).

3.3.2.6 Site Specific Immobilization Strategies

To provide a better control of the protein orientation during the immobilization process, different site-specific immobilization techniques have been proposed. This approach could also be divided in noncovalent and covalent strategies.

Site specific and non-covalent immobilization

Most of these approaches exploit complementary affinity interactions between biomolecules. They rely on specific interactions occurring in biological process [39] such as biotine/streptavidin interactions or Protein A or G towards antibodies.

Polyhistidine. This method is based on affinity between polyhistidine and divalent metal cations such as Ni^{2+}, Cu^{2+} and Zn^{2+}. Recombinant proteins exhibiting a $(His]_6$ tag at the C- or N-terminus can be immobilized on metal treated surfaces via a chelating agent such as nitriloacetic acid (NTA). This immobilization is reversible (Fig. 3.30). This strategy has been employed to develop an electrochemical biosensor, by immobilizing horse heart Cytochrome c on gold electrode surface [44].

Biotin-(strept)Avidin interaction. This approach relies on one of the strongest noncovalent bonds. The specificity of the interaction allows uniform orientation of immobilized proteins. It is based on the interaction between strept(avidin) and biotin (Fig. 3.31). An example of application is given by Xia et al. [46]. They have developed a sensor surface, by immobilizing antibodies directed against Alzheimer using Biotin-(strept)]Avidin interaction. This non-covalent bonding permits repeated SPR measurements on the same surface and thereby improves the reproducibility of the methods and enhances the sample throughput. They have been able to measure Alzheimer disease biomarkers (amyloid peptides) in cerebrospinal fluid (CSF) samples from AD patients.

Protein A/Protein G-mediated Immobilization. The immobilization of antibodies by use of protein A/protein G relies on the specific interaction with the constant region [Fc Fragment] of the antibodies. Using this immobilization method ensures that the binding site of the antibody, located on the Fab variable region, remains available for specific antigen capture [32].

DNA-Directed Immobilization. This strategy has been developed to convert DNA arrays into protein arrays using DNA hybridization. In this method, the complementary DNA conjugates to protein will be immobilized by specific interaction between the DNA grafted on the microarray and the DNA-protein conjugate. However, it is important to note that the incorporation of oligonucleotides into large proteins is still poorly developed [32].

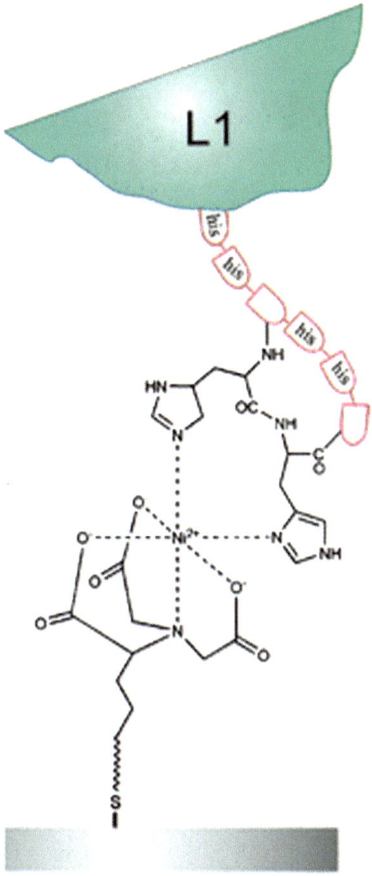

Fig. 3.30 Immobilization of a protein-His-Tag via NTA complex. Reprinted with permissions from [45]

Site specific and covalent immobilization

Click chemistry. In order to control an oriented and covalent coupling of a protein to a substrate, a unique chemical group or sequence could be introduced in the protein at a site-specific location. Several bio-orthogonal chemistries can lead to an oriented covalent immobilization of proteins. In this context, the 'click chemistry' concept becomes very popular. This chemistry rely on; (i) mild reaction conditions, (ii) insensitivity to oxygen and water, (iii) ability to use water as reaction medium, and the formation of a stable product under physiological conditions. In recent years several works include, copper or ring-strain catalyzed azide—alkyne cycloadditions, Staudinger-ligations, Diels—Alder cycloadditions, thiol—ene additions, and oxime formation [48].

Targeting the Carbohydrate Moiety. Many glycoproteins, including antibodies, hormones, and enzymes, can be immobilized through their carbohydrate moiety without affecting their biological activity, since the carbohydrate region is generally located in areas not involved in specific activity. Moreover, coupling of antibodies

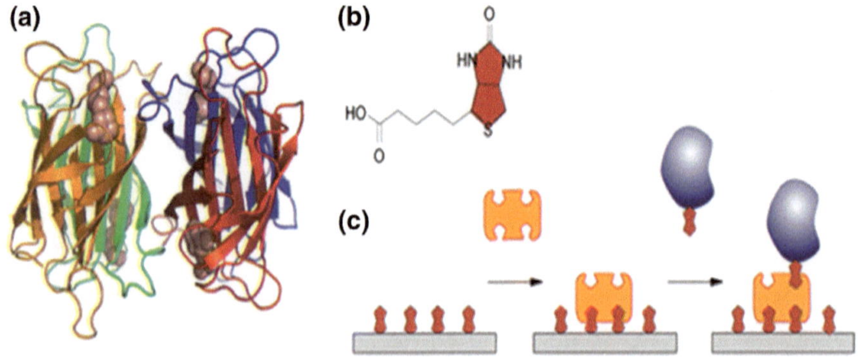

Fig. 3.31 **a** Streptavidine, **b** biotin, **c** immobilization of a protein relying on biotinstreptavidin interaction. Reprinted with permission from [47]

via carbohydrate moiety, located on the Fc region, was proven to enhance steric accessibility of the binding sites [32, 49].

3.3.3 Coatings for Proteins Anti Fouling

In this chapter, we will present the key strategies for preparing antifouling coating. Proteins solution containing fibrinogen or bovine serum albumin [BSA] or lysozyme are often employed to evaluate the antifouling properties of the coating. However, it is also important to evaluate the resistance toward highly complex mixture such as blood, plasma or human serum.

3.3.3.1 Poly(Ethylene Glycols) PEG and Poly[Ethylene Oxide] PEO Based Coating

PEG is one of the most widely used polymers to prevent proteins adsorption. The antifouling properties of PEG have been widely reported in the literature [50]. A range of groups have shown that by independently varying the molecular weight and chain density of PEG layers the protein resistance can be tuned accordingly. Indeed, a decrease of proteins adsorption is observed by increasing the grafting density or the chain length [30, 51, 52]. Layers with increasing grafting density pass from a single grafted chain (mushroom) regime to a brush regime with deformed polymer coils [30]. The effects of the polyethylene glycol (PEG) density, chain length, on proteins adsorption have been studied by using a complex mixture of proteins (lysozyme, human serum albumin (HSA), IgG and lactoferrin). For shorter chains a high density of PEG surface attachment sites is required to prevent proteins adsorptions. If the initial density of functional groups is low, the PEG chains should be longer to obtain a protein resistance effect. The development of PEG coatings by attachment of PEGs from aqueous solution onto an amine interlayer (or an alternative equivalent amine base layer) provides a more convenient and versatile route for the fabrication of protein resistant PEG coatings [53]. High molecular weight PEG molecules grafted on SiO2 surfaces showed adsorption resistance of proteins from human serum evaluated with two different techniques: ellipsometry and QCM-D.

In another work, silane-PEGs with either carboxylic acid (silane-PEG-COOH, SPC) or biotin have been grafted on stainless steel surfaces (Fig. 3.32) [54]. Stainless Steel is a commonly used metal in industrial and medical settings, where it becomes exposed to organic and biological agents that may result, for example, in biofouling and microbial-induced corrosion. In this work the silanes-PEGs formed less than 10 Å thick overlayers with close to 90% surface coverage. Recently, it has been demonstrated that a coating by PEO stearate on a Dyneon THV microchip prevents the adsorption of proteins such as BSA and trypsin inhibitor and allow their analysis by electrokinetic mode [55].

Another approach rely on amphiphillic triblock PEO-bl-PPO-bl-PEO copolymers consisting of central hydrophobic propylene oxide (PPO) segment flanked by two hydrophilic PEO chains has been also evaluated on gold surfaces by SPR. An ultra-low adsorption of single protein solutions of fibrinogen, BSA, and complex mixtures of proteins has been obtained [56]. The same effect has been observed on stainless steel surface coated with the triblock copolymer a poly(ethylene oxide)-poly(propylene oxide)-poly(ethylene oxide) (PEO-PPO-PEO). The non-specific biofouling of fibronectine and biotine is reduced by 70%. Similar reductions in protein adsorption have been reported, for example, by Yang et al. [57] and Harder et al. [58]. Y. Yang and his colleagues coated stainless steel surface with a poly (ethylene oxide)-poly(propylene oxide)-poly(ethylene oxide) (PEO-PPO-PEO) triblock copolymer, and showed that the coated stainless steel surfaces were capable of significantly reducing the adsorption of bovine serum albumin (BSA) [57]. Another triblock copolymer structures composed by a central polypropylene sulfide flanked by two external PEG chains (PEG-bl-PPS-bl-PEG) grafted on gold surface exhibit anti-proteins adsorptions towards highly complex media such as human serum measured by SPR [59] (Figs. 3.22 and 3.33).

Another type of multidentate copolymeric has been proposed (Fig. 3.34). They are composed by three different monomers that have been randomly copolymerized in order to obtain a macro- molecule bearing grafted PEG side chains and providing a non-fouling property when exposed to human serum [61].

3.3.3.2 Oligo[Ethylene Glycols] SAMs Based Coating

Regarding OEG-Sam coating the typical structure is R1-$(CH2)_n$-R2. R1 is an hydrophilic moiety at the ω-position, (CH2)n is a nonpolar alkane chain and R2 is a group on α-position that is chemically reactive to substrate. Chapman et al., pioneered research on proteins adsorption [i.e. fibrinogen, a large (340 kD) blood

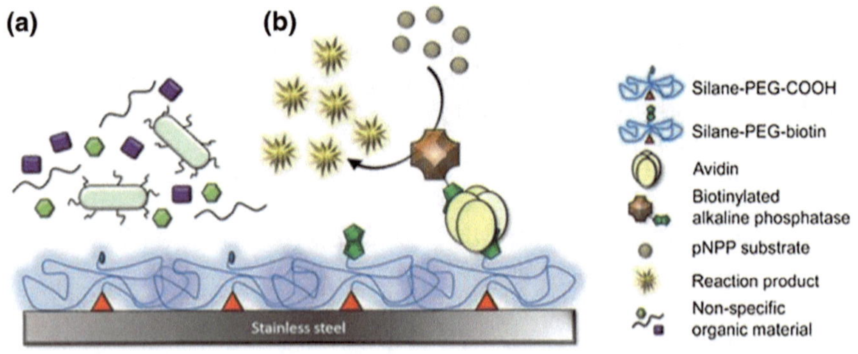

Fig. 3.32 Schematic illustration of silane-PEG-modified Stainless Steel. **a** silane-PEG-COOH overlayer resists non-specific biofouling of both bacteria and proteins. **b** silane-PEG-biotin overlayer enables selective functionalization of SS via avidin-biotin bridges. Reprinted with permission from [54]

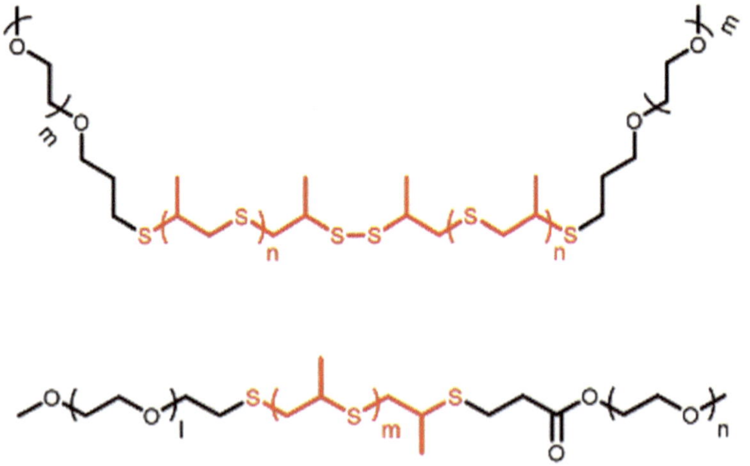

Fig. 3.33 Molecular structure of two variants of sulfur-based poloxamer analogues (PEG-bl-PPS-bl-PEG triblock copolymers). Reprinted with permission from [60]

plasma proteins and lysozyme, a small protein (14 kD), onto ω-functional SAMs. They showed that the best results were obtained when functional groups share the following properties: containing polar functional groups, incorporating hydrogen bond accepting groups without any hydrogen bond donating groups, and having no net charge [62]. The same groups has also evaluated the ability of SAM to resist to model protein adsorption such as fibrinogen, RNase, lysozyme and carbonic anhydrase, when incorporating oligo(ethylene glycols) (EG) groups into alkenethiols. [63]. The results obtained showed that by incorporating oligomer of EGn groups [n = 6], the mixed SAM completely resist to model proteins

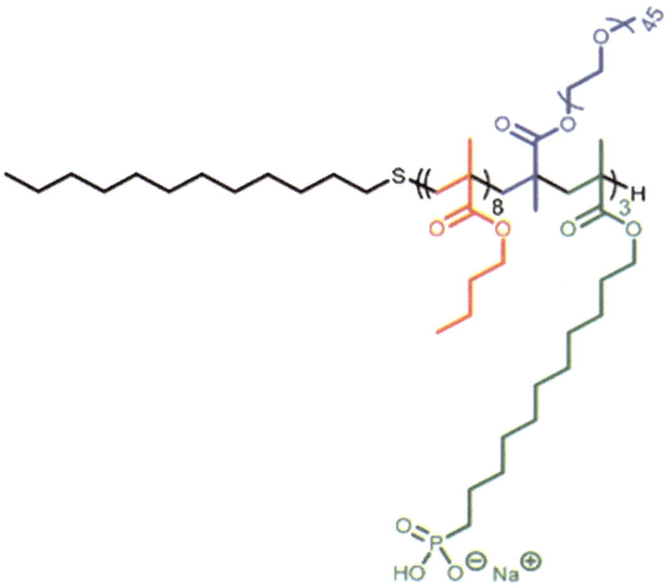

Fig. 3.34 Molecular structure of the PEG-polyalkyl phosphonate terpolymer. Reprinted with permission from Ref. [60]

adsorption. The surface inertness towards proteins could be due to a reduction of the number of hydrogen bond donors.

3.3.3.3 Zwitterionic SAMs Based Coating

Zwitterionic-based materials, such as phosphoryl choline, phosphobetaine, sulfobetaine, and carboxybetaine-based polymers or surfaces have been shown to greatly reduce nonspecific protein adsorption. To explain their repellent activity towards proteins, the hypothesis is that zwitterionic surfaces interact with water molecules creating thereby a hydration layer. This layer prevents protein adhesion on the surface. Several studies demonstrated the resistance of phosphorycholine (PC) SAM to proteins adsorption [64, 65]. In this work protein adsorption has been investigated on gold chip terminated by PC-SAMs by an SPR sensor. The results showed a dramatic decrease of two model proteins adsorption; Fibrinogen and BSA. A recent study, investigated polyurethane membranes modified with zwitterionic polymers such as 2-methacryloyloxyethyl phosphorylcholine (PU-MPC) showed a significant reduction of plasmatic proteins adsorption (Fig. 3.35) [66].

Recently, brushes produced by the copolymerization, on Si3N4 surfaces, of a standard zwitterionic sulfobetaine monomer have demonstrated their antifouling properties towards fibrinogen (Fig. 3.36) [67]. In this study, they showed a proof of principle with biotin-functionalized brushes on Si3N4 that combine excellent

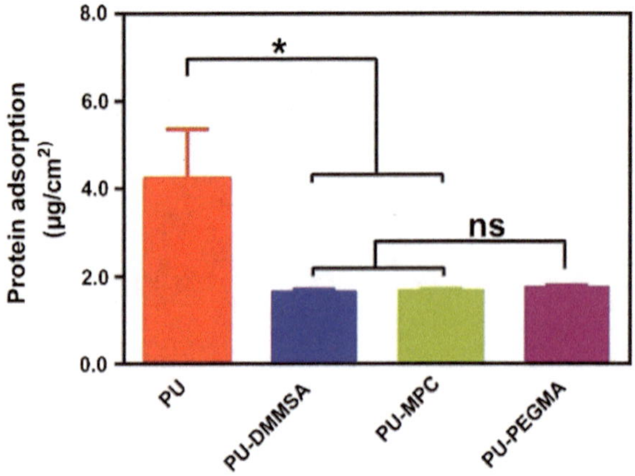

Fig. 3.35 Reduction of proteins adsorption after coating with zwitterionic polymers, methacryloyloxyethyl phosphorylcholine (PU-MPC). Reprinted with permission from [66]

antifouling properties toward fibrinogen with specific avidin binding (Fig. 3.37). The avidin binding could allow in the next step bioreceptors attachment.

Zwitterionic SAM carboxybetaine methacrylate and acrylamide surfaces have also demonstrated by surface plasmon resonance (SPR) measurements their ability to reduce fibrinogen from buffer solution [68] and BSA from undiluted human blood plasma [69] non-specific adsorption. In addition, it has been possible to develop an antibody-functionalized poly carboxybetaine acrylamide gold surface platform to detect a biomarkers of carcinoma, Activated leukocyte cell adhesion molecule (ALCAM), in blood plasma using a sensitive surface plasmon resonance (SPR) sensor [69].

3.3.4 *Anti Bacterial Coatings*

Biomedical device-associated infection by bacteria is one of the most common and problematic complications faced worldwide [70]. Numerous epidemiological studies have shown that the most common bacteria causing hospital infections are Pseudomonas aeruginosa, Staphylococcus aureus, Enterococci and Escherichia coli [71, 72].

Bacteria could be considered as living colloids that interact easily with surfaces. This is due to their small size, low density, surface charge, and variable degrees of

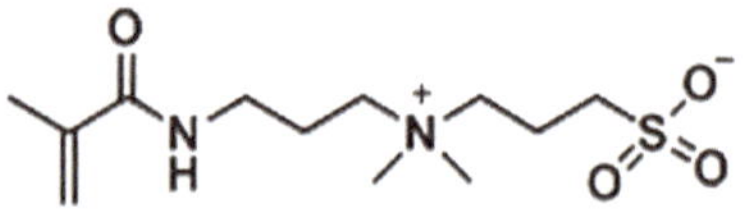

Fig. 3.36 Standard Monomer of a standard zwitterionic sulfobetaine. Reprinted with permission from [67]

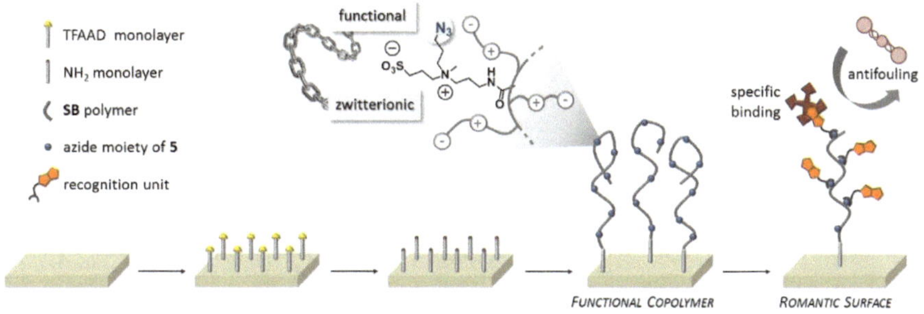

Fig. 3.37 Schematic representation of functional antifouling polymer brushes as obtained from the copolymerization of a clickable monomer with a standard SB monomer. Reprinted with permission from [67]

cell surface hydrophobicity. Their adhesion in surfaces could leads to a growth and production of an extracellular polymeric matrix and eventually to biofilm. Within this biofilm bacteria are more resistant to usual treatments by antimicrobial agents, which in turn increase health problems. Therefore, it is of paramount importance to prevent or control microbial attachment in the surface to avoid proliferation of adherent bacteria and biofilm formation. To prevent pathogenic biofouling on surfaces specific surface chemistries could be employed. The bacteria adhesion is influenced by the hydrophobicity and the charge of the external membrane of bacteria but also by the physico-chemical properties of the materials and its topography. Therefore, to prevent bacteria adhesion in surfaces by modifying surface chemistry two approaches could be proposed; repel or bacteria killing (Fig. 3.38). Repelling prevent bacteria attachment by steric hindrance or charged anionic polymers leading thereby to a repulsive electrostatic force between the negatively charged bacteria and the surface.

Killing surface could be prepared by using polymers interacting with the bacteria membranc and cause its disruption or inactivate microbial cell activity [73–75]. Another approach relies on the incorporation of antibiotics on the polymers [76].

3.3.4.1 Polymers for Bacteria Repelling

For Bacteria repelling surfaces that are coated with non-charged hydrophilic polymers the reduced attachment of bacteria could be due to a steric hindrance phenomenon [78]. The most widely described repelling polymers is the poly (ethylene glycol) (PEG). The bacteria repelling by PEG could be explained by several hypothesis such as; hydrophilicity, high polymer flexibility, large excluded volume, steric hindrance effect [53]. It has also been demonstrated that the longer is the PEG chain the higher is the resistance of the bacterial attachment [79]. PEG molecules bearing terminal hydroxyl, amino and sulfonates groups were tested against Escherichia Coli (E. Coli) and Staphyloccus Epidermidis (S. Epidermidis).

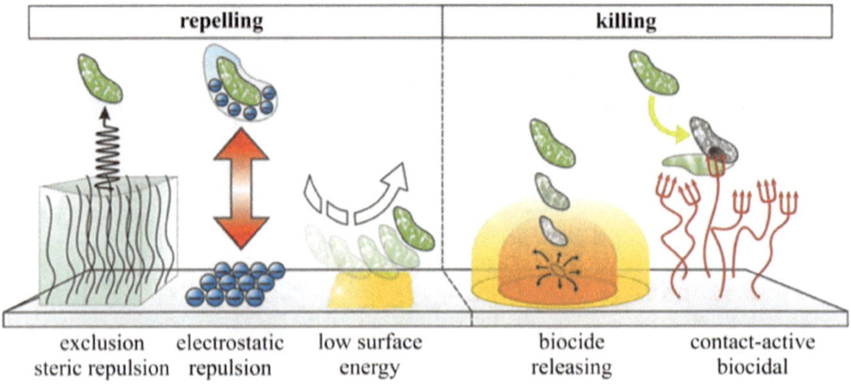

Fig. 3.38 Repelling and killing bacteria surfaces. Reproduced with permissions from [77]

Results obtained showed that the best repelling effect is obtained for both bacteria with PEG- terminated by sulfonates [79].

Several authors also obtained a significant reduction of bacteria adhesion (70 to 95%) with poly[ethyleneterephtalate] (PET) surfaces modified by poly[ethylene oxide] (PEO) [80]. Glass and silica surfaces have also been modified with a covalent grafting of PEO brush. They showed that for the most hydrophobic strain the repelling is less effective [81, 82]. Thus, a dramatic reduction of attachment was obtained with S. Epidermidis, S. Aureus, S. Salivarius and E. Coli. Other materials that reduce bacterial adhesion are based on surfactants such as hydroxyapatite and chitosan or biological molecules such as albumin and heparin. Friedman and co-workers coated TiO2 surfaces with BSA using carbodiimide chemistry. A recent study performed on titanium surfaces sprayed with calcium phosphate incorporating

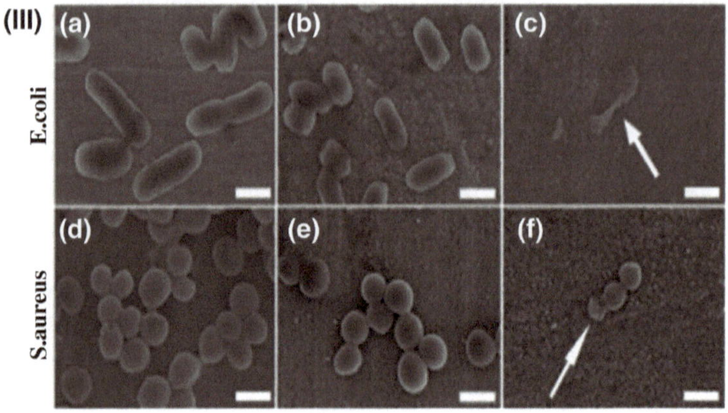

Fig. 3.39 SEM images of E. Coli and S. Aureus (c and h) when silver (Ag) is incorporated into calcium phosphate coatings on Titanium. Reprinted with permission from [83]

silver showed a strong reduction of E. Coli and S. Aureus attachment [83] (Fig. 3.39).

3.3.4.2 Polymers for Bacteria Killing

Killing surfaces could be obtained with cationic polymers. These polymers are able to interact with bacteria surfaces leading to their disruption [73].

Non-natural cationics polymers

The first study has been proposed by Klibanov and coworkers. They developed antimicrobial surfaces which involve a covalent coating of amino-modified glass slides with long hydrophobic polycationic chains i.e. poly(4-vinyl-N-alkylpyridinium bromide). These derivatized surfaces become bactericidal since they kill up to 99% of deposited bacteria such as Tiller et al. [84]. The bactericidal mechanism of the surface-attached polycationic chains probably involves their penetration into the bacterial membranes, leading to cell damage and death [85].

Another approach to provide antimicrobial properties to medical device surfaces involves their silanization with quaternary ammonium-containing silane agents. Microfibrillated cellulose surfaces have been modified with alkoxysilane octadecyldimethyl [3-trimethoxysilylpropyl]-ammonium chloride rendering the material highly microbiocidal [86]. The quaternary ammonium functionalized surfaces have a high positive surface charge exerting thereby a strong adhesive force on negatively charged bacteria, presumably causing membrane leakage and cell death [87–89]. These surfaces have been tested for antibacterial activity against S. Aureus, E. Coli and P. aeruginosa. Results obtained showed that more than 99% of S. aureus or E. coli bacteria initially exposed to the material were killed after a 24 h contact. The killing efficiency of these surface was found to be smaller toward P. aeruginosa [95%] compared to E. coli and S. aureus. A similar approach has been described recently when glass was modified by triethoxysilane terminated with a quaternary ammonium [90]. In this study, bacterial suspensions of S. Aureus or E. coli were sprayed at coated and uncoated slides. After 24 h, the number of colonies for both strains decreased; 92% and 71% for S. aureus and E. coli respectively.

Another important class of polymeric biocides is based on quaternary ammonium compounds. Recently, new random copolymers, where rather small fractions of reactive units such as carboxyl groups or epoxides are copolymerized with quaternary ammonium biocidal units, covalently attached or electrostatically bound onto the polymeric chains. Two series of copolymers, poly(4-vinylbenzyl chloride-co-acrylic acid), P(VBC-co-AAx), and poly(sodium 4-styrenesulfonate-co-glycidyl methacrylate), P[SSNa-co-GMAx], were synthesized via free radical copolymerization and further modified by the incorporation of biocidal units either covalently (4-vinyl benzyl dimethylhexadecylammonium chloride) or electrostatically bound (cetyltrimethylammonium 4-styrenesulfonate, SSAmC16). Antimicrobial tests also revealed that the cross-linked membranes presented strong antimicrobial activity against S. aureus and P. aeruginosa [91]. Another study showed that hydrophilic monomer acrylic acid (AA), at low contents, with the covalently attached bacteriostatic group vinyl benzyl dimethylhexadecylammonium chloride (VBCHAM) in the copolymer P(AA-co-VBCHAM88), resulted in a high bacteriostatic activity against P. aeruginosa and E. faecalis (6 log reduction in certain cases) [92]. This performance of the coating materials, evidencing their strong antimicrobial activity suggests that such materials are promising for antimicrobial applications [93].

Natural cationics polymers

Chitosan is a polycationic polymer that can be employed alternatively to the non-natural polycationic polymers described above. Because of its non-toxicity, biodegradability, biocompatibility and antibacterial activity chitosan has attracted attention for application in the food and pharmaceutical industries [78]. Chitosan is the N-deacetylated derivative of chitin. Chitosan inhibits the growth of a wide variety of bacteria, including E. Coli and S. Aureus [94]. The exact mechanism of the antimicrobial action of chitin, chitosan, and their derivatives is still unknown, but it has been suggested that an interaction between positively charged chitosan molecules and negatively charged bacteria cell membranes could occur leading thereby to membrane disruption [94].

Because of their practical importance, microbiological applications of zwitterionic polymers and their derivatives have been studied in depth in recent years, resulting in new mechanistic understandings and novel molecular designs. Recently, a coating by chitosan–iron oxide on graphene oxide nanocomposite hydrogel has been developed. The chitosan–iron oxide coated graphene oxide nanocomposite hydrogel films displayed significant antimicrobial activities against both Gram-positive and Gram-negative bacterial strains, such as methicillin-resistant Staphylococcus aureus, Staphylococcus aureus, and Escherichia coli, and also against the opportunistic dermatophyte Candida albicans [95].

3.3.4.3 Nanostructured Surfaces

Somme recent development showed that nanostructured surfaces could prevent surface colonization by bacteria. A recent work demonstrated the ability of a highly ordered arrays of surface nanopillars from cicada wing surface elicits bacterial membrane disruption (Fig. 3.40) [96, 97].

3.3.5 Conclusion

To develop lab on chip or sensor, it is of paramount importance to control proteins immobilization in order to obtain the best bio-functionalized surface in term of grafting density and bioreceptors orientation. A large variety of immobilization techniques has been described in this chapter. However, the most widely used

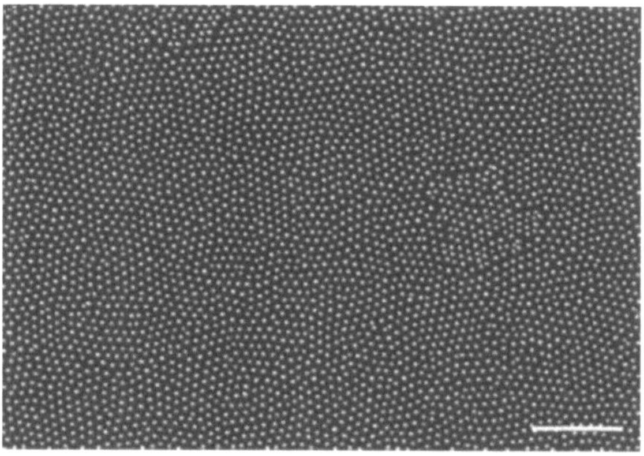

Fig. 3.40 Scanning electron micrograph of the upper surface of the forewing, at 25,000 × magnification. The surface consists of an array of nanoscale pillars, with approximately hexagonal spacing. Scale bar is 2 μ m. Reprinted with permission from [96]

approach for the immobilization of bioreceptors, to optimize biorecognition and resist to non-specific grafting (antifouling), is covalent attachment. For specific proteins attachment, the SAM technique is majority used directly or indirectly on the host substrate. To reach a high sensitive bioactive surface, all the steps for SAM protocols need to be extensively characterized, to optimize the efficiency of the biografting, and the reproducibility finely evaluated. In parallel, many developments of functionalizable antifouling coatings towards proteins and bacteria have been proposed. We demonstrated in this review the large variety of surface chemistry developed from polyethylene glycol to self-assembled monolayers (neutral or zwitterionic) to prevent unspecific adsorption. For antibacterial surfaces, similar surface chemistry is proposed. However, two different purposed could be attained; repel or bacteria killing. Repelling prevent bacteria attachment by steric hindrance or charged anionic polymers. Killing surface could be prepared by using polymers interacting (electrostatic interactions) with the bacteria membrane and cause its disruption or inactivate microbial cell activity. At least a recent and promising alternative has been proposed by designing nano-structured surface with nanopillars may be a promising route to minimize bacterial contamination/infection.

3.4 Packaging Technologies for Fluidics and Biodevices (Yeong-Eun Yoo)

3.4.1 Introduction

In Sect. 3.2, several fabrication processes are described for micro/nano structures like channels or pillars using various materials. These fabricated micro/nano structures are open along the structures on the substrate, so a capping process with a cover plate is needed to seal the open structures to deliver the fluidic samples without overflow or leakage. Sealing the surface structures like fluidic channels or chambers depends on the size and shape of the structures, materials of the parts, bonding strength required, etc.

For mass production of device based on fluidic channel, injection molding using thermoplastic resin may be the most competitive process due to its excellent productivity, good dimensional accuracy and low cost, etc. Although injection molding process is now available for plastic substrate with nano or micro-structures on its surface, sealing the fluidic channel is still a big hurdle.

The major goal of sealing the fluidic channel is to bond the substrates hard, hermetically but still precisely to avoid deforming the channel. For the bonding of two substrates to seal the fluidic channel, the substrate mostly needs to be altered chemically or mechanically on the surface. Heating or applying a solvent softens the substrate especially in the skin region (Fig. 3.41), which results in inter-diffusion of the material and bonding at the interface of the substrates.

A dry or wet adhesive may be used between the substrates for sealing. This adhesive bonding might be simple and fast to use, but have some issues like clogging by overflow of the adhesive during the coating or pressing process, misalign of the adhesive layer to the channel substrate or leakage path formed by porous structures in the adhesive layer (Fig. 3.42).

Clamping two substrates mechanically is another method for sealing of the fluidic channel device as shown in Fig. 3.43. It looks quite simple conceptually, but it is not easy to seal the fluidic channel hermetically due to the deformation of the substrate caused by point load for clamping.

Generally, more heating, solvent or adhesive is applied, higher strength of bonding can be obtained, but the surface structures are more deformed. The structures order of millimeters may not be affected significantly by the small amount of deformation or change at the interface (Fig. 3.44a). In case of sealing the structures less than hundreds of micrometers, however, the deformation or interference on the structures may be comparable to or even larger than the scale of the

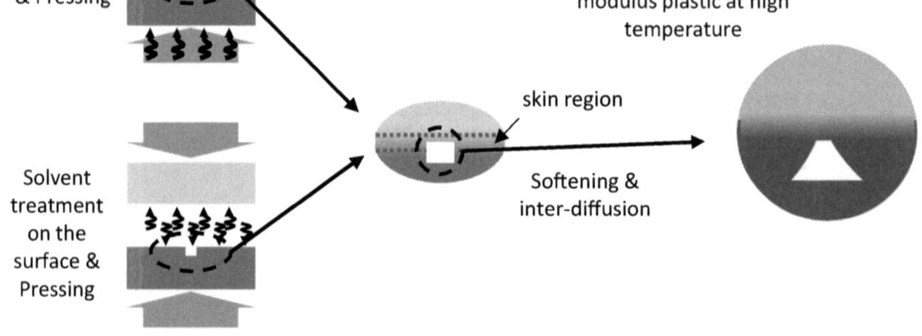

Fig. 3.41 Schematics on sealing of fluidic channel thermoplastic substrate by thermal or solvent bonding and the channel deformation issue

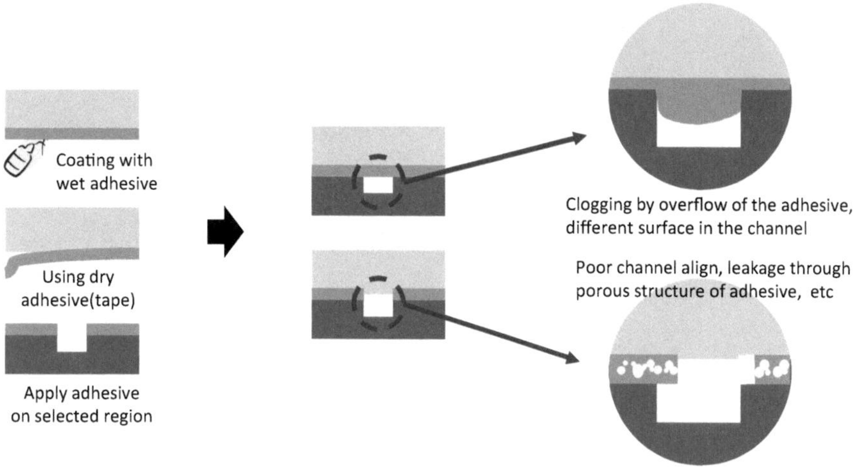

Fig. 3.42 Schematics on sealing of fluidic channel thermoplastic substrate by adhesive bonding and some of issues such as channel clogging, leakage or poor channel align

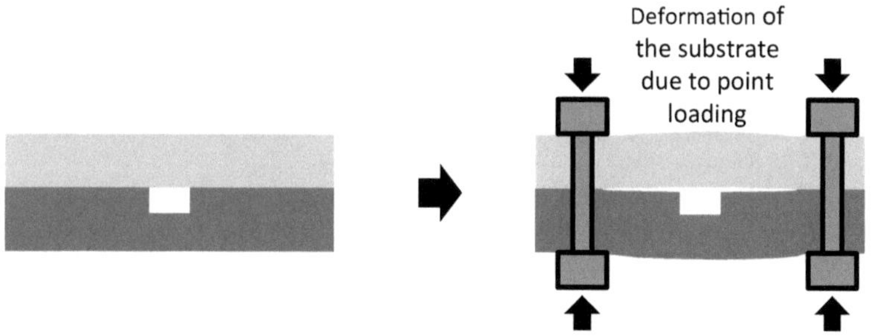

Fig. 3.43 Schematics on sealing of fluidic channel thermoplastic substrate by mechanical joining and leakage issue due to substrate deformation caused by mechanical clamping

structure itself (Fig. 3.44b). This significant influence on the channel requires that a very precisely controlled bonding method be used to enhance the bonding strength and the dimensional stability of the micro-structures at the same time, especially for polymeric substrates.

Beside the issues mentioned above, restriction on heat, mechanical stress, UV radiation or solvent during the sealing process are additional issues on sealing of fluidic channel especially for devices employing biomaterials. Most biomaterials like antibody, other proteins or lipid are degradable or susceptible to losing their functionality when exposed to thermal, chemical or mechanical stress, UV radiation etc. Considering most bonding require at least one of processes such as heating, solvent treatment, pressing or UV exposure processes, it should be designed very carefully not to damage the biomaterials.

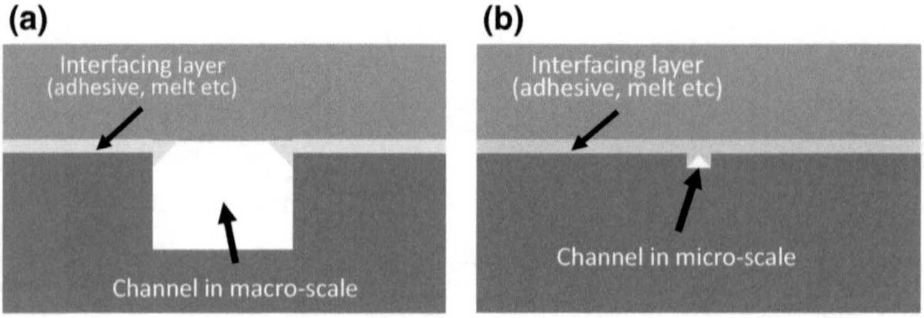

Fig. 3.44 Influence of the bonding layer on a fluidic channel **a** in order of millimeter **b** in order of micrometer

Although diverse sealing methods have been proposed for nano or micro-fluidic device, it is still challenging task for many cases. In this chapter, some sealing methods are reviewed mainly based on typical bonding processes for plastics such as thermal bonding, adhesive bonding, ultra-sonic welding and solvent bonding. A sealing by direct contact using PDMS, which is mostly used in laboratory, is reviewed briefly.

3.4.2 Sealing by Direct Contact Based on PDMS

PDMS is one of the materials widely used for packaging of micro-fluidic device because it can contact the other surface conformally due to its low elastic modulus, typically less than 1 MPa [18], and provides consequently good sealing even through solid to solid contact. The bond strength for PDMS device depends on the characteristics of the PDMS such as the mixing ratio of the curing agent or the degree of cure [16, 98]. Higher mixing ratio of the curing agent makes the PDMS have lower elastic modulus and contact the other surface more conformally, which results in better adhesion. The partially cured PDMS can also contact the other surface better because the incompletely cured PDMS has lower elastic modulus compared to the fully cured one. Although PDMS/PDMS or PDMS/other surface can bond together well reversibly, PDMS's or PDMS and some materials can also bond irreversibly when the surface is modified by surface treatment such as oxygen plasma. The oxidized surface of PDMS by exposing to oxygen plasma has silanol groups (Si–OH) temporarily, less than 30 min approximately in the air, and can form covalent siloxane bonds (Si–O–Si) when facing two surfaces tightly before silanol groups disappear [99, 100]. The PDMS and Si-based materials can also form covalent bond when the surface is oxidized by exposing to oxygen plasma [16, 98, 100, 104].

3.4.3 Adhesive Bonding

Bonding with an adhesive is one of general methods for sealing the fluidic channel. For this bonding method, the two parts are bonded through a 3rd material such as thermoset resin, UV curable resin or thermoplastic resin. The adhesive is coated to thin layer usually on the flat surface and cured or solidified after facing to press two substrates. The adhesive layer is cured or solidified in different ways depending on the material. The thermoset adhesive such as epoxy resin or thermo-reactive polymers is cured by heating up to curing temperature of the resin [101]. The UV curable resin is cured by UV light with appropriate wavelength to the resin much faster than the thermoset resin. These thermoset and UV curable resin are mostly liquid during the process to be coated on a substrate with a simple process. The thermoplastic adhesive layer previously coated on the substrate like thermoplastic film is heated up over the glass transition temperature of the adhesive and cooled down for re-solidification after facing and pressing two substrates. In general, the glass transition temperature of the thermoplastic adhesive is lower than of the substrate to reduce the deformation of the substrate or fluidic channel during the bonding process. The bonding process using adhesive is relatively simple, but one of the issues is to maintain the channel dimension precisely. The adhesive layer coated on a part is squeezed to flow by a compressive force on the layer for bonding. The squeezing flow of the adhesive during the bonding process may result in overflow into the micro-channel and blocking or narrowing the channel (Fig. 3.42). This blocking or narrowing the channel may depend on the thickness of the adhesive layer, channel size, viscosity of the adhesive, the design of the channel array or the surface structure, etc. Some solutions to reduce this blocking problem by the overflow have been proposed such as employing additional guiding channel for overflow adhesive [102].

3.4.4 Sealing Fluidic Channel by Thermal Bonding

For bonding two thermoplastic substrates directly without additional materials like adhesives, the substrates should be heated up to be softened and pressed after facing target surfaces to enhance the inter-diffusion at the interfacing zone of the mating surfaces. The bonding strength and sealing performance enhance as the substrate is heated to higher temperature and pressed with higher pressure. The dimensional stability of the fluidic channel and the substrate, however, is influenced negatively due to the larger deformation caused by lower modulus of the material at higher temperature and higher stress. The deformation of the fluidic channel or substrate and bonding strength increases for a same pressing condition as the temperature of the substrate increases and consequently the modulus of the thermoplastic decreases. Low temperature of substrate and low pressing pressure usually results in good dimensional stability, but poor bonding strength. High temperature and high pressing pressure usually results in poor dimensional stability, but good bonding strength. Some studies tried to enhance the dimensional stability and bonding strength at the same time by balancing the temperature for the substrate and pressure for pressing. Some other methods have tried to enhance the bonding strength, sealing performance and dimensional stability such as matching thermoplastic substrates with different glass transition temperature (T_g), modifying the substrate to have varying T_g through thickness or activating the surface of the substrate to have functional chemical groups [103, 104].

It may be a good match to use a high T_g of fluidic channel substrate and a low T_g of cover plate with lower T_g for good dimensional stability and bonding strength when the temperature for the bonding process is set to between the glass transition temperatures of the fluidic channel substrate and the cover plate (Fig. 3.45).

The thermoplastic substrate can be pre-softened or have lower T_g at the skin layer by solvent, UV radiation or plasma treatment on the surface [103, 105], which may cause degradation of the polymer (Fig. 3.46). The thickness of low T_g layer by these treatments is reported to be very thin, 50 nm for example [105]. Using the substrate with thin low T_g or pre-softened layer on the surface, the process temperature can be set to near low T_g to restrict the deformation within skin layer, which may be negligible for the channel, without bulk deformation of the fluidic channel or substrate itself. The bonding strength for this bonding configuration still can be enhanced significantly due to the active diffusion in the softened low T_g layer even at low temperature and higher pressing pressure considering the high modulus of bulk substrate and the channels [106].

Surface modifications to have functional group such as −COOH or $-NH_3$, etc. by chemical reaction or plasma treatment also have been studied to enhance affinity of the two surfaces and bonding strength consequently. Some studies reported that the bonding strength enhanced compared to bare substrates, but not as much as bonding with softened skin layer by solvent exposure, etc. [104, 105].

As shown in Fig. 3.47, materials such as metal or carbon absorbing irradiation energy by laser or microwave and releasing heat are applied in skin region of the substrates to soften the skin region preferentially [107].

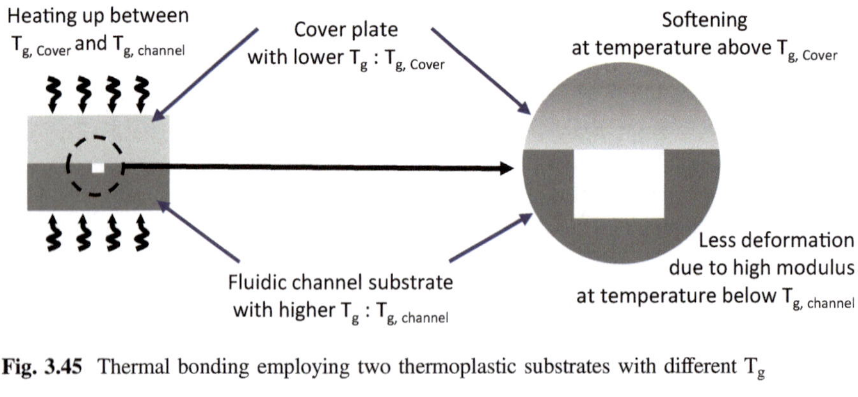

Fig. 3.45 Thermal bonding employing two thermoplastic substrates with different T_g

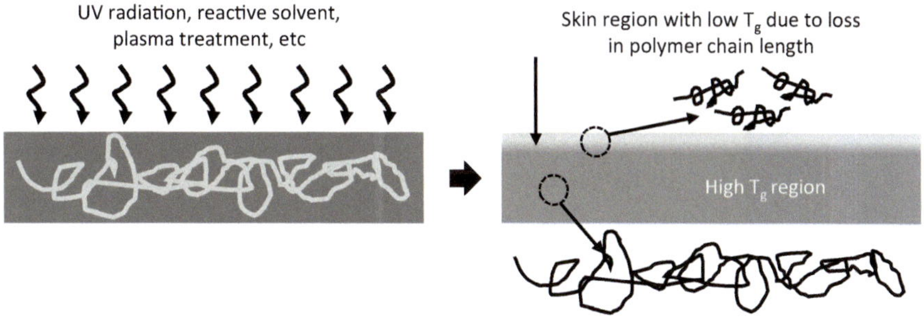

Fig. 3.46 Modification of skin region of the thermoplastic substrate to have lower T_g

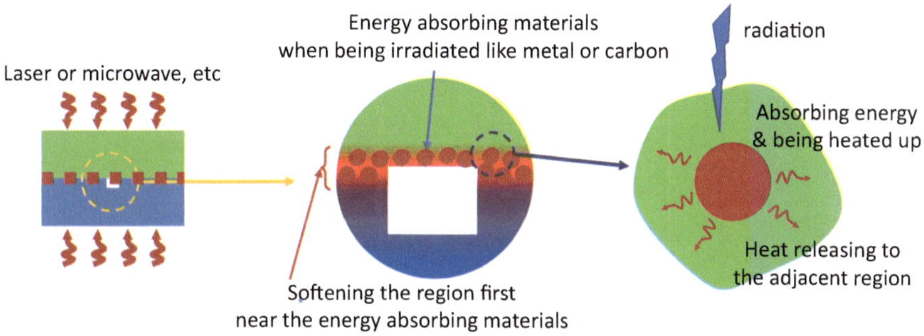

Fig. 3.47 Softening the skin region preferentially employing metal or carbon materials absorbing energy and releasing heat

In a typical thermal bonding process by heating and pressing, the substrates are heated from the other surface of the facing surface while the skin region on the facing surface is needed to be softened. This configuration for bonding in the press results in excessive heating on the back surface of the substrate (Fig. 3.48). If the bonding surface is set to required temperature, then the back surface of the substrate will have higher temperature and severe deformation. This will become worse for a thicker substrate or a substrate with lower thermal conductivity. As an alternative process, heating the substrate from the facing surface may be effective approach to heat and soften the skin region locally (Fig. 3.49).

3.4.5 Solvent Bonding

When a thermoplastic substrate is exposed to a solvent, the substrate becomes softened from the surface under treatment and the mating substrates are bonded by pressing thereafter under a certain pressure. The bonding based on solvent may be efficient in dimensional stability of the fluidic channel and bonding strength because

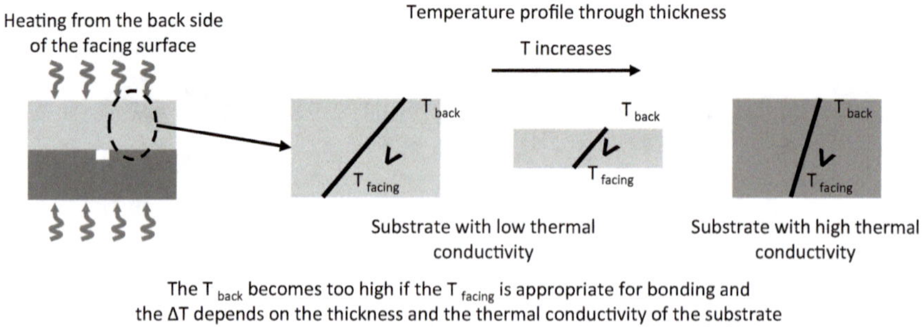

Fig. 3.48 Temperature profile in the substrate for a thermal bonding process heating & pressing in the press at the same time

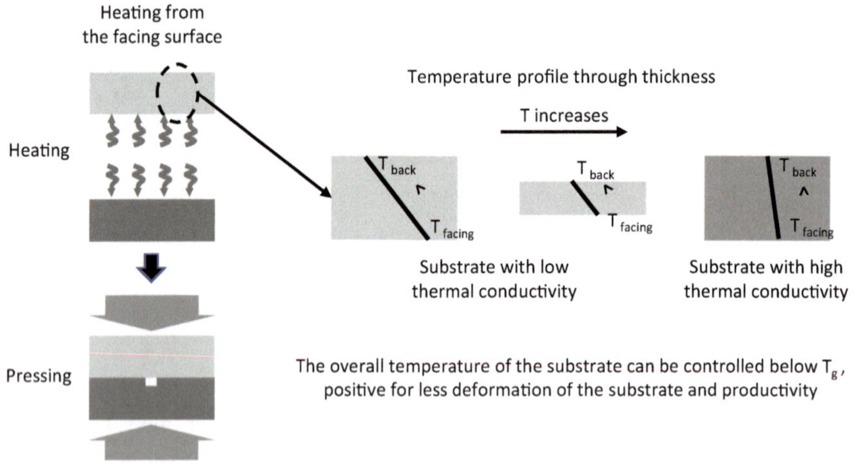

Fig. 3.49 A thermal bonding process heating from the facing surface and the temperature profile

the solvent may soften only the limited skin region if the process and solvent involving are controlled properly. For optimization of solvent bonding, the thermoplastic material and solvent should be selected carefully first because the thermoplastics react differently depending on the solvents and its concentration. For example, Acetone dissolves PMMA (Poly (methyl methacrylate)) quite well but not COC (Cyclic Olefin Copolymer). The softening also depends on the phase of the solvent (liquid or vapor), exposure time or temperature when applied to the surface. If a thermoplastic substrate is exposed to a reactive solvent to the thermoplastic in liquid phase for long time, the substrate may be softened too deeply, resulting in severe deformation during the pressing process. When the surface of the fluidic channel is exposed to the solvent, the fluidic channel might be clogged due to the swelling of the channel surface by the solvent and this clogging may become worse for the channel in smaller scale [108]. The bonding processes using solvent have been studied in various way to enhance the dimensional stability and bonding strength. To prevent channel deformation due to excessive solvent, a solvent less reactive with the substrate was selected to avoid softening the substrate too much [105]. In the other study [109], cyclohexane vapor was used to reduce the penetration of the solvent into the COC(Cyclic Olefin Copolymer) substrate and deformation consequently. Additional DUV exposure thereafter induces a cross-linking reaction in the softened layer at the interface which results in higher bonding strength. The fluidic channel was filled with sacrificial material like wax before being exposed to solvent not to contact with solvent [110]. After bonding two substrates, the wax was eliminated through the channel by heating the substrates over melting temperature of the wax.

Another issue on solvent bonding is chemical contamination because the biological substances in the fluidic channel may lose their functionalities due to the solvent [111].

3.4.6 Ultrasonic Welding

Ultrasonic welding is one of bonding processes widely used for plastic parts. Ultrasonic welding uses high frequency vibration (15 $\sim$ 70 kHz, typically 20 kHz for plastic welding) generated by piezoelectric transducer and transmitted to the target part via booster and horn. The ultrasonic welding equipment consists of a generator, a converter (piezoelectric transducer or similar one), a booster and a horn as shown in Fig. 3.50.

The transmitted vibration energy to the workpiece is concentrated into a structure, called welding bid, for localized melting at the contact point (Fig. 3.51). The welding bid is usually designed to have sharp edges to contact counterpart at only restricted areas and to be melted locally by friction or plastic deformation of the welding bid. The size and shape of the welding bid are designed by considering the thickness of welding layer or the volume for melt at the interface with the counterpart for welding. Typical size of the welding bid for conventional plastic part is in the order of hundreds of micro-meters, but this is too large for direct welding on micro-structure (Fig. 3.52).

For many micro-fluidic devices, many of flow channels are less than a hundred of micro-meter but the land or ridge for the channels is larger than a millimeter. Therefore, for most ultrasonic welding structure for micro-fluidic, the welding bid contacts the macro region to be welded and cover the micro-channels. If the ridge of the channel is too small to apply welding bid in order of hundreds micrometer, then the welding bids are required to be applied to other regions, practically apart from the channels, on the device (Fig. 3.53). In this case, the channels are sealed by mechanical contact between the channel ridge and cover plate, which may cause leak in case of incomplete contact due to a deflection of the plate or difference in

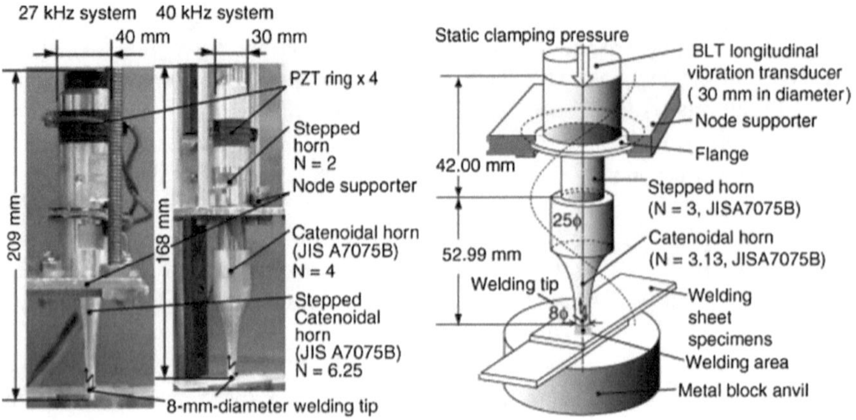

Fig. 3.50 Some configurations of 27, 40 and 67 kHz longitudinal vibration systems for ultrasonic plastic welding machine [112]

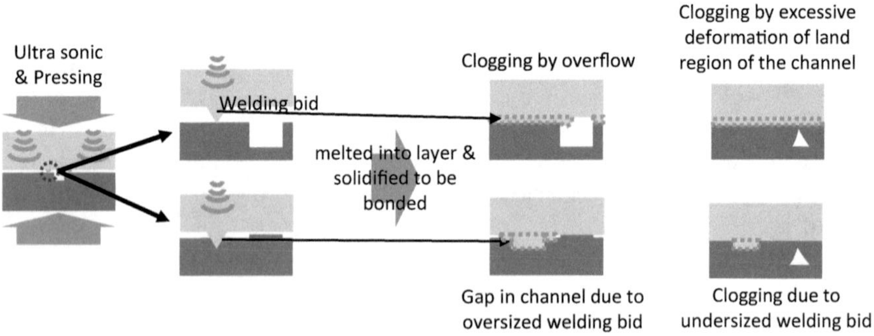

Fig. 3.51 Some configurations of channel substrate and cover plate with welding bid and issues

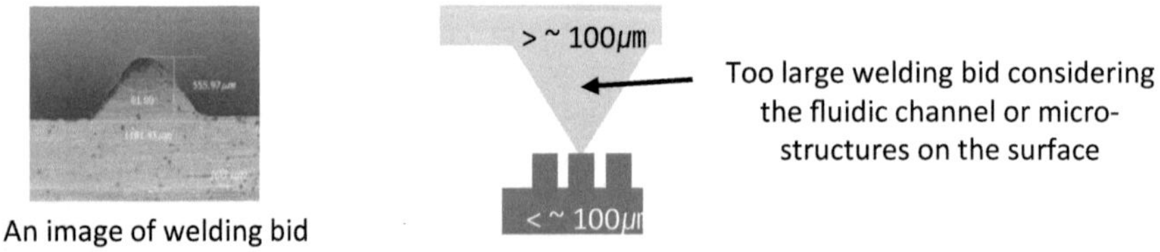

Fig. 3.52 An image of welding bid and an issue in the size of the welding bid and channel

height of the ridge, etc. As shown in Figs. 3.51 and 3.53, too large welding bid results in a thick melting layer and consequently larger channel or gap between the cover plate and the ridge. Too small welding bid may result in insufficient bonding strength or channel collapse due to severe deformation of the ridge.

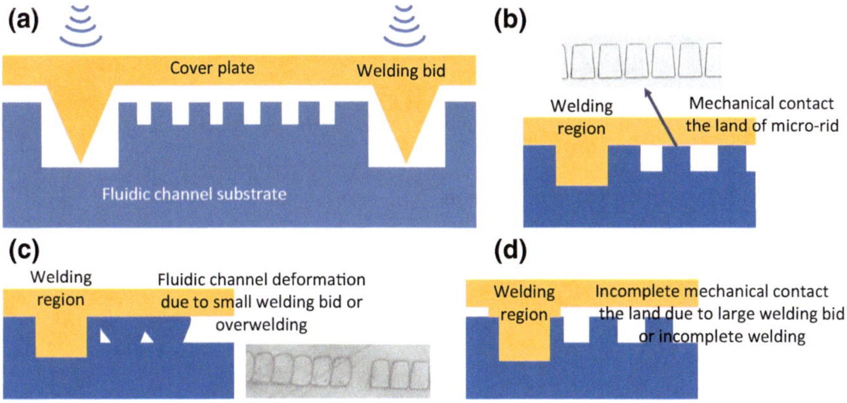

Fig. 3.53 A design of fluidic channel platform with micro-ridge for ultrasonic welding and typical results depending on the welding bid design and process conditions

References

1. Y. Shang, Y. Zeng, Y. Zeng, Integrated microfluidic lectin barcode platform for high performance focused glycomic profiling. Sci. Rep. **6**, 20297 (2016)
2. E. Primiceri, M.-S. Chiriaco, R. Rinaldi, G. Maruccio, Cell chips as new tools for cell biology—results, perspectives and opportunities. Lab Chip **13**, 3789 (2013)
3. L.N. Abdulkadir, K. Abou-El-Hossein, A.I. Jurmare, P.B. Odedeyi, M.M. Liman, T.A. Olaniyan (2018) Ultra-precision diamond turning of optical silicon-a review. Int. J. Adv. Manuf. Technol. Published online: 20 January 2018 (2018)
4. J. Nestler, A. Morschhauser, K. Hiller, T. Otto, S. Bigot, J. Auerswald, H.F. Knapp, J. Gavillet, T. Gessner, Polymer lab-on-chip systems with integrated electrochemical pumps suitable for large-scale fabrication. Int. J. Adv. Manuf. Technol. **47**, 137 (2010)
5. P. Li, D. Zdebski, H.H. Langen, A.M. Hoogstrate, J.A.J. Oosterling, R.H. Munnig Schmidt, A.M. Allen, Micromilling of thin ribs with high aspect ratios. J. Micromech. Microeng. **20** (11), 115013 (2010)
6. D.L. Zariatin, G. Kiswanto, T.J. Ko, Investigation of the micro-milling process of thin-wall features of aluminum alloy 1100. Int. J. Adv. Manuf. Technol. **93**, 2625 (2017)
7. E.G. Mintegi, Micromilling technology: a global review. (2017). https://www.slideshare.net/endika55/micromilling-technology-a-global-review. Accessed 17 July 2017
8. E. Kuram, B. Ozcelik, Micro milling, in *Modern Mechanical Engineering, Materials Forming, Machining and Tribology*, zd. by J.P. Davim (Springer, Berlin Heidelberg, 2014) p. 325
9. D.J. Guckenberger, T.E. de Groot, A.M.D. Wan, D.J. Beebe, E.W.K. Young, Micromilling: a method for ultra-rapid prototyping of plastic microfluidic devices. Lab Chip **15**, 2364 (2014)
10. X. Cheng, Z. Wang, K. Nakamoto, K. Yamazaki, A study on the micro tooling for micro/nano milling. Int. J. Adv. Manuf. Technol. **53**, 523–533 (2011)
11. H. Weule, V. Hüntrup, H. Tritschler, Micro-cutting of steel to meet new requirements in miniaturization. CIRP Ann. **50**(1), 61 (2001)
12. X. Liu, R.E. DeVor, S.G. Kapoor, An analytical model for the prediction of minimum chip thickness in micromachining. J. Manuf. Sci. Eng. **128**(2), 474 (2006)

13. E. Vazquez, C.A. Rodríguez, A. Elías-Zúñiga, J. Ciurana, An experimental analysis of process parameters to manufacture metallic micro-channels by micro-milling. Int. J. Adv. Manuf. Technol. **51**, 945 (2010)
14. T. Wu, K. Cheng, R. Rakowski, Investigation on tooling geometrical effects of micro tools and the associated micro milling performance. Proc IMechE Part B: J Eng. Manuf. **226**(9), 1442 (2012)
15. J. Steigert, S. Haeberle, T. Brenner, C. Müller, C.P. Steinert, P. Koltay, N. Gottschlich, H. Reinecke, J. Rühe, R. Zengerle, J. Ducree, Rapid prototyping of microfluidic chips in COC. J. Micromech. Microeng. **17**, 333–341 (2007)
16. J.C. McDonald, G.M. Whitesides, Poly(dimethylsiloxane) as a material for fabricating microfluidic devices. Acc. Chem. Res. **35**(7), 491 (2002)
17. A. San-Miguel, H. Lu, Microfluidics as a tool for *C. elegans* research, WormBook, ed. The C. elegans Research Community, WormBook (2013)
18. M.A. Eddings, M.A. Johnson, B.K. Gale, Determining the optimal PDMS-PDMS bonding technique for microfluidic devices. J. Micromech. Microeng. **18**(6), 067001 (2008)
19. S. Mohanty, D.J. Beebe, G. Mensing, PDMS connectors for macro to microfluidic interfacing. Chips & Tips (*Lab on a Chip*): 23 October 2006. http://blogs.rsc.org/chipsandtips/2006/10/23/pdms-connectors-for-macro-to-microfluidic-interfacing/?doing_wp_cron=1510476786.5287001132965087890625. Accessed 12 October 2017 (2006)
20. J. Greener, W. Li, D. Voicu, E. Kumacheva, Reusable, robust NanoPort connections to PDMS chips. Chips & Tips (*Lab on a Chip*): 08 August 2008. http://blogs.rsc.org/chipsandtips/2008/10/08/reusable-robust-nanoport-connections-to-pdms-chips/. Accessed 12 October 2017
21. J. Wang, W. Chen, J. Sun, C. Liu, Q. Yin, L. Zhang, Y. Xianyu, X. Shi, G. Hu, X. Jiang, A microfluidic tubing method and its application for controlled synthesis of polymeric nanoparticles. Lab Chip **14**, 1673 (2014)
22. Y. Temiz, R.D. Lovchik, G.V. Kaigala, E. Delamarche, Lab-on-a-chip devices: How to close and plug the lab? Microelec. Eng. **132**(25), 156 (2015)
23. A. Mathur, S.S. Roy, M. Tweedie, S. Mukhopadhyay, S.K. Mitra, J.A. Mc Laughlin, Characterisation of PMMA microfluidic channels and devices fabricated by hot embossing and sealed by direct bonding. Curr. Appl. Phys. **9**, 1199–1202 (2009)
24. Y.-E. Yoo, T.H. Kim, D.-S. Choi, H.-J. Lee, S.J. Choi, S.K. Kim, Study on Molding of a Nanostructured Plastic Plate and Its Surface Properties. Jap. J. Appl. Phys. **48**, 06FH07 (2009)
25. D.-H. Kim, M.-H. Kang, Y.H. Chun, Development of a new injection molding technology: Momentary mold surface heating process. J. Injection Molding Technol. **5**(4), 229 (2001)
26. D. Yao, B. Kim, Development of rapid heating and cooling systems for injection molding applications. Polymer Eng. Sci. **42**(12), 2471 (2002)
27. T. Saito, I. Satoh, Y. Kurosaki, A New concept of active temperature control for an injection molding process using infrared radiation heating. Polymer Eng. Sci. **42**(12), 2418 (2002)
28. J.A. Chang, S.C. Chen, J.C. Cin, Rapid mold temperature control on micro-injection molded parts with high aspect ratio micro-features, Antec 2016, 1275 (2006)
29. Y.E. Yoo, T.H. Kim, D.S. Choi, S.M. Hyun, H.J. Lee, K.H. Lee, S.K. Kim, B.H. Kim, Y.H. Seo, H.G. Lee, J.S. Lee, Injection molding of a nanostructured plate and measurement of its surface properties. Curr. Appl. Phys. **9**(2), e12 (2009)
30. H. Vaisocherova, E. Brynda, J. Homola, Functionalizable low-fouling coating for label-free biosensing in complex biological media. Anal. Bioanal. Chem. **407**, 3927–3953 (2015)
31. O. Seitz, P.G. Fernandes, R. Tian, N. Karnik, H.C. Wen, H. Stiegler, R.A. Chapman, E.M. Vogel, Y. Chabal, Control and stability of self-assembled monolayers under biosensing conditions. J. Mat. Chem. **21**, 4384–4392 (2011)
32. F. Rusmini, Z. Zhong, J. Feijen, Protein Immobilization Strategies for Protein. Biochips. Biomacromolecules **8**, 1775–1789 (2007)
33. T.M. Blattler, S. Pasche, M. Textor, H.J. Griesser, High salt stability and protein resistance of poly[L-lysine]-g-poly[ethylene glycol]copolymers covalently immobilized via aldehyde

plasma polymer interlayers on inorganic and polymeric substrates. Langmuir **22**, 5760–5769 (2006)
34. Z. Zhang, S. Chen, S. Jiang, Dual-functional biomimetic materials: non-fouling poly [carboxybetaine] with active functional group for proteins immobilization. Biomacromol **7**, 3311–3315 (2006)
35. H. Chen, C. Zhao, M. Zhang, Q. Chen, J. Ma, J. Zheng, Molecular understanding and structural-based design of polyacrylamides and polyacrylates as antifouling materials. Langmuir **32**, 3315–3330 (2016)
36. X. Zeng, Z. Shen, R. Mernaugh, Recombinant antibodies and their use in biosensors. Anal. Bioanal. Chem. **402**, 3027–3038 (2012)
37. S. Balamurugan, A. Obubuafo, S. Soper, D. Spivak, Surface immobilization methods for aptamer diagnostic applications. Anal. Bioanal. Chem. **390**, 1009–1021 (2008)
38. A. Ulman, Formation and structure of self-assembled monolayers. Chem. Rev. **96**, 1533–1554 (1996)
39. E.S. Redeker, D.T. Ta, D. Cortens, B. Billen, W. Guedens, P. Adriaensens, Protein engineering for direct immobilization. Bioconjugate Chem. **24**, 1761–1777 (2013)
40. J.A. Camarero, New development for the site-specific attachment of proteins to surfaces. Biophys. Rev. Lett. **1**, 1–28 (2006)
41. N. Stephanopoulos, M.B. Francis, Choosing effective proteins bioconjugation strategy. Nat. Chem. Biol. **7**, 876–884 (2011)
42. M. Ammar, C. Smadja, D. Tandjidora D., M. Azzouz, J. Vigneron, A. Etcheberry, M. Taverna, E. Dufour-Gergam, Chemical engineering of self-assembled Alzheimer's peptide on silanized silicon surface. Langmuir. **30**, 5863–5872 (2014)
43. M. Ammar, C. Smadja, L.Giang Thi Phuong, J. Vigneron, A. Etcheberry, M. Taverna, E. Dufour-Gergam, A new controlled concept of immune-sensing platform, for specific detection of Alzheimer's biomarkers. Biosens. Bioelectron. **40**, 329–395 (2013)
44. F. Schröper, A. Baumann, A. Offenhausser, D. Mayer, Direct electrochemistry of novel affinity tag immobilized recombinant horse heart cytochrome *c*. Biosens. Bioelectron. **34**, 171–177 (2012)
45. J. Fick, T. Wolfram, F. Belz, S. Roke, Surface-specific interaction of the extracellular domain of protein L1 with nitrilotriacetic acid-terminated self-assembled monolayers. Langmuir **26**, 1051–1056 (2010)
46. N. Xia, L. Liu, M.G. Harrington, J. Wang, F. Zhou, Regenerable and Simultaneous Surface Plasmon Resonance Detection of Aβ[1–40] and Aβ[1–42] Peptides in Cerebrospinal Fluids with Signal Amplification by Streptavidin Conjugated to an N-Terminus-Specific Antibody. Anal. Chem. **82**, 10151–10157 (2010)
47. C. You, M. Bhagawati, A. Brecht, J. Piehler, Affinity capturing for targeting proteins into micro and nanostructures. Anal. Bioanal. Chem. **393**, 1563–1570 (2009)
48. J.-F. Lutz, Copper-free azide-alkyne cycloadditions: new insights and perspectives. Angew. Chem. Int. Ed. **47**, 2182–2184 (2008)
49. G. Fleminger, E. Hadas, T. Wolf, B. Solomon, Oriented immobilization of periodate-oxidized monoclonal antibodies on amino and hydrazide derivatives of Eupergit C. Appl. Biochem. Biotechnol. **23**, 123–137 (1990)
50. I. Banerjee, R.C. Pangule, R.S. Kane, Antifouling coatings: recent developments in the design of surfaces that prevent fouling by proteins, bacteria and marine organisms. Adv. Mater. **23**, 690–718 (2011)
51. L.D. Unsworth, H. Sheardown, J.L. Brash, Protein resistance of surfaces prepared by sorption of end-thiolated poly[ethylene glycol] to gold: effect of surface chain density. Langmuir **21**, 1036–1041 (2005)
52. L.D. Unsworth, H. Sheardown, J.L. Brash, Protein resistant poly[ethylene oxide]-grafted surfaces: chain density-dependant multiple mechanisms of action. Langmuir **24**, 1924–1929 (2008)

53. P. Kingshott, H. Thissen, H.J. Griesser, Effects of cloud-point grafting, chain length, and density of PEG layers on competitive adsorption of ocular proteins. Biomaterials **23**, 2043–2056 (2002)
54. V. Hynninen, L. Vuori, M. Hannula, K. Tapio, K. Lahtonen, T. Isoniemi, E. Lehtonen, M. Hirsimäki, J.J. Toppari, M. Valden, V.P. Hytönen, Improved antifouling properties and selective biofunctionalization of stainless steel by employing heterobifunctional silane-polyethylene glycol overlayers and avidin-biotin technology. Sci. Rep. **6**, 1–12 (2016)
55. N. Aboud, D. Ferraro, M. Taverna, S. Descroix, C. Smadja, N.T. Tran, Dyneon THV, a fluorinated thermoplastic as a novel material for microchip capillary electrophoresis. Analyst. **141**, 5776–5783 (2016)
56. Y. Chang, W.L. Chu, W.Y. Chen, J. Zheng, L. Liu, R.C. Ruaan, A. Higuchi, A systematic SPR study of human plasma protein adsorption behavior on the controlled surface packing of self-assembled poly[ethylene oxide] triblock copolymer surfaces. J. Biomed. Mater. Res. A. **93**, 400–408 (2010)
57. Y. Yang, P.G. Rouxhet, D. Chudziak, J. Telegdi, C.C. Dupont-Gillain, Influence of poly [ethylene oxide]-based copolymer on protein adsorption and bacterial adhesion on stainless steel: modulation by surface hydrophobicity. Bioelectrochemistry Amst. Neth. **97**, 127–136 (2014)
58. P. Harder, M. Grunze, R. Dahint, G.M. Whitesides, P.E. Laibinis, Molecular Conformation in Oligo[ethylene glycol]- Terminated Self-Assembled Monolayers on Gold and Silver Surfaces Determines their Ability To Resist Protein Adsorption. J. Phys. Chem. B. **102**, 426–436 (1998)
59. L.M. Feller, S. Cerritelli, M. Textor, J.A. Hubbell, S.G.P. Tosatti, Influence of Poly [propylene sulfide-block-ethylene glycol] Di- and Triblock Copolymer Architecture on the Formation of Molecular Adlayers on Gold Surfaces and Their Effect on Protein Resistance: A Candidate for Surface Modification in Biosensor Research. Macromolecules **38**, 10503–10510 (2005)
60. C. Blaszykowski, S. Sheikh, M. Thompson, Biofluids. A survey of state-of-the-art surface chemistries tominimize fouling from human and animal. Biomater. Sci. **3**, 1335–1370 (2015)
61. V. Zoulalian, S. Zürcher, S. Tosatti, M. Textor, S. Monge, J.J. Robin, Self-Assembly of Poly (ethylene glycol)−Poly(alkyl phosphonate) Terpolymers on Titanium Oxide Surfaces: Synthesis, Interface Characterization, Investigation of Nonfouling Properties, and Long-Term Stability. Langmuir **26**, 74–82 (2010)
62. R.G. Chapman, E. Ostuni, L. Yan, G.M. Whitesides, Preparation of Mixed Self-Assembled Monolayers [SAMs] That Resist Adsorption of Proteins using the reaction of amines with a SAM that presents interchain carboxylic anhydride groups. Langmuir **16**, 6927–6936 (2000)
63. R.G. Chapman, E. Ostuni, S. Takayama, R.E. Holmin, L. Yan, G.M. Whitesides, Surveying for surfaces that resist the adsorption of proteins. J. Am. Chem. Soc. **122**, 8303–8304 (2000)
64. S.F. Chen, J. Zheng, L.Y. Li, S.Y. Jiang, Strong resistance of phosphorylcholine self-assembled monolayers to protein adsorption: insights into nonfouling properties of zwitterionic materials. J. Am. Chem. Soc. **127**, 14473–14478 (2005)
65. S.F. Chen, L.Y. Liu, S.Y. Jiang, Strong Resistance of Oligo[phosphorylcholine] Self-Assembled Monolayers to Protein Adsorption. Langmuir **22**, 2418–2421 (2006)
66. P. Liu, T. Huang, P. Liu, S. Shi, Q. Chen, L. Li, J. Shen, Zwitterionic modification of polyurethane membranes for enhancing the anti-fouling property. J. Colloid Interface Sci. **15**, 91–101 (2016)
67. S.C. Lange, E. Van Andel, M.M.J. Smulders, H. Zuilhof, Efficient and tunable three-dimensional functionalization of fully zwitterionic antifouling surface coating. Langmuir **32**, 10199–10205 (2016)
68. Z. Zhang, M. Zhang, S. Chen, T.A. Horbett, B.D. Ratner, S. Jiang, Blood compatibility of surfaces with superlow protein adsorption. Biomaterials **29**, 4285 (2008)
69. H. Vaisocherová, W. Yang, Z. Zhang, Z. Cao, G. Cheng, M. Piliarik, J. Homola, S. Jiang, Ultralow Fouling and Functionalizable surface chemistry based on a zwitterionic polymer

enabling sensitive and specific protein detection in undiluted Blood Plasma. Anal. Chem. **80**, 7894–7901 (2008)
70. J.W. Costerton, P.S. Stewart, E.P. Greenberg, Bacterial biofilms: a common cause of persistent infections. Science **284**, 1318–1322 (1999)
71. N.B. Jaballah, A. Bouziri, K. Mnif, A. Hamdi, A. Khaldi, W. Kchaou, Epidemiology of hospital-acquired bloodstream infections in a Tunisian pediatric intensive care unit: A 2-year prospective study. Am. J. Infect. Control **35**, 613–618 (2007)
72. R. Tan, J. Liu, M. Li, J. Huang, J. Sun, H.J. Qu, Epidemiology and antimicrobial resistance among commonly encountered bacteria associated with infections and colonization in intensive care units in a university affiliated hospital in Shanghai. Microbiol. Immunol. **47**, 87–94 (2014)
73. L. Ferreira, A. Zumbuehl, Non-leaching surfaces capable of killing microorganisms on contact. J. Mater. Chem. **19**, 7796–7806 (2009)
74. F. Hui, C. Debiemme-Chouvry, Antimicrobial N-halamine polymers and coatings: a review of their synthesis, characterization, and applications. Biomacromol **14**, 585–601 (2013)
75. R. Li, P. Hu, X. Ren, S.D. Worley, T.S. Huang, Antimicrobial N-halamine modified chitosan films. Carbohydr. Polym. **92**, 534–539 (2013)
76. M.L.W. Knetsch, L.H. Koole, New strategies in the development of antimicrobial coatings: the example of increasing usage of silver and silver nanoparticles. Polymers **3**, 340–366 (2011)
77. F. Siedenbiedel, J.C. Tiller, Antimicrobial polymers in solution and on surfaces: overview and functional principles. Polymers **4**, 46–71 (2012)
78. N. Gour, K.X. Ngo, C. Vebert-Nardin, Anti-infectious surfaces achieved by polymer modification. Macromol. Mater. Eng. **299**, 648–668 (2014)
79. K.D. Park, Y.S. Kim, D.K. Han, Y.H. Kim, E.B.H. Lee, H. Suh, K.S. Choi, Bacterial adhesion on PEG modified polyurethane surfaces. Biomaterials **19**, 851–859 (1998)
80. N.P. Desai, S.F.A. Hossainy, J.A. Hubbell, Surface-immobilized polyethylene oxide for bacterial repellence. Biomaterials **13**, 417–420 (1992)
81. A. Roosjen, H.J. Kaper, H.C. Van der Mei, W. Norde, H.J. Busscher, Inhibition of adhesion of yeasts and bacteria by poly[ethylene oxide]-brushes on glass in a parallel plate flow chamber. Microbiology **149**, 3239–3246 (2003)
82. A. Roosjen, H.J. Kaper, H.C. Van der Meir, W. Norde, Microbial adhesion to poly[ethylene oxide] brushes: influence of polymer chain length and temperature. Langmuir **25**, 10949–10955 (2004)
83. Z. Geng, R. Wang, X. Zhuo, Z. Li, Y. Huang, L. Ma, Z. Cui, S. Zhu, Y. Liang, Y. Liu, H. Bao, X. Li, Q. Huo, Z. Liu, X. Yang, Incorporation of silver and strontium in hydroxyapatite coating on titanium surface for enhanced antibacterial and biological properties. Mater. Sci. Eng., C **71**, 852–861 (2017)
84. J.C. Tiller, S.B. Lee, K. Lewis, A.M. Klibanov, Polymer surfaces derivatized with poly [vinyl-N-hexylpyridinium] kill airborne and waterborne bacteria. Biotechnol. Bioengen. **79**, 465–471 (2002). https://doi.org/10.1002/bit.10299
85. J.C. Tiller, C.J. Liao, K. Lewis, A.M. Klibanov, Designing surfaces that kill bacteria on contact. Proc. Natl. Acad. Sci. USA. **98**, 5981–5985 (2001)
86. M. Andresen, P. Stenstad, T. Moretro, S. Langsrud, K. Syverud, L.S. Johansson, P. Stenius, Non leaching antimicrobial films prepared from surface-modified microfibrillated cellulose. Biomacromol **8**, 2149–2155 (2007)
87. A.J. Isquith, E.A. Abbot, P.A. Walters, Surface-bonded antimicrobial activity of an organosilicon quaternary ammonium chloride. Appl. Microbiol. **24**, 859–863 (1972)
88. G. Harkes, J. Dankert, J. Feijen, Growth of uropathogenic Escherichia coli strains at solid surfaces. J. Biomater. Sci. Polym. Ed. **3**, 403–418 (1992)
89. B. Gottenbos, H.C. Van der Mei, F. Klatter, P. Nieuwenhuis, H.J. Busscher, In vitro and in vivo antimicrobial activity of covalently coupled quaternary ammonium silane coatings on silicone rubber. Biomaterials **23**, 1417–1423 (2002)

90. M.J. Saif, J. Anwar, M.A. Munawar, A novel application of quaternary ammonium compounds as antibacterial hybrid coating on glass surfaces. Langmuir **25**, 377–379 (2009)
91. D. Druvari, N.D. Koromilas, G.C. Lainioti, G. Bokias, G. Vasilopoulos, Polymeric quaternary ammonium-containing coatings with potential dual contact-based and release-based antimicrobial activity. ACS Appl. Mater. Interfaces. **8**, 35593–35605 (2016)
92. E. Kougia, M. Tselepi, G. Vasilopoulos, G.C. Lainioti, N.D. Koromilas, D. Druvari, G. Bokias, A. Vantarakis, J.K. Kallitsis, Evaluation of antimicrobial efficiency of new polymers comprised by covalently attached and/or electrostatically bound bacteriostatic species. Based on Quaternary Ammonium Compounds. Molecules **20**, 21313–21327 (2015)
93. E.I. Rabea, M.E.T. Badawy, C.V. Stevens, G. Smagghe, W. Steurbaut, Chitosan as antimicrobial agent: applications and mode of action. Biomacromol **4**, 1457–1465 (2003)
94. A. Konwar, S. Kalita, J. Kotoky, D. Chowdhury, Chitosan–Iron Oxide Coated Graphene Oxide Nanocomposite Hydrogel: A Robust and Soft Antimicrobial Biofilm. ACS Appl. Mater. Interfaces. **8**, 20625–20634 (2016)
95. E.P. Ivanova, J. Hasan, H.K. Webb, V.K. Truong, G.S. Watson, J.A. Watson, V.A. Baulin, S. Pogodin, J.Y. Wang, M.J. Tobin, C.N. Löbbe, R.J. Crawford, Natural bactericidal surfaces: mechanical rupture of pseudomonas aeruginosa cells by cicada wings. Small **8**, 2489–2494 (2012)
96. S. Pogodin, J. Hasan, V.A. Baulin, H.K. Webb, V.K. Truong, H.P. Nguyen, V. Boshkovikj, C.J. Fluke, G.S. Watson, J.A. Watson, R.J. Crawford, E.P. Ivanova, Biophysical model of bacterial cell interactions with nanopatterned cicada wing surfaces. Biophysical J. **104**, 835–840 (2013)
97. M.A. Unger, H.-P. Chou, T. Thorsen, A. Scherer, S.R. Quake, Monolithic microfabricated valves and pumps by multilayer soft lithography. Science **288**, 5463 (2000)
98. P. Rezai, P.R. Selvaganapathy, G.R. Wohl, Plasma enhanced bonding of polydimethylsiloxane with parylene and its optimization. J. Micromech. Microeng. **21**, 065024 (2011)
99. S.K. Sia, G.M. Whitesides, Microfluidic devices fabricated in poly (dimethylsiloxane) for biological studies. Electrophoresis **24**(21), 3563 (2003)
100. M.-E. Vlachopoulou, A. Tserepi, P. Pavli, P. Argitis, M. Sanopoulou, K. Misiakos, A low temperature surface modification assisted method for bonding plastic substrates. J. Micromech. Microeng. **19**, 015007 (2009)
101. H.-Y. Chen, A.A. McClelland, Z. Chen, J. Lahann, Solventless adhesive bonding using reactive polymer coatings. Anal. Chem. **80**, 4119 (2008)
102. H.H. Ruf, T. Knoll, K. Misiakos, R.B. Haupt, M. Denninger, L.B. Larsen, P.S. Petrou, S.E. Kakabakos, E. Ehrentreich-Förster, F.F. Bier, Biochip-compatible packaging and micro-fluidics for a silicon opto-electronic biosensor. Microelec. Eng. **83**, 1677 (2006)
103. H. Shinohara, J. Mizuno, S. Shoji, Low-Temperature polymer bonding using surface hydrophilic treatment for chemical/bio microchips, in *Solid State Circuits Technologies*, ed. by J.W. Swart, (InTechOpen), Chap. 22, p. 445 (2010)
104. H. Yu, Z.Z. Chong, S.B. Tor, E. Liu, N.H. Loh, Low temperature and deformation-free bonding of PMMA microfluidic devices with stable hydrophilicity via oxygen plasma treatment and PVA coating. RSC Adv. **5**, 8377 (2015)
105. L. Brown, T. Koerner, J.H. Horton, R.D. Oleschuk, Fabrication and characterization of poly (methylmethacrylate) microfluidic devices bonded using surface modifications and solvents. Lab Chip **6**, 66 (2006)
106. M.I. Mohammed, K. Quayle, R. Alexander, E. Doeven, R. Nai, S.J. Haswell, A.Z. Kouzani, I. Gibson, Improved manufacturing quality and bonding of laser machined microfluidic systems. Procedia Technology **20**, 219 (2015)
107. J.W. Lai Pik, Carbon nanotube microwave-assisted thermal bonding of plastic micro biochip. Dissertation, The Hong Kong Polytechnic University (2010)
108. H. Klank, J.P. Kutter, O. Geschke, CO2-laser micromachining and back-end processing for rapid production of PMMA-based microfluidic systems. Lab Chip **2**, 242 (2002)
109. D.A. Mair, M. Rolandi, M. Snauko, R. Noroski, F. Svec, J.M.J. Frechet, Room-temperature bonding for plastic high-pressure microfluidic chips. Anal. Chem. **79**, 5097 (2007)

110. R.T. Kelly, T. Pan, A.T. Woolley, Phase-changing sacrificial materials for solvent bonding of high-performance polymeric capillary electrophoresis microchips. Anal. Chem. **77**(11), 3536 (2005)
111. D. Figeys, Y.B. Ning, R. Aebersold, A microfabricated device for rapid protein identification by microelectrospray ion trap mass spectrometry. Anal. Chem. **69**(16), 3153 (1997)
112. J. Tsujino, M. Hongoh, M. Yoshikuni, H. Hashii, T. Ueoka, Welding characteristics of 27, 40 and 67 kHz ultrasonic plastic welding systems using fundamental- and higher-resonance frequencies. Ultrasonics **42**, 131 (2004)

Chapter 4
Mechanics for Fluidics and Bio-Devices

Luis Guillermo Villanueva, Annalisa De Pastina and Magalie Faivre

Abstract In this chapter, we firstly present mechanical elements which are essential components of Lab-On-Chip devices as they can provide sensing, mixing, pumping and controlled delivery of small fluidic volumes. Microvalves, crucial for the on-chip implementation of complex fluidic patterns, are discussed first. Micropumps, typically built as a collection of microvalves operated in sequence, are subsequently described. Finally, a short overview of nanomechanical biosensors, which have demonstrated great capabilities in label-free sensing applications, is provided. Particular attention will be paid to mechanical sensing in liquids, focusing on limitations and novel techniques. In a second part, we will focus on concentrating, focusing, trapping, sorting and single cell biomechanical characterization which are critical steps in various biomedical applications such as diagnostics, therapeutics, drug screening and cell biology. Microfluidic technologies propose attractive engineered microenvironments for the manipulation and the study of cellular mechanics compatible with high throughput. The different approaches available to act on cells and perform mechanical phenotyping in microsystems are detailed and discussed.

Keywords Microvalves · Micropumps · Nanomechanical sensors · Cell manipulation · Cell biomechanics · Passive microfluidics · Microfluidic deformability assays

---

L. G. Villanueva · A. De Pastina
EPFL-STI-IGM-NEMS, Station 9, ME D2 2726, CH-1015, Lausanne, Switzerland
e-mail: guillermo.villanueva@epfl.ch

M. Faivre (✉)
Institut des Nanotechnologies de Lyon INL-UMR5270, CNRS, Université de Lyon, Université Lyon 1, Villeurbanne 69622, France
e-mail: magalie.faivre@univ-lyon1.fr

G. Barbillon et al. (eds.), *Engineering of Micro/Nano Biosystems*, Microtechnology and MEMS, https://doi.org/10.1007/978-981-13-6549-2_4

4.1 Introduction

Mechanics and biology seem at first view rather remotely related fields but as demonstrated in this chapter, mechanics, fluidics and biology can be nicely combined to control fluids, to manipulate and sort cells and to perform biodetection. In the following sections, microcomponents for fluidics, passive microfluidics and mechanics for biodevices will be successively reviewed:

Macroscale mechanical pumps and valves can be used for fluid injection and distribution in microchannels. However, their miniaturization is finally desirable to get fully integrated micro/nano biosystems that are portable and more autonomous. As detailed in Sect. 4.2 of this chapter, a large variety of microvalves and micropumps were proposed and developed for this purpose.

The development of atomic force microscopy and of micro/nano-electromechanical systems (M/NEMS) in the last three decades clearly demonstrated the high sensitivity of micro/nano-mechanical structures such as cantilever micro/nano-beams, micro/nano-bridges, membranes… and of acoustic microdevices to added mass, to mechanical stress or/and to tiny forces. Their potential for biodetection began to be investigated around year 2000 and some typical examples of biomechanical detection are reviewed in the last part of Sect. 4.2.

Micro/nano bio system can also take benefit of mechanics for cell manipulation, sorting and analysis owing to the similar size of fluidic microchannels and of cells, and to the variability of cell density and mechanical properties. The various ways to apply passive microfluidics to cells manipulation and to perform mechanical characterization of cells by the way of microfluidics are described in Sects. 4.3 and 4.4.

4.2 Mechanical Microcomponents for Fluidics (Annalisa De Pastina and Luis Guillermo Villanueva)

Micro- and nano-electromechanical systems (M-NEMS) are defined as electrically transduced continuum mechanical structures, such as beams, strings, plates, or membranes, with dimensions in the order of micrometers. What started being a very efficient solution for inertial sensing in the automotive industry has expanded to numerous different applications, including playing a fundamental role in miniaturized microfluidic devices [1, 2]. Indeed, mechanical elements are essential components of Lab-On-Chip devices as they can provide sensing, mixing, pumping and controlled delivery of small fluidic volumes.

This part of the chapter focusses on *Microvalves*, *Micropumps*, and *Nanomechanical biosensors*, as they are the main types of components required for fully integrated microfluidic systems.

- *Microvalves* control routing, timing, and separation of fluids within a microfluidic device and are crucial for the on-chip implementation of complex fluidic patterns.
- *Micropumps* generate temporal and volumetric flow and allow reducing the amount of external hardware necessary to operate a microfluidic device.
- *Nanomechanical biosensors* have demonstrated great capabilities in label-free sensing applications, such as detection of viruses and single molecules [3].

In the following paragraphs microvalves are discussed first, followed by micropumps, since many micropumps are built as a collection of microvalves operated in sequence. These two broad categories are further subdivided into several types. Each type inherently possesses many strengths and weaknesses which must be considered when choosing an appropriate solution for a given application. Finally, a short overview of nanomechanical biosensors operating in static and dynamic mode is provided. Particular attention will be paid to mechanical sensing in liquids, focusing on limitations and novel techniques.

4.2.1 Mechanical Microvalves

Microvalves can be divided in three major types: active (internal and external), and passive. *Active* microvalves have in common an actuating element (the different kinds are analyzed below) to stop or allow the flow. Active microvalves can be classified depending on whether the actuator is external (e.g. built-in modular or pneumatic) or the actuator is integrated on the microfluidic chip using MEMS-based technologies (e.g. magnetic, electric, piezoelectric or thermal actuators). Internal valves offer high performances, but also high complexity and fabrication costs. In contrast, external valves are cheaper and simpler, but the final solution is bulky, not viable for miniaturization. *Passive* valves, sometimes called check-valves, do not include actuation, they open to forward pressure showing a diode-like behavior.

Microvalves' applications include flow regulation, on/off switching and sealing of liquids, gases or vacuum. The desired characteristics of an ideal microvalve include no leakage flow, reduced dead volume, reduced power consumption, large pressure resistance, fast response time, insensitivity to contamination, ability to operate both with liquids and gases, low complexity, low fabrication cost, and disposability [4]. Unfortunately, up to date there is no valve technology or design that simultaneously satisfies all these requirements. Therefore, the most appropriate type of microvalve will change according to the demands of a specific application.

4.2.1.1 Active External

External active microvalves are actuated by the aid of external systems such as built-in modular or pneumatic means. This approach suffers, logically, from difficulties in miniaturization, given the fact that additional components are required, like air/vacuum pumps or solenoids. However, external microvalves exhibit no leakage flow even at high input pressures and excellent performances in on/off switching or sealing.

Two main categories can be distinguished, namely *modular* and *pneumatic* external valves. *Modular* external microvalves include built-in [5] or rotary configurations [6]. Microvalves actuated by an external *pneumatic* system are distinguished between membranes [7] and in-line [8].

Quake et al. presented an *in-line pneumatic valve* system consisting of a crossed-microfluidic PDMS channel block, sealed to glass or elastomer substrate, as shown in Fig. 4.1a. An actuation channel is filled by pressurized gas allowing a thin membrane to deflect and close the bonded fluidic channel. Two different configurations have been demonstrated and compared, as presented in Fig. 4.1b, c.

The push-down configuration (Fig. 4.1b) consists of 100 μm wide and 10 μm thick channels with a minimum actuation pressure of 40 kPa to seal the fluidic channel [8]. The second valve generation, push-up configuration (Fig. 4.1c), allows an increase fluidic channel height up to 55 μm, and a lower valve actuation pressure of 15 kPa [9].

The push-up membrane was further improved by Pandolfi and Ortiz. They demonstrated that the channel closure can be assisted by out-of-plane buckling of the membrane caused by controlled chemical swelling in silicone oil. This strategy allowed them to decrease the actuation pressure down to 8 kPa [10].

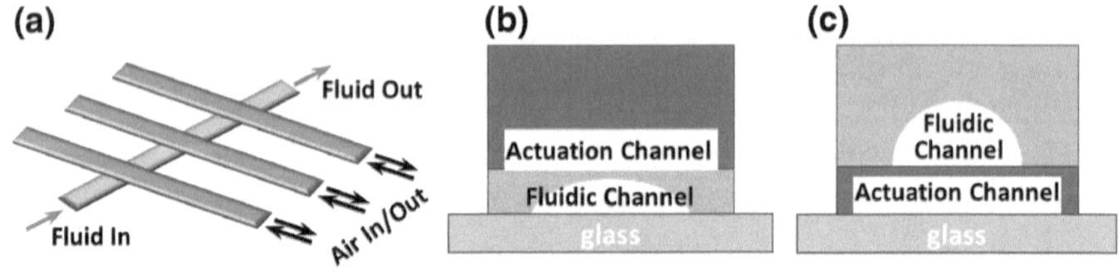

Fig. 4.1 **a** 3-D representation of three in-line valves with individual control proposed in [8]. **b** Push-down valve geometry. A curved membrane of variable thickness is deflected by applying pressure to close the channel underneath. This configuration requires high aspect ratio fluidic channels. **c** Push-up valve configuration. A membrane of uniform thickness is deflected when pressure is applied in the actuation channel to close the fluidic channel on top. The pressure applied in this case is not dependent of the aspect ratio of the fluidic channel

4.2.1.2 Active Internal

As opposed to the previous type of valves, it is also possible to have the actuation mechanism integrated on-chip. In those cases, the moving component is usually a membrane. Various actuation principles can be adopted to actuate mechanical moving parts and the actuators are integrated in the valve structure itself.

Mechanical active microvalves have two main disadvantages: unavoidable leakage flow and relatively high cost due to their complicated structure. While leakage flow is a critical feature for on/off switching applications, it is less critical for flow regulation. Therefore, micromachined active microvalves are usually preferred for gas or selected liquid regulation, even though numerous on/off switching applications have been reported [11]. Due to their elevated cost, internal active microvalves are mainly reusable and are usually implemented in non-disposable devices.

The following paragraphs provide a brief analysis of mechanical active microvalves, classified according to their actuation principle. Bear in mind that we do not intend to do a thorough review of these actuation principles, just introducing them succinctly.

Electromagnetic

A wire carrying a current in the presence of a magnetic field will experience the Lorentz force $\vec{F}_{em} = (\vec{I} \times \vec{B})L$, where $\vec{F}_{em}$ is the electromagnetic force, $\vec{I}$ is the current through the wire, $\vec{B}$ is the magnetic field and L is the wire length. The magnitude of the force, $|\vec{F}_{em}|$, scales down with dimensions $\propto l^{\alpha}$, with α between 1 and 2, if we consider an external magnetic field, and depending on whether we take constant current, voltage or power [12]. In any case, this scaling shows that for very small sizes the force is compromised.

Electromagnetic microvalves operate at low voltages and are able to generate large actuation forces (see Fig. 4.2a). However, electromagnetic actuators require an external magnetic field, thus rendering them difficult to miniaturize, even though in some cases they have been shown to be integrated [13]. As current needs to be sent through the wires, electromagnetic microvalves exhibit high power consumption and high heat dissipation [14].

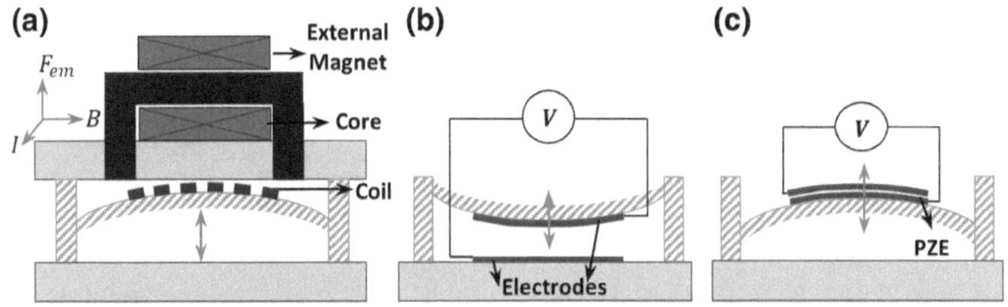

Fig. 4.2 General schematic of three different actuation principles for active internal microvalves: **a** electromagnetic, **b** electrostatic, and **c** piezoelectric

In 1979 Terry et al. proposed the first active micromachined valve, using a solenoid plunger, which was physically connected to a 100 nm thick nickel membrane. The normally closed microvalve constituted a component of an integrated gas chromatography system [15].

Electrostatic

Electrostatic actuation is based on the Coulomb attraction force between oppositely charged plates. Assuming a system with two parallel plates and a voltage difference (V) is set between them, the force takes the form: $F_{es} = -\frac{1}{2}\frac{\varepsilon_0\varepsilon_r A V^2}{x^2}$ where $\varepsilon_0\varepsilon_r$ is the dielectric constant, A the electrode area, and x the electrode spacing. Due to the scaling of the electrostatic force, $F_{es} \propto l^0$, the miniaturization of electrostatic actuators is much simpler than electromagnetic ones.

In electrostatic microvalves, one of the electrodes to give rise to the force is typically located on the membrane, whereas the second one is typically placed on the substrate, within the microchannel to be closed or opened, as presented in Fig. 4.2b. This type of valves has a fast response time and low power consumption in the order of 1 mW, since almost no current flows through the system. The main limitation is the small achievable displacements, usually in the order of few micrometers.

Modifications of the basic schematic shown in Fig. 4.3 have been proposed in several ways. The efficiency of the system can be improved using S-shaped membranes that move back and forth (laterally) instead of up and down [16]. Different materials have been used for the fabrication of the actuation membrane: flexible materials as polymide [17], requiring lower actuation voltages, and rigid silicon membrane [18, 19], preferred for high pressure control.

Piezoelectric

Piezoelectricity is the property of a material to respond with a mechanical deformation (ε_j) to an applied voltage. We consider a piezoelectric layer of width w_{PZE}, thickness t_{PZE}, Young's modulus E_{PZE} and piezoelectric coefficients d_{ij}. Applying an electric voltage across the thickness of the piezoelectric layer (V) causes a deformation in the piezoelectric layer and therefore a force (assuming a structure working in pure elongation), $F_{PZE} = E_{PZE} \cdot d_{33} \cdot w_{PZE} \cdot L_{PZE} \frac{V}{t_{PZE}}$ and scales as l. Although large force and fast response time are achieved through piezoelectric actuators, small strokes and complexity in the fabrication are challenging issues.

The drawback of small strokes has been overcome via different alternatives as the hydraulic amplification of the piezoelectric deflection [20], stacked piezoelectric

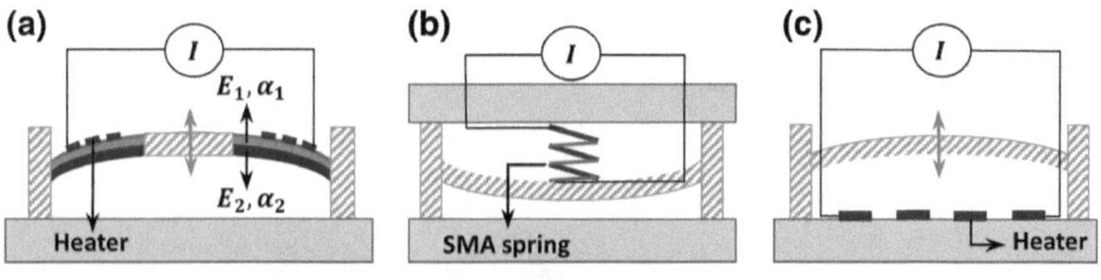

Fig. 4.3 Schematic representation of three different thermo-mechanical actuation principles: **a** bimetallic, **b** shape memory alloy, and **c** thermopneumatic actuation

disks [21] or piezo bimorphs, presented in Fig. 4.2c (i.e. structures working in flexion instead that elongation) [22, 23].

For example, Rogge et al. proposed a hydraulic transmission system, which is able to translate a small-scale deflection of the piezodisk into a larger movement of a valve membrane, showing an increase of the membrane displacement up to 25 times, achieving a valve stroke of 50 μm [24].

Thermal

Thermo-mechanical actuation is based on the expansion caused by a change in temperature in either the mechanical structure (bimetallic effect and shape memory alloys) or its surroundings (thermopneumatic). In general, thermal actuation provides large strokes and large actuation forces, but this comes with a slow response time and a high power consumption.

Bimetallic

Bimetallic actuation, schematically represented in Fig. 4.3a, is based on the difference of thermal expansion coefficients of multiple materials bonded together. When a multilayer stack of dissimilar materials, not necessarily metals, is subjected to temperature changes (ΔT), thermal stresses (σ_i) are induced in each material: $\sigma_i = E_i \cdot \alpha_i \cdot \Delta T$, where E_i is the Young's modulus of material i. This stress distribution translates into bending of the structure [25]. Among thermal microvalves, bimetallic actuated ones exhibit the lowest power consumption.

Shape Memory Alloy (SMA)

Shape memory alloy (SMA) actuated microvalves make use of the shape memory effect in SMAs such as Au/Cu, In/Ti and Ni/Ti. These materials can be easily deformed at room temperature, but they recover their pre-deformed shape when heated up above a phase transition temperature, which is characteristic of each material. This phase transition results in mechanical deformation that is used for actuation [26], as shown in Fig. 4.3b.

As an example, we can take the case of Kohl et al. who developed normally open gas microvalves actuated by microfabricated SMA thin films of NiTiPd. The main components are a polyimide membrane, a spacer and the SMA device. They demonstrate control of pressure differences below 250 kPa, corresponding to a gas flow of 360 mL·min^{-1} with a stroke of about 20 μm [27].

Thermopneumatic

Thermopneumatic actuation is based on thermally induced volume change and/or phase change of fluids sealed in a cavity with at least one compliant wall (Fig. 4.3c). For liquids, the thermally induced pressure increase is related to the thermal expansion coefficient, and the temperature increase. Since the actuation force is given by $F_{tp} = P \cdot A$, the thermopneumatic force is related to the surface A of the moving membrane, thus it scales as the square of the size. Thermopneumatic actuation combined with soft materials is very promising to achieve large actuation force and large stroke, with however a large power consumption. An example is reported from Takao et al. [28], presenting a thermo-pneumatic in-channel PDMS microvalve applicable to integrated blood test system. The microvalve consists of two separated chambers and a PDMS diaphragm actuated by pneumatic pressure controlled by an integrated Au microheater.

Bistable

As opposed to what we have seen till now, this refers more to a design choice, rather than to an actuating mechanism. Power consumption is one of the major factor of interest when choosing the suitable microvalve for a specific application. Any of the actuation mechanisms shown till now might need to have a continuous application of power to keep the microvalves open or closed. This limitation can be overcome through the design of bistable structures that only require power during the transition between two stable positions, regardless of the actuation principle (Fig. 4.4).

Bistable microvalves have been reported using thermal buckling of membranes [29], pneumatic switching actuation [30], and combination of multiple actuation principles [31, 32].

An example of thermal buckling is reported by Goll et al. [29]. The authors use a polyimide membrane which buckles due to a compressive stress induced by thermal treatment and mechanical loading. Because of the bistability of the microvalve, only a short pressure rise and a short pressure drop are obtained by controlling an electric current through a resistive heater in the actuator chamber.

4.2.1.3 Passive

While active valves control flow rate through pressure differences and have complex structures due to their various actuation principles, passive valves (check valves) only open to forward pressure and have simple structures, showing diode-like characteristics. The valving efficiency of passive valves is relatively poor, since the performance of these check valves depends on input pressure. This lack of efficiency results in leakage flows at low pressure. Despite this drawback,

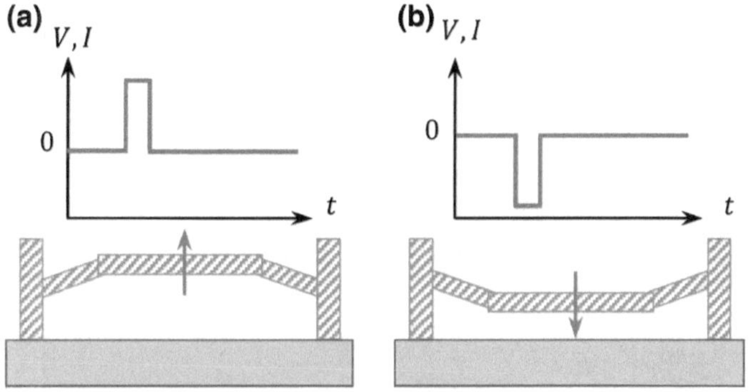

Fig. 4.4 Schematic representation of the actuation principle of bistable microvalves

most passive microvalves are incorporated as check valves in inlets and outlets of reciprocal displacement micropumps. The mechanical moving component can be constituted of a cantilever (flap type), a membrane or a ball, as shown in Fig. 4.5.

In case of disposable microfluidics, irreversible *burst valves* are highly preferred, due to their single-use nature and simple fabrication [33].

Flap and Membrane

Flap and membrane valves usually have a mechanical component that forms a seal with the valve seat, allowing fluid flow only in one direction and preventing reverse flow.

Voldman et al. successfully developed a silicon flap microvalve, placed at the end of a silicon nozzle, to control biochemical reactions of two compounds [34]. However, the use of a rigid material as silicon can represent a limiting factor, since stiction between the moving parts can occur, and even slight contaminations can affect the integrity of the sealing. Such limitations can be overcome by the use of soft materials like PDMS or hydrogels, which are not only flexible, but also cheaper and biocompatible [35]. On the other hand, PDMS, polymide and hydrogels tend to leak liquid and gases due to their high permeabilities. Therefore, parylene is often chosen as valve material, because of its low permeability to liquids and low Young's modulus. As an example, Kim et al. reported in [36] the fabrication of two different configurations of parylene micromechanical check valves: cantilever-type flap valve and bridge-type membrane valve. In both cases, one valve controls the flow in one direction, while the other controls the flow in the opposite direction.

The main difference between the flap-type and the membrane-type configuration lies in the relative movement between the fluid and the mechanical part of the valve. For the cantilever-type flap valve, the flow going through the valves produces rotating motion of the valves and induces turbulences in the flow around the valve opening ends. For the bridge-type valve, the flow can go through the valve with a symmetrically distributed velocity profile.

Ball

Ball valves allow the control of a unidirectional flow. The working principle is very similar to flap and membrane valves, but the mechanical moving element in this case is a microsphere.

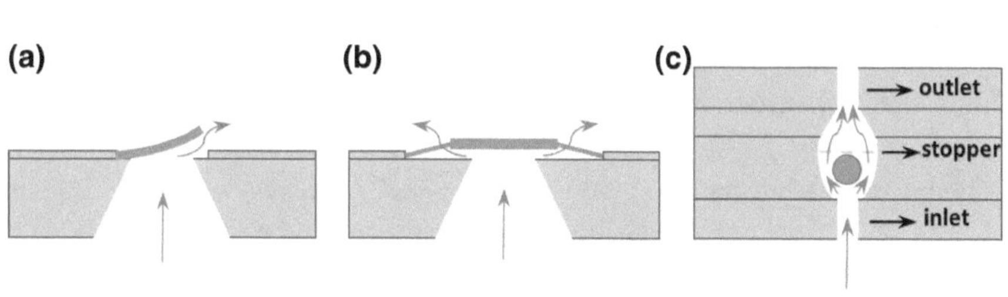

Fig. 4.5 Simple representation of passive microvalves, where the mechanical moving component can be either a cantilever (**a**), a membrane (**b**), or a ball (**c**). In case of ball microvalves a stopper is needed in order to keep the ball confined in his cavity and allow the valve opening. The blue arrows represent the direction of the flow

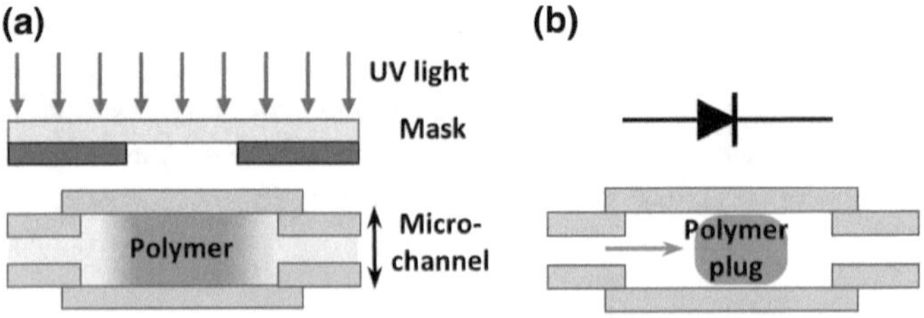

Fig. 4.6 **a** Schematic of photopolymerization method for the fabrication of mobile piston in microchannels. **b** The resulting diode-like polymer plug is shown

Ball valves historically find their main application in heart valve prosthesis. In 1954, Hufnagel et al. treated patients with aortic insufficiency with insertion of an acrylic ball valve into the descending aorta [37], while in September 1960, Starr-Edwards performed the first successful mitral valve replacement using a caged-ball valve [38, 39].

Miniaturization of these valves allowed their use as passive check valves in micropump structures, as will be described in Sect. 4.2.2 Mechanical Micropumps.

In-line Polymerized Gel Mobile Structure

In-line polymerized gel microvalves are a particular example of ball valves. They use in situ UV or visible light induced photopolymerization to create gel plugs and interrupt liquid flows, working as diode-like mechanical components (Fig. 4.6a and b).

Photopolymerization is induced by selectively exposing the chip to UV light, in order to define mobile pistons inside the microchannels. After UV exposure, the unwanted polymer can be flushed away with organic solvents while sealing or confining surfaces prevent the moving part from flowing out when pressure is applied. The main advantage of this technique is the simple fabrication that does not require any sacrificial material or mechanical assembly.

Hasselbrink et al. used photopolymerization of Teflon-like polymer to fabricate micropistons in microfludic channels, demonstrating sealing pressures up to 30 MPa [40]. Another example is given by Seidemann et al. who fabricated SU8 photoepoxy triangular check valve, anchored to a S-shaped spring [41].

4.2.2 Mechanical Micropumps

Micropumps generate temporal and volumetric flow and allow to reduce the amount of external hardware necessary to operate a microfluidic device. They find application in many different fields such as microfluidic transport, microelectronic cooling, micropropulsion for space exploration, dispensing therapeutics, and single cells monitoring [42–45].

In essence, pumps are smart combination of valves (both active and passive) that put fluid in motion. Pumps may be classified based on the applications they serve, the materials from which they are constructed, the liquids they handle, and even their orientation in space.

We adopt here a more basic classification system according to the way by which micropumps induce fluid flow and pressure [44, 46]. This classification is thus related to the pump itself and is independent of any consideration external to the pump or even to the materials from which it is constructed. Under this classification criterion, all pumps are divided into two major categories: dynamic micropumps and displacement micropumps.

Displacement micropumps exert pressure forces on the working fluid through one or more moving boundaries. Mechanical microvalves are typically used as a constitutive part of displacement pumps.

Dynamic micropumps continuously add energy to the working fluid in order to increase its pressure directly or indirectly through its momentum. Momentum added to the fluid in a displacement pump can subsequently be converted into pressure by the action of an external fluidic resistance.

Several parameters need to be taken into account in order to optimize the micropump performances. Some of them are reliability, power consumption, cost, miniaturization, biocompatibility, and efficiency. This latter concept, pump efficiency η, is defined as the ratio of the power delivered to the fluid by the pump P_{pump} and the power required to drive the pump actuator $P_{actuator}$. In an ideal pump, P_{pump} and $P_{actuator}$ are identical as no losses exist. In a real pump three sources of losses can be distinguished: fluid leakage losses (volumetric efficiency η_V), frictional losses (mechanical efficiency η_M), and losses due to imperfect pump construction (hydraulic efficiency η_H). Therefore, the total pump efficiency can be written as follows: $\eta = \eta_V \eta_M \eta_H = P_{pump}/P_{actuator}$ [42].

4.2.2.1 Displacement

Displacement micropumps can operate in a periodic or aperiodic manner. Periodic displacement micropumps give a continuous motion to the fluid, and can be divided into reciprocating and rotary types, depending on the nature of movement carried out by the pressure-producing mechanical element. Aperiodic displacement pumps do not typically involve the movement of any mechanical parts and so they are not going to be studied here, but the interested reader can further read about them in [44].

Reciprocating: Diaphragm

The majority of reported micropumps are reciprocating displacement pumps in which the moving surface is a diaphragm, also called membrane pumps or diaphragm pumps. Alternatively the moving mechanical element can also be constituted by a piston [44].

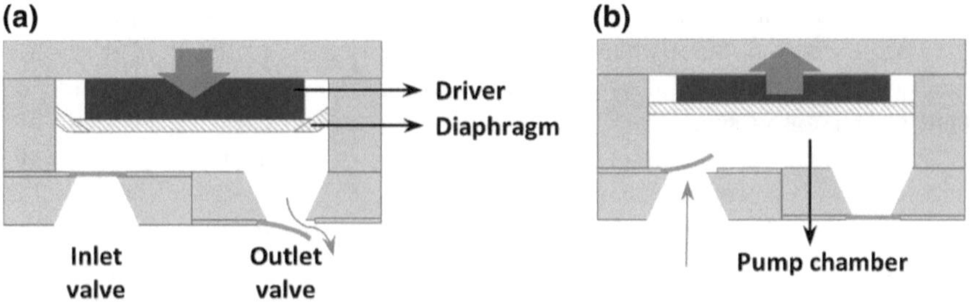

Fig. 4.7 Structure and operation of a typical reciprocating displacement micropump. During operation the driver acts on the pump diaphragm to alternately increase and decrease the pump chamber volume. Fluid is forced out of the pump chamber through the outlet check valve, during the discharge stroke (**a**). During the suction stroke fluid is drawn in through the inlet check valve due to the pump chamber expansion (**b**)

The basic components of diaphragm pumps are shown in Fig. 4.7, and are a pump chamber, an actuator mechanism or driver and two passive check valves. Richter et al. have demonstrated that the maximum pressure p_{max} produced by reciprocating displacement micropumps is inversely dependent of the fluid compressibility κ, consequently these devices are generally capable of achieving higher pressures with liquid-phase working fluids better than with gas phase [47]. Complications may arise due to the presence of air bubbles in the working fluid, which would increase its compressibility, decreasing p_{max}. Therefore, susceptibility to bubbles is a significant problem for reciprocating displacement micropumps.

The most common method for fabricating micropumps is micromachining of silicon combined with glass bonding layers. The first silicon micropump was proposed by Van Lintel et al. [48] in 1988, and consisted of a single-chamber reciprocating displacement silicon micropump, driven via piezoelectric disk actuator placed on top of a glass diaphragm. While most micropumps presented in literature have a single pump chamber, a few micropumps have been reported with multiple chambers. One example is given by Smith et al. who fabricated a reciprocating displacement micropump with three chambers in series, piezoelectrically actuated and requiring no valve to rectify the flow [49]. Also, Shoji et al. proposed a reciprocating displacement microvalve with two pump chambers arranged in parallel, showing a continuous and rippleless flow in the pump output [50].

Actuation within the pump has been reported to be of many different types, including every single case analyzed in the previous paragraph. Thus, the same advantages and limitations discussed for microvalves are still valid for micropumps.

Pump diaphragm material must be chosen according to the frequency range of actuation and fluid properties. For micropumps driven by low frequency and/or low force actuators, a low elastic modulus diaphragm material, such as silicone rubber, generally allows to maximize the pump stroke volume, thus positively impacting device performances [51]. However, the high permeability of soft polymers to liquids may be a concern depending of the application. On the other hand, the fast mechanical response of a stiff diaphragm generally yields the best performance in case of drivers operating at high frequencies. For example silicon and glass are the most common diaphragm materials in piezoelectric driven reciprocating displacement micropumps.

Rotary

A small number of microscale displacement pumps employ rotary elements such as gears or vanes for generating fluidic flow. Based on different design concepts, rotary micropumps make use of either viscous forces or pressure forces to carry out the pumping action [52].

Ashouri et al. proposed a novel high-efficiency prototype microfluidic pump that uses magnetic properties of a ferrofluid in both pumping and valving mechanisms, presented in Fig. 4.8. A cylindrical magnetic piston covered by ferrofluid functions as a revolving element which sweeps the perimeter of the cylindrical pumping chamber. The piston is externally actuated by a motorized off-center permanent magnet. Ferrofluid is used to cover the gaps between the magnetic piston and the channel walls, also serves as a separating plug between the inlet and the outlet of the chamber preventing recirculation of the pumped fluid inside the chamber. This novel revolving piston design eliminates the need for an inlet valve, while a ball passive valve is placed at the outlet of the pump [53].

Key issues for such pumps include ensuring the immiscibility of the ferrofluidic plug and liquid being pumped, degradation of the ferrofluid over time, and the need to incorporate an external controller for the magnet.

Another interesting example of rotary displacement micropump is the one proposed by Chou et al. They used a microfabricated rotary pump to demonstrate active mixing of biological entities, overcoming their slow diffusion process [54]. Because microfluidic devices operate at low Reynolds number, many fluidic

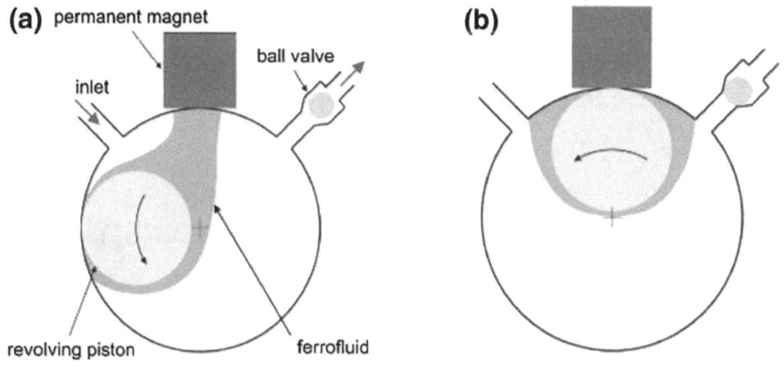

Fig. 4.8 Cross section and working principle of rotary micropump proposed by Ashouri et al. When the revolving piston is sweeping the larger sector between the inlet and the outlet ports (**a**) the counter clockwise displaced volume of the working fluid will be pushed into the outlet port. When instead the magnetic revolving piston is confined to the small sector between the inlet and the outlet ports (**b**) the check valve located at the outlet will prevent the fluid from flowing reversely from the outlet port to the inlet port [53]

operations are limited by diffusion in laminar flow. An active mixing apparatus takes much smaller area and thus saves the amount of time required and of precious reagents used, compared to a long narrow channel.

4.2.2.2 Dynamic

Dynamic micropumps require continuous addition of energy to the working fluid, in order to increase its pressure directly or indirectly through its momentum. Momentum added to the fluid in a displacement pump can then be converted into pressure by the action of a change in the fluidic resistance. These pumps generate a continuous flow without pulsation, and are simpler to fabricate than displacement micropumps. Typically, dynamic micropumps have no mechanical moving parts, and are based on non mechanical physical principle, which does not belong to the intent of this paragraph [44]. However, dynamic micropumps based on ultrasonic flexural plate waves and surface acoustic waves have been reported, which allow for directed flow in simple microchannels with no valves or other moving parts.

Acoustic

The use of mechanical actuation for fluidic pumping requires the use of sophisticated and often costly fabrication processes to develop the device and its micromachined moving parts.

Tovar and Lee demonstrated a novel and simple method for pumping and mixing fluids within a microfluidic system using an acoustic energy source, namely a piezoelectric (PZT) buzzer. Such device, termed lateral cavity acoustic transducer (LCAT), is based on the principle of acoustic microstreaming: air bubbles trapped in lateral cavities adjacent to the microfluidic channels are excited by an external acoustic field, which causes the air/liquid interface to vibrate. This energy is then passed to the bulk liquid and produces a net force perpendicular to the bubble interface and out the end of the cavity, as shown in Fig. 4.9. The pumping flow rate generated by the cavities is controlled through adjusting the amplitude and frequency at which the PZT is driven and also by determining the optimal bubble

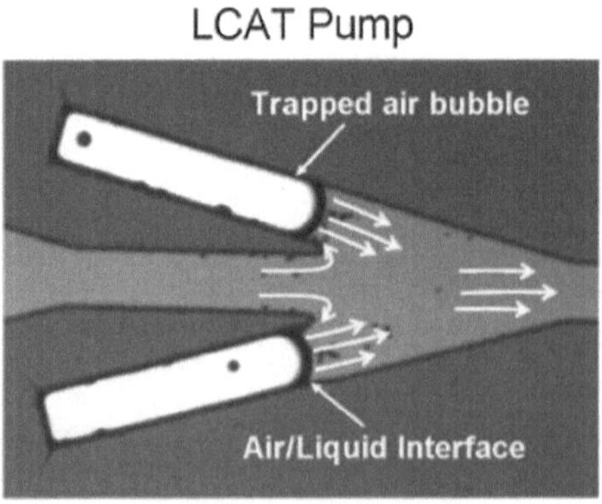

Fig. 4.9 Micrograph of the lateral cavity acoustic transducer micropump design. The microfluidic channel is 100 μm wide [55]

trapping angle. The maximum flow rate was obtained with an angle of 15° between the lateral cavity and the microfluidic channel [55].

Mixing of beads and cells into downstream microchannels has also been demonstrated by simply placing the air bubble cavity perpendicular to the microfluidic channel, which results in a drag force deflecting the particle in the direction perpendicular to the bulk flow [56].

The main advantage that the LCAT device has over other pumping methods is the ease of operation and fabrication. However, the classical configuration of closed channel networks can present several disadvantages. In pumping biological molecules and cells there is concern over reagent loss by adhesion to the wall, and clogging of small channels. Also, the pressure required to move a liquid confined into a microfluidic channel scales inversely with the channel dimension [57]. This means that the power of the pumps has to be increased by the same factor the channel size is reduced, which complicates the scaling and the integration into a portable system.

An alternative to the common closed microfluidic networks is given by the *surface acoustic waves* (SAW) technology [43], that will be discussed in more detail in the following paragraph, and that can be applied to fluidic volumes from few picoliters up to several microliters. A representative example of SAW microfluidic actuation is provided by Guttenberg et al. [58], who proposed a microfluidic device for on-chip PCR, operating at a planar open surface. In such device the fluid is transported on a piezoelectric substrate, and in order to prevent evaporation of the PCR reagents at high temperatures the fluidic sample is enclosed in droplets of mineral oil that can be ultrasonically actuated along the surface plane. Actuation of small droplets on the surface of a SAW chip is caused by the effect of acoustic streaming. This phenomenon appears when intensive sound fields are travelling through a liquid. The viscosity of the medium attenuates the amplitude of the acoustic waves during its propagation, which leads to a pressure gradient in the liquid. If the gradient is large enough, the fluid starts to move in the same direction as the sound wave. The RF power threshold for the movement strongly depends on the contact angle of the droplets. Therefore, the distribution and the motion of the fluid on the surface is controlled mainly by the surface free energy, which makes surface contamination a sensitive concern in SAW microfluidic actuation.

4.2.3 Mechanical Biosensors

Micro- and Nano-mechanical biosensors are mechanical transducers with micro- and nano-sized moving parts. Their fabrication relies on the standard semiconductor processing techniques, enabling for efficient batch production. Nanomechanical biosensors are usually cantilever-shaped, but can also include doubly-clamped beams, membranes, and SAW or BAW (bulk acoustic waves) devices.

Robust and well-established electrical and optical sensors still dominate the field of biological detection in terms of market share. However, in the past two decades

Micro and Nano Mechanical systems (N/MEMS) have been receiving a lot of attention at the research level. The main reason for this development of N/MEMS in life-science applications (BioMEMS) has been to address the *mechanical nature* of many fundamental biological processes. A clear example is cell mechanical properties such as stiffness or viscoelasticity, which represent a valid diagnostic biomarker for several pathologies like cancer, malaria and sickle cell anemia among others [59, 60]. In addition, continuous advances in micro- and nano-fabrication techniques allow the size of mechanical devices to be comparable to the size of some biomolecules, thus resulting in a highly sensitive *mechanical response* and in outstanding *mass resolution* [61–63]. Furthermore, N/MEMS can exhibit extremely high *mechanical compliance*. This enables the measurement of pN forces generated by biological interactions, by translating them into a mechanical displacement, enhancing *force responsivity* [64, 65].

Mechanical biosensors can operate in *static* or *dynamic* mode, depending on whether quasi-static deflection or shifts in the resonance frequency are monitored and thus related to the biological event under study.

4.2.3.1 Static Mode Biosensors

Static-mode biosensors are usually cantilever shaped and typically measure surface stress or surface forces induced by biological entities such as single-base DNA, pathogens or protein biomarkers [66, 67]. Other approaches consider the detection of forces at the tip of cantilevers, using a configuration similar to that of an atomic force microscope (AFM) [68, 69].

Common surface stress biosensors consist of microcantilevers with one active side functionalized to attach a monolayer of receptors that exhibit high affinity to the target molecules, while the opposite passive side is inert. The devices are placed in a fluid cell where the sample solution is delivered. The binding of target molecules to the receptors causes a surface stress change that induces a quasi-static deflection of the cantilever (see Fig. 4.10).

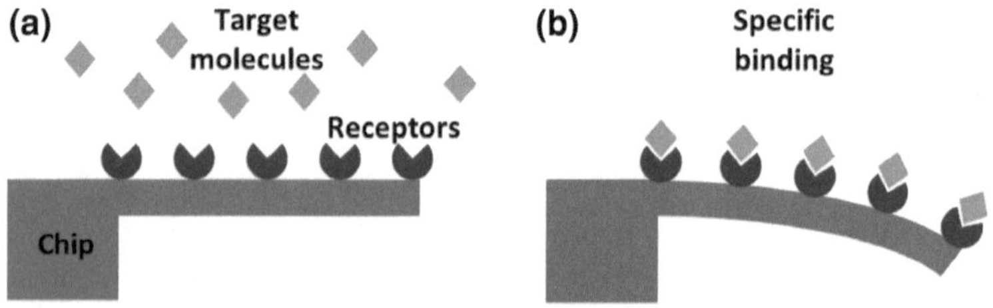

Fig. 4.10 Working principle of a static mode biosensor. Binding of target molecules induces a surface stress, which leads to a quasi-static deflection of the cantilever. This bending can be measured with optical or electrical displacement detection methods

The stress induced cantilever deflection is real time monitored via optical or electrical displacement detection methods. The surface stress variation can be evaluated through the change of curvature the microcantilevers undergo when the sample solution flows over them. According to the Stoney's equation $\Delta\kappa = 6\frac{(1-v)}{Eh^2}\Delta\sigma$, where κ is the curvature, v and E the Posson's ratio and the Young's modulus of the material respectively, h the thickness of the cantilever and $\Delta\sigma$ the surface stress change. The Stoney's equation is only valid under the assumption of plates with unclamped edges. Therefore, in order to take into account the clamped edge of microcantilevers configuration, a correction factor needs to be considered [70, 71].

These devices have shown remarkable label-free capabilities [72], however deflection is highly susceptible to non-specific binding, temperature and flow fluctuations. Very stable flows, together with differential measurements of non-functionalized cantilever help to partially circumvent spurious deflections [67].

4.2.3.2 Dynamic Mode Biosensors

Dynamic mode biosensors are mechanical devices which vibrate close to the resonance frequency of the device f_r proportional to the square root of the ratio between stiffness k and mass m of the structure. An analyte landing or binding on the resonator surface induces a change of its mass (and stiffness [73]), which is transduced into a shift of the resonance frequency, as seen in Fig. 4.11.

Micro- and Nano-mechanical resonators have demonstrated great capabilities in life-science label-free mass sensing applications, such as detection of viruses and single molecules [3, 63, 74]. Mass responsivity is defined as the change in resonance frequency due to a change in mass. $\Re = \frac{\partial f_r}{\partial m} \approx -\frac{1}{2}\frac{f_r}{m}$ [3, 75]. In order to achieve the highest mass responsivity, one can reduce the mass or increase the

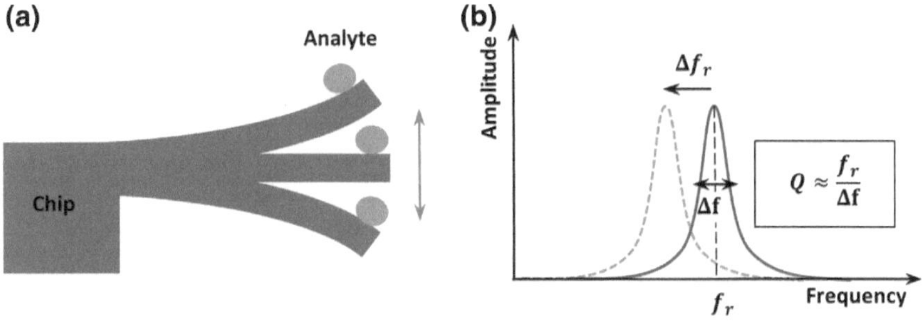

Fig. 4.11 Working principle of a dynamic mode mechanical mass sensor. An analyte landing on the free-end surface of a singly-clamped beam resonator induces a change of its mass and stiffness, which is transduced into a shift of the resonance frequency Δf_r. The accuracy of the measurement of the resonance frequency is defined by the quality factor Q, related to the full width of the half maximum of the resonant peak Δf

frequency (or both). The minimum detectable mass by a mechanical resonator is given by the ratio between the noise and the responsivity, and it is typically given by: $\delta m_{min} \approx \frac{\delta f_{min}}{\Re} = -\frac{1}{2}\frac{\delta f_{min}}{f_r}m$. Using Robbins formula [76], we can then write: $\delta m_{min} \approx -\frac{1}{2}m\frac{1}{Q}\frac{Noise}{Signal}$. Thus, it is clear that a large quality factor, Q, is necessary to optimize the sensing performance. Q determines the slope of the phase versus frequency curve of the device near resonance. It can also be seen as the ration between energy stored and energy lost per cycle of the resonator: $Q = \frac{E_{stored}}{E_{lost}}$.

In order to attain large quality factors, one needs to reduce the energy loss. This typically implies to work in vacuum [66]. However, when tackling real time biosensing applications, like protein or DNA detection, the immersion of the resonator into fluids is inevitable since most biological processes naturally occur in a *fluidic environment*. In this scenario, fluid viscous damping causes a drop of Q of several orders of magnitude.

Detection in Liquids

The main strategies developed to exploit nanomechanical sensor performances for biological detection consist in (i) capture in fluidic phase and detection in vacuum, (ii) measurements in humid environment, and (iii) continuous operations in liquid focused on enhancing the quality factor [66, 67]. The first one will not be detailed here as the operation of the device is not performed in liquid.

Humid environment

Certain biological species like bacteria can grow and survive in humid environment. In this case, even though viscous damping is larger than in vacuum, it does not reach the levels of full immersion in liquid, thus the quality factor remains acceptable. An example of this biodetection is the growth monitoring of Escherichia coli using cantilevers coated with nutritive layers (sugar) and operation in air with 100% humidity [77].

Real-Time monitoring in liquid

There are many other cases, e.g. the detection of analytes, which require full fluidic immersion while the measurement is running. This, as announced before, poses a practical issue since viscous damping degrades Q. The solution for this has been, up to date, limited to a handful of experimental techniques. One example is the use of acoustic modes (QCM and SAW). Another option, considering flexural devices, is to use larger frequencies, higher vibrational modes [78], individual actuation and detection, vertical nanopillars [79–81] and suspended microchannels [82].

Quartz Crystal Microbalance

An interesting device from the performance point of view is the quartz crystal microbalance (QCM). It consists of a quartz disc with electrodes that allow for the excitation of a shear acoustic mode, which has an inherent low viscous interaction with the surrounding fluid (Fig. 4.12) [83]. QCMs are used now in a range of scientific disciplines, such as chemistry, biology, and material sciences, demonstrating also detection of biomolecules and bacteria [84]. The resonance frequencies of these devices are usually in the range of 10–50 MHz. The thickness of the crystal determines its resonant frequency, but at higher frequencies the devices become

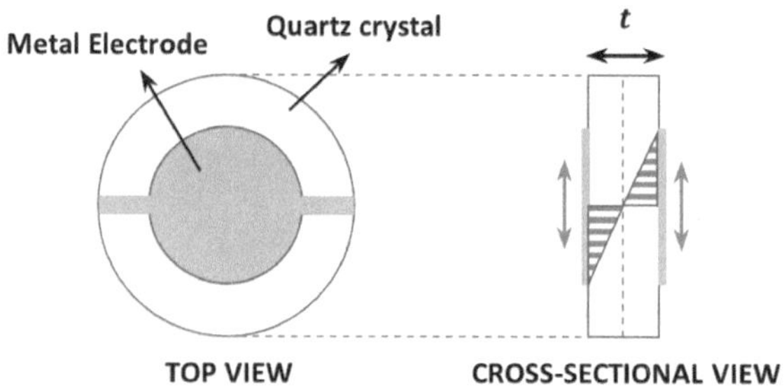

Fig. 4.12 Schematic of Quartz Crystal Microbalance (QCM). A quartz crystal disk oscillates in thickness-shear mode when the metal electrodes deposited on each side of the disk are connected to an alternate voltage signal, generating a bulk shear acoustic wave (BAW)

very thin and too fragile. Because of the device thickness in the range of millimeters, the sensitivity of QCM is limited [84]. The lowest limit of detection of QCM in liquid has been achieved in the work by Cho et al. They demonstrated nanomolar sensitivity for continuous analyte monitoring of C-reactive protein, a biomarker for coronary heart disease, using an indirect-competitive assay [85].

Surface Acoustic Wave Devices

Another example of acoustic sensors are SAW devices, which have been described in the previous paragraph for their capabilities in microfluidic actuation and micro-objects manipulation like mixing, pumping, merging and sorting among other applications [43]. SAW devices have also demonstrated to represent a cost-effective technology for label-free detection of biomolecular interactions, allowing highly sensitive real time detection of biomolecules such as DNA, protein and bacteria [86, 87]. Recently, Senvely and coworkers proposed a low cost innovative system capable of analyzing few picoliters of biological cell samples at high frequencies. Tumor cells are trapped inside microcavities along the SAW device surface. Through interaction with the samples, the resultant SAW reaching the output electrode exhibits a phase delay which is related to the cell mechanical stiffness [88].

SAW devices generate and detect acoustic waves using interdigital transducers (IDT) on the surface of a piezoelectric crystal, typically lithium niobate (LiNbO3) or lithium tantalate ($LiTaO_3$). As shown in Fig. 4.13, an input IDT converts an electrical signal into an acoustic wave through direct piezoelectric effect. On the opposite side, an output IDT converts the acoustic wave back into an electrical signal, through inverse piezoelectric effect. The surface acoustic wave travels along the space between the IDTs, known as the delay-line. The acoustic energy stays confined close to the surface of the PZE crystal in a depth range close to the acoustic wavelength, regardless of the thickness of the complete substrate. For this reason, the wave is potentially very sensitive towards any change on the surface, such as mass loading, viscosity and conductivity changes [86]. To permit the use of a SAW device as a biosensor, analyte-specific molecules are immobilized on the device surface to catch analyte molecules from the sample solution, as shown in Fig. 4.13. Analytes binding to the immobilized receptor molecules will influence the velocity of the SAW and hence the output signal measured by the system electronics [86].

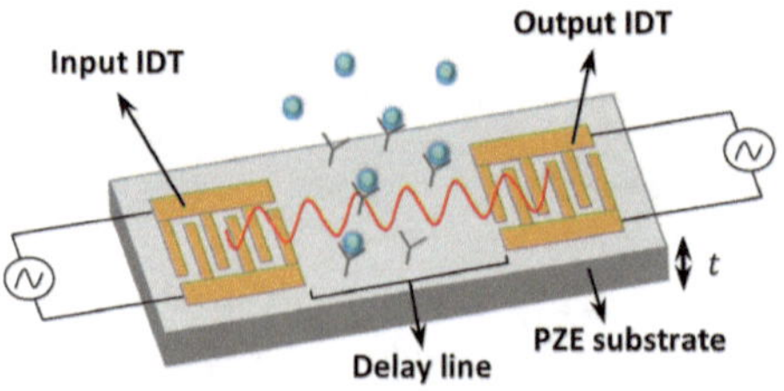

Fig. 4.13 Schematic of a Surface Acoustic Wave sensor (SAW). An input interdigitated electrode (IDT) sitting on a piezoelectric substrate converts an electrical signal into an acoustic wave. An output IDT converts the acoustic wave back into an electrical signal. The surface acoustic wave travels along the space between the IDTs, known as the delay-line. Analyte binding to the immobilized receptors will influence the velocity of the SAW and hence the output signal measured by the system electronics

SAW devices can be found to be as thin as few hundreds of micrometers and their typical frequencies are much larger than those of QCMs. However, the acoustic waves are not purely shear, and thus radiate compression waves into the liquid, which induces larger damping than in the case of shear modes, but lower than in the case of flexural [87].

Suspended Microchannel Resonators

As we have seen above, the minimum detectable mass is proportional to the mass of the device. In order to minimize said parameter, the optimum choice is to work with *flexural devices*, which can reach nanoscale dimensions and masses in the order of picograms, potentially allowing single molecule mass sensitivities. However, flexural devices exhibit out-of-plane vibrations, therefore their energy loss is much larger compared to acoustic resonators.

An important contribute to mechanical biosensing in liquids came from Manalis' group in 2003, when they presented a device inspired to the millimeter scale densitometer developed by Stabinger and co-workers in 1967 [89]. Suspended microchannel resonators (SMRs) are resonant mass sensors for continuous biomolecular detection in a pico-liter fluid volume [82].

SMRs consist of a hollow resonant structure containing embedded microfluidic channels. The fluid to be analyzed flows inside the resonator, thus viscous damping is almost completely suppressed, and mass loading is minimized. Real time analysis of fluidic samples is made possible while the whole device can be kept in dry environment or even in vacuum. SMRs have been successfully employed for several biosensing applications such as measuring single cell mass, density, volume, deformability, viscosity and monitoring cell growth. Quality factors up to 15,000 and very good mass sensitivities down to 1 ag in 1 kHz bandwidth have been demonstrated [90–95]. However, one of the major limitations of resonators with embedded fluidics (SMCs) is the costly and rather challenging fabrication process, with many steps, which include the use of aggressive chemicals.

Two modalities of operation are possible with SMRs devices: flow-through detection and affinity-based capture, shown in Fig. 4.14.

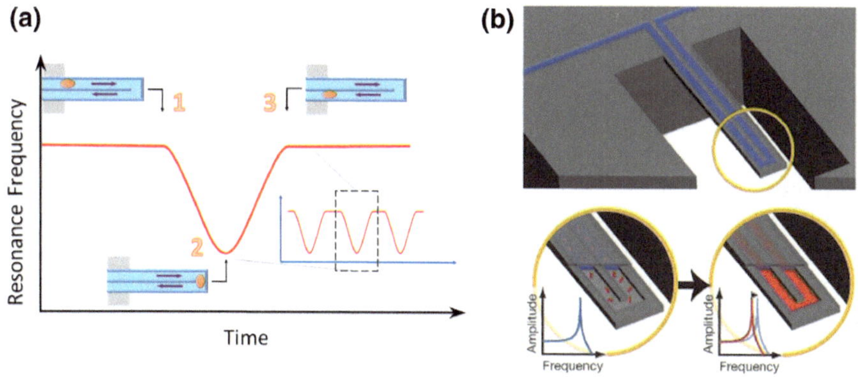

Fig. 4.14 Sensing principle of suspended microchannel resonators. **a** Flow-through detection: a particle flowing inside the embedded microfluidic channel induces a shift in the resonance frequency of the device which is related to the particle mass and position along the structure. **b** Affinity-based capture: target analytes are captured by molecular recognition on a bio-functionalized surface of the device. This induces a shift in the resonance frequency due to mass loading effect [95]

In the *flow-through detection method* a particle travelling inside the microfluidic channel induces a shift in the resonance frequency of the device, which is a function of the position of the particle along the resonating structure, and of the particle density, mass and volume. This approach has been mainly used to study the development of cells along time as a function of different conditions. In this case, the maximum flow rate is determined by the *minimum particle residence time*, which yields sufficient averaging to achieve a detectable signal above the noise floor.

In the *affinity-based capture method* target analytes are immobilized by molecular recognition on the bio-functionalized surface of the microfluidic channel. The functionalization can provide high specificity of capture, reducing false positives that can be caused by particles with similar mass. In this case, the added mass signal from immobilized target analytes can be averaged over a period of time longer than that for free-particle transit through the microchannel. Therefore, in contrast with the flow-through detection method, affinity-capture based SMRs devices are not limited by the analyte minimum residence time. Instead, the sensitivity is determined by the kinetics of diffusion and binding, which can be controlled through geometry and flow-rate.

The main limitation of this type of devices is the throughput, defined as the sample volume that can be analyzed per unit of time [96]. High throughputs can be obtained through pre-concentration of the analyte solution and device parallelization, as recently demonstrated by Cemark and co-workers [91].

4.3 Passive Microfluidics (Magalie Faivre)

4.3.1 Introduction

Microfluidic diagnostic tools provide numerous advantages over conventional macroscale methods in terms of sample volume, low cost, portability and integration in a complete chain of analysis in Lab-On-a-Chip devices. Among the key features of fluidic microsystems is the possibility to structure the geometry at a scale comparable to that of cells, the ability to control precisely the cellular environment and to perform analysis at the single-cell level while dealing with large populations. The high heterogeneity of cell samples, containing many different species at very different abundance levels, is one of the great challenges in clinical research and biomedical engineering. Concentrating, focusing, trapping and sorting these cells are thus critical steps in various biomedical applications including (i) diagnostics, (ii) therapeutics and (iii) cell biology, where the least abundant cells can be the most important. Indeed, some domains such as oncology, stem cells research or infectious disease, have raised great interest in identifying, accumulating and counting various types of "rare cells"—cells present as a very small subpopulation in a large amount of surrounding cells—or bacteria in blood [97]. Rare cells such as Circulating Tumor Cells (CTCs) may be difficult to isolate as large volumes of blood are required to be relevant [98], imposing a very high throughput of the separation technique. For example, HIV diagnostic and theranostic rely on the separation and count of human T-lymphocytes (CD4+) from whole blood [99, 100]. Non-invasive prenatal diagnosis can also be performed by separating rare fetal cells such as Nucleated Red Blood Cells (NRBCs) from peripheral maternal blood [101]. For example, CTCs [102], stem cells [103] and malaria-infected Red Blood Cells (RBCs) [104] would highly benefit from rapid, efficient and low-cost separation microdevices.

Microfluidic cell manipulation and sorting methods can be divided in two: active and passive approaches. Active methods are based on the application of an external field to achieve particle manipulation, (dielectrophoresis, acoustophoresis, magnetophoresis, etc.…) whereas passive techniques rely on the channel geometry and associated hydrodynamic forces to operate on cells. Although it may be considered as a passive sorting approach, the immunological-based separation, which exploits the presence of antibodies at the surface of the cells of interest, will not be discussed here. The different label-free passive techniques rely on several cells intrinsic physical properties such as size, shape, deformability or density to manipulate and sort them. Passive label-free techniques generally require fewest preparation steps than active ones, making it a highly attractive approach for cell manipulation. For detailed information on active techniques, affinity-based methods and their comparison with passive ones, the readers are suggested to refer to specific chapters of the present book or the following review papers from Xuan et al. [105], Bhagat et al. [106], Shields IV et al. [107], Autebert et al. [108] and Gosset et al. [109].

In this section, we aim at introducing and discussing the different approaches available to perform concentration, trapping, manipulation and sorting of cells and their applications. We propose to analyze many passive techniques for cell manipulation, concentration and isolation in microsystems, detailing their mode of action and trying to evaluate their performances to allow comparison. We document potential applications, discuss the main advantages and limitations of the different approaches, and attempt to outline the main remaining challenges in this fast evolving field.

4.3.2 *Microfiltration*

A commonly employed technique to isolate cells is microfiltration. Most of cell sorting devices using microfiltration rely on both size and deformability to define the pore size and the type of filtration. In the following, we will detail various approaches with one common feature: the use of micropatterns to induce specific behavior of particles flowing in a microfluidic device.

4.3.2.1 Size Exclusion

The use of micropatterns (pillars, weirs, wells, etc.…) is very attractive for size and deformability-based cell sorting or capture. This approach is especially well adapted to the microsystem format where characteristic dimensions are comparable to the typical size of cells. Size-based microfiltration faces many challenges including the heterogeneity of size within a population of cell and membrane clogging. Four types of microfilters have been implemented for whole blood filtration [110]: membrane, weir, pillar and cross-flow (Fig. 4.15a).

Membrane microfilters are composed of well-defined pores, designed to prevent the passage of cells below a critical size. More than 90% recovery of lung cancer cells from spiked blood samples, with a relatively high processing throughput of 10 mL/h, was recently demonstrated using a highly porous polydimethylsiloxane (PDMS) membrane filter-based microdevice [114].

Weir-type microfiltration has been used to produce cell-free blood plasma [115, 116] for diagnostics. The microdevice has been designed such as the sample would flow through a planar groove, small enough to retain all blood cells (Fig. 4.15a). However, very small amount of plasma—typically several tens of nanoliters—was produced due to the use of capillary-driven flow [116]. The obtained filtrate was tested negative for significant levels of hemoglobin and cellular debris.

Pillar-type microfiltration implements micro-obstacles which spacing is chosen adequately to retain cells above a given critical size (Fig. 4.15a). Such design has been successfully applied to the isolation of fetal nucleated RBCs from cord blood using rows of pillars with decreasing spacing [117]. However, it is unsuitable for

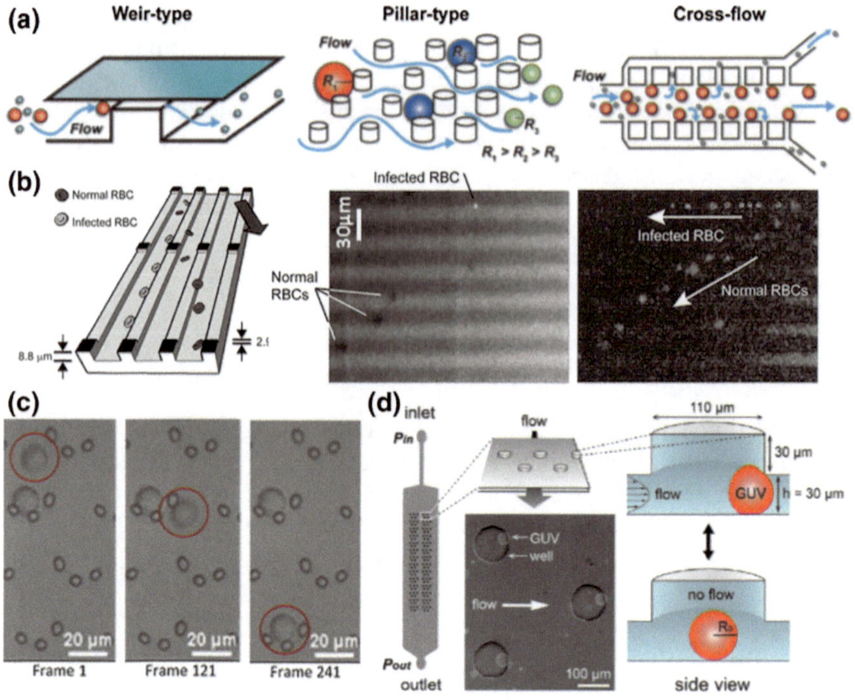

Fig. 4.15 **a** Some examples of implementation of the microfiltration technique: weir-type, pillar-type filters or cross-flow (Reproduced from [109], copyright The Authors (2010) with permission of Springer). **b** Application of the microfiltration to the isolation of malaria-infected RBCs from healthy ones (Reproduced with permission from [111]). **c** The use of crescent-shaped isolation wells has been reported to trap CTCs (Reproduced from [112], copyright (2010) with permission from Elsevier). **d** Example of deformability-based microfluidic device implementing indentations etched in the channel ceiling to trap particles (Reproduced from [113] with permission of The Royal Society of Chemistry)

prenatal diagnosis due to the relatively low throughput achieved (0.35 mL/h) and the high probability of clogging, the cells being trapped in the direction of the flow.

Cross-flow microfiltration was developed to reduce clogging and filter saturation. The working principle is similar, except that the flow is perpendicular to the array of obstacles or weir filter (see Fig. 4.15a). Cross-flow filtration has been used by Sethu et al. to achieve separation of 97% of White Blood Cells (WBCs) from whole blood [118] at a flow rate of 5 μL/min. A similar approach has been implemented by Han et al. [111] to isolate mechanically impaired malaria-infected RBCs, rigidified by the presence of the parasite, from healthy ones. The cells are flown in a channel with several parallel trenches which walls have been dimensioned such as deformable cells can squeeze from one trench to another but stiffer ones cannot. A flow is applied at an angle with respect to the trenches (Fig. 4.15b), allowing deformable healthy RBCs to follow the flow and be collected on one side of the outlet whereas the parasited RBCs keep linear trajectories and are collected on the other side of the outlet [111]. Reported studies on mechanical filtration microsystems [119] show limited efficiency in the target cell isolation at high throughput. Recent published work reports another approach to overcome clogging: the use of low frequency oscillating flow [120]. They achieve collection of target polystyrene particles with 100% efficiency and demonstrate the separation of CTCs from whole blood (50 fixed MDA-MB-231 cells in 200 μL of blood processed at 200 μL/h).

4.3.2.2 Isolation Wells

Microsytems implementing micropatterns can also be employed to perform cell trapping. Located in the path of the flow, those micropatterns forming cavities are designed such as only one cell gets trapped in it. Such platforms offer the possibility to perfuse different reagents to the individually captured cells and to study their response, which can be highly attractive for drug screening assays in the framework of personalized medicine applications. Di Carlo et al. [121] demonstrate the capture of single HeLa cells using arrays of physical U-shaped hydrodynamic trapping structures. Their microsystem allows both the culture of individual adherent cells and the dynamic control of fluid perfusion with uniform environments. With a similar approach, Tan et al. [112] have demonstrated the capture of CTCs (issued from 8 different types of cancer: breast and gastric carcinomas, tongue and pharynx squamous carcinomas, etc…) in blood using arrays of crescent-shaped isolation wells and their subsequent release for further analysis (Fig. 4.15c). Typically, 100 cells suspended in 1 mL of blood were successfully flown at a pressure of 5 kPa in the device and performances with very low count of cancer cells (typically 1–3 cancer cells in 1 mL of buffer) were evaluated at 81%. Recently, Yamada et al. [113] have reported a clever approach to trap and release giant unilamellar vesicles (GUVs) in a thin and wide microfluidic channel, as they cross indentations etched in the channel ceiling (Fig. 4.15d). The particles being squeezed to flow into the shallow channel, partially relax when entering a well with a larger height, leading to their capture. They demonstrate that GUVs, whose diameter is slightly larger than the channel height, can be trapped and subsequently released by flowing the outer fluid beyond a critical velocity [113]. They report a size-dependency of the critical flow velocity necessary to untrap the GUVs.

In this review we introduced different label-free passive manipulation techniques, which characteristics are gathered for comparison in Table 4.1.

4.3.2.3 Deterministic Lateral Displacement

The implementation of microstructures in a microfluidic device can also be employed to continuously sort cells without clogging. Deterministic lateral displacement (DLD) is a size-sorting technique where cells navigate through an array of posts [122]. Contrary to the previous techniques, there is no retention of cells on the chip as both the cells of interest and the rest of the sample are flowing separately

Table 4.1 Label-free passive microfluidic manipulation and separation techniques

Method	Principle	Properties exploited	Sample	Performances	Flow Rate	Ref.
Microflitration	Size exclusion	Size, deformability	Whole blood*	>97% depletion of WBCs	5 μL/min	[118]
Wells	Cell trapping	Size, deformability	CTCs in PBS[a]	80% CTCs isolation	5 kPa	[112]
DLD[b]	Flow through posts array	Size, deformability, shape	Whole blood	>99% WBCs isolation	0.3 nL/s	[122]
Hydrophoresis	Flow alterations	Size, deformability, density	~2% hematocrit blood	210-fold enrichment of WBCs	1 μL/min	[123]
Gravity	Cell sedimentation	Size, density	3 and 20 μm bead suspension	~100% isolation of 2 populations	1.2 mL/hr	[124]
Hydrodynamic filtration	Streamline modification	Size, shape	Liver cells	~100% hepatocytes isolation	50 μL / min	[125]
PFF[c]	Streamline modification	Size	0.13% hematocrit blood	80% RBCs isolation	20 μL/hr	[126]
Inertia in straight channels	Lift forces	Size	E. coli in diluted blood	300-fold enrichment of bacteria	18 μL/min	[127]
Inertia in curved channels	Lift and Dean drag forces	Size	Beads suspension	Sorting rate of 1 g/hr	1.5 mL/min	[128]
Fahraeus effect	Hydrodynamic focusing	Size, deformability	16% hematocrit blood	24% plasma recovery	200 μL/hr	[129]
Leukocyte margination	Steric margination	Size, deformability	Whole blood	94% NCs[d] isolation	5 nL/s	[130]
Bifurcation law	Hydrodynamic force	Size, deformability	Whole blood	25% volume of plasma isolated	4 μL/min	[131]

*Whole blood corresponds to 45% hematocrit

[a]PBS: Phosphate Buffer Saline, [b]DLD: Deterministic Lateral Displacement, [c]PFF: Pinch Flow Fractionation, [d]NCs: Nucleated Cells

out of the system. At low Reynolds number, an object entering a streamline will tend to follow it, thus having a deterministic behavior. When encountering an obstacle, the behavior will change according to the particle size. Indeed, if the particle has a typical size smaller than a critical hydrodynamic diameter D_c, it will follow the main stream line, circumventing the obstacle. In the contrary, if the object size is larger than D_c, the object will bump into the obstacle which will lead to a shift in stream line (Fig. 4.16a). The critical hydrodynamic diameter D_c is related to the inter-posts distance d and the period of the array N according to the following equation [132]:

$$D_c = \sqrt{\frac{N}{3}} \cdot \frac{d}{N}, \tag{4.1}$$

The influence of the geometrical parameters on D_c was carefully evaluated by Inglis et al. [135] and have been used to fractionate undiluted whole blood samples [122], separate parasites from human blood [136] and isolate cancer cells [137]. By tuning the posts size, their spacing, as well as the shift between adjacent post rows, Huang et al. [138] demonstrated that D_c could be tuned precisely enough to sort 0.8, 0.9 and 1.03 μm in diameter particles upon their size. Obstacles can be any shape: cylindrical, rectangular or triangular [127]. Tottori et al. [139] used a DLD microfluidic device to sort viable Jurkat cells from nonviable apoptotic ones, the latter being smaller due to shrinkage and fragmentation. However, such microsystem is much more complicated than it seems as one need to take into account the deformability of the cells. Actually, DLD is a size-based separation technique but it is also sensitive to shape (Fig. 4.16b) and rigidity (Fig. 4.16c) of the target cell as shown by Beech et al. [133] on RBCs. Finally, thermally tunable DLD has been demonstrated [134] recently. Indeed, by implementing poly-N-isopropylacrylamide (poly-NIPAM)—a thermo-sensitive hydrogel known to be swollen or shrinked, respectively below and above a critical temperature—pillar arrays, it is possible to tune at will the size and spacing of the posts, by adapting the temperature as illustrated Fig. 4.16d.

4.3.3 Hydrophoresis

The introduction of micropatterns in the microfluidic channel can finally be exploited to perform cell focusing using a phenomenon referred to as hydrophoresis. This technique relies on the formation of a lateral pressure gradient within a microfluidic channel due to the presence of an array of slanted obstacles on the microchannel floor and/or ceiling; the apparition of associated micro-vortices altering the flow [140] and leading to the focalization of particles (Fig. 4.17a). Using this phenomenon, Choi et al. have developed a device that exploits differences in both the size and deformability of cells for passive sorting [123]. This principle was used to demonstrate a 210-fold enrichment of WBCs from RBCs with

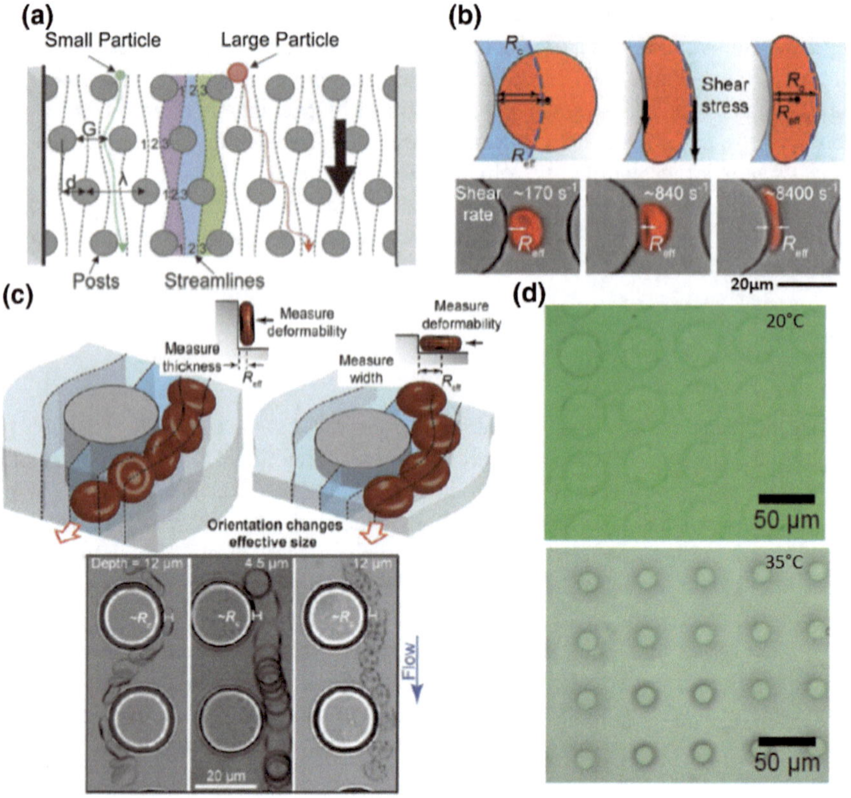

Fig. 4.16 **a** Deterministic lateral displacement approach is a size-based separation method relying on the posts size, the distance between them, as well as the relative shift be bigger ones are deflected on one side of the device (Reproduced from [122], copyright (2006) National Academy of Sciences, U.S.A.). DLD have also been demonstrated to be sensitive to cell **b** shape and **c** rigidity (Reproduced from [133] with permission of The Royal Society of Chemistry). **d** poly-NIPAM pillar arrays allow thermally tunable DLD (Reproduced from [134] with permission from Chemical and Biological Microsystems Society)

a throughput of 4×10^3 cells/s. More recently, the same group has reported the recovery of 98.5% of CTCs spiked in blood (1 MCF-7: 7500 blood cells) at a flow rate of 44 μL/min [141].

Similarly, a device with successive ridges along the channel ceiling has been shown to focus and sort cells [143, 144]. This microsystem contains an array of herringbone shaped slanted grooves in the ceiling of a microfluidic channel that passively focuses cells to specific location of the channel upon their density, thus acting their separation. They successfully sort a mixture of two bead populations whose density difference is as small as 0.1 g/cm^3. With a different design, Sollier et al. [142] present a ridgeless microvortex, introducing a sudden geometrical

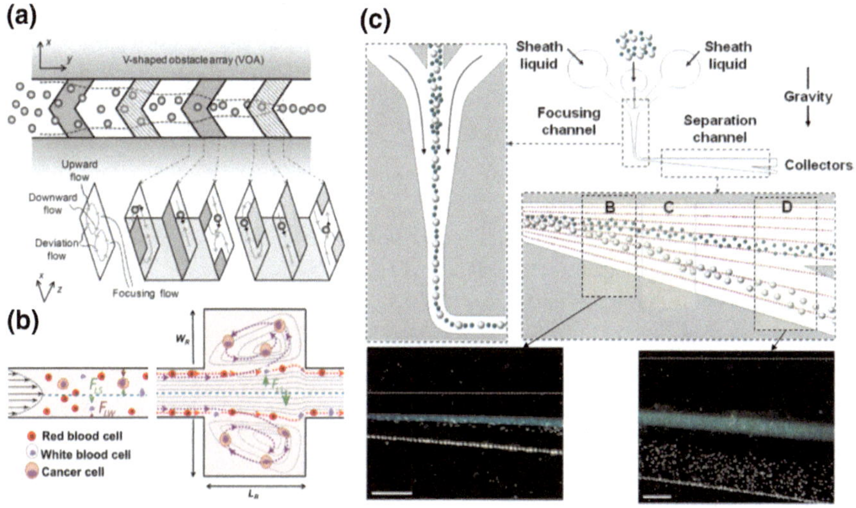

Fig. 4.17 **a** Hydrophoresis manipulation techniques include ridge-based microvortex (Reproduced with permission from [140] Copyright (2008) American Chemical Society) and **b** microvortex using a sudden geometric expansion (Reprinted from [142] with permission of The Royal Society of Chemistry). **c** Size-based particle sorting has been demonstrated using a gravitational force with hydrodynamic amplification (Reproduced with permission from [124], copyright (2007) American Chemical Society)

expansion in a microfluidic channel to create predictable recirculations to capture CTCs from blood as illustrated on Fig. 4.17b.

4.3.4 *Gravity*

Huh et al. [124] have demonstrated the size-based particle sorting using gravity with hydrodynamic amplification. Briefly, the randomly distributed particle suspension is injected and tightly focused in one dimension by sheath flows into the device parallel to gravity. After flowing around a 90° turn, the stream of particles flows in a divergent channel normal to gravity and are allowed to sediment at a speed depending on their size, their density relative to that of the suspending fluid and the external viscosity (Fig. 4.17c). Sorting of objects with different size is amplified by the asymmetrical geometry of the widening, creating a downward-angled flow. This sedimentation-based microsystem has been successfully demonstrated with polystyrene particles and polydispersed droplets, with typical flow rates of 1 mL/h. However this technique has only been demonstrated with the separation of objects with very large size difference and waits to be applied to cell sorting. Gravity is also employed in centrifugal microfluidic where microsystems developed on disc, also known as Lab-On-a-Disc, are spinned in order to both move the fluid inside the chip and perform separation [145].

4.3.5 Hydrodynamics

Several techniques will be discussed in the next section due to their similarities in terms of operating principles. These label-free passive manipulation techniques rely on the behavior of particles in a laminar flow, at low Reynolds number, where the center of mass of the particle follow the associated fluid streamline.

4.3.5.1 Hydrodynamic Microfluidic Filtration

Hydrodynamic filtration was introduced by Yamada et al. [146] for size-based separation in a microfluidic device. In this technique the sample is injected in a main channel having multiple side branches. Small amount of liquid is repeatedly drained through the side branches, hence concentrating and aligning all the cells along the sidewalls of the main microchannel (Fig. 4.18a). The size difference positions small particles closer to the sidewall than larger ones, allowing the former to be filtered out earlier. The size threshold is determined by the relative flow rate in the main and side channels, which depend on the flow resistance of each branch and may be adjusted by changing the pressure drop applied to each channel. Yamada et al. applied this technique to the selective enrichment of leukocytes from 10-folds diluted blood at 20 μL/min [146] and liver cells in culture medium at $\sim 2 - 3 \times 10^6$ cells/mL at 50 μL/min with a throughput higher than 2×10^5 cells/min [125]. Recently Aoki et al. [147] applied the principle of hydrodynamic filtration to the focusing of 5 μm particles along the centerline of a main microchannel. They used multiple loop side channels to drain the liquid out upstream, before re-injecting it into the main channel from both sides simultaneously, serving as sheath flow to focus the particles.

4.3.5.2 Pinched Flow Fractionation

Pinched flow fractionation (PFF) is a label-free hydrodynamic technique for size-based separation introduced by Yamada et al. [148]. This method relies on the use of a sheath buffer injected such that the particulate content of the sample is pushed against one wall in a pinched segment of the channel. Due to the difference in size, the center of mass of small cells are located on streamlines closer to the wall than that of bigger ones. This difference in lateral position is significantly amplified by the parabolic velocity profile in Poiseuille flow. As the channel widens after the pinched segment, the particles of different size are rapidly dispersed perpendicular to the main flow, allowing them to be collected into distinct outlets (Fig. 4.18b). Unless

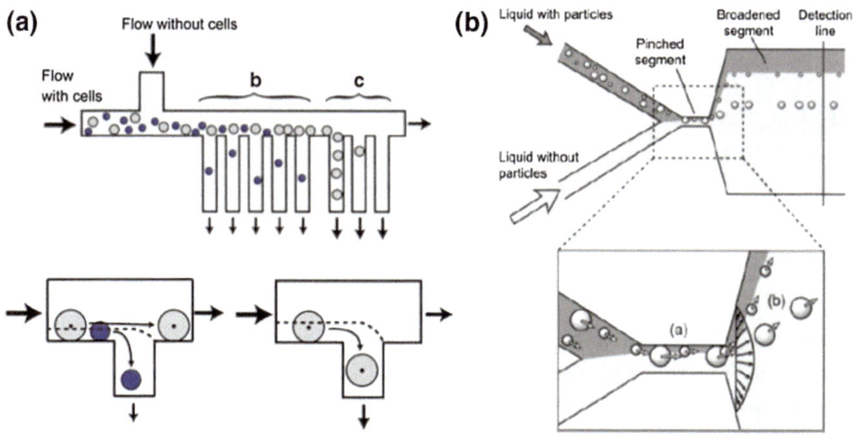

Fig. 4.18 **a** Hydrodynamics filtration (Reproduced from [125], copyright Springer Science + Business Media, LLC (2007) with permission of Springer) and **b** Pinched flow fractionation (Reproduced from [148], copyright (2004) with permission of Springer) are hydrodynamics techniques exploiting the difference in position of the particles center of mass relative to the channel wall according to their size

other separation techniques relying on the device geometry (size and spacing of microstructures, radius of curvature of the channel, shape of micropatterns…), cells of varying sizes can be separated effectively in the same device by adjusting the flow rates of the sample and sheath buffer. Takagi et al. [126] have improved the separation efficiency by playing with the hydrodynamic resistance of the different outlets. They have shown enrichment of RBCs from a 0.3% diluted blood suspension, at a flow rate of 20 μL/h, with a separation efficiency above 90% [126]. Higher flow rates are usually avoided as inertial forces acting on the particles tend to affect their motion, and thus the sorting efficiency. Recently Lu et al. [149, 150] have demonstrated a continuous shape-based separation of spherical and peanut-shaped rigid particles of equivalent spherical diameters. This technique, referred to as elasto-inertial pinched flow fractionation, exploits the dependence of the lift on the shape of a particle suspended in viscoelastic fluids, to amplify the separation induced by the pinched flow fractionation microdevice [149].

4.3.6 *Inertial Forces*

In microchannels, inertia is usually neglected as flow is considered laminar in absence of turbulence (i.e. for Reynolds number Re above 2000). However, for Re typically ranging from 1 to 100, inertial effects can become significant and perturb the particles trajectories, which no longer follow streamlines. These inertial forces can therefore be exploited to precisely manipulate the cell position within these systems. For example, in a circular pipe, initially randomly distributed particles tend to concentrate at an equilibrium position corresponding to 0.6 times the pipe radius [151]. The number and location of the particles equilibrium positions may be tuned by controlling the channel geometry [128, 152]. For more detailed explanations, an interesting review by Di Carlo highlights the physics and the applications of this technique toward high-throughput cell separation [153].

4.3.6.1 Inertial Forces in Straight Channels

Lateral deviation of cells—away from the walls—in a straight microchannel occurs because of the competition between two inertial lift forces: the shear-induced and the wall-induced lift forces [154]. As a result, particles will move to specific equilibrium positions within the channel cross-section, according to their size (Fig. 4.19a). Di Carlo et al. were among the first to investigate the inertial focusing of objects in straight microchannels [128]. They observed that initially uniformly distributed microbeads of 9 μm in diameter were focused along the centers of each of the four walls (see Fig. 4.19b) as the Reynolds number was increased (Re = 90), leading to local particle concentration. Using this principle, Bhagat et al. have demonstrated continuous filtration of particles from a suspension [155]. The same phenomenon has been implemented in microsystem in order to achieve controlled encapsulation of single cells into microdroplets [156] and sheathless focusing of cells for high throughput flow cytometry [157].

Recently, Di Carlo's group [158] studied inertial focusing of particles in non-rectangular cross-section channels, such as semi-circular or triangular shapes. They demonstrated that they can modulate the location and number of focusing positions depending on the cross-section shapes and Reynolds number. By connecting channels with different cross-sectional shapes, they reported the focusing of microparticles to a single stream with 99% purity [158].

4.3.6.2 Inertial Forces in Curved Channels

In a curved channel, the curvature induces an additional force referred to as the Dean drag force. Indeed higher momentum fluid at the center of the channel moves lower momentum fluid near the walls, resulting in two counter-rotating vortices—referred to as the Dean vortices—normal to the main flow [159]. The Dean drag force will concentrate particles on specific flow lines at the expense of others. Di Carlo et al. [128] observed that the four equilibrium positions observed in straight channels, were reduced to two in a symmetric sinusoidal channel and to a single equilibrium position in an asymmetric channel (Fig. 4.19b). They demonstrated that cells could be differentially focused and sorted based on size under laminar flow, with a high throughput (1.5 mL/min), using a serpentine pattern (Fig. 4.19c). Later, the same group demonstrated the focusing and sorting of cells using a spiral channel [160]. Wu et al. demonstrated the isolation of bacteria from diluted blood with a

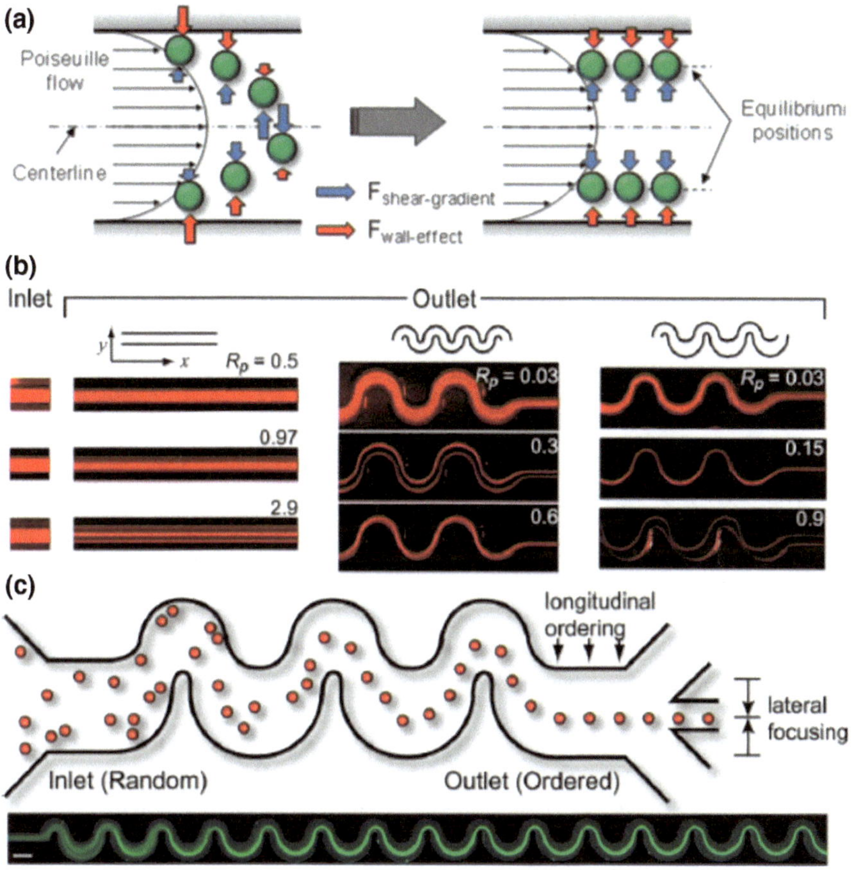

Fig. 4.19 **a** Inertial forces can be exploited to manipulate the cell position within the channel (reproduced from [154] with permission of The Royal Society of Chemistry). **b** In a straight channel, particles tend to be focused along the centers of each of the four walls, whereas in a curved channel, the number of equilibrium positions is reduced to two. **c** Di Carlo's group demonstrated the focusing of particles using an asymmetric sinusoidal channel. R_p represents the Reynolds number associated with the particle and includes parameters describing both the particle and the channel through which it is translating. (Reproduced with permission from [128], copyright (2007) National Academy of Sciences, USA)

purity above 99% and a recovery rate of 62%, at a flow rate of 15 μL/min [127]. Dean Flow was used to separate two types of neuronal cells with average diameters of 8 and 15 μm respectively, with an efficiency of 80% and a throughput of the order of 10^6 cells/min [161]. More recently, Bhagat et al. managed to isolate 90% of tumor cells in a 20% hematocrit blood sample spiked with MCF7 breast cancer cells, at 4 mL/h maximum flow rate [162]. The use of a spiral channel presenting a trapezoidal cross section has been demonstrated to further improve the performance of such device to focus and separate cells [163, 164].

Finally, as inertia-based focusing methods exploit the particularities of flow at high flow rate (i.e. large Re), it may offer a high throughput, which makes it suitable for high-speed applications such as flow cytometry. On the other hand, this technique may be limited to only dilute samples, the efficiency being deteriorated by cell-cell interactions.

4.3.7 Biomimetic

The concept of biomimetic techniques for cell manipulation consists in biologically inspired designs exploiting cell behavior observed in vivo, most of the time in blood vessels.

4.3.7.1 Fahraeus Effect

One of the most famous hemodynamic phenomenon of the microcirculation is the Fahraeus effect [165, 166]. For instance, in veins with a diameter smaller than 300 μm, RBCs tend to drift away from the high shear regions adjacent to the vessel wall, resulting in their accumulation at the center of the blood vessel. Thus leaving a "cell-free" plasma layer adjacent to the vessel wall [167]. Faivre et al. [129] used a geometric constriction to enhance the size of the "cell-free" layer (Fig. 4.20a). They report the recovery of 24% of the initial blood plasma from a 16% hematocrit blood suspension (dilution 1:3), at a working flow rate of 200 μL/h [129]. Key parameters of this approach include the sample viscosity—which can be tuned by dilution—the flow rate, the cells volume and deformability, as well as the shape and size of the geometric constriction. Later, Sollier et al. [168] has improved the performances of such devices by adding geometrical singularities such as ear-cavities or corner-edges. They reached a recovery of 10.7% of the plasma, at a flow rate of 100 μL/min, from a blood sample diluted at 1:20. The extracted plasma presented a purity similar to that of centrifugation methods [168].

4.3.7.2 Leukocyte Margination

As the large number of RBCs migrate towards the center of the blood vessel, mechanical collisions between RBCs and WBCs—much larger and stiffer than RBCs—displace the leucocytes near the walls. This phenomen is referred to as the leukocyte margination. Shevkoplyas et al. [169] demonstrated the possibility to sort WBCs from blood by exploiting leukocyte margination. Briefly, whole blood is flown through multiple symmetric and asymmetric bifurcations (Fig. 4.20b). Most of the RBCs flown in the wider outlet, while marginalized leukocytes entered the narrower outlet, leading to nearly 10-fold enrichment of WBCs. Jaggi et al. [172] increased the throughput (in the range of μL/min) by using higher microchannels.

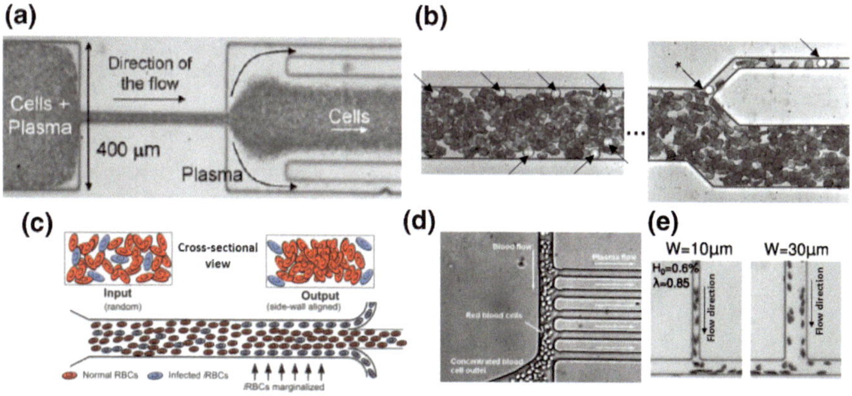

Fig. 4.20 Biomometic approaches include **a** the Fahraeus effect (Reprinted from [129], copyright (2006), with permission from IOS Press), **b** the leukocyte margination (Reprinted with permission from [169], copyright (2005) American Chemical Society) which has also been applied to the isolation of **c** mechanically impaired malaria-infected RBCs (Reproduced from [170] with permission of the Royal Society of Chemistry), **d** the bifurcation law (Reproduced from [131] with permission of the Royal Society of Chemistry) and **e** its inverse in the case of very low hematocrit (Reprinted from [171], copyright (2016), with permission from Elsevier)

Recently, Hou et al. [170] took advantage of the greater rigidity of malaria-infected RBCs (iRBCs) to perform separation of parasited cells in microsystem based on the same principle as illustrated Fig. 4.20c. They claim nearly 75% of separation efficiency for early stage iRBCs, where deformability is only slightly altered compare to healthy RBCs, and more than 90% for late stage iRBCs, with flow rates of 5 µL/min.

Finally, an automated biomimetic microfluidic device made for sorting nucleated cells population from unprocessed whole blood sample was presented [130]. This device is composed of successive triangular channel expansions that mimic the post-capillaries venules expansions in human body. The authors report an extraction of 90% of the nucleated cells with 45-fold enrichment in concentration, at the rate of 5 nL/s [130]. The physics underlying the leukocyte margination, i.e. the "crowding" effect of RBCs on WBCs, favors high RBCs concentration. This technique is thus compatible with the use of whole blood samples, hence avoiding additional sample preparation steps.

4.3.7.3 Bifurcation Law

When a microchannel splits into two channels, a particle will follow different streamlines according to its initial position. In the case of a bifurcation, the flow velocity difference between the two branches decides of the particle trajectory. For example, RBCs flowing at a microchannel split will predominantly choose the daughter channel with the higher flow rate. This effect is referred to as the

Zweifach-Fung effect—also known as the bifurcation law—and has been exploited by Yang et al. [131] in a microsystem to achieve plasma extraction from a blood sample (Fig. 4.20d). Very recently, Shen et al. [171] shown that the inverse of the Zweifach-Fung effect can occur if the hematocrit is small enough. Indeed, in such conditions, they observed that RBCs flow preferentially in the daughter channel with the lower flow rate (Fig. 4.20e), leading to hematocrit in that branch even greater than in the parent vessel [171]. They explain this result by a complex organization of the RBCs prior to the bifurcation and they highlight the extreme dependence of RBCs transport on channel geometry and cell mechanical properties.

4.3.8 Conclusion

In this review we introduced different label-free passive manipulation techniques, which characteristics are gathered for comparison in Table 4.1.

Here the term passive refers to microfluidic devices relying predominantly on channel geometry and/or associated hydrodynamic forces, whereas active describes microsystems using an external force field to function. We chose not to discuss here immunological techniques, where antibody-antigen associations are exploited to achieve cell capture, separation or labelling. The extensive literature in the field underlines the strong interest drawn by label-free passive techniques to manipulate, concentrate, trap, isolate or sort cells using these approaches. Many of the methods presented are at the proof-of-concept stage and few have demonstrated their performances on actual clinical samples. Although they present the advantage on active techniques to be easily manufacturable and to provide cells with minimum residence time under harmful conditions, they still need to overcome some limitations such as low selectivity and low throughput. Advances on parallelization and other scale-up approaches are required especially when rare cells (e.g. CTCs in blood) are involved, as CTCs levels can be as low as 1 cell/mL in early stages. Another drawback of these techniques relying on microscale fluid dynamics, is the need to integrate precise and autonomous flow handling solutions. Finally, the sample preparation procedure often associated may impact the performances of the manipulation due to user variations. Therefore building a microfluidic device that require minimal sample preparation and reducing the dimensions of the apparatus associated with the chip remains a major challenge to be commercially viable.

The selection of the adapted method will depend on the properties of the target cells regarding those of the rest of the sample, and on the needs of the application (efficiency, reliability, simplicity, throughput…). Current microfluidics technology has already demonstrated that it can be used to separate and analyze diseased or rare cells from a sample, more rapidly, with more sensitivity and in a more controlled way. Ultimately, the goal is to develop a portable diagnostic tool—the so-called Lab-On-a-Chip—integrating a whole chain of complex processes needed to perform the desired analysis. Such devices will have tremendous applications in cell biology research, diagnosis, drug screening and personalized medicine.

4.4 Mechanics for Biodevices

4.4.1 Introduction

Biological cell mechanical properties are mainly dictated by the characteristics of its cytoskeleton, which is an elaborate network of polymer fibers [173]. One of the primary applications of the mechanical phenotyping is the identification of pathological cells. Indeed various pathologies and modifications of cell states are reported to influence cell mechanical properties. For example, variation of Red Blood Cells (RBCs or erythrocytes) deformability is associated with malaria [174] and sickle cell anemia [175], White Blood Cells (WBCs) deformability varies in sepsis [176] and acute respiratory distress syndrome [177], and increase deformability of metastatic cancer cells [178] and drop of stiffness during stem cell differentiation [179] have been observed. A better understanding of cell deformability and its interactions with physical environment may imply enormous developments in disease diagnostics, therapeutics and drug screening assays. Although conventional techniques allowing the quantification of cellular mechanical properties, such as atomic force microscopy (AFM), micropipette aspiration and optical tweezers [180] are well-established, they require sophisticated equipments and skilled personal and most importantly are low throughput (of the order of several tens of cells per day). Due to a match between cellular length scales and typical sizes accessible by microfabrication methods, microfluidic technologies propose attractive engineered microenvironments for the study of cellular mechanics compatible with high throughput. The present review begins with an overview of the sub-cellular organization contributing to overall deformability and mechanical response of cells. We then describe recent advances in microfluidic assays for single cell biomechanical characterization. Finally, limitations of current systems and ways in which they can be improved are discussed.

4.4.2 Biomechanics of Cells

Cell mechanical behavior is inextricably related to their intra-cellular composition and architecture, particularly the cytoskeleton. The cytoskeleton is a complex 3D network of biopolymers which primarily defines cell shape and mechanical response to physical stresses and plays a central role in mechanotransduction, mitosis and migration [173]. It is mainly composed of actin microfilaments, intermediate filaments and microtubules. Actin microfilaments are linear polymers that prevent overall cell deformation in response to external stimuli and can polymerize into helicoidal filaments and depolimerize within minutes to facilitate cell motion. Intermediate filaments form an intracellular intricated network connecting the nucleus to the plasma membrane and other organelles. They provide cell structural integrity and resist tension at strain levels where actin networks do not retain their structural integrity [180]. Microtubules consist in tubulin polymers that contribute to the transport of intra-cellular components within the cell and resist compression. Both the composition and the organization of these combined elements contribute to the cell mechanical phenotype [173]. The interaction of a cell with its environment can also dramatically influence its mechanical properties. Indeed when spread on a surface, adherent cells exhibit a different cytoskeleton conformation than when in suspension, leading to differences in mechanical response to solicitations [181]. In the case of adherent cells, the polymerization/depolymerization of cytoskeleton components in response to both intracellular and extracellular stimuli may significantly alter the global cell mechanical signature. This is especially noticeable during biological processes such as cell division, or cell-cell contacts where extensive cytoskeletal remodeling occurs.

The cell membrane contributes also to the cell mechanical properties due to its connections to the cytoskeleton. It passes strains and stresses on the intracellular tensile and compressive components. For small deformations, membrane viscoelasticity can contribute to overall cell mechanical behavior. It is especially relevant when cells are in suspension, which is the case in flow-based approaches, where viscoelastic characteristics of the membrane plays a role in the cell distribution within the channel [182] (see Sect. 4.3). The presence of the nucleus and other organelles can contribute locally to the mechanical properties of the cell according to their physical size. High-resolution mechanical mapping of flattened cells using AFM presents different rigidity, depending if the measurement has been performed over the nucleus or over cytoplasmic and cytoskeletal areas [183].

The RBC membrane possesses a unique structure responsible of their remarkable mechanical properties (deformability, elasticity, shear resistance, etc.…) allowing them to go through the small capillaries of the microcirculation. RBCs have an elastic 2D mesh-like spectrin cytoskeleton anchored to the internal side of a lipid bilayer [184]. The membrane of the erythrocyte encloses a cytoplasm made of hemoglobin—allowing the transportation of oxygen through the organism—which viscosity at 25 °C is 10 mPa.s. Together with the cytoplasmic fluid and the absence of nucleus and other organelles, the membrane controls the overall deformability of the cell. Under normal conditions, the membrane deforms at constant area and exhibits both elastic and viscous behaviors. The extension of the membrane is governed by the lipid bilayer, which tends to resist area expansion. The shear resistance of the membrane is directly linked to the density of spectrin and thus, to the cytoskeleton [185]. RBCs have been extensively studied in the past fifty years in order to characterize mechanical phenotypes associated with both healthy and pathological states.

4.4.3 Microfluidic Mechanical Characterization of Cells

A wide variety of experimental biophysical assays have been used to probe deformability of living cells. For example AFM and other indentation techniques allow monitoring the cell deformation in response to a force applied through physical compression or indentation of a probe and can be used to extract cell elastic and viscoelastic properties. In micropipette aspiration, either a portion or a whole cell is aspired into the opening of a micropipette while monitoring the extension of the membrane, leading to the evaluation of the viscoelastic properties locally or at the whole-cell level, respectively. Unlike AFM and micropipette aspiration, optical tweezers are attractive mechanical assays because they do not necessarily require mechanical contact with the target cells. Optical tweezers uses a highly focused laser beam to create a 3D light gradient exerting a force on the cell, relying on the dielectric contrast with the surrounding media. Although these conventional techniques are well-established, they are extremely low throughput (several tens of cells per day typically), which makes it unadapted to diagnostics and biomedical applications, where millions of cells per tens of minutes need to be processed. This section highlights emerging microfluidic techniques addressing this throughput challenge.

4.4.3.1 Optical Stretcher

A microfluidic version of the optical tweezers has been developed by Guck et al. [186] and is referred to as optical strecher. It consists in a two-beam laser trap optimized to serially deform isolated cells flowing through a microfluidic channel (Fig. 4.21a). The cells are being individually stretched at forces on the order of 200–500 pN, which is still not sufficient to induce large deformation. They found that deformability, probed using optical forces, is sensitive enough to detect changes between non-malignant (MCF-10), cancerous (MFC-7) and even metastatic (modMCF-7) human breast epithelial cells [186]. Mechanical phenotyping of cells without any modifications or contact was demonstrated with relatively high throughput (1 cell/min) despite the fact that we are still far from the intended throughput. The same group demonstrated the measurement of the elasticity of malaria infected RBCs (iRBCs) using the optical stretcher [187]. The device combines the selectivity and sensitivity of single-cell elasticity measurements with a throughput that is higher than conventional single-cell techniques. One drawback of this technique is the exposure of cells to the laser beam during stretching.

4.4.3.2 Electro-Deformation

Transient deformation of cells in a high-frequency electric field (Fig. 4.21b) is reported to allow quantitative measurements of the shear elastic modulus and viscosity of plasma membranes of healthy RBCs (hRBCs) [188]. It relies on the principles detailed in Chap. 6. Using analytical models or numerical simulations, estimation of healthy RBCs (hRBCs) shear modulus $\mu = 6.1 \times 10^{-6}$ N/m and membrane viscosity $\eta_m = 3.4 \times 10^{-7}$ N · s/m are obtained and present a good agreement with the micropipette results of Evans et al. [190]. Electro-deformation

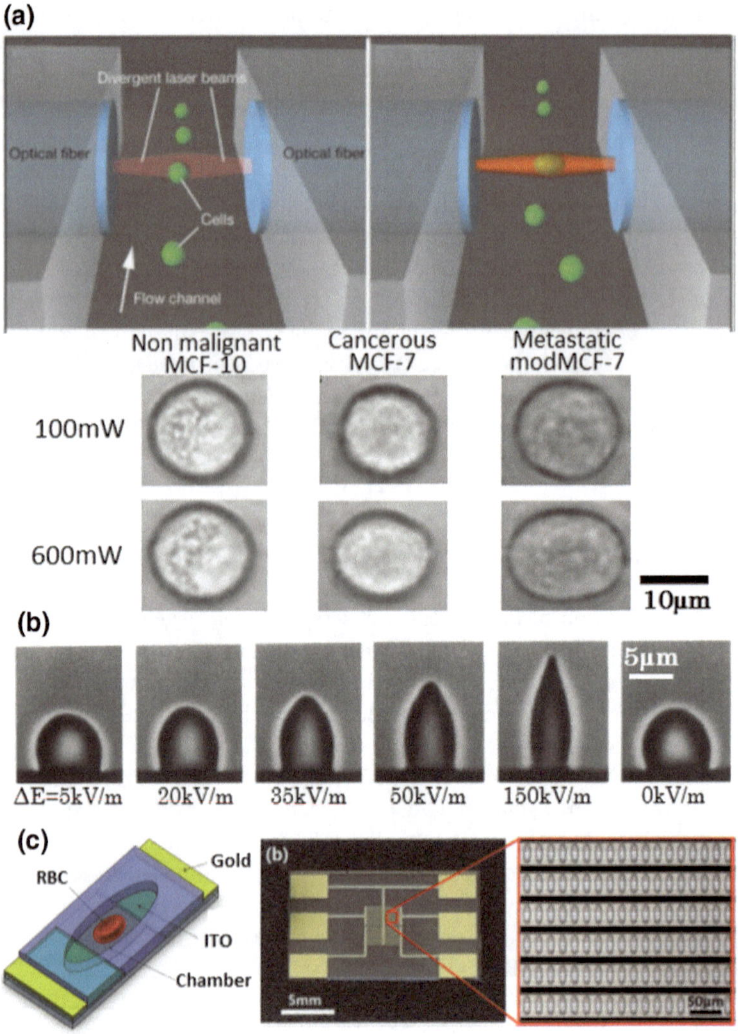

Fig. 4.21 **a** Principle of the optical stretcher developed by Guck et al. Flowing cells get stretched when intercepting the laser beam. Correlation between elongation and metastatic characteristic of breast cell lines has been observed. (Reproduced from [186], copyright (2005) by the Biophysical society, with permission from Elsevier). **b** Electro-deformation of a RBC; cell elongation increases with the applied electric field (reproduced from [188], with permission from Elsevier). **c** Schematic view and actual pictures of the single-cell microchamber array chip allowing mechanical characterization with improved throughput (Reproduced from [189] with the permission of AIP Publishing)

has also been used to quantify the deformability of individual eukaryotic cells [191, 192]. Recently, in order to improve the throughput, a single-cell microchamber array chip was proposed to trap and deform individual RBCs [189] (Fig. 4.21c). One drawback of electro-deformation is the necessity to suspend the cells in a buffer of low conductivity, which add sample preparation steps prior to analysis and lead to rapid aging of the cells [188].

4.4.3.3 Compression Experiments

The discrimination of two different cell types (neutrophiles and fibroblasts) known to have different cytoskeletal structures has been reported using microfluidic compression experiments [193]. The method relies on the observation of the deformation resultant of the cell compression by an actuated flexible membrane (Fig. 4.22a). The relaxation time—the characteristic time necessary for the cells to recover their equilibrium shape as compression is stopped—is used to discriminate two cell lines. Changes in the mechanical response due to disruption of the cellular actin network using Cytochalasin D—an inhibitor of actin polymerization—is also highlighted [193]. With a similar approach, Kim et al. [194] used difference in the bulge generation associated with cell compression, to identify cancerous (MCF-7) from non-malignant breast cells (MCF-10A). Indeed upon cell compression by a flexible membrane and beyond a given strain, cellular membrane expands and small bulges form on the peripheral cell membrane. They explain these bulge formations where cytoskeleton detaches from the membrane bilayer, by a lower amount of actin present in cancerous cells than in normal ones [194].

Compression experiments with micro-electro-mechanical systems (MEMS), to probe living cells stiffness, have also been reported in the literature [196, 197]. They usually rely on the controlled displacement of a moving actuator compressing the cell against a fixed part. For example, Barazani et al. [197] developed a MEMS relying on an AC driven electrothermal micro-actuator to compress single yeast cells against a reference back spring. The motion of the reference spring divided by the cell deformation provides the cell stiffness relative to the reference spring constant. They report an average stiffness of the yeast cells of 9.3 ± 3.1 Nm^{-1} and an average force necessary to rupture their membrane of 0.47 ± 0.1 μN. Gnerlich et al. [196] have worked on a similar MEMS and they have reported the compression of NIH 3T3 mouse fibroblasts by moving the actuator by 4 μm for 90 s using a calibrated piezo-driver. They have measured the associated forces decreased during the compression from 38 to 5 nN, which is well described by a viscoelastic cell model. Finally, we can cite the work from Lafitte et al. [198] presenting the utilization of silicon nano-tweezers (SNT) to probe the mechanical properties of biological samples (DNA and cells). Typically, one tip of the tweezer can be alternatively open and closed to compress the cell and the other one is fixed and senses the absolute force applied. The mechanical response of the cells can be deduced in real-time from the changes of the resonance frequency of the SNT, knowing that the resonance response changes according to the added rigidity and

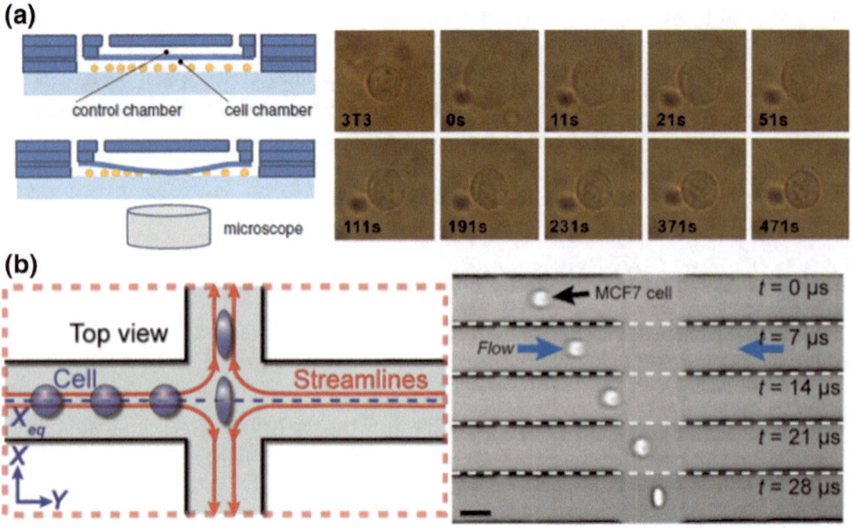

Fig. 4.22 **a** Schematic of the principle of compression experiements. Sequence of relaxation of a 3T3 cell after compression is stopped (Reproduced from [193], copyright The authors (2010), with permission of Springer). **b** Schematic view and sequence of deformation of a focused cell being stretched when entering the extensional flow region (Reproduced with permission from [195])

viscosity of the cell. However, the use of MEMS to characterize the mechanical phenotype of cells over a large cell population remains to be demonstrated.

4.4.3.4 Fluid-Based Deformability Assays

Perhaps the simplest method to measure cellular mechanical properties while achieving high-throughput is by monitoring cell deformations as they flow through microfluidic channels. Cellular deformability can be challenged either by the walls of the device or by the associated shear stress without any contact, thus limiting the impact of cell membrane adhesion with channel boundaries. Microfluidic deformability-based passive separation techniques can also be used to characterize mechanical phenotype of cells. Such techniques have been detailed in Sect. 4.3 and therefore will not be addressed again here.

Fluid stretching. RBCs are highly deformable and their deformability can therefore be challenged using shear stress generated by microfluidic channels wider than the typical size of the cell. Forsyth et al. [199] reported three different dynamic behaviors of RBCs in a pressure-driven flow, according to the shear rate: stretching, tumbling and recoiling. Upon chemical treatment with glutaraldehyde, which is known to rigidify both cell membrane and cytoplasm, RBCs presented lower deformation and increased occurrence of tumbling, which is a motion typical of rigid bodies. Lee et al. [200] reported the monitoring of RBC deformation in a hyperbolic converging microchannel. They showed that the extensional flow is more efficient to induce cell deformation than a shear flow, as RBCs presented deformation index of 0.51 and 0.29 at 3 Pa respectively. Such approach was used to detect the decrease in deformability of RBCs following heat treatment.

Gosset et al. [195] proposed an automated microfluidic tool using inertial focusing to uniformly deliver cells to a stretching extensional flow where cells are deformed at high strain rates (Fig. 4.22b). They characterized the deformability of populations of leukocytes and malignant cells in pleural fluid samples. They predicted disease state in patients with cancer and immune activation with a sensitivity of 91% and a specificity of 86% at high throughput (2000 cells/s). Both lymphocyte activation and stem cell pluripotency were found to be associated with increased deformability [195].

Constriction channels. Constriction channels with dimensions smaller than the typical size of a single cell provide a simple and powerful tool to probe cell deformability. Several types of geometric restrictions have been reported in the literature to quantify cell mechanical phenotype, using however the same type of readout (transit time, deformation index and relaxation time). For example Selby et al. [201] reported the use of a well-defined geometrical constriction ranging from 8 to 2 μm in width and controlled pressure gradients in order to discriminate healthy and malaria-iRBCs. They report that as parasites develop within the host cell, the increase of RBCs rigidity prevent their passage through smaller constrictions. Indeed Fig. 4.23a illustrates the blockage of a 6 μm wide channel by schizontes-iRBCs, whereas a hRBC (in red) manages to deform past the barricade. Using the same approach, Suresh [180] probed the mechanical properties of human pancreatic cells flowing in a microfluidic channel at 0.5 μL/min (Fig. 4.23b). By studying transit velocity measurements, shape change and relaxation at the exit of the narrowing, they connected changes in cell deformability and motility [180]. The distinction between non-malignant and cancerous cells flowing in similar geometries was demonstrated using entry time [202, 203]—defined as the time necessary for the cell to deform and enter completely into the microchannel—and deformation index measurements [203].

Flow of cells in 16 parallel channels implementing multiple constrictions has also been reported (Fig. 4.23c) [204]. The study revealed that transit time scales indeed with cell size and that simultaneous transit events occurring across neighboring constrictions can influence transit time [204]. A comparable geometry was used to quantify the effect of taxol—an anti-cancer drug—on the transit time of tumor cells [207] through the serial constrictions. The results show that when travelling through the first constriction, cells treated with taxol undergo longer transit times than untreated cells, however for subsequent transits the difference was smaller between the two groups [207]. Rosenbluth et al. [208] measured single-cell transit times of white blood cell populations passing through in vitro capillary networks, allowing to observe several channels in parallel. They demonstrate clinical relevance of their geometry in sepsis and leukostasis, showing increased transit time with patient samples relative to control ones. Their results also revealed that chemotherapy reduce leukemia cell deformability [209]. Herricks et al. [210] reported that ring-iRBCs get blocked upstream whereas hRBCs reached positions further downstream in converging channels. The dynamic behavior of RBCs flowing through arrays of 15 μm pillars, either slowly through 5–2 μm wide gaps or rapidly along 10 μm wide channels has been studied [211]. They observed that mechanically impaired RBCs—either due to heat treatment or issued from malaria-iRBCs or from patients with hereditary spherocytosis—tend to accumulate in narrow slits more frequently than normal RBCs, hence demonstrating their increased rigidity. The same approach was used by Preira et al. [205] to explore the effect of actin organization and miosin II on the deformability of leukocytes using pore-based deformability measurement (Fig. 4.23d). They highlighted that cell stiffness

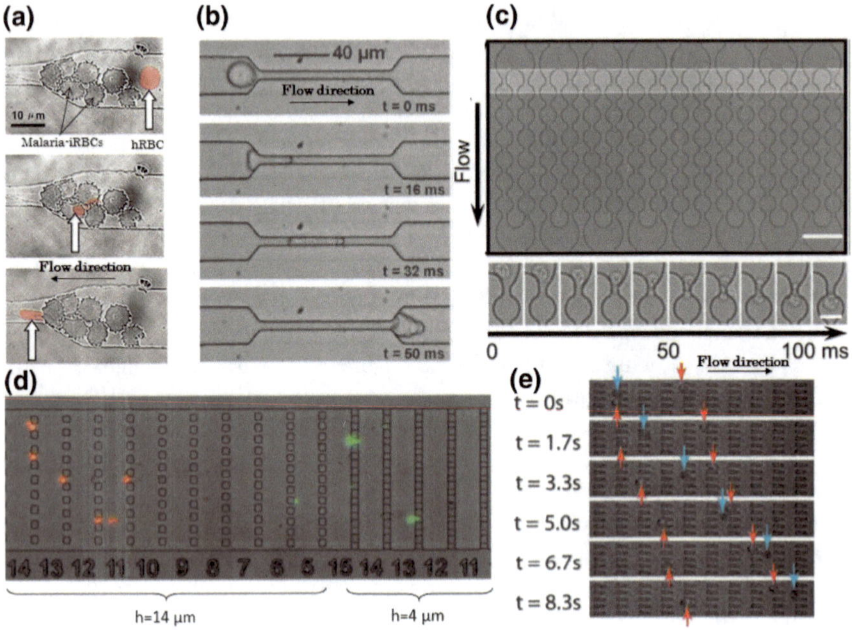

Fig. 4.23 **a** Video-microscopic images showing malaria-iRBCs being stuck at the entry of a geometric narrowing, whereas a helathy cell (in red) managed to deform past the blockage (Reproduced with permission from [201], copyright (2003) National Academy of Sciences, U.S. A.). **b** Sequence of deformation of a human pancreatic tumor cell flowing through a constriction (Reproduced from [180], copyright (2007) with permission of Elsevier). **c** Image of the multi-constriction array, which consist in 16 channels each implementing a series of 5 constrictions. Sequence of deformation of a single HL-60 cell during flow through a 5 μm constriction; transit time of 80 ms is measured (Reproduced from [204] with permission of the Royal Society of Chemistry). **d** Illustration of the deformability-based separation. Stiff THP-1 cells (treated by glutaraldehyde) labelled in red are trapped in larger pores than compliant THP-1 cells (treated by Latrunculin-A) labelled in green (Reproduced from [205] with permission of the Royal Society of Chemistry). **e** Time-lapse images of ring stage-iRBCs (red arrows) and hRBCs (blue arrows) flowing in the device; the hRBC travels faster than the iRBC (Reproduced from [206] with permission of the Royal Society of Chemistry)

depends mainly on actin organization, whereas shape is related to miosin II. Similarily, Bow et al. [206] have monitored dynamic mechanical responses of 10^3–10^4 individual RBCs using converging or diverging obstacles (Fig. 4.23e). They have experimentally demonstrated that differences in RBC deformability, demonstrated by differences in transit velocity, are enhanced in geometries with sharper corners, for a given pressure difference. The influence of an anti-malaria drug (Artesunate) on the dynamic deformability of ring-infected iRBCs was also evaluated [212]. After Artesunate treatment, they observed a 50% decrease in the transit velocity of iRBCs whereas only 10% in speed reduction is observed for hRBCs.

A limitation to constriction channels however is the influence of size and adhesiveness on cells behavior, which makes it difficult to emphasize the role of deformability alone. Indeed, recent studies trying to take the cell size into account have been reported in the literature [213]. They showed that cell size significantly influences travel times and that for a given volume, stiffer cells have longer transit times than compliant ones. However, efforts still need to be made to characterize adhesion and/or friction between channel walls and cell membrane, for example by varying the walls coating. Nyberg et al. [204] have reported variability in transit time measurements as the PDMS device age after plasma treatment; they attribute such discrepancies to modifications in channel surface properties. Another drawback of this approach is the need to adapt precisely the size of the constriction to dimensions of cells of interest to avoid clogging or deform significantly the cells.

Pressure drop measurements. Abkarian et al. [214, 215] proposed a high-speed microfluidic approach to measure dynamical pressure-drop variations associated with the flow of individual cells in confined geometries. The chip is composed of two identical channels (picture in Fig. 4.24a); the cells suspension is injected in the upper channel whereas a solution of contrast agent is injected in the lower one. The two channels merge on the right side of the chip and the contrast agent allow the visualization of the co-flow virtual interface. In absence of cells flowing in the upper channel, the virtual interface is centered, hence traducing the equilibrium between the pressures in the two channels. When a deformable cell enters the upper channel, an additional hydrodynamic resistance is added to the channel resistance, leading to changes in the local pressure profile, thus inducing a displacement of the virtual interface. Thanks to an appropriate calibration, the deflation of the interface can be converted into a pressure drop measurement. The authors have shown that the amplitude of pressure drop depends on both the volume and the deformability (graph in Fig. 4.24a) of the cells under study. Indeed, pressure-drop measurements are able to discriminate WBCs and RBCS, as well as healthy and chemically rigidified RBCs [214]. Khan et al. [203] applied this approach to the mechanical characterization of healthy and cancerous brain cells. The two cell lines were differentiated using entry time and to a little extend pressure drops, rather than on speeds and elongations. They observed that brain tumor cells have a longer entry time, thus suggesting that the well-spread idea that cancer cells are more deformable than benign cells may not apply to brain cancer cells [203].

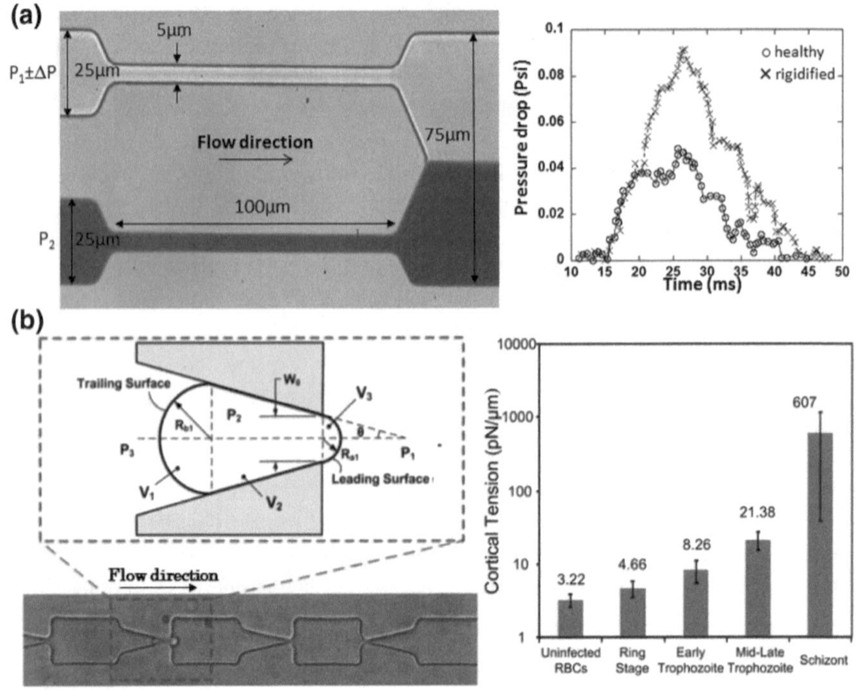

Fig. 4.24 **a** Pressure drop measurements rely on the geometry of the device. The monitoring of the displacement of the virtual interface in the co-flow region of the chip is correlated to the pressure drop associated with the flow of a cell in the upper channel. Pressure drop versus time for (×) a hRBC and (o) a rigidified RBC after 0.001% glutaraldehyde treatment. **b** Microfluidic adaptation of the micropipette aspiration technique. RBCs are deformed when passing the funnel constriction and the associated geometric model. Measured cortical tension of hRBCs and malaria-iRBCs in various stages of infection (reproduced from [216] with permission of the Royal Society of Chemistry)

Microfluidic micropipette aspiration. Guo et al. [216] proposed an adaptation of the conventional micropipette aspiration technique into microfluidic format. Figure 4.24b presents the geometry of the microfluidic micropipette aspiration device implementing multiple funnel-shaped constrictions with openings ranging from 5 down to 1 μm. As a cell is flowing in the device, it gets trapped at the nozzle of a funnel, the pressure difference is manually increased until the RBC is allowed to squeeze past the restriction. The threshold pressure is used to determine the intrinsic stiffness of each cell. Using the same mathematical model as in conventional pipette aspiration technique, i.e. considering the RBC as a liquid-drop with constant volume, they managed to extract the cell cortical tension. Malaria-iRBCs by early (ring) through late (schizont) developmental stages of the parasite were shown to be 1.5 to 200-fold stiffer than uninfected cells (graph in Fig. 4.24b) as already reported in the literature [217]. However, the rectangular cross-section of the geometry can rise concerns about the validity of applying conventional micropipette models to retrieve cellular mechanical properties.

4.4.4 Conclusion

Although they present many advantages such as possible automation, absence of labeling and compatibility with high throughput, each microfluidic mechanical characterization technique has some drawbacks. In optical stretching, cells are exposed to the power of the laser which may damage them and electro-deformation require a change in suspending media; such low conductivity buffer may accelerate aging of the cells. Multi-steps soft lithography processes are needed to implement the flexible membrane in compression experiments. A disadvantage of constriction channels is the persistent possibility of device clogging. Indeed, as the section of the constriction is smaller than the typical size of the cells, one need to adapt carefully the width of the constriction. A group proposed recently an improvement with the development of a constriction channel with tunable section [218, 219]. One of the walls of the constriction was replaced by a thin membrane, which deflection is controlled via external pressure. In case of channel obstruction by larger particles or cellular aggregates, the increase of the applied pressure causes the augmentation of the constriction cross-section through deflection of the membrane, thus removing the obstruction.

Up to recently, the various techniques presented above presented the same limitation in terms of throughput because of the use of high-speed video-microscopy for monitoring cellular mechanical responses such as the deformation index. Indeed, even when using automated post-processing routines, the huge amount of image data to analyze imposes computational power and remains time consuming. In replacement of high-speed camera, electrical impedance measurements has been reported in the literature to detect RBCs [220] and cancer cells [213] deformation in constriction channels, with a throughput of 100 cells/s. In 2015, Jochen Guck's group has proposed an automated real-time deformability cytometer (RTDT) able to perform continuous cell mechanical characterization of large populations (above 100 000 cells) with throughput greater than 100 cells/s [221, 222]. Their system consists in the real-time measurement of the deformation associated with the passage of cells into a constriction channel. They report that RT-DC is sensitive to cytoskeletal alterations and can distinguish cell-cycle phases, track stem cell differentiation into distinct lineages and identify cell populations in whole blood by their mechanical fingerprints. More recently, Deng and Chung [223] proposed a similar device allowing nearly real-time analysis of transient deformation of cells hitting a wall at a T-junction. Their automated process allow the mechanical characterization of cells at a throughput of 2000 cells/s. Finally, the last challenge that still need to be addressed is the translation of raw parameters such as entry time or deformation index into cellular intrinsic mechanical characteristics such as Young's modulus or cortical tension. Indeed, intrinsic deformability parameters are independent of the setup (viscosity, flow speed, etc.…) and allow the comparison of data obtained with various techniques and by various research groups. Therefore there is a strong need for theoretical framework and mechanical modeling of the cellular response in order to extract intrinsic mechanical properties.

References

1. T. Vilkner, D. Janasek, A. Manz, Micro total analysis systems. Recent developments. Anal. Chem. **76**, 3373–3386 (2004). Accessed 01 June 2004
2. H.A. Stone, A.D. Stroock, A. Ajdari, Engineering flows in small devices. Ann. Rev. Fluid Mech. **36**, 381–411 (2004)
3. A. Boisen, S. Dohn, S.S. Keller, S. Schmid, M. Tenje, Cantilever-like micromechanical sensors. Rep. Progr. Phys. **74**, 036101 (2011). Accessed 01 Mach 2011
4. G.T. Kovacs, *Micromachined Transducers Sourcebook*, 1st edn. (McGraw-Hill Science/Engineering/Math, Boston, 1998)
5. K.W. Oh, C. Park, K. Namkoong, A world-to-chip microfluidic interconnection technology with dual functions of sample injection and sealing for a multichamber micro PCR chip, in *18th IEEE International Conference on Micro Electro Mechanical Systems, MEMS 2005* (2005), pp. 714–717
6. K.I.T.H.K. Nkashima, 10-way micro switching valve chip for multi-directional flow control (2003)
7. M.T. Taylor, P. Nguyen, J. Ching, K.E. Petersen, Simulation of microfluidic pumping in a genomic DNA blood-processing cassette. J. Micromech. Microeng. **13**, 201 (2003)
8. M.A. Unger, H.-P. Chou, T. Thorsen, A. Scherer, S.R. Quake, Monolithic microfabricated valves and pumps by multilayer soft lithography. Science, **288**, 113–116 (2000). Accessed 07 April 2000
9. V. Studer, G. Hang, A. Pandolfi, M. Ortiz, W.F. Anderson, S.R. Quake, Scaling properties of a low-actuation pressure microfluidic valve. J. Appl. Phys. **95**, 393–398 (2004) 01 Jan 2004
10. A. Pandolfi, M. Ortiz, Improved design of low-pressure fluidic microvalves. J. Micromech. Microeng. **17**, 1487 (2007)
11. K.W. Oh, C.H. Ahn, A review of microvalves. J. Micromech. Microeng. **16**, R13 (2006)
12. Mechanics over micro and nano scales (Springer, New York, 2011)
13. B. Bae, N. Kim, H. Kee, S.-H. Kim, Y. Lee, S. Lee et al. Feasibility test of an electromagnetically driven valve actuator for glaucoma treatment. J. Microelectromech. Syst. **11**, 344–354 (2002)
14. K.W. Oh, A. Han, S. Bhansali, C.H. Ahn, A low-temperature bonding technique using spin-on fluorocarbon polymers to assemble microsystems. J. Micromech. Microeng. **12**, 187 (2002)
15. S.C. Terry, J.H. Jerman, J.B. Angell, A gas chromatographic air analyzer fabricated on a silicon wafer. IEEE Trans. Electron Devices **26**, 1880–1886 (1979)
16. K. Sato, M. Shikida, An electrostatically actuated gas valve with an S-shaped film element. J. Micromech. Microeng. **4**, 205 (1994)
17. M. Patrascu, J. Gonzalo-Ruiz, M. Goedbloed, S.H. Brongersma, M. Crego-Calama, Flexible, electrostatic microfluidic actuators based on thin film fabrication. Sens. Actuat. A Phys. **186** 249–256 (2012)
18. W. van der Wijngaart, H. Ask, P. Enoksson, G. Stemme, A high-stroke, high-pressure electrostatic actuator for valve applications. Sens. Actuat. A Phys. 100, 264–271 (2002). Accessed 01 Sep 2002

19. S. Messner, J. Schaible, H. Sandmaier, and R. Zengerle, Three-way silicon microvalve for pneumatic applications with electrostatic actuation principle, Microfluid. Nanofluid. **2**, 89–96 (2005). Accessed 28 July 2005
20. T. Ninomiya, Y. Okayama, Y. Matsumoto, X. Arouette, K. Osawa, N. Miki, MEMS-based hydraulic displacement amplification mechanism with completely encapsulated liquid. Sens. Actuat. A: Phys. **166**, 277–282 (2011)
21. E.-H. Yang, C. Lee, J. Mueller, T. George, Leak-tight piezoelectric microvalve for high-pressure gas micropropulsion. J. Microelectromech. Syst. **13**, 799–807 (2004)
22. T. Goettsche, J. Kohnle, M. Willmann, H. Ernst, S. Spieth, R. Tischler et al. Novel approaches to particle tolerant valves for use in drug delivery systems. Sens. Actuat. A Phys. **118**, 70–77 (2005). Accessed 31 Feb 2005
23. P. Ivaldi, J. Abergel, M.H. Matheny, L.G. Villanueva, R.B. Karabalin, M.L. Roukes et al., 50 nm thick AlN film-based piezoelectric cantilevers for gravimetric detection. J. Micromech. Microeng. **21** (2011)
24. T. Rogge, Z. Rummler, W.K. Schomburg, Polymer micro valve with a hydraulic piezo-drive fabricated by the AMANDA process. Sens. Actuat. A Phys. **110**, 206–212 (2004). Accessed 01 Feb 2004
25. H. Jerman, Electrically activated normally closed diaphragm valves. J. Micromech. Microeng. **4**, 210 (1994)
26. D.J. Leo, Engineering analysis of smart material systems, in *Engineering Analysis of Smart Material Systems* (Wiley, 2007), pp. i–xiv
27. M. Kohl, D. Dittmann, E. Quandt, B. Winzek, Thin film shape memory microvalves with adjustable operation temperature. Sens. Actuat. A Phys. **83**, 214–219 (2000). Accessed 22 May 2000
28. H. Takao, K. Miyamura, H. Ebi, M. Ashiki, K. Sawada, and M. Ishida, A MEMS microvalve with PDMS diaphragm and two-chamber configuration of thermo-pneumatic actuator for integrated blood test system on silicon. Sens. Actuat. A Phys. **119**, 468–475 (2005). Accessed 13 April 2005
29. C. Goll, W. Bacher, B. Büstgens, D. Maas, W. Menz, W.K. Schomburg, Microvalves with bistable buckled polymer diaphragms. J. Micromech. Microeng. **6**, 77 (1996)
30. A. Chen, T. Pan, Manually operatable on-chip bistable pneumatic microstructures for microfluidic manipulations. Lab Chip **14**, 3401–3408 (2014). Accessed 31 July 2014
31. C. Megnin, J. Barth, M. Kohl, A bistable SMA microvalve for 3/2-way control. Sens. Actuat. A Phys. **188**, 285–291 (2012)
32. B. Yang, B. Wang, W.K. Schomburg, A thermopneumatically actuated bistable microvalve. J. Micromech. Microeng. **20**, 095024 (2010)
33. A.K. Au, H. Lai, B.R. Utela, A. Folch, Microvalves and micropumps for BioMEMS. Micromachines **2**, 179–220 (2011). Accessed 24 May 2011
34. J. Voldman, M.L. Gray, M.A. Schmidt, An integrated liquid mixer/valve. J. Microelectromech. Syst. **9**, 295–302 (2000)
35. J. Kim, J. Baek, K. Lee, Y. Park, K. Sun, T. Lee et al., Photopolymerized check valve and its integration into a pneumatic pumping system for biocompatible sample delivery. Lab Chip **6**, 1091–1094 (2006). Accessed 26 July 2006
36. G.-H. Feng, E.S. Kim, Micropump based on PZT unimorph and one-way parylene valves. J. Micromech. Microeng. **14**, 429 (2004)
37. C.A. Hufnagel, W.P. Harvey, P.J. Rabil, T.F. McDermott et al. Surgical correction of aortic insufficiency. Surgery **35**, 673–683 (1954)
38. A. Starr, M.L. Edwards, Mitral replacement: clinical experience with a ball-valve prosthesis. Ann. Surg. **154**, 726–740 (1961)
39. Prosthetic Heart Valves: Practice Essentials, Background, Design Features 35 (2017). Accessed 09 March 2017
40. E.F. Hasselbrink, T.J. Shepodd, J.E. Rehm, High-pressure microfluidic control in lab-on-a-chip devices using mobile polymer monoliths. Anal. Chem. **74**, 4913–4918 (2002). 01 Oct 2002

41. V. Seidemann, S. Bütefisch, S. Büttgenbach, Fabrication and investigation of in-plane compliant SU8 structures for MEMS and their application to micro valves and micro grippers. Sens. Actuat. A Phys. **97–98**, 457–461 (2002), 2002/04/01
42. A. Nisar, N. Afzulpurkar, B. Mahaisavariya, A. Tuantranont, MEMS-based micropumps in drug delivery and biomedical applications, Sens. Actuat. B Chem. 130, pp. 917–942, 2008/03/28/ 2008
43. G. Destgeer, H.J. Sung, Recent advances in microfluidic actuation and micro-object manipulation via surface acoustic waves. Lab Chip (2015). Accessed 28 May 2015
44. D.J. Laser, J.G. Santiago, A review of micropumps, J. Micromech. Microeng. **14**, R35 (2004)
45. R. Hebden, A.M. Baker, L. Stenmark, J. Köhler, J.L. Moerel, W. Halswijk, Towards the development of a monopropellant microrocket engine using MEMS technology. ResearchGate
46. J.K. Igor, P.M. Joseph, C. Paul, C.H. Charles, Classification of pumps, in *Pump Handbook*, Fourth edn (McGraw Hill Professional, Access Engineering, 2008)
47. M. Richter, R. Linnemann, P. Woias, Eurosensors XIRobust design of gas and liquid micropumps. Sens. Actuat. A: Phys. **68**, 480–486 (1998). Accessed 15 June 1998
48. H.T.G. Van Lintel, F.C.M. Van de Pol, S. Bouwstra, A piezoelectric micropump based on micromachining of silicon. Sens. Actuat. **15**, 153–167 (1988)
49. J.G. Smits, Proceedings of the 5th International Conference on Solid-State Sensors and Actuators and Eurosensors IIIPiezoelectric micropump with three valves working peristaltically. Sens. Actuat. A Phys. **21**, 203–206 (1990). Accessed 01 Feb 1990
50. S. Shoji, S. Nakagawa, M. Esashi, Proceedings of the 5th International Conference on Solid-State Sensors and Actuators and Eurosensors IIIMicropump and sample-injector for integrated chemical analyzing systems. Sens. Actuat. A Phys. **21**, 189–192 (1990). Accessed 01 Feb 1990
51. C. Yamahata, F. Lacharme, Y. Burri, M.A.M. Gijs, A ball valve micropump in glass fabricated by powder blasting. Sens. Actuat. B Chem. **110**, 1–7 (2005). Accessed 30 Sep 2005
52. B. Cetin, R. Salemmilani, D. Li, Microfluidic Rotary Pump, in *Encyclopedia of Microfluidics and Nanofluidics*, ed. by D. Li (Springer, New York, 2013), pp. 1–6
53. M. Ashouri, M.B. Shafii, A. Moosavi, H. Amiri Hezave (2015) A novel revolving piston minipump. Sens. Actuat. B Chem. **218**, 237–244. Accessed 31 Oct 2015
54. H.-P. Chou, M.A. Unger, S.R. Quake, A microfabricated rotary pump. Biomed. Microdevices **3**, 323–330 (2001)
55. A.R. Tovar, A.P. Lee, Lateral cavity acoustic transducer. Lab Chip **9**, 41–43 (2009). Accessed 07 Jan 2009
56. M.V. Patel, A.R. Tovar, A.P. Lee, Lateral cavity acoustic transducer as an on-chip cell/particle microfluidic switch. Lab Chip **12**, 139–145 (2012)
57. J.P. Brody, P. Yager, R.E. Goldstein, R.H. Austin, Biotechnology at low Reynolds numbers. Biophys. J. **71**, 3430–3441 (1996). Accessed 12 1996
58. Z. Guttenberg, H. Müller, H. Habermüller, A. Geisbauer, J. Pipper, J. Felbel et al., Planar chip device for PCR and hybridization with surface acoustic wave pump. Lab Chip **5**, 308–317 (2005). Accessed 22 Feb 2005
59. M. Plodinec, M. Loparic, C.A. Monnier, E.C. Obermann, R. Zanetti-Dallenbach, P. Oertle et al. The nanomechanical signature of breast cancer. Biophys. J. **104**, 321a (2013). Accessed 29 Jan 2013
60. Y. Nematbakhsh, C.T. Lim, Cell biomechanics and its applications in human disease diagnosis. Acta Mech. Sin. **31**, 268–273 (2015). Accessed 12 May 2015
61. J. Tamayo, P.M. Kosaka, J.J. Ruz, A. San Paulo, M. Calleja, Biosensors based on nanomechanical systems. Chem. Soc. Rev. **42**, 1287–1311 (2013)
62. M. Calleja, P.M. Kosaka, Á.S. Paulo, J. Tamayo, Challenges for nanomechanical sensors in biological detection. Nanoscale **4**, 4925–4938 (2012). Accessed 27 June 2012

63. A. Gupta, D. Akin, R. Bashir, Single virus particle mass detection using microresonators with nanoscale thickness. Appl. Phys. Lett. **84**, 1976–1978 (2004). Accessed 15 March 2004
64. G. Villanueva, F. Perez-Murano, M. Zimmermann, J. Lichtenberg, J. Bausells, "Piezoresistive cantilevers in a commercial CMOS technology for intermolecular force detection. Microelectron. Eng. **83**, 1302–1305, April–Sep 2006
65. G. Villanueva, J.A. Plaza, J. Montserrat, F. Perez-Murano, J. Bausells, Crystalline silicon cantilevers for piezoresistive detection of biomolecular forces. Microelectron. Eng. **85**, 1120–1123 (2008)
66. J. Tamayo, P. M. Kosaka, J.J. Ruz, Á.S. Paulo, M. Calleja, Biosensors based on nanomechanical systems. Chem. Soc. Rev. **42**, 1287–1311 (2013). Accessed 16 Jan 2013
67. J.L. Arlett, E.B. Myers, M.L. Roukes,Comparative advantages of mechanical biosensors. Nat. Nanotechnol. **6**, 203–215, (2011). Accessed April 2011
68. G. Tosolini, G. Villanueva, F. Perez-Murano, J. Bausells, Silicon microcantilevers with MOSFET detection. Microelectron. Eng. **87**, 1245–1247, May–Aug 2010
69. G. Villanueva, J. Montserrat, F. Perez-Murano, G. Rius, J. Bausells, Submicron piezoresistive cantilevers in a CMOS-compatible technology for intermolecular force detection. Microelectron. Eng. **73–4**, 480–486 (2004)
70. J. Tamayo, J. J. Ruz, V. Pini, P. Kosaka, and M. Calleja, Quantification of the surface stress in microcantilever biosensors: revisiting Stoney's equation. Nanotechnology **23**,475702 (2012)
71. V. Pini, J.J. Ruz, P.M. Kosaka, O. Malvar, M. Calleja, J. Tamayo, How two-dimensional bending can extraordinarily stiffen thin sheets. Sci. Rep. **6**. Accessed 11 July 2016
72. J. Fritz, M.K. Baller, H.P. Lang, H. Rothuizen, P. Vettiger, E. Meyer et al., Translating biomolecular recognition into nanomechanics. Science **288**, 316–318 (2000)
73. D. Ramos, J. Tamayo, J. Mertens, M. Calleja, L.G. Villanueva, A. Zaballos, Detection of bacteria based on the thermomechanical noise of a nanomechanical resonator: origin of the response and detection limits. Nanotechnology **19**, 035503 (2008). Accessed 23 Jan 2008
74. E.A. Corbin, F. Kong, C.T. Lim, W.P. King, R. Bashir, Biophysical properties of human breast cancer cells measured using silicon MEMS resonators and atomic force microscopy. Lab Chip **15**, 839–847 (2015). Accessed 22 Jan 2015
75. K.L. Ekinci, M.L. Roukes, Nanoelectromechanical systems. Rev. Sci. Instrum. **76**, 061101 (2005). Accessed 01 June 2005
76. M. Sansa, E. Sage, E.C. Bullard, M. Gely, T. Alava, E. Colinet et al., Frequency fluctuations in silicon nanoresonators. Nat. Nanotechnol. **11**, 552 (2016)
77. K.Y. Gfeller, N. Nugaeva, M. Hegner, Rapid biosensor for detection of antibiotic-selective growth of escherichia coli. Appl. Environ. Microbiol. **71**, 2626–2631 (2005). Accessed 01 May 2005
78. M.K. Ghatkesar, T. Braun, V. Barwich, J.-P. Ramseyer, C. Gerber, M. Hegner et al., Resonating modes of vibrating microcantilevers in liquid. Appl. Phys. Lett. **92**, 043106 (2008). Accessed 28 Jan 2008
79. E. Oesterschulze, P. Kehrbusch, B. Radzio, E. A. Ilin, A. Thyssen, J. W. Deitmer et al., Tailoring the interface of hybrid microresonators in viscid fluids enhances their quality factor by two orders of magnitude. Lab on a Chip **12**, 1316–1319 (2012). Accessed 07 March 2012
80. M. Melli, G. Scoles, M. Lazzarino, Fast detection of biomolecules in diffusion-limited regime using micromechanical pillars. ACS Nano **5**, 7928–7935 (2011). Accessed 25 Oct 2011
81. D. Ziegler, P.D. Ashby, Encased cantilevers for ultra-low-noise force spectroscopy of proteins and ligand receptor complexes. Biophys. J. **102**, 579a (2012). Accessed 31 Jan 2012
82. T.P. Burg, S.R. Manalis, Suspended microchannel resonators for biomolecular detection. Appl. Phys. Lett. **83**, 2698–2700 (2003). Accessed 29 Sep 2003
83. M. Rodahl, F. Höök, A. Krozer, P. Brzezinski, B. Kasemo, Quartz crystal microbalance setup for frequency and Q-factor measurements in gaseous and liquid environments. Rev. Sci. Instrum. **66**, 3924–3930 (1995). Accessed 01 July 1995

84. S.K. Vashist, P. Vashist, Recent advances in quartz crystal microbalance-based sensors. J. Sens. e571405 (2011), 19 Sep 2011
85. N. Kim, D.-K. Kim, Y.-J. Cho, Development of indirect-competitive quartz crystal microbalance immunosensor for C-reactive protein. Sens. Actuat. B Chem. **143**, 444–448 (2009). Accessed 04 Dec 2009
86. K. Länge, B.E. Rapp, M. Rapp, Surface acoustic wave biosensors: a review. Anal. Bioanal. Chem. **391**, 1509–1519 (2008). Accessed 12 Feb 2008
87. N.G. Durmuş, R.L. Lin, M. Kozberg, D. Dermici, A. Khademhosseini, U. Demirci, Acoustic-based biosensors, in *Encyclopedia of Microfluidics and Nanofluidics* ed. by D. Li (Springer, New York, 2015), pp. 28–40
88. S.U. Senveli, Z. Ao, S. Rawal, R.H. Datar, R.J. Cote, O. Tigli, A surface acoustic wave biosensor for interrogation of single tumour cells in microcavities **16**, 163–171 (2015). Accessed 15 Dec 2015
89. Oscillating U-tube, in *Wikipedia* (2016)
90. S. Byun, S. Son, D. Amodei, N. Cermak, J. Shaw, J. H. Kang et al., Characterizing deformability and surface friction of cancer cells. Proc. Natl. Acad. Sci. **110**, 7580–7585 (2013). Accessed 07 May 2013
91. N. Cermak, S. Olcum, F.F. Delgado, S.C. Wasserman, K.R. Payer, M.A. Murakami et al., High-throughput measurement of single-cell growth rates using serial microfluidic mass sensor arrays. Nat. Biotechnol. (2016). Advance online publication. Accessed 05 Sep 2016
92. A.K. Bryan, V.C. Hecht, W. Shen, K. Payer, W.H. Grover, S.R. Manalis, Measuring single cell mass, volume, and density with dual suspended microchannel resonators. Lab on a Chip **14**, 569–576 (2013). Accessed 23 Dec 2013
93. S. Olcum, N. Cermak, S.C. Wasserman, K.S. Christine, H. Atsumi, K.R. Payer et al., Weighing nanoparticles in solution at the attogram scale. Proc. Natl. Acad. Sci. **111**, 1310–1315 (2014). Accessed 28 Jan 2014
94. M.F. Khan, S. Schmid, P.E. Larsen, Z.J. Davis, W. Yan, E.H. Stenby et al., Online measurement of mass density and viscosity of pL fluid samples with suspended microchannel resonator. Sens. Actuat. B Chem. **185**, 456–461 (2013). Accessed Aug 2013
95. T.P. Burg, M. Godin, S.M. Knudsen, W. Shen, G. Carlson, J.S. Foster et al., Weighing of biomolecules, single cells and single nanoparticles in fluid. Nature **446**, 1066–1069 (2007). Accessed 26 April 2007
96. J.L. Arlett, M.L. Roukes, Ultimate and practical limits of fluid-based mass detection with suspended microchannel resonators. J. Appl. Phys. **108**, 084701 (2010). Accessed 15 Oct 2010
97. J. El-Ali, P.K. Sorger, K.F. Jensen, Cells on chips. Nature **442**, 403–411 (2006)
98. M. Cristofanilli, G.T. Budd, M.J. Ellis et al., Circulating tumor cells, disease progression, and survival in metastatic breast cancer. N. Engl. J. Med. **351**, 781791 (2004)
99. W.R. Rodriguez, N. Christodoulides, P.N. Floriano, S. Graham, M. Mohanty, M. Dixon, M. Hsiang, T. Peter, S. Zavahir, I. Thior, D. Romanovicz, B. Bernard, A.P. Goodey, B.D. Walker, J.T. McDevitt, A microchip CD4 counting method for HIV monitoring in resource-poor settings. PLoS Med **2**(7), 663–672 (2005)
100. X. Cheng, D. Irimia, M. Dixon et al., A microfluidic device for practical label-free CD4+ T cell counting of HIV-infected subjects. Lab Chip **7**(2), 170–178 (2007)
101. R. Huang, T.A. Barber, M.A. Schmidt, R.G. Tompkins, M. Toner, D.W. Bianchi, R. Kapur, W.L. Flejter, A microfluidics approach for the isolation of nucleated red blood cells (NRBCs) from the peripheral blood of pregnant women. Prenat. Diagn. **28**, 892899 (2008)
102. J. den Toonder, Circulating tumor cells: the grand challenge. Lab Chip **11**(3), 375–377 (2011)
103. R. David, M. Groebner, W.M. Franz, Magnetic cell sorting purification of differentiated embryonic stem cells stably expressing truncated human CD4 as surface marker. Stem Cells **23**(4), 477–482 (2005)

104. P. Gascoyne, C. Mahidol, M. Ruchirawat, J. Satayavivad, P. Watcharasit, F. Becker, Microsample preparation by dielectrophoresis: isolation of malaria. Lab Chip **2**(2), 70–75 (2002)
105. X. Xuan, J. Zhu, C. Church, Particle focusing in fluidic devices. Microfluid. Nanofluid. (2010). https://doi.org/10.1007/s10404-010-0602-7
106. A.A.S. Bhagat, H. Bow, H. Wei Hou, S. Jin Tan, J. Han, C. Teck Lim, Microfluidic for cell separation. Med. Biol. Eng. Comput. (2010). https://doi.org/10.1007/s11517-010-0611-4
107. C.W. Shields IV, C.D. Reyes, G.P. Lpez, Microfluidic cell sorting: a review of the advances in the separation of cells from debulking to rare cell isolation. Lab Chip (2015). https://doi.org/10.1039/c4lc01246a
108. J. Autebert, B. Coudert, F.C. Bidard, J.Y. Pierga, S. Descroix, L. Malaquin, J.L. Viovy, Microfluidic: an innovative tool for efficient cell sorting. Methods (2012). https://doi.org/10.1016/j.ymeth.2012.07.002
109. D.R. Gosset, W.M. Weaver, A.J. Mach, S.C. Hur, H.T. Kwong Tse, W. Lee, H. Amini, D. Di Carlo, Label-free cell separation and sorting in fluidic systems. Anal. Bioanal. Chem. (2010). https://doi.org/10.1007/s00216-010-3721-9
110. H. Ji, V. Samper, Y. Chen, C. Heng, T. Lim, L. Yobas, Silicon-based microfilters for whole blood cell separation. Biomed. Microdevices **10**, 251–257 (2008)
111. J. Han, B. Hansen, P. Abgrall, Continuous-flow deformability-based cell separation. Patent No. US20110081674 A1 (2011)
112. S.J. Tan, R.L. Lakshmi, P. Chen, W.T. Lim, L. Yabos, C.T. Lim, Versatile label free biochip for the detection of circulating tumor cells from peripheral blood in cancer patients. Biosens. Bioelectron. (2010). https://doi.org/10.1016/j.bios.2010.07.054
113. A. Yamada, S. Lee, P. Bassereau, C.N. Baroud, Trapping and release of Giant Unilamellar Vesicles in microfluidic wells. Soft Matter **10**, 5878–5885 (2014)
114. X. Fan, C. Jia, G. Li, H. Mao, Q. Jin, J. Zhao, A microfluidic chip integrated with a high-density PDMS-based microfiltration membrane for rapid isolation and detection of circulating tumor cells. Biosens. Bioelectron. **71**, 380–386 (2015)
115. J.P. Brody, T.D. Osborn, F.K. Forster, P. Yager, A planar microfabricated fluid filter. Sens. Actuat. A Phys. **54**, 704–708 (1996)
116. T.A. Crowley, V. Pizziconi, Isolation of plasma from whole blood using planar microfilters for lab-on-a-chip applications. Lab Chip (2005). https://doi.org/10.1039/b502930a
117. H. Mohamed, J.N. Turner, M. Caggana, Biochip for separating fetal cells from maternal circulation. J. Chromatogr. A **1162**(2), 187–192 (2007)
118. P. Sethu, A. Sin, M. Toner, Microfluidic diffusive filter for apheresis (leukapheresis). Lab Chip **6**, 83–89 (2006)
119. P. Wilding, L.J. Kricka, J. Cheng, G. Hvichia, M.A. Shoffner, P. Fortina, Integrated cell isolation and polymerase chain reaction analysis using silicon microfilter chambers. Anal. Biochem. (1998). https://doi.org/10.1006/abio.1997.2530
120. Y. Yoon, S. Kim, J. Lee, J. Choi, R.W. Kim, S.J. Lee, O. Sul, S.B. Lee, Clogging-free microfluidics for continuous size-based separation of microparticles. Scientific reports (2016). https://doi.org/10.1038/srep26531
121. D. Di Carlo, L.Y. Wu, L.P. Lee, Dynamic single cell culture array. Lab Chip **6**, 1445–1449 (2006)
122. J.A. Davis, D.W. Inglis, K.J. Morton, D.A. Lawrence, L.R. Huang, S.Y. Chou, J.C. Sturm, R.H. Austin, Deterministic hydrodynamics: taking blood apart. Proc. Natl. Acad. Sci. U.S.A **103**(40), 14779–14784 (2006)
123. S. Choi, S. Song, C. Choi, J.K. Park, Sheathless focusing of microbeads and blood cells based on hydrophoresis. Small **4**, 634–641 (2008)
124. D. Huh, J.H. Bahng, Y. Ling, H. Wei, O.D. Kripfgans, J.B. Fowlkes, J.B. Grotberg, S. Takayama, Gravity-driven microfluidic particle sorting device with hydrodynamic separation amplification. Anal. Chem. **79**, 1369–1376 (2007)

125. M. Yamada, K. Kano, Y. Tsuda, J. Kobayashi, M. Yamato, M. Seki, T. Okano, Microfluidic devices for size-dependent separation of liver cells. Biomed. Microdevices **9**(5), 637–645 (2007)
126. J. Takagi, M. Yamada, M. Yasuda, M. Seki, Continuous particle separation in a microchannel having asymmetrically arranged multiple branches. Lab Chip **5**, 778 (2005)
127. Z. Wu, B. Willing, J. Bjerketorp, J.K. Jansson, K. Hjort, Soft inertial microfluidics for high throughput separation of bacteria from human blood cells. Lab Chip **9**(9), 1193–1199 (2009)
128. D. Di Carlo, D. Irimia, R.G. Tompkins, M. Toner, Continuous inertial focusing, ordering and separation of particles in microchannels. Proc. Natl. Acad. Sci. U.S.A **104**(48), 18892–18897 (2007)
129. M. Faivre, M. Abkarian, K. Bickraj, H.A. Stone, Geometrical focusing of cells in a microfluidic device: an approach to separate blood plasma. Biorheology **43**(2), 147–159 (2006)
130. A. Jain, L.L. Munn, Biomimetic postcapillary expansions for enhancing rare blood cell separation on a microfluidic chip. Lab Chip **11**, 2941–2948 (2011)
131. S. Yang, A. Undar, J.D. Zahn, A microfluidic device for continuous, real time blood plasma separation. Lab Chip **6**(7), 871–880 (2006)
132. J.P. Beech, Microfluidics separation and analysis of biological particles. Ph.D. Thesis, Lund University, Sweeden (2011), https://tegen.ftf.lth.se/files/theses/beech2011.pdf
133. P. Beech, S.H. Holm, K. Adolfsson, J.O. Tegenfeldt, Sorting cells by size, shape and deformability. Lab Chip **12**, 1048–1051 (2012)
134. N. Tottori, Y. Sakurai, T. Nisisako, Y. Tanagida, T. Hatsuzawa, Thermally tunable deterministic lateral displacement through hydrogel micro pillar arrays. in *Proceedings of 20th International Conference on Miniaturized Systems for Chemistry and Life Sciences*, 9–13 October 2016, Dublin, Ireland (2016), pp. 140–141
135. D.W. Inglis, J.A. Davis, R.H. Austin, J.C. Sturm, Critical particle size for fractionation by deterministic lateral displacement. Lab Chip **6**, 655–658 (2006)
136. S.H. Holm, J.P. Beech, M.P. Barrett, J.O. Tegenfeldt, Separation of parasites from human blood using deterministic lateral displacement. Lab Chip **11**(7), 1326–1332 (2011)
137. Z. Liu, F. Huang, J. Du, W. Shu, H. Feng, X. Xu, Y. Chen, Rapid isolation of cancer cells using microfluidic deterministic lateral displacement structure. Biomicrofluidics **7**, 11801 (2013)
138. L.R. Huang, E.C. Cox, R.H. Austin, J.C. Sturm, Continuous particle separation through deterministic lateral displacement. Science **304**, 987–990 (2004)
139. N. Tottori, T. Nisisako, J. Park, Y. Tanagida, T. Hatsuzawa, Separation of viable and nonviable mammalian cells using a deterministic lateral displacement device. Biomicrofluidics **10**, 014125 (2016)
140. S. Choi, J.K. Park, Sheathless hydrophoretic particle focusing in a microchannel with exponentially increasing obstacle arrays. Anal. Chem. **80**, 3035–3039 (2008)
141. B. Kim, J.K. Lee, S. Choi, Continuous sorting and washing of cancer cells from blood cells by hydrophoresis. Bio. Chip J. (2015). https://doi.org/10.1007/s13206-016-0201-0
142. E. Sollier, D.E. Go, J. Che, D.R. Gosset, S. O'Byrne, W.M.M. Weaver, N. Kummer, M. Rettig, J. Goldman, N. Nickols, S. McCloskey, R.P. Kulkarni, D. Di Carlo, Size-selective collection of circulating tumor cells using Vortex technology. Lab Chip **14**, 63–77 (2014)
143. C.H. Hsu, D. Di Carlo, C. Chen, D. Irimia, M. Toner, Microvortex for focusing, guiding and sorting of particles. Lab Chip **8**, 2128–2134 (2008)
144. S. Song, S. Choi, Inertial modulation of hydrophoretic cell sorting and focusing. App. Phys. Let. **104**(7), 074106 (2014)
145. M. Madou, J. Zoval, G. Jia, H. Kido, J. Kim, N. Kim, Lab on a CD. Annu. Rev. Biomed. Eng. **8**, 601–628 (2006)
146. M. Yamada, M. Seki, Hydrodynamic filtration for on-chip particle concentration and classification utilizing microfluidics. Lab Chip **5**(11), 1233–1239 (2005)
147. R. Aoki, M. Yamada, M. Yasuda, M. Seki, In-channel focusing of flowing microparticles utilizing hydrodynamic filtration. Microfluid. Nanaofluid. **6**, 571–576 (2009)

148. M. Yamada, M. Nakashima, M. Seki, Pinched flow fractionation: continuous size separation of particles utilizing a laminar flow profile in a pinched microchannel. Anal. Chem. **76**(18), 5465–5471 (2004)
149. X. Lu, X. Xuan, Elasto-inertial pinched flow fractionation for continuous shape-based particle separation. Anal. Chem. **87**, 11523–11530 (2015)
150. X. Lu, X. Xuan, Continuous microfluidic particle separation via elasto-inertial pinched flow fractionation. Anal. Chem. **87**(12), 6389–6396 (2015)
151. G. Segre′, A. Silberberg, Radial particle displacements in poiseuille flow of suspensions. Nature **189**, 209–210 (1961)
152. D.R. Gossett, D. Di Carlo, Particle focusing mechanisms in curving confined flows. Anal. Chem. **81**, 8459–8465 (2009)
153. D. Di Carlo, Inertial microfluidics. Lab Chip **9**(21), 3038–3046 (2009)
154. J.S. Park, S.H. Song, H.I. Jung, Continuous focusing of microparticles using inertial lift force and vorticity via multi-orifice microfluidic channels. Lab Chip **9**(7), 939–948 (2009)
155. A.A.S. Bhagat, S.S. Kuntaegowdanahalli, I. Papautsky, Inertial microfluidics for continuous particle filtration and extraction. Microfluid. Nanofluid. **7**(2), 217–226 (2009)
156. J.F. Edd, D. Di Carlo, K.J. Humphry, S. Koester, D. Irimia, D.A. Weitz, M. Toner, Controlled encapsulation of single-cells into monodisperse picolitre drops. Lab Chip **8**, 1262–1264 (2008)
157. S.C. Hur, H.T. Tse, D. Di Carlo, Sheathless inertial cell ordering for extreme throughput flow cytometry. Lab Chip (2010). https://doi.org/10.1039/b919495a
158. J. Kim, J. Lee, C. Wu, S.M. Nam, D. Di Carlo, W. Lee, Inertial focusing in non-rectangular cross-section microchannels and manipulation of accessible focusing position. Lab Chip (2016). https://doi.org/10.1039/c5lc01100k
159. A. Berger, L. Talbot, L.S. Yao, Flow in curved pipes. Ann. Rev. Fluid Mech. **15**, 461–512 (1983)
160. A. Russom, A.K. Gupta, S. Nagrath, D. Di Carlo, J.F. Edd, M. Toner, Differential inertial focusing of particles in curved low-aspect-ratio microchannels. New J. Phys. **11**, 75025 (2009)
161. S.S. Kuntaegowdanahalli, A.A.S. Bhagat, G. Kumar, I. Papautsky, Inertial microfluidics for continuous particle separation in spiral microchannels. Lab Chip **9**, 2973–2980 (2009)
162. A.A.S. Bhagat, H.W. Hou, L.D. Li, C.T. Lim, J. Han, Dean flow fractionation (DFF) Isolation of circulating tumor cells (CTCs) from blood, in *Proceedings of 15th International Conference on Miniaturized Systems for Chemistry and Life Sciences*, October 28–November 1 2012, Okinawa, Japan (2012), pp. 524–526
163. G. Guan, L. Wu, A.A. Bhagat, Z. Li, P.C. Chen, S. Chao, C.J. Ong, J. Han, Spiral microchannel with rectangular and trapezoidal cross-sections for size based particle separation. Sci. Rep. **3**, 1475 (2013)
164. M.E. Warkiani, G. Guan, K.B. Luan, W.C. Lee, A.A. Bhagat, P.K. Chaudhuri, D.S. Tan, W. T. Lim, S.C. Lee, P.C. Chen, C.T. Lim, J. Han, Slanted spiral microfluidics for the ultra-fast, label-free isolation of circulating tumor cells. Lab Chip **14**, 128–137 (2014)
165. A.R. Pries, K. Ley, P. Gaehtgens, Generalization of the fahraeus principle for microvessel networks. Am. J. Physiol-Heart C **251**(6), H1324–H1332 (1986)
166. H.L. Goldsmith, G.R. Cokelet, P. Gaehtgens, Robin fahraeus: evolution of his concepts in cardiovascular physiology. Am. J. Physiol. **257**(Heart Circ. Physiol. 26):H1005–H1015 (1989)
167. Goldsmith HL (1971) Red cell motions and wall interactions in tube flow, in *Federation Proceedings*, vol. 30, pp. 1578–1588
168. E. Sollier, H. Rostaing, P. Pouteau, Y. Fouillet, J.L. Achard, Passive microfluidic devices for plasma extraction from whole human blood. Sens. Actuat. B-Chem. **141**, 617–624 (2009)
169. S.S. Shevkoplyas, T. Yoshida, L.L. Munn, M.W. Bitensky, Biomimetic autoseparation of Leukocytes from whole blood in a microfluidic device. Anal. Chem. **77**(3), 933–937 (2005)

170. H.W. Hou, A.A.S. Bhagat, A.G. Lin CHong, P. Mao, K.S. Wei Tan, J. Han, C.T. Lim, Deformability based cell margination—a simple microfluidic design for malaria-infected erythrocyte separation. Lab Chip **10**, 2605–2613 (2010)
171. Z. Shen, G. Coupier, B. Kaoui, B. Polack, J. Harting, C. Misbah, T. Podgorski, Inversion of hematocrit partition at microfluidic bifurcations. Microvasc. Res. **105**, 40–46 (2016)
172. R.D. Jaggi, R. Sandoz, C.S. Effenhauser, Microfluidic depletion of red blood cells from whole blood in high-aspect-ratio microchannels. Microfluid. Nanofluid. **3**, 47–53 (2006)
173. B. Alberts, A. Johnson, J. Lewis, M. Raff, K. Roberts, P. Walter, The cytoskeleton, in *Molecular Biology of the Cell*, 5rd edn. ed. Alberts (Garland Science, New York, 2008)
174. J.P. Mills, M. Diez-Silva, D.J. Quinn, M. Dao, M.J. Lang, K.S.W. Tan, C.T. Lim, G. Milon, P.H. David, O. Mercereau-Puijalon, S. Bonnefoy, S. Suresh, Effect of plasmodial RESA protein on deformability of human red blood cells harboring Plasmodium falciparum. Proc. Natl. Acad. Sci. U.S.A **104**(22), 9213–9217 (2007)
175. G.A. Barabino, M.O. Platt, D.K. Kaul, Sickle cell biomechanics. Ann. Rev. Biomed. Eng. **12**, 345–367 (2010)
176. M. Nishino, H. Tanaka, H. Oqura, Y. Inoue, T. Koh, K. Fujita, H. Sugimoto, Serial changes in leukocyte deformability and whole blood rheology in patients with sepsis or trauma. J. Trauma **59**(6), 1425–1431 (2005)
177. A.D. van der Meer, A.A. Poot, J. Feijen, I. Vermes, Analyzing shear stress-induced alignment of actin filaments in endothelial cells with a microfluidic assay. Biomicrofluidics **4** (1), 011103 (2010)
178. S. Suresh, J. Spatz, J.P. Mills, A. Micoulet, M. Dao, C.T. Lim, M. Beil, T. Seufferlein, Connectiones between single-cell biomechanics and human disease states: gastrointestinal cancer and malaria. Acta Biomater. (2005). https://doi.org/10.1016/j.actbio.2004.09.001
179. R.D. Gonza′lez-Cruz,V.C. Fonseca, E.M. Darling, Cellular mechanical properties reflect the differentiation potential of adipose-derived mesenchymal stem cells. Proc. Natl. Acad. Sci. U.S.A **109**(24), E1523 (2012)
180. S. Suresh, Biomechanics and biophysics of cancer cells. Acta Biomater. **3**, 413–438 (2007)
181. S.M.A. Haghparast, T. Kihara, J. Miyake, Distinct mechanical behavior of HEK293 cells in adherent and suspended states. Peer J (2015). https://doi.org/10.7717/peerj.1131
182. S.C. Hur, N.K. Henderson-MacLennan, E.R. McCabe, D. Di Carlo, Deformability-based cell classification and enrichment using inertial microfluidics. Lab Chip **11**, 912–920 (2011)
183. E.M. Darling, Force scanning: a rapid, high-resolution approach for spatial mechanical property mapping. Nanotechnology **22**, 175707 (2011)
184. N. Mohandas, E. Evans, Mechanical properties of the red cell membrane in relation to molecular structure and genetic defects. Ann. Rev. Biophys. Bio. **23**, 787–818 (1994)
185. R.E. Waught, P. Agre, Reductions of erythrocyte membrane viscoelastic coeffients reflect sprectrin deficiencies in hereditary spherocytosis. J. Clim. Invest. **81**, 133–141 (1988)
186. J. Guck, S. Schinkinger, B. Lincoln, F. Wottawah, S. Ebert, M. Romeyke, D. Lenz, H.M. Erickson, R. Ananthakrishnan, D. Mitchell, J. Ka¨s, S. Ulvick, C. Bilby, Optical deformability as an inherent cell marker for testing malignant transformation and metastatic competence. Biophys. J. **88**,3689–3698 (2005)
187. J.M.A. Mauritz, T. Tiffert, R. Seear, F. Lautenschlger, A. Esposito, V.L. Lew, J. Guck, C.F. Kaminskia, Detection of Plasmodium falciparum-infected red blood cells by optical stretching. J. Biomed. Opt. **15**(3), 030517 (2010)
188. H. Engelhart, E. Sackmann, On the measurement of shear elastic moduli and viscosities of erythrocyte plasma membranes by transient deformation in high frequency electric fields. Biophys. J. **54**, 495–508 (1988)
189. I. Doh, W.C. Lee, Y.H. Cho, A.P. Pisano, F.A. Kuypers, Deformation measurement of individual cells in large populations using a single-cell microchamber array chip. Appl. Phys. Lett. **100**, 173702 (2012)
190. E. Evans, R. Skalak, *Mechanics and thermodynamics of biomembranes* (Florida CRC Press, Boca Raton, 1980)

191. L.A. MacQueen, M.D. Buschmann, M.R. Wertheimer, Mechanical properties of mammalian cells in suspension measured by electro-deformation. J. Micromech. Microeng. **20**, 065007 (2010)
192. J. Chen, M. Abdelgawad, L.M. Yu, N. Shakiba, W.Y. Chien, Z. Lu, W.R. Geddie, M.A.S. Jewett, Y. Sun, Electrodeformation for single cell mechanical characterization. J. Micromech. Microeng. **21**, 054012 (2011)
193. G. Du, A. Ravetto, Q. Fang, J.M.J. den Toonder, Cell types can be distinguished by measuring their viscoelastic recovery times using a microfluidic device. Biomed. Microdevices **13**, 29–40 (2010)
194. Y.C. Kim, S.J. Park, J.K. Park, Biomechanical analysis of cancerous and normal cells based on bulge generation in a microfluidic device. Analyst **133**, 1432–1439 (2008)
195. D.R. Gosset, H.T.K. Tse, A.S. Lee, Y. Ying, A.G. Lindgreen, O.O. Yang, J. Rao, A.T. Clark, D. Di Carlo, Hydrodynamic stretching of single cells for large population mechanical phenotyping. Proc. Natl. Acad. Sci. U.S.A **109**, 7630–7635 (2012)
196. M. Gnerlich, S.F. Perry, S. Tatic-Lucic, A submersible piezoresistive MEMS lateral force sensor for a diagnostic biomechanics platform. Sens. Actuat. A-Phys. **188**, 111–119 (2012)
197. B. Barazani, S. Warnat, A. Fine, T. Hubbard, MEMS squeezer for the measurement of single cell rupture force, stiffness change, and hysteresis. J. Micromech. Microeng. **27**, 025002 (2017)
198. Lafitte N, Guillou H, Kumemura M, Jalabert L, Fujii T, Fujita H Collard D, Integrated MEMS platform with silicon nanotweezers and open microfluidic device for real-time and routine biomechanical probing on molecules and cells, in *35th Annual International Conference of the IEEE Engineering in Medicine and Biology Society* (EMBS 2013), Osaka, Japan (2013), pp. 148–151
199. A.M. Forsyth, J.D. Wan, W.D. Ristenpart, H.A. Stone, The dynamic behavior of chemically "stiffened" red blood cells in microchannel flows. Microvasc. Res. **80**, 37–43 (2010)
200. S.S. Lee, Y. Yim, K.H. Ahn, S.J. Lee, Extensional flow-based assessment of red blood cell deformability using hyperbolic converging microchannel. Biomed. Microdevices **11**, 1021–1027 (2009)
201. J.P. Shelby, J. White, K. Ganesan, P.K. Rathod, D.T. Chiu, A microfluidic model for single-cell capillary obstruction by Plasmodium falciparum infected erythrocytes. Proc. Natl. Acad. Sci. U.S.A **100**(25), 14618–14622 (2003)
202. H.W. Hou, Lee G.Y.H. LiQS, A.P. Kumar, C.N. Ong, C.T. Lim, Deformability study of breast cancer cells using microfluidics. Biomed. Microdevices **11**, 557–564 (2015)
203. Z.S. Khan, S.A. Vanapalli, Probing the mechanical properties of brain cancer cells using a microfluidic cell squeezer device. Biomicrofluidics **7**, 011806 (2013)
204. K.D. Nyberg, M.B. Scott, S.L. Bruce, A.B. Gopinath, D. Bikos, T.G. Mason, J.W. Kim, H. S. Choig, A.C. Rowat, The physical origins of transit time measurements for rapid, single cell mechanotyping. Lab Chip (2016). https://doi.org/10.1039/c6lc00169f
205. P. Preira, V. Grandne, J.M. Forel, S. Gabriele, M. Camara, O. Theodoly, Passive circulating cell sorting by deformability using microfluidic gradual filter. Lab Chip **13**, 161–170 (2013)
206. H. Bow, I.V. Pivkin, M. Diez-Silva, S.J. Goldfless, M. Dao, J.C. Niles, S. Suresh, J. Han, A microfabricated deformability-based flow cytometer with application to malaria. Lab Chip **11**, 1065–1073 (2011)
207. M. Mak, D. Erickson, A serial micropipette microfluidic device with applications to cancer cell repeated deformation studies. Integr. Biol. **5**, 1374–1384 (2013)
208. M.J. Rosenbluth, W.A. Lam, D.A. Fletcher, Analysing cell mechanics in hematologic diseases with microfluidic biophysical flow cytometry. Lab Chip **8**, 1062–1070 (2008)
209. W.A. Lam, M.J. Rosenbluth, D.A. Fletcher, Chemotherapy exposure increases leukemia cell stiffness. Blood **109**(8), 3505–3508 (2007)
210. T. Herricks, M. Antia, P.K. Rathod, Deformability limits of Plasmodium falciparum-infected red blood cells. Cell. Microbiol. **11**(9), 1340–1353 (2009)
211. J. Picot,P.A. Ndour, S.D. Lefevre, W. El Nemer, H. Tawfik, J. Galimand, L. Da Costa, J.A. Ribeil, M. de Montalembert, V. Brousse, B. Le Pioufle, P. Buffet, C. Le Van Kim, O.

Français, A biomimetic microfluidic chip to study the circulation and mechanical retention of red blood cells in the spleen. Am. J. Hematol. (2015). https://doi.org/10.1002/ajh.23941
212. S. Huang, A. Undisz, M. Diez-Silva, H. Bow, M. Dao, J. Han, Dynamic deformability of Plasmodium falciparum-infected erythrocytes exposed to artesunate in vitro. Integr Biol **5**, 414–422 (2013)
213. A. Adamo, A. Sharei, L. Adamo, B. Lee, S. Moa, K.F. Jensen, Microfluidics-based assessment of cell deformability. Anal. Chem. **84**, 6438–6443 (2012)
214. M. Abkarian, M. Faivre, H.A. Stone, High-speed microfluidic differential manometer for cellular-scale hydrodynamics. Proc. Natl. Acad. Sci. U.S.A **103**(3), 538–542 (2006)
215. M. Abkarian, M. Faivre, R. Horton, K. Smistrup, C.A. Best-Popescu, H.A. Stone, Cellular-scale hydrodynamics. Biomed. Mater. **3**, 034011 (2008)
216. Q. Guo, S.J. Reiling, P. Rohrbach, H.S. Ma, Microfluidic biomechanical assay for red blood cells parasited by Plasmodium falciparum. Lab Chip **12**, 1143–1150 (2012)
217. B.M. Cooke, N. Mohandas, R.L. Coppel, The malaria-infected red blood cell: structural and functional changes. Adv. Parasit. **50**, 1–86 (2001)
218. S.B. Huang, Z. Zhao, D.Y. Chen, H.C. Lee, Y.N. Luo, T.K. Chiu, J.B. Wang, J. Chen, M.H. Wu, A clogging-free microfluidic platform with an incorporated pneumatically-driven membrane-based active valve enabling specific membrane capacitance and cytoplasm conductivity characterization of single cells. Sens. Actuat. B Chem. **190**, 928–936 (2014)
219. W. Beattie, X. Qin, L. Wang, H. Ma, Clog-free cell filtration using resettable cell traps. Lab Chip **14**, 2657–2665 (2014)
220. Y. Zheng, E. Shojaei-Baghini, A. Azad, C. Wang, Y. Sun, High-throughput biophysical measurement of human red blood cells. Lab Chip **12**, 2560–2567 (2012)
221. O. Otto, P. Rosendahl, A. Mietke, S. Golfier, C. Herold, D. Klaue, S. Girardo, S. Pagliara, A. Ekpenyong, A. Jacobi, M. Wobus, N. Tpfner, U.F. Keyser, J. Mansfeld, E. Fischer-Friedrich, J. Guck, Real-time deformability cytometry: on-the-fly cell mechanical phenotyping. Nat. Methods (2015). https://doi.org/10.1038/nmeth.3281
222. M. Xavier, P. Rosendahl, M. Herbig, M. Krater, D. Spencer, M. Bornhauser, R.O.C. Oreffo, H. Morgan, J. Guck, O. Otto, Mechanical phenotyping of primary human skeletal stem cells in heterogeneous populations by real-time deformability cytometry. Integr. Biol. **8**, 616–623 (2016)
223. Y. Deng, A.J. Chung, Next generation deformability cytometry: fully automated, high-throughput and near real-time cell mechanotyping, in *Proceedings of 20th International Conference on Miniaturized Systems for Chemistry and Life Sciences*, 9–13 Oct 2016, Dublin, Ireland (2016), pp. 148–149
224. J. Chen, Y. Zheng, Q.Y. Tan, E. Shojaei-Baghini, Y.L. Zang, J. Li, P. Prasad, L.D. You, X. Y. Wu, Y. Sun, Classification of cell types using a microfluidic device for mechanical and electrical measurement on single cells. Lab Chip **11**, 3174–3181 (2012)

Chapter 5
Optics and Fluidics

Amos Chungwon Lee, Jinhyun Kim, Jinseong Noh, Howon Lee and Sunghoon Kwon

Abstract The use of light in biotechnology has reversely enlightened the path for development in biosciences. In this chapter, the basic optical principles used in micro/nano systems for biological applications are explained and their applications are followed. This chapter is comprised of two parts: integration of optics in biodevices and optofluidics. The former part describes optics, fluorescence, and plasmonics that are integrated in micro/nano systems and related technology and discuss their current and future applications. The latter part describes fabrication and manipulation methods that are enabled by optics.

Keywords Optics · Fluorescence · Plasmonics · Optofluidics · Photopolymerization · Nanocomposites · Laser · Optomechanics · Optical tweezers

5.1 Introduction

Optical technologies, which have brought revolutionary impact to micro/nano systems, are widely incorporated in current biotechnologies. Along with fluidic technologies, by which biomolecules and biological specimen can be handled in simple manner, optical systems enable not only simple synthesis and fabrication of complicated matters at the micro/nano scale, but also manipulation of small objects that are intangible otherwise. Through merging the fields of fluidics and optics, biotechnology has seen remarkable progress in various fields including medicine, chemistry, engineering, and biology. In this chapter, we describe various principles and examples of different biotechnologies that utilize optical technologies, demonstrating the uses, advantages, and potential of optics and fluidics in

A. C. Lee (✉) · J. Kim · J. Noh · H. Lee · S. Kwon
Seoul National University, Seoul, Republic of Korea
e-mail: amoslee89@gmail.com

S. Kwon
e-mail: skwon@snu.ac.kr

G. Barbillon et al. (eds.), *Engineering of Micro/Nano Biosystems*,
Microtechnology and MEMS, https://doi.org/10.1007/978-981-13-6549-2_5

micro/nano scale biotechnology. The topics were categorized into integration of optics in biodevices and optofluidics. The optofluidics part was categorized into fabrication and manipulation using optics. Through specific cutting-edge examples that were provided in the script, the readers should get a sense how the integration of optics and fluidics is currently used and possibly will be used in the future.

5.2 Integration of Optics in Biodevices (Amos Chungwon Lee, Jinhyun Kim, Jinseong Roh, Howon Lee, and Sunghoon Kwon)

5.2.1 Introduction and Basics

The integration of optics and microfluidics has brought significant advances in micro/nano-systems, broadly ranging from biosensors to particle manipulation techniques [1–5]. In this section of this book, we categorized cutting-edge technologies that utilize optofluidics in three: synthesis, manipulation, and detection. Synthesis part of this chapter will introduce photopolymerization and its uses in microfluidic environments. The manipulation part will mostly be consisted of optomechanics in microfluidic environments, suggesting the bridge between the microfluidic systems to generic bioassays. The detection part of this chapter will address the detection methods that were enabled by optics in micro and nano-systems. This chapter was written to address current technologies as well as fundamental knowledge of the basics of optics in micro/nano-systems.

Since matter can also modulate or change light, which in turn gives us useful information about that matter, we can apply this phenomenon for high resolution and sensitive signal detection. A direct example is the object that we see. We can see this object since it absorbs and reflects light. In this section, we will study the detection methods in optofluidics.

Light is useful in observing materials in two aspects. First, light is one of the most abundant sources that can generate energy. Second, we can easily manipulate light with lens, filter, and many other optical devices. Therefore, we can easily configure the observing condition by modifying light. Third, in visible light wavelength, the light carries relatively low energy and barely modifies the target material. In other words, light can preserve the integrity of biomaterial we want to observe.

Optical microscopy is widely used as a device in order to observe small biomaterials in biology. We can simply observe biomaterials by illuminating the bright field and observing the transmitted light (bright-field microscopy). The absorption of light by the sample causes the contrast in the image, which we can observe and recognize this sample. Although the bright-field image suffers from low contrast in biomaterial, we even can clearly observe the boundary of bacterial cells and their shape. Jungil Choi et al. (2013) have used this simple bright field image to observe bacterial cell division and ultimately reduce the antibiotic susceptibility testing (AST) time [6]. They used Agarose gel to immobilize bacteria and diffused antibiotics using microfluidic channel (Fig. 5.1a). After time-lapse observation of bacteria cell division, they could accurately obtain MICs (minimal inhibitory concentrations) and reduced the time of AST about 6 folds.

The image of an object is very complicated and shows specific characteristics of this object. This leads to a new concept of using shape as unique identifying character of an object. If we manage to make such shape physically unclonnable, we can fundamentally block counterfeiting. Bae et al. (2015) used photopolymerization discussed in Fig. 5.1 to generate particle followed by silica coating and dry to induce a wrinkle [7, 8]. The biomimetic wrinkle pattern was similar to human fingerprints so that the random generated pattern contained unique codes. This means that the wrinkled microparticle can be generated with low-cost and scalable manner to be used as unique ID of an object. Furthermore, after developing algorithms to decode wrinkles from image obtained with a portable microscope attached to a smartphone, they expect to utilize this particle in real world.

There are many other interesting techniques and types of microscopes, which will not be covered here. However, all the optical microscopies have fundamental drawback. Since light has wave property, all the optical microscopies have a resolution limit called Abbe diffraction limit.

$$\mathrm{d} = \frac{\lambda}{2\mathrm{n}\sin\theta} \tag{5.1}$$

In green light wavelength (500 nm), diffraction limit is about 0.25 um, which is smaller than the size of biological cells (1–100 um) but is bigger than that of proteins (1–100 nm). In other words, we cannot see protein and many other objects smaller than 500 nm. Let's now explore some useful techniques to overcome these limits.

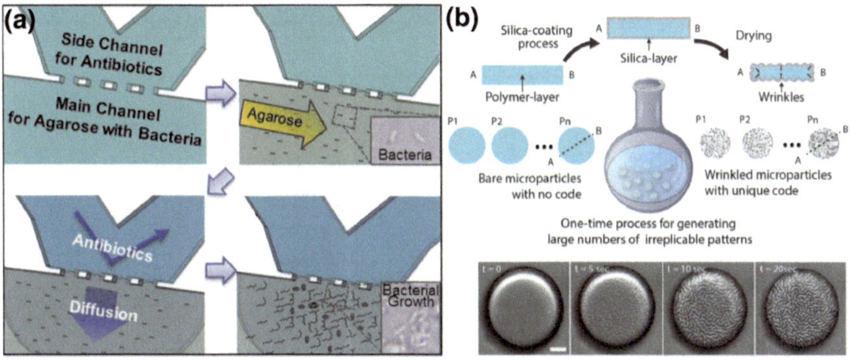

Fig. 5.1 **a** A device to rapidly measure AST by observing bacterial cell division. (With the permission from RSC publishing group.) **b** Overall procedure to generate physically unclonnable wrinkles (With the permission from Wiley)

5.2.2 *Fluorescence*

Fluorescence is one of the most widely tools to detect biological material below diffraction limit of light observation system. It is emission of light from a matter, which has absorbed light. When light passes through the matter, light is absorbed by this matter. In Quantum mechanical sense, quantum energy of the photon is transferred to the molecule and electron of the atom is excited. After electron undergoes non-radiative relaxation where small energy of excited electron is dissipated in the form of vibration, molecule then emits remaining energy in the form of photon (Fig. 5.2). This is the principle of fluorescence and therefore has longer wavelength than absorbed light.

Since wavelength of emitted fluorescence is longer that excitation light, excitation light can now be filtered out by short pass-filter, which in turn enables observation of signal (fluorescence) without background. With fluorescence molecules bound to a detection analyte, since it can bind to the analyte in molecular scale, the theoretical observation limit is below single molecule [9]. Even though diffraction of emitted fluorescent light remains same, fluorophore (fluorescent chemical compound) can label (or attached to) the target molecule, so we can now observe and quantify the analyte as well as image how this analyte is spatially distributed. Furthermore, efforts to obtain various color fluorophores leads to a development of a variety of fluorescent protein like Fig. 5.3a. We can now detect and observe small biological materials such as protein in relatively non-destructive, full color way.

Fluorescence signal detection technology can also take advantage of fluidic system. Since most biological materials are dissolved in liquid (even cells are filled with cytoplasm), a direct application would be multiplexed quantification of target molecules. An example is DNA microarray [10]. A DNA microarray is an array of DNA probe spots bound to a solid surface. Each probe has DNA sequence

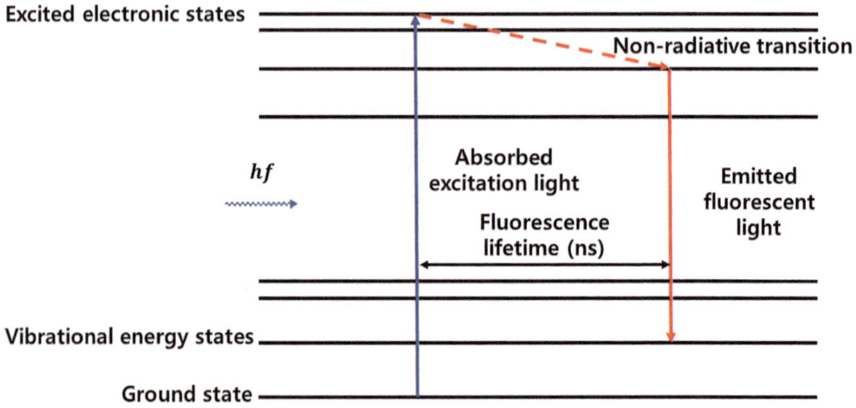

Fig. 5.2 Energy-level diagram showing the absorption and emission of light in fluorescence

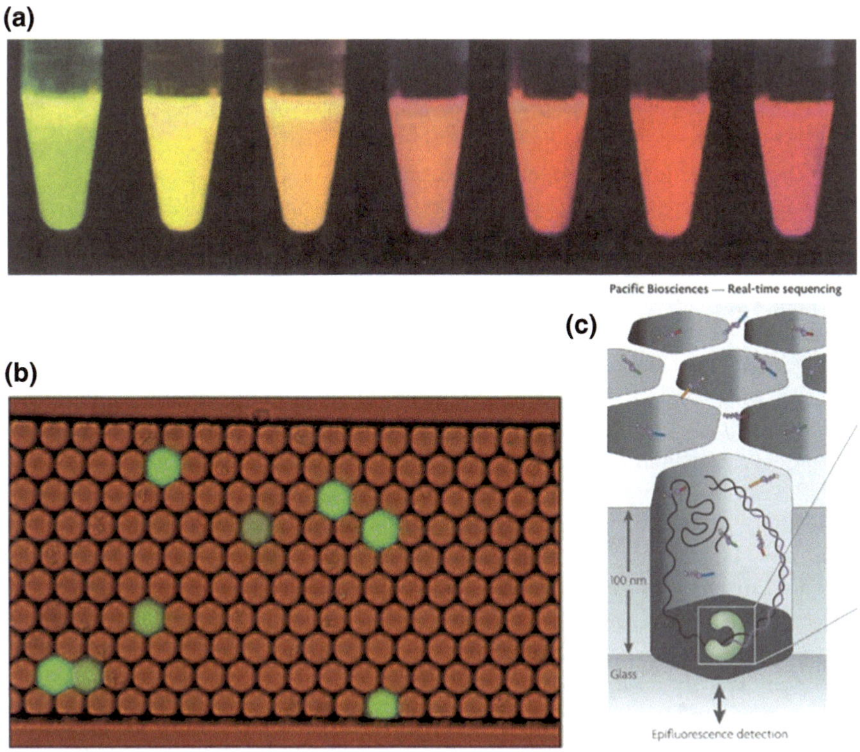

Fig. 5.3 **a** fluorescence image of various color fluorescent proteins (from left, mHoneydew, mBanana, mOrange, tdTomato, mTangerine, mStrawberry, mCherry) (With the permission from Springer). **b** Fluorescence microscope image of droplets. (With the permission from RSC). **c** Schematic of real-time DNA sequencing platform in Pacific Bioscience (With the permission from Springer)

complementary to target cDNA, and samples are prepared by reverse transcription of sample mRNA followed by labeling with fluorescence. The brighter spot on DNA microarray is, the more abundant target mRNA is, and we can quantify a variety of mRNA expressions in parallel. Even more, if we make DNA microarray chip to target cancer related gene, we can easily diagnose cancer status of a patient and Point of Care (POC) is not that far.

Multiplexing can also be achieved by emulsion droplets. Since each droplet in microfluidics can act as an isolated independent micro-reactor, droplet provides significantly reduced sample volume as well as high-throughput parallelization with low cost [11]. Also, microfluidic system can be designed to manipulate droplets so that fusion and splitting [12] as well as to sort the droplet can be realized [13]. Figure 5.3b shows the image of fluorescence signals in droplets. By detecting this fluorescence mark, we can sort each cells according to fluorophore target molecules of the cells.

Fluidic environment provides free movement of molecules dissolved in the fluid. This means fluorescence probe also moves freely in the fluid. Pacific Biosciences exploited this freely moving fluidic environment combined with fluorescent-labeled dNTP to perform real-time DNA sequencing from single DNA molecules [14]. Figure 5.3c describes the scheme of one DNA sequencing [15]. First, fluidic environment enables fluorescent-labeled dNTP move freely in reactor, so that no change of solution is needed. Second, fluorescence molecule is excited by zero-mode waveguide, which substantially reduces the observation volume to the surface of the polymerization reaction. A zero-mode waveguide is an optical waveguide that guides light energy into a very small volume, smaller than the wavelength of the light. If fluorescence-labeled dNTP approaches polymerase to be polymerized, then fluorescent dye is excited and this represents just the real-time sequencing at single molecular scale.

Flourescence molecules also can be attached to biomolecules observe DNA nanostructures using FRET efficiency analysis. Forster resonance energy transfer or FRET describes a mechanism of energy transferring between two chromophores. Depending on the distance between the two chromophore, the emitted fluorescence will differ. For example, Choi et al. has described reconfigurable DNA accordion rack, which changes its shape according to the input DNA locks [16]. Through measuring the FRET efficiency between the two dyes, Choi et al. has described configurability of DNA nanostructures.

As you can see in this section, one can construct many more complicated and desired platform by simply designing the microfluidic environment combined with fluorescence detection. Although fluorescence can theoretically be used to detect single molecule, it is hard to obtain single molecule imaging. This is because (1) fluorescence molecules are dissolved but not attached to target molecule and create background noise. (2) Fluorescence also is a visible light so that we can obtain blurred image of emitted fluorescence. There are many researches about enhancing resolution of fluorescence microscopy, which is beyond the scope of this book. But interested readers can study more deeply about advanced techniques of fluorescence microscopy.

5.2.3 Surface Plasmon and Surface-Enhanced Raman Scattering

Fluorescence, in principle, measures the emission of light from a material after absorption of light to detect a material. By contrast, another light-matter interaction that can be used to detect a material is the scattering of light [17–20]. Especially, Raman spectroscopy has some powerful advantages to study and detect a material. First, in contrast to the emission spectroscopy, Raman spectrometry does not require labels (for example, fluorophore) for detection. Since attachment or labeling of fluorophores is not only laborious, but also can interact with or modify the target molecules and other detecting system. Second, Raman spectroscopy can measure vibrational, rotational, and other low-frequency modes in a molecule, which can be a fingerprint of that molecule. Third, Raman spectrometry does not suffer from rapid photobleaching commonly observed in fluorescence spectrometry. Briefly, Raman spectroscopy can be easy, label-free, and non-destructive ways of detecting materials.

Scattering is a physical process where light is forced to deviate from a trajectory they travel due to non-uniformity is that material. Among them, Raman scattering is inelastic scattering of photon, i.e. the scattered photon has a different wavelength from incident photon (Fig. 5.4). This energy difference corresponds to the quantum energy difference between vibrational modes of these molecules. Since vibrational modes are specific to the chemical bonds of the molecule, the spectrum of scattered photon provides the fingerprint of this molecule [17].

Even though Raman spectroscopy has great advantages, Raman scattering is an extremely inefficient process compared to fluorescence (Its scattering cross-section is smaller than that of fluorescence over fourteen orders of magnitude folds [18]. This means that we should amplify Raman scattering signal in order to increase the sensitivity of the detection. Luckily, there is a phenomenon called SERS (Surface-Enhanced Raman scattering). When light is incident to metal nanoparticle, if incident frequency of light meets the nanoparticles' surface plasmon resonance condition, it has been known that this surface plasmon induce strong electromagnetic fields near nanoparticle, which significantly amplifies the Raman scattering signals of nearby analyte [21]. In other words, there are two conditions to achieve SERS: (1) specific structure of metal nanoparticle, (2) target analyte should be sufficiently close to that nanoparticle to enable chemical interaction between them. If we manage to achieve these conditions, we can detect materials in single molecule resolution.

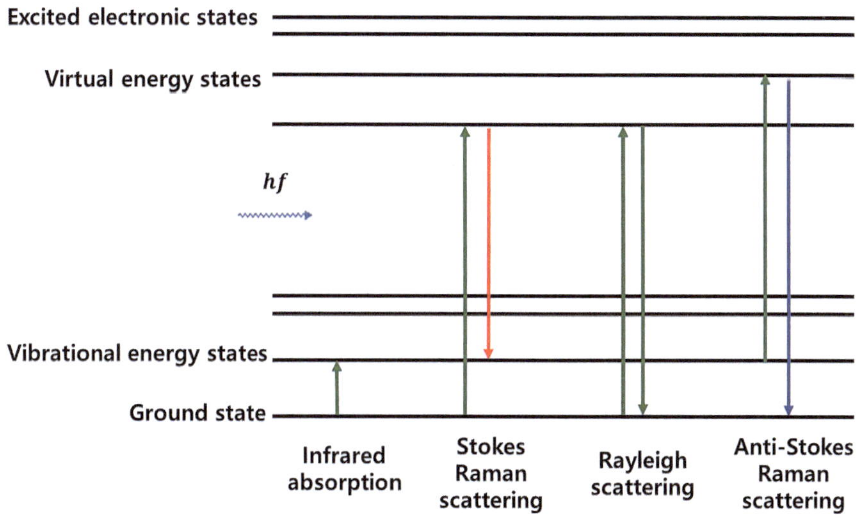

Fig. 5.4 Energy-level diagram showing various scattering events

There are many efforts to achieve SERS satisfying above two conditions. What's important is that signal enhancement should not only be high and uniform but also reproducible. Selena Chan et al. [22] produced efficient and reliable SERS substrates by coating silver onto nanoporous silicon and achieved seven orders of magnitude Raman enhancement (Fig. 5.5a). D. Lim et al. (2011) produced gold

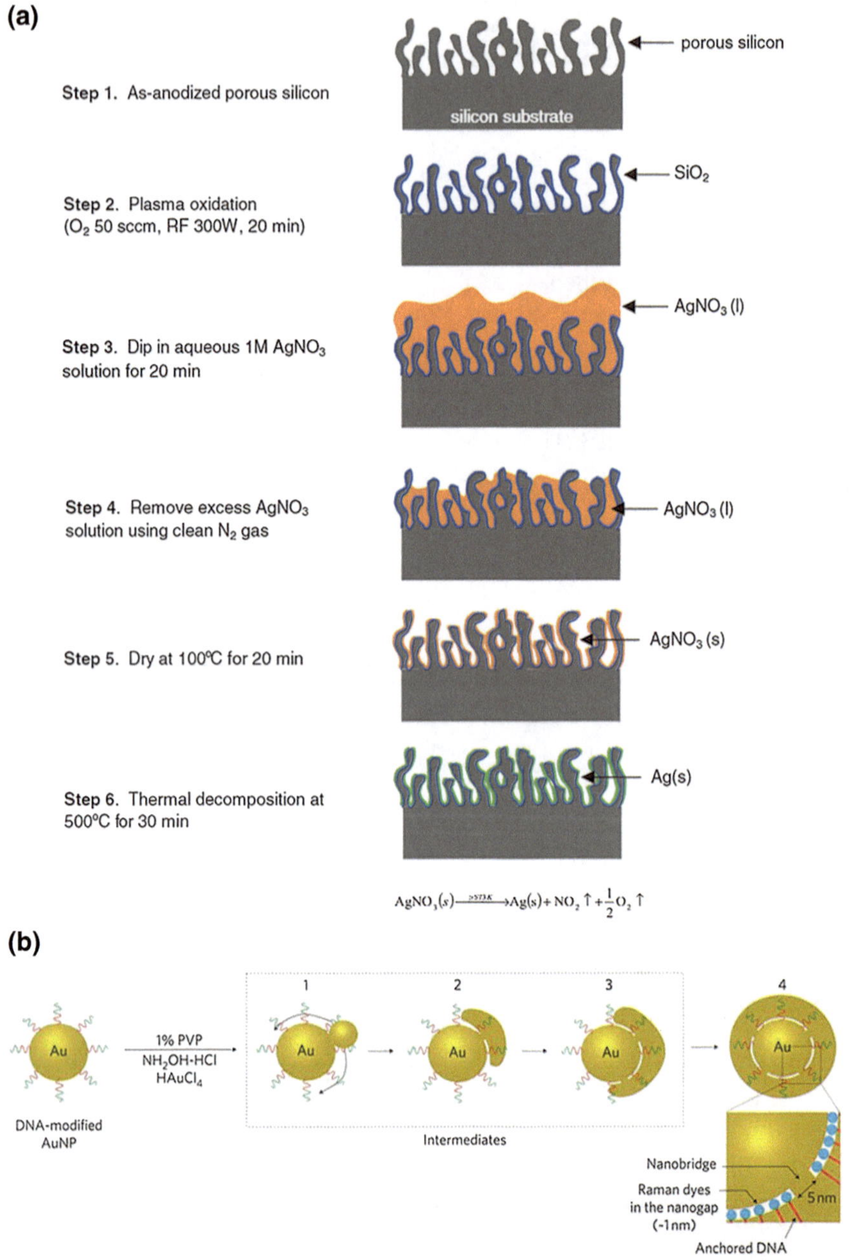

$$AgNO_3(s) \xrightarrow{\geq 573K} Ag(s) + NO_2 \uparrow + \frac{1}{2} O_2 \uparrow$$

Fig. 5.5 **a** Fabrication process of metal-coated nanoporous silicon (With the permission from Wiley). **b** Schematic of making gold nanobridged nanogap particles (With permission of Springer)

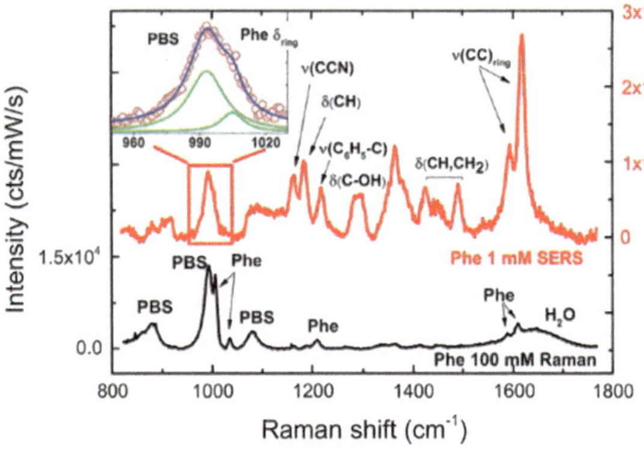

Fig. 5.6 Using LIQUISOR, which employs gold nanorods, capped with Cetryltrimethlyammonium Bromide and dispersed in deionized water, Fazio et al. was able to detect SERS spectrum of Phenylalanine (red line) compared to the solution phase Raman of Phe 100 mM in PBS (black line). The enhancement factor was around $\sim 10^5$ (With permission of Springer)

nanobridged nanogap particles by anchoring DNA and obtained reproducible and sensitive (up to 10 fM for the detection limit) SERS signal (Fig. 5.5b) [23].

Figure 5.6 depicts Raman and SERS spectra of Phenylalanine. A strong difference between these two spectra is observed. Indeed, an enhancement factor was found and equal to 10^5. Aside from researches mentioned above, there are many other trials to increase performance of SERS and is under active research [24]. SERS is a promising part of optical detection and is expected that its applications will continuously be extended.

5.3 Optofluidics (Amos Chungwon Lee, Jinhyun Kim, Jinseong Roh, Howon Lee, and Sunghoon Kwon)

5.3.1 Micro/Nano-Structure Synthesis and Fabrication Using Optics

5.3.1.1 Photopolymerization

Light has been widely used in micro and nano-systems in fabricating processes, such as photolithography [25]. In microsystems for biological applications, the synergetic integration of optics and fluidics brought significant advances in particle fabrication and in situ polymerization for bioassays. Photodegradable polymers or photocurable monomers, which are used in these processes, are thus gaining interests in bio-industry for their controllability and simplicity [5, 26–30]. There have been a wide variety of efforts to apply this uniqueness of photopolymers in research fields ranging from 3D printing to tissue engineering.

According to Roger Philips, photopolymerization can be categorized into two major classes of chemical reactions [25]. The first class involves photochemically active cross-linking groups that can act as the intermediate excited state. The second class involves an excited photoinitiator molecule to interact with monomer to induce chain reaction to form polymers. The figure above illustrates the second class, where the photoinitiators are split into two or more parts when light source is exposed. The radicals or ions thereof react with the monomers to form polymers (Fig. 5.7). These photocurable materials with different properties can be chosen for specific applications.

We designed this section of Optofluidics to concern synthesis using optofluidics. Specifically, photopolymerization, nanocomposites, and their biological applications will be discussed.

Photopolymerization itself has been widely used as a tool for biological and biochemical applications. In this chapter, we will open optofluidics in micro/nano-systems with photopolymerization used for biological applications. Specifically, we will concentrate on the optofluidic lithography systems and its applications.

Patterning the shapes or the structures with photopolymerization can be applied to a broad range of micro/nano-systems such as particle manipulation, actuators, and assay system using the particles and the actuators [2, 31–38]. Scientists and researchers have developed numerous ways to construct particles or structures with desired shape and function. To realize desired shapes, methods for patterning the light exposure are widely used. The patterned light exposure would activate the curing agents in desired regions. In 2007, Chung and colleagues reported optofluidic maskless lithography (OFML) system for real-time synthesis of photopolymerized microstructures in microfluidic channels [39] (Fig. 5.8).

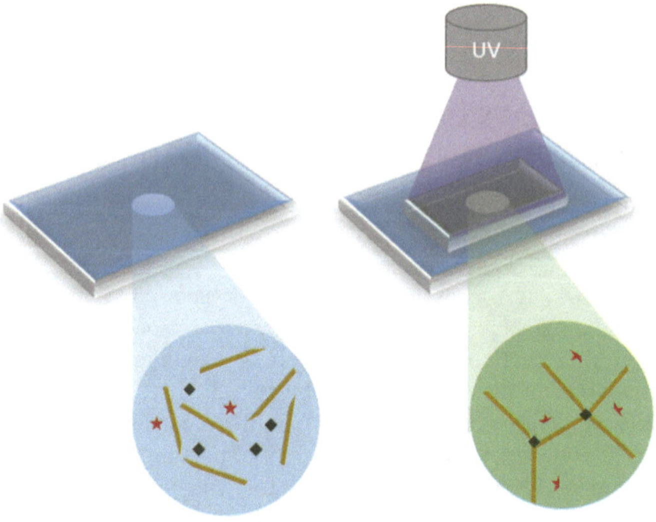

Fig. 5.7 Photopolymerization. Oligomers are suspended for photopolymerization (Left). When exposed under Ultraviolet light, photopolymerization initiators or curing agents are usually photocleaved, linking the oligomers (Right). The star-shaped particles indicate curing agents while rods and squares are oligomers and monomers to be polymerized

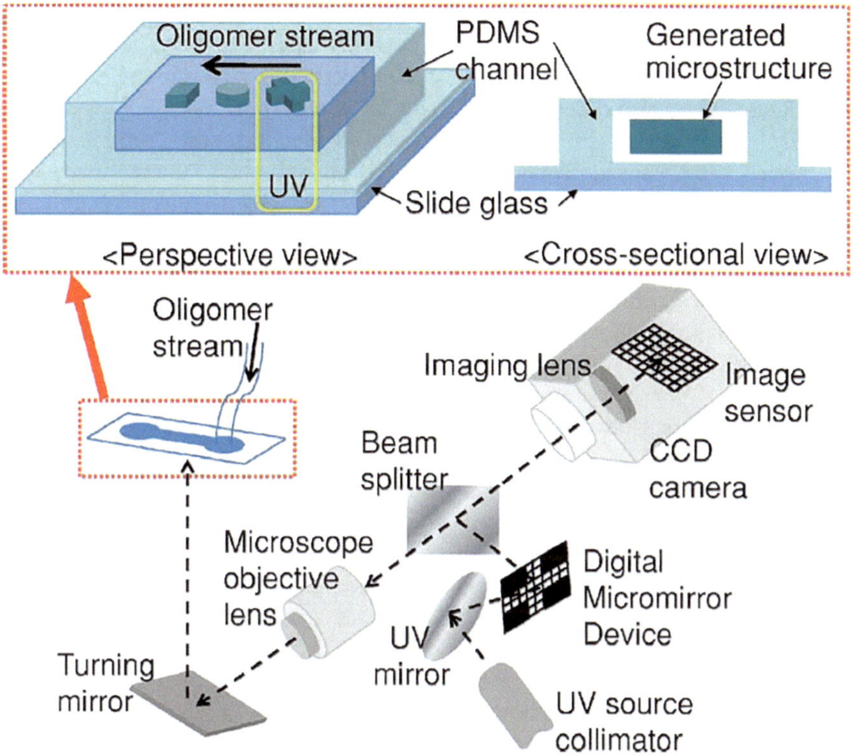

Fig. 5.8 Optofluidic maskless lithography setup by Kwon et al. UV source collimator collimates the UV lights that are modulated by digital micromirror device. Beam splitter splits the modulated beam for image processing and patterning photopolymers (With permission of American Institute of Physics)

From the UV source, the light is collimated to prevent light diffraction and guide the light rays to the digital micromirror device. Contrast to the traditional photolithography, in which photomasks were used for patterning, digital micromirror device can be controlled with a processor in real-time. The OFML system developed here guides UV light that is selectively reflected to the digital micromirror to the beam splitter. The UV light is then refracted through microscope objective lens to pattern the desired shape in desired scale. They report that they were successful in synthesizing polymeric microstructures with a single pixel resolution of 1.54 μm by 1.54 μm at 0.1 s exposure time. Using such in situ photopolymerization techniques, liquid-phase photocurable materials can be patterned to form anchored and unanchored polymeric structures, widely ranging from valves to micro particles.

For example, Kwon's group from Seoul National University reported guided and fluidic self-assembly of microstructures using railed microfluidic channels. The microstructures were synthesized by in situ OFML system.

Inside the fluidic channels fabricated with soft lithography, photopolymerization could be utilized to fabricate particles in desired shapes (Fig. 5.9). The ultraviolet light, which cures the photopolymers, is modulated via digital micromirror device. The digital micromirror device (DMD) is an optical semiconductor device that utilizes digital light processing module, enabled by the progression in micro-electro-mechanical technology. As the micro-units of these devices are modulated using DMD, the collimated UV light rays can be selectively reflected onto the substrates. Because the micromirrors are dynamic, materials can be cured in different shapes. Another advantage of digital micromirrors is to easy control spatially and temporally the desired to-be-exposed region. Spatial and temporal resolution can be revolutionary because traditional photolithography can be difficult and complex when synthesizing micro-scale structures dynamically. Using this phenomenon, Kwon's group from Seoul National University fabricated microparticles with different shapes on a micro-rail in situ [40, 41]. On to the railed microfluidic channel, they used UV-DMD system to fabricate particles in different shapes. The particles were used in self-assembly, allowing complex microstructures assembled in fluid. Even though the particles were not different in their functions, the assembly of such different population of particles promises complex functional micro-system, which can be applied to different areas in biotechnology. The most significant point of this development is that parallel fabrication of micro/nano-devices are possible through this fluidic self-assembly.

Not only two-dimensional particles can be assembled, but also three-dimensional fabrication of heterogeneous microstructures can be fabricated using soft membrane deformation and OFML system (Fig. 5.10). Seung Ah Lee et al. proposed a method for high-throughput generation of three-dimensional

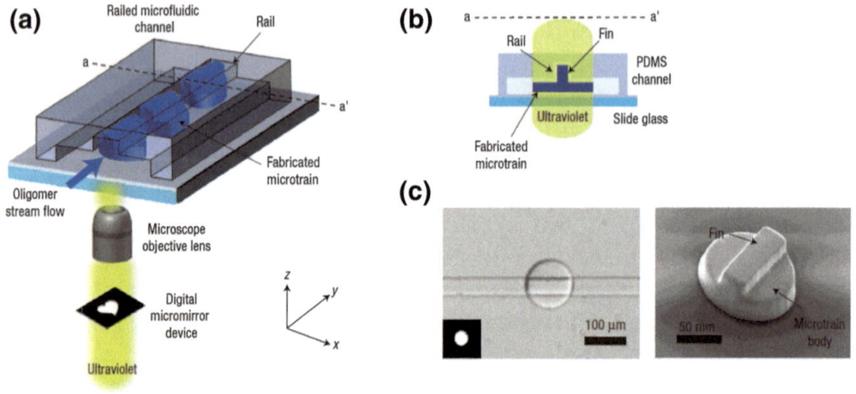

Fig. 5.9 **a** Schematic of in situ photopolymerization in microfluidic channels. **b** A sectional view of the system. **c** An example of a microparticle flowing with the rail (With permission of Springer group)

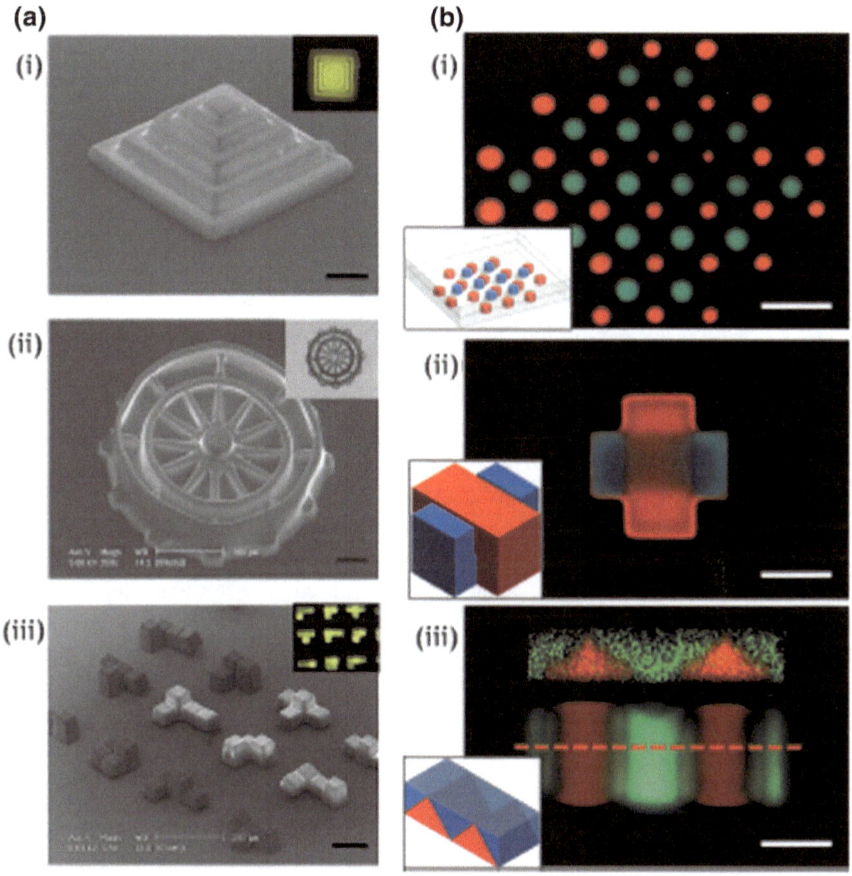

Fig. 5.10 **a** Scanning Electron Microscopy (SEM) images of fabricated 3 dimensional microparticle structures **b** A colorful view of the 3-dimensional microparticles (With permission of RSC group)

microstructures using a membrane mounted microfluidic channel [42]. If the OFML system described above can be utilized to create 2D structures in different layers, 3D microstructures can be fabricated. The group used a soft membrane to control the thickness of each layer, which allows large area projection lithography with high vertical resolution. Using microfluidics, the photocurable material resin can be switched, producing heterogeneous composition of each and every layer of the three dimensional structure. Kwon's group demonstrated three-dimensional patterning of different types of cells in a hydrogel for a microfluidic platform. This demonstration showed that three-dimensional structures made by OFML can be utilized to study co-culture of cells and cell-to-cell interaction, and perhaps cell-to-tissue interactions. This simple and high-throughput patterning of 3D microstructures would have an impact to produce heterogeneous biomaterials with various shapes.

The same group also reported fine-tuned gray-scale OFML for three-dimensional freeform shape microstructure fabrication. On top of the three dimensional fabrication, they utilized gray shade levels, that are modulated through multiple reflections of light in the DMD. Another implementation of photopolymerization in microfluidic device is a synthesis of membrane-type microvalves in rectangular microfluidic channels via seal photopolymerization (Fig. 5.11). Microfluidic devices have revolutionized bioassays, as the fluid mechanics in micro world have characteristics that can be utilized in bioassays. This work implemented photopolymerization to develop the limitations of microfluidic valves that can control the flow in integrated microfluidic circuits. Microvalves usually are made of membranes that can be seal the microchannel as the gas pressure pushes out the membrane. The idea of this work is that through controlled polymerization enabled by photocurable resin, the channels can be sealed via polymerization by photolithography, producing a structure matching the shape of the deformed membrane curvature [43]. This piece of art has enabled accurate sealing of microvalves, which are widely used in cell sorting, fluid control, and other biological assays.

Using in situ optofluidic photopolymerization, three-dimensional fluidic self-assembly by axis translation is also possible. Chung et al. used two dimensionally fabricated micro-components in railed microfluidics to produce three-dimensional structures (Fig. 5.12).

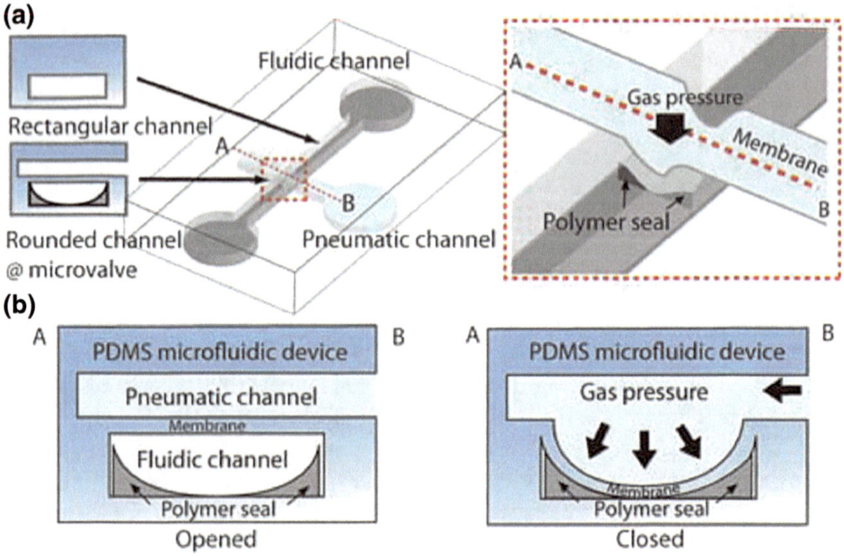

Fig. 5.11 **a** Schematic of fabrication of the polymer seals for microfluidic system with microvalves. **b** Sectional views of the system (With permission of RSC group)

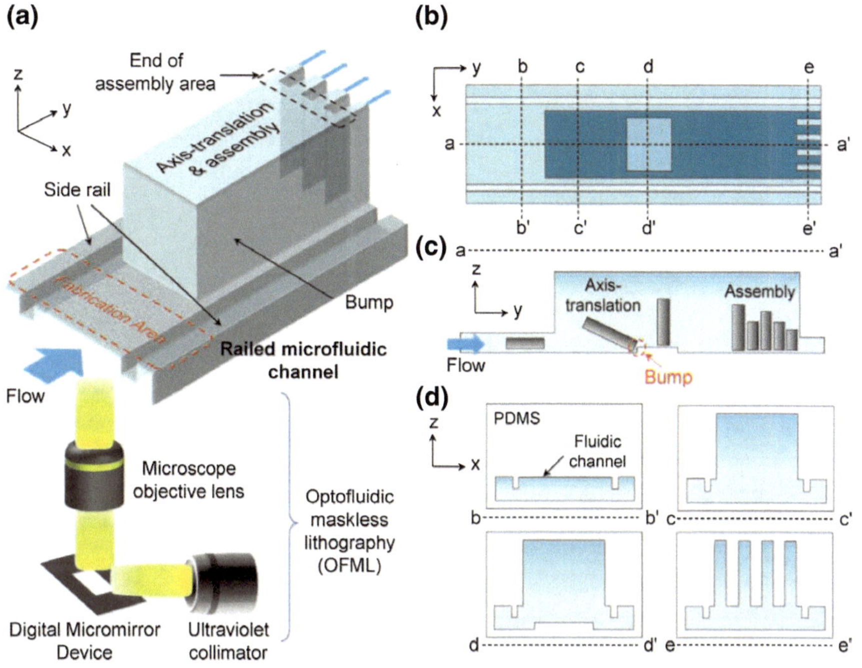

Fig. 5.12 **a** Schematic of in situ photopolymerization in microfluidic channels. **b** A sectional view of the system. **c** An example of a microparticle flowing with the rail. **d** cross sectional view of the microchannel according to its location (With permission of Wiley group)

Using OFML, desired 2D structure was cured in the fabrication area, as indicated in the figure. The 2D structure is then raised by the premade bump for axis-translation and assembly [44] (Fig. 5.13).

Conclusion

In this section, we have introduced methods that are possible with the optofluidic structure synthesis alone. 2D and 3D structures can be synthesized and assembled using the photopolymerization and variety of bioassays or mimetics of biological phenomena, which can be performed by utilizing the advantages of optofluidic synthesis methods.

5.3.1.2 Nanocomposites

The great advantage of in situ photopolymerization is that the photocurable resin can incorporate other materials [2, 36–38]. Materials, such as DNA, antibody, magnetic nanoparticles, and more can be encapsulated in situ by photopolymerization. The various shapes of particles and structures can provide advantage in biological applications in many ways. Nanocomposites are the photocurable

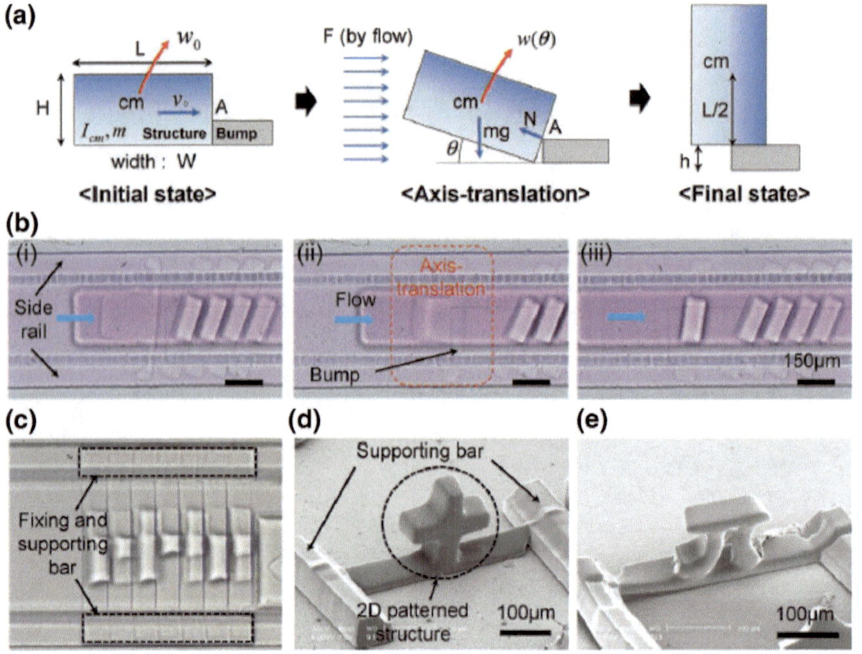

Fig. 5.13 **a** A schematic of axis-translation of the microparticles. **b** Top view of microparticles with axis-translation. **c–e** Examples of microparticles that are axis-translated (With permission of Wiley group)

polymer structures encapsulating nano-sized materials. This section will mainly be consisted of nanocomposites that can be or are being utilized in biological applications.

Optofluidic encapsulation of nano-materials as well as micro-materials can be applied in many ways. Song et al. has described liquid-capped encoded microcapsules for multiplex assays which can encapsulate thousands of liquids inside a core-shell structures, which are assembled onto a microchip for diagnosis and examination [45]. Chung et al. introduced optofluidic encapsulation and manipulation of silicon microchips using image processing based optofluidic maskless lithography and railed microfluidics [41, 42] (Fig. 5.14). Specifically, this work showed the integration of a fluidic channel, encapsulation, fin fabrication, guide channels, and assembly. This platform would be a platform to manipulate microscale devices and self-assemble in order to form large-scale systems. This would not only be applied to the field of display, but also to the particle based biosensors.

Structures at micron-scale, such as silicon chips, can be encapsulated in the photocurable polymers. Interestingly, if photolithography technique such as OFML, which uses DMD, is employed, nano-materials can be encapsulated in even smaller particles or structures. For example, Hyoki Kim et al. have used superparamagnetic colloidal nanocrystal clusters (CNCs), which are comprised of single-domain

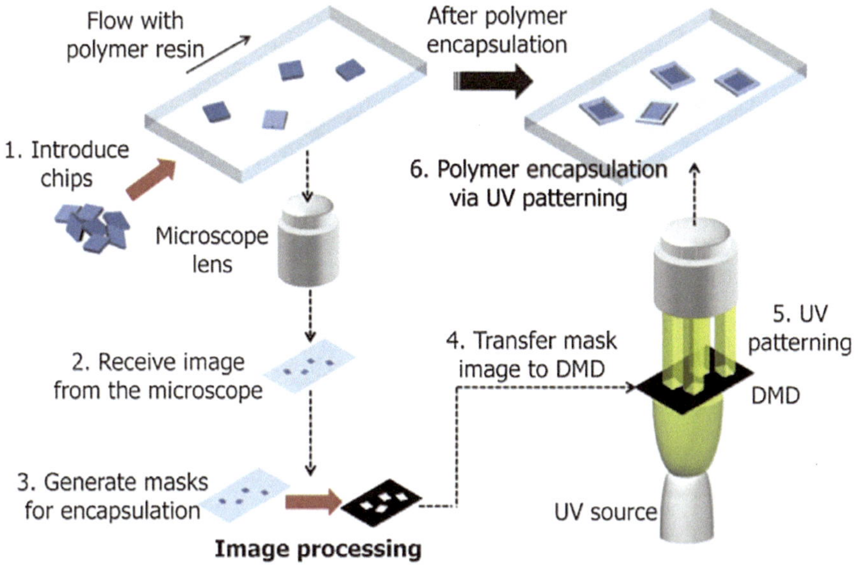

Fig. 5.14 Scheme of coating the particles with polymers with image processing and dynamically controlling the DMD (With the permission from the RSC group)

magnetite nanocrystals (Fig. 5.15). These microparticles, which contain nanocrystals inside silica shells, can form chain-like structures along the external magnetic field lines. When aligned in chain, the color of the light diffracted from the chain is determined by Bragg diffraction theory. Kim et al. have employed these characteristics to cure the resin mixed with superparamagnetic CNCs [46, 47] (Fig. 5.17). When desired external magnetic field is applied, the CNC chains will form to diffract desired color (Fig. 5.16). Because the color would be removed when the external magnetic field is suppressed, the photocurable resin could be cured in order to fix the CNC chain. The fixed chain would fix the structural color of the cured material, thus given the name: M-ink. M-ink can be used in color-coding the particles that are used in biological applications. Applying the unique characteristic of the optofluidic synthesis methods enabled M-ink nanocomposite.

One of the advantages of this nanocomposite is that the particles can rotate when different external magnetic forces are applied to the fixed magnetic axes of the particle. The rotation can provide faster and larger surface contact inside the solution, possibly increasing the efficiency of different bioassays inside the sample fluid.

The same group has applied this idea in multiplexed DNA detection. High-throughput and multiplexed bioassays are realized through optofluidically synthesized magnetic particles [48]. These encoded microparticles allowed to identify distinct codes on assay particles in the pool of diverse samples. The figure shows the reflection micrographs of three different particles with different DNA oligomer probes and multiplexed analysis based on DNA oligomer target

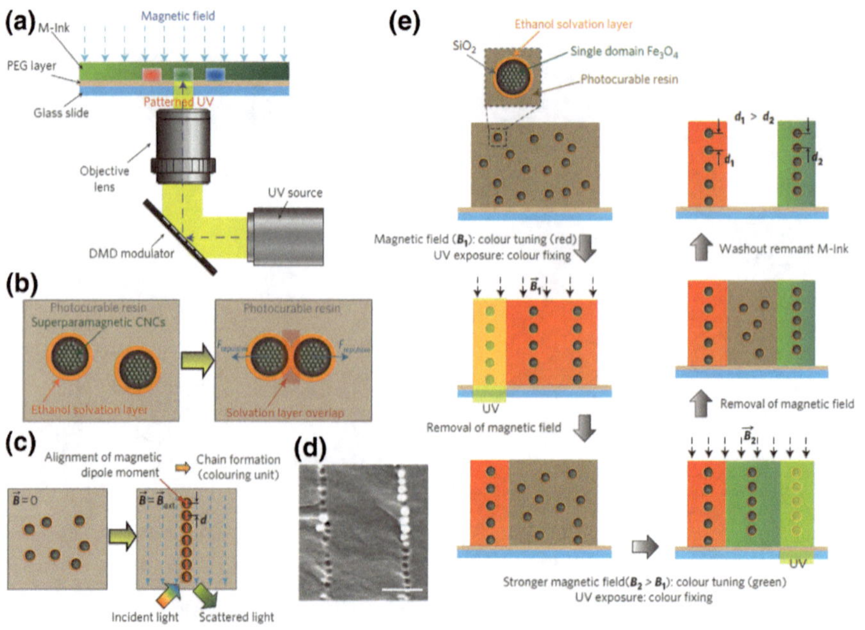

Fig. 5.15 **a** Schematic of the fabrication of the M-ink microparticles. **b** The underlying principle of M-ink particles. **c** Alignment of magnetic dipole moment. **d** SEM image of the aligned magnetic beads. **e** A scheme of multi-color tuning of the microparticle (With the permission from the Springer)

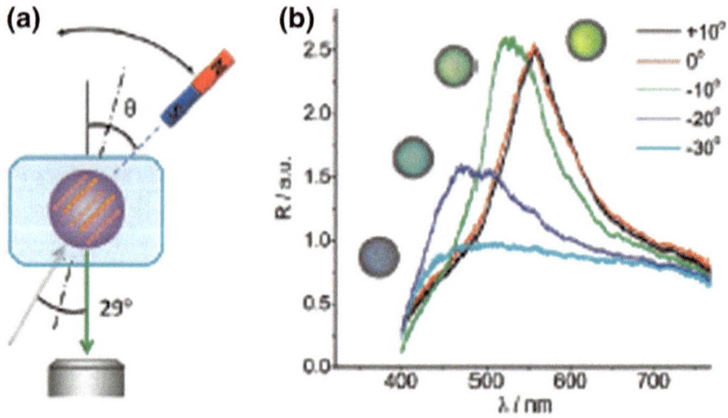

Fig. 5.16 **a** Scheme of the experimental setup for studying the angular dependence of the diffraction property of the magnetochromatic microspheres. **b** Reflection spectra and corresponding digital photos recorded from a single microsphere at different tilting angles (With the permission from the Wiley Online Library)

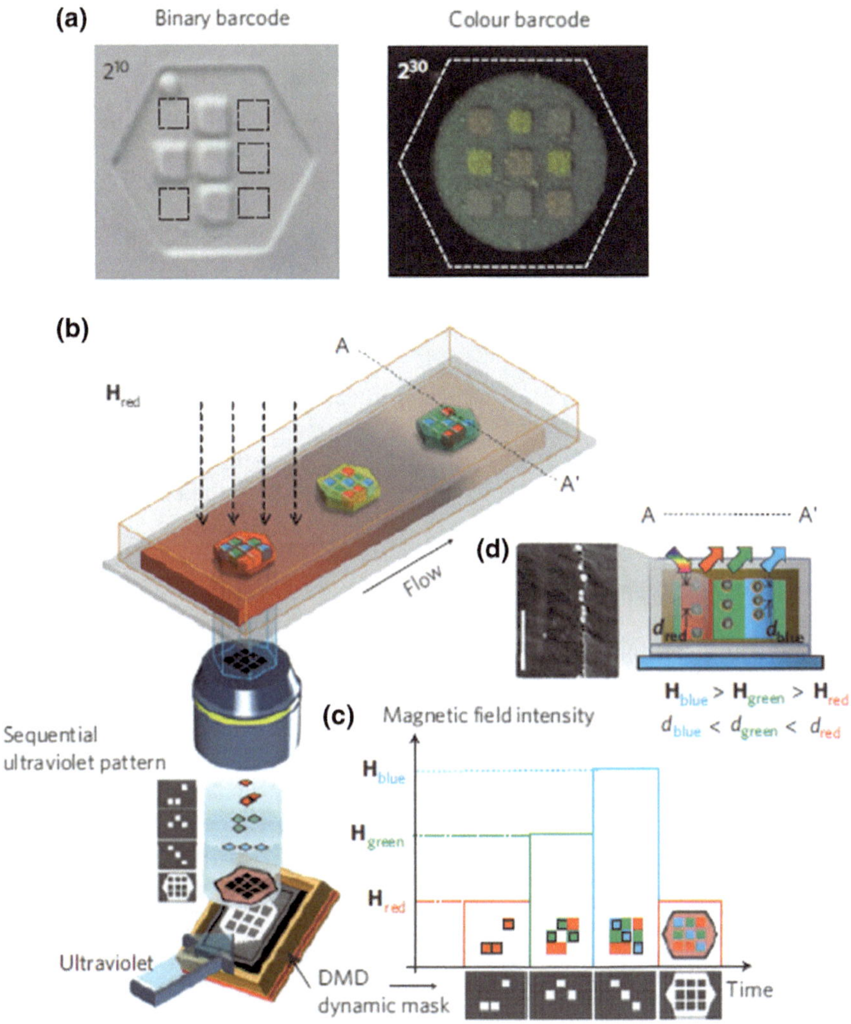

Fig. 5.17 **a** Using nanocomposite system, microparticles with binary barcodes and color barcodes can be fabricated. **b** The scheme of in situ fabrication of the magnetic particles. **c** Examples of fabricated colorful microparticles. **c** Color coding example by controlling magnetic field intensity. **d** The relationship of the distance between the CNCs and the color (With the permission from the Springer)

hybridization. The code region and probe region are segregated, for detection of codes for hybridization assays. Through this platform, the group has developed a novel method that can encode billions of different codes. The optofluidic manner of synthesizing such nanocomposites can be scalable, high-throughput and straightforward, opening a new possibility to new multiplexed bioassays.

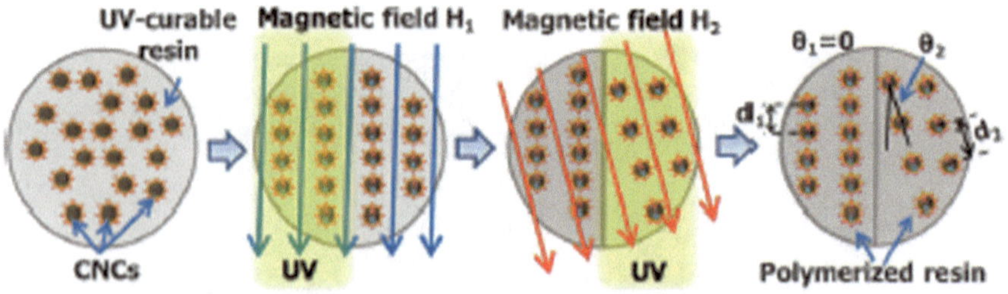

Fig. 5.18 Lithographic compartmentalization of an emulsion template (With permission of RSC group)

Magnetic particles can also be compartmentalized using this optofluidic lithography technique [49] (Fig. 5.18).

The group sequentially exposed the particle, changing the direction of the applied magnetic field. The resulting particle cured has two different compartments, of which structure colors are different.

Regular magnetic nanocomposite microspheres can be made using droplet microfluidics [50] (Fig. 5.19). A microfluidic valve controlled by gas pressure can insert M-ink reservoir inside the stream of mineral oil to be encapsulated. The M-ink microsphere droplet is then transferred to a wider channel, where the magnetic field is modulated and photopolymer composites are cured.

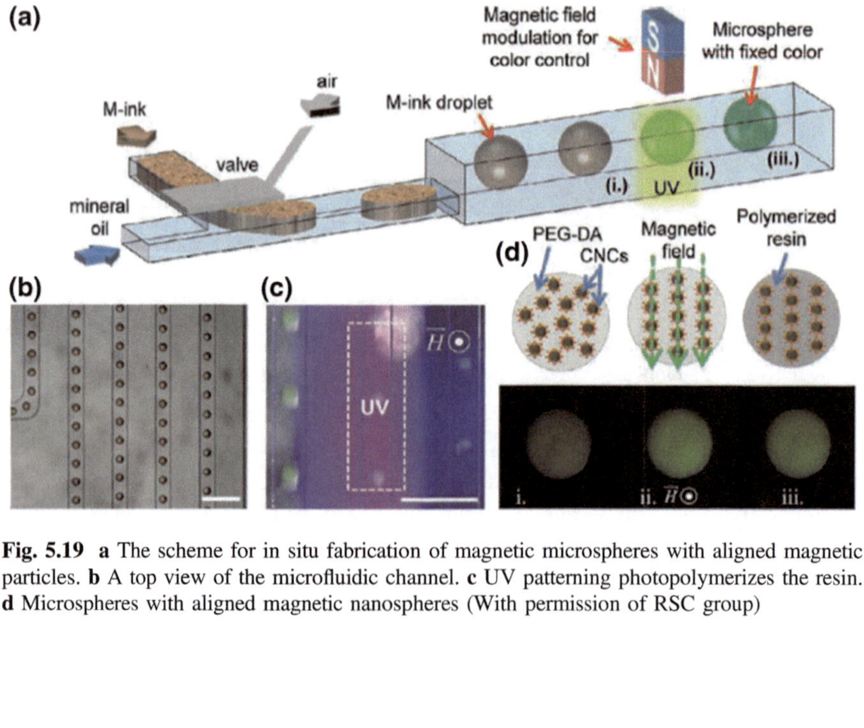

Fig. 5.19 **a** The scheme for in situ fabrication of magnetic microspheres with aligned magnetic particles. **b** A top view of the microfluidic channel. **c** UV patterning photopolymerizes the resin. **d** Microspheres with aligned magnetic nanospheres (With permission of RSC group)

We introduced some examples of magnetic nanocomposites synthesized optofluidically, which can form structural colors for barcoding. Barcoding enables a wide range of high-throughput bioassays in scalable and less expensive manner. However, the magnetic force such as translational magnetic force or magnetic torque can be utilized. Chung et al. have introduced an in situ fabrication and actuation of polymer magnetic microstructures [51] (Fig. 5.20).

Photocurable material is mixed with superparamagnetic CNCs and dispersed inside a PDMS channel. The PDMS channel has an opening so that the material is in contact with the bare glass. When cured, the photocurable material will form an anchor. When the cured object is anchored, and when the right magnetic field is applied, the cured nanocomposite will actuate. The movements of these actuators could be expected when the magnetic torque is calculated.

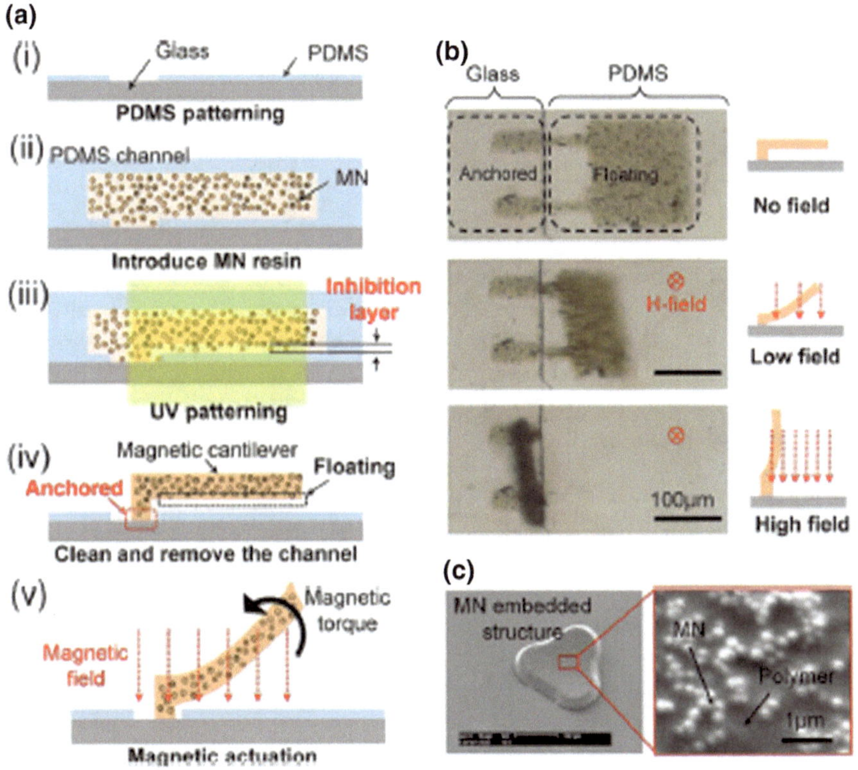

Fig. 5.20 **a** Schematic of fabrication of the magnetic microactuators. **b** A microscopic view of the microactuators. **c** SEM images of the magnetic nanoparticle embedded structures (With permission of Springer)

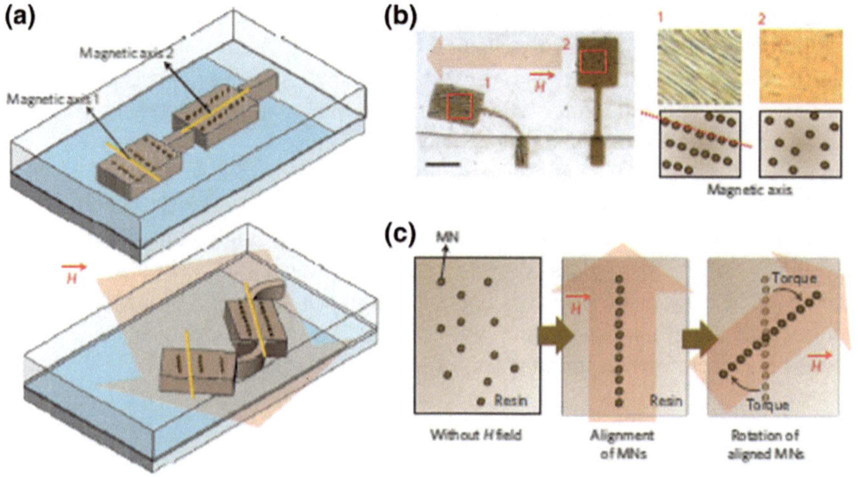

Fig. 5.21 **a** Schematic of multi-axis magnetic microactuators. **b** A microscopic view of the microactuators. **c** The principle of the microactuation (With permission of Springer)

The same phenomenon applies to magnetic actuators developed by Kim et al. Magnetic anisotropy can be programmed easily and readily when using photopolymerization [52] (Fig. 5.21). Patterns can be generated via DMD modulator to produce versatile patterns. Polymeric micro-components are widely used in MEMS devices, but components that can actuate complex motions have not been developed. Using this technique, many complex static microstructures can be fabricated. The microstructures synthesized can be programmed in actuation when the right sequential magnetic axes are formed by external magnetic field, which is applied to the superparamagnetic CNC chains.

Conclusion

Nanocomposites that can be easily generated using optofluidic synthesis methods are attractive in that their shapes, colors, magnetic properties, biological properties, and chemical properties can be modulated using diverse strategies. The field will continue to progress, and more biological and chemical assays would be possible.

5.3.1.3 Biological Applications

In this section, we will look into some of the biological applications of particles or structures synthesized in optofluidic manner.

A good example for applying optofluidic photopolymerization technique is a demonstration given by Kim et al. [53] (Fig. 5.22). They have developed a high resolution polymer fabrication method based on the OFML system for in situ charge selective nanoporous hydrogel fabrication in a glass microchannel.

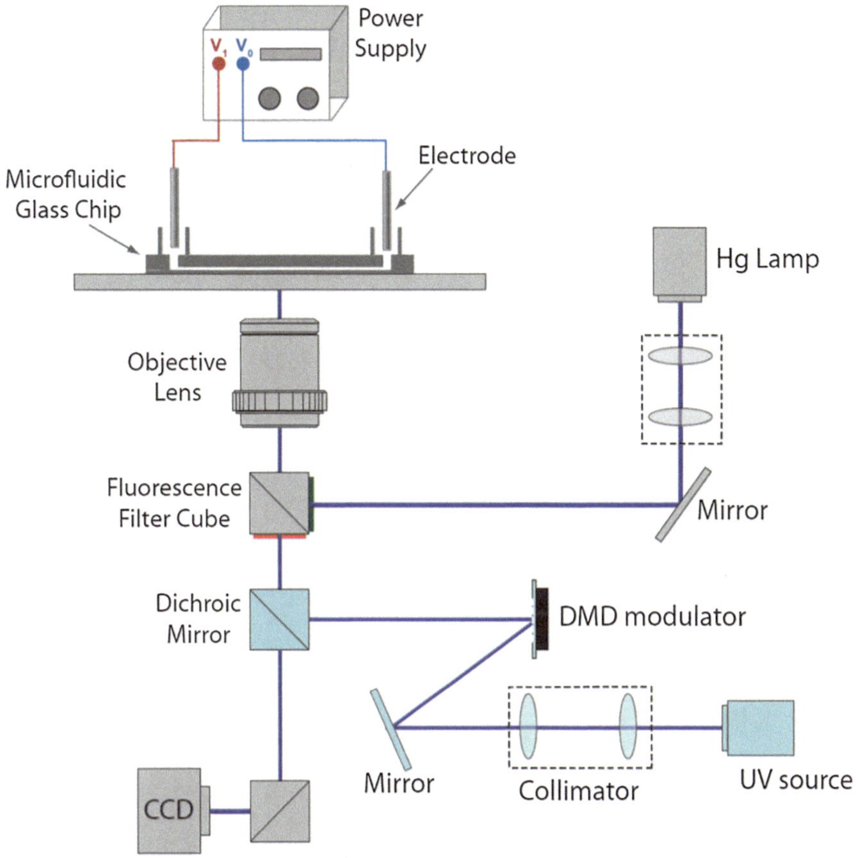

Fig. 5.22 Schematic illustration of instrumentation for charge selective preconcentration, A genatively charged hydrogel is produced by OFML system. Spatial light modulator (DMD) inside the system produces arbitrary UV light pattern dynamically (With permission of American Institute of Physics)

Using photopolymerization, they added micro-components of poly-AMPS [(2-acrylamido-2-methyl-1-propanesultonic acid)]. This component forms negatively charged nanoporous hydrogel and acts as cation-selective filter for biological sample preconcentration.

Another example includes generation of microcups using the same OFML setup. These microcups can act as 3D hydrogel cell microcarriers that can be manipulated easily. The authors show that different microcarriers are labeled with different fluorescence and shapes, which can encode the each cell carriers. One can additionally imagine nanocomposite microcarriers that have different extrinsic properties. In that manner, the optofluidically-synthesized microcarriers can be an excellent technique to address cell-based bioassay in high throughput manner [54] (Fig. 5.23).

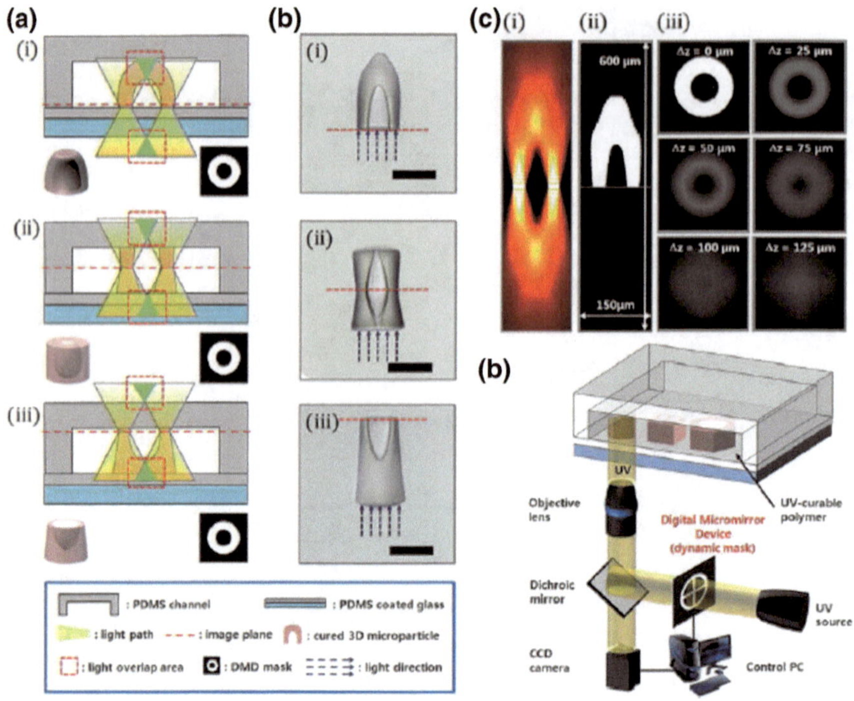

Fig. 5.23 Fabrication of microcups. **a** Schematic illustration of the approach used to manufacture 3D microcups with 3 different morphologies with arbitrary control of focus depth. **b** Bright field images of fabricated 3D microcups with 3 different shapes. All the 3D microstructures are extruded from a same ring-shaped mask pattern. All scale bars indicate 100 mm. **c** Simulation of light propagation. (i) A cross-sectional view of calculated z-dependent intensity distribution of light projected from a ring-shaped object. (ii) Estimation of the particle shape by determining the gel point. (iii) In-plane light intensity distribution; different values of Dz are displayed. **d** The OFML system used for creating 3D microstructures. The OFML system is composed of two parts: the focus-tunable channel and the optical system (With permission of RSC group)

In addition to cell carriers, these hydrophobic oligomers synthesized in situ can function as picoliter droplet carriers [55] (Fig. 5.24). Hydrophobic oligomer, TMPTA, [trimethyloporpane triacrylate] is cured using OFML setup. Then, after washing out the hydrophobic oligomer resin, Hydrophilic oligomer, PEG-DA [polyethylene glycol) using these liquid handlers can be used in multiplexed bioassays in small volume. Multiplexing the liquid in these micro-liquid handlers could be applied to different types of biological assays that require small volumes.

Lee et al. have developed polymer based chemical delivery to multichannel capillary patterned cells [56] (Fig. 5.25). The group used acyl homoserine lactone (AHL) as a model chemical released to the cells. AHL-laden hydrogels were synthesized through photopolymerization. This is an example of nanocomposite that combines the photopolymers with nano-biochemistry. The nanoporous hydrogel will release AHL slowly, affecting the *E. coli* that is prepared inside the

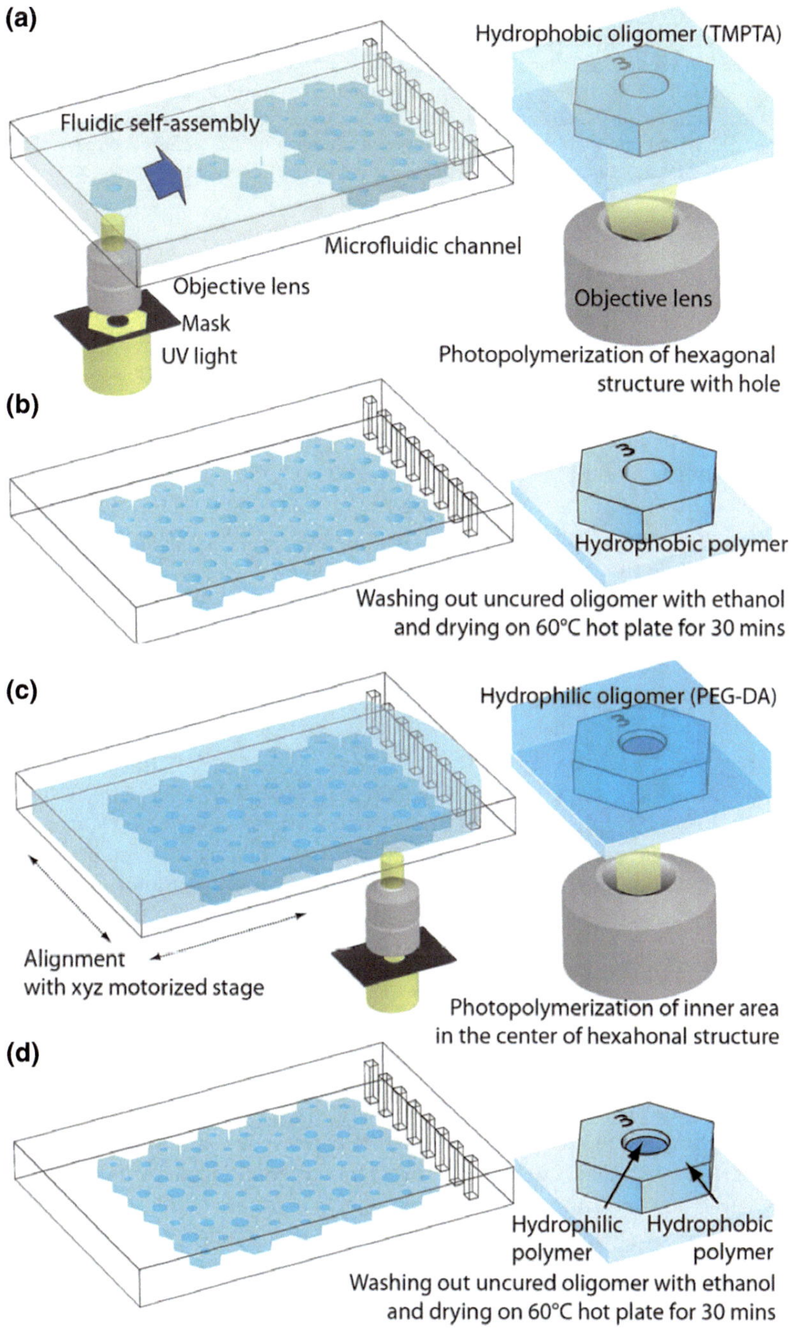
(a)
Hydrophobic oligomer (TMPTA)
Fluidic self-assembly
Microfluidic channel
Objective lens
Mask
UV light
Objective lens
Photopolymerization of hexagonal structure with hole
(b)
Hydrophobic polymer
Washing out uncured oligomer with ethanol and drying on 60°C hot plate for 30 mins
(c)
Hydrophilic oligomer (PEG-DA)
Alignment with xyz motorized stage
Photopolymerization of inner area in the center of hexahonal structure
(d)
Hydrophilic polymer
Hydrophobic polymer
Washing out uncured oligomer with ethanol and drying on 60°C hot plate for 30 mins

◂**Fig. 5.24** Processes for fabricating the amphiphilic microcarriers in a microfluidic channel: **a** Injection of hydrophobic oligomer (TMPTA) into a microfluidic channel followed by UV illumination in a hexagonal shape with a center hole for the hexagonal outer structure. This is followed by fluidic self-assembly of the outer structure against the channel anchors. **b** Washing the channel out with ethanol and drying on a 60 °C hot plate for 30 min. **c** Injecting hydrophilic oligomer (PEG-DA) into the same microchannel and illuminating UV light in a circular pattern to fill the hole in the hexagonal structure. **d** Washing out uncured oligomer and drying on a 60 °C hot plate for 30 min (With permission of Springer)

shallow channel. The advantage of this method is that the reacting channels are chemically isolated flow through environment.

Another example of the particles can be a QR code particle developed by Han et al. [57] (Fig. 5.26). In their paper, they report polymer microtaggant patterned with QR code.

So far, we have looked at generation of structures, structural components, and particles using optofluidic lithography. In situ photopolymerization can realize multiplex and high throughput bioassays that were impossible. However, more techniques that can actually control and handle these millions of small reactors. We want to conclude this chapter by introducing a method of handling these particles.

The method is called “partipetting” which mixes the terminology particles and pipetting [58, 59] (Fig. 5.27). The method allows scalable multiplexed bioassay in a single swap as demonstrated in the figure. The encoded particles, which we already have introduced above, can be oriented in each well for further analysis. The assays can be done on top of the platform and decoded for identifying the micro-scale reactors. We expect this field to grow more, as there are infinitely different strategies possible for convenient, fast, and low-cost bioassays using optofluidic fabrication. For example, if partipetting is used, particles with certain shapes can be assembled into the corresponding well with the indentations with reciprocal shape. This method was defined as hierarchical shape-by-shape assembly of microparticles for micromete-scale viral delivery of two different genes [60]. Also this technique has been applied to point-of-care immunoassay device called ELIPatch [61]. ELIPatch is a thumbnail-sized patch with immunospot array for multiplexed protein detection from human skin surface.

Summary

In this chapter, we introduced synthesis using optofluidic phenomena. We first described photopolymerization and tools that can use this phenomenon. Then, we looked deeply into nanocomposites that can be synthesized using in situ photopolymerization and some biological applications of these systems. By integrating the advantages of optics and microfluidics, optofluidic synthesis synergistically can provide various useful tools to the scientists and engineers.

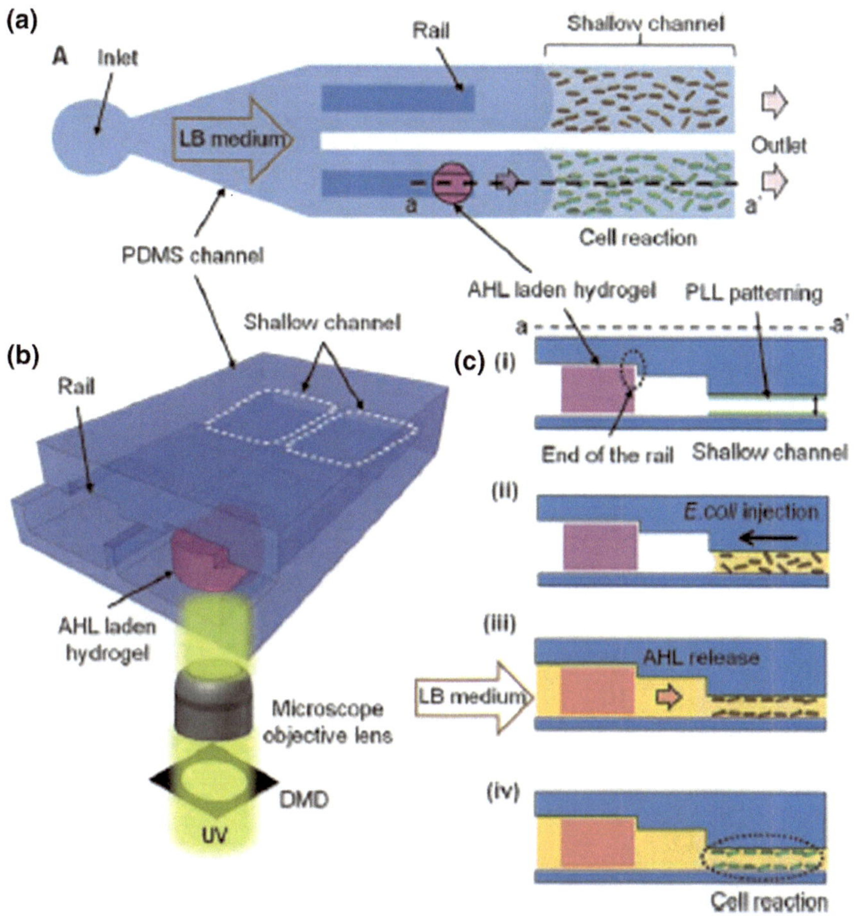

Fig. 5.25 **a** Top view diagram showing on-chip polymer based chemical delivery. **b** 3D diagram showing the control channel and the test channel containing the loaded hydrogel. The hydrogel contains AHL, which reacts with genetically modified E. coli cells. **c** The microchannel has three heights. The rail layer used to immobilize the polymer is 60 mm, the fluidic channel below the rail layer is 40 mm, and the shallow channels containing the capillary patterned cells are 20 mm. (i) Cross-sectional view cut at a–a0 from (**a**). PLL is pre-patterned in the shallow channel area. (ii) E. coli cells are injected and tether in the shallow channel. Surface tension prohibits the cells and growth medium from flooding into the deep channel. (iii) LB medium is introduced from the inlets into the deep channel where the chemical laden hydrogel is stacked. AHL is released from the hydrogel, exposing patterned cells to AHL. (iv) The cells' synthetic quorum sensing circuit is activated causing the E. coli to produce GFP (With the permission from RSC publishing group)

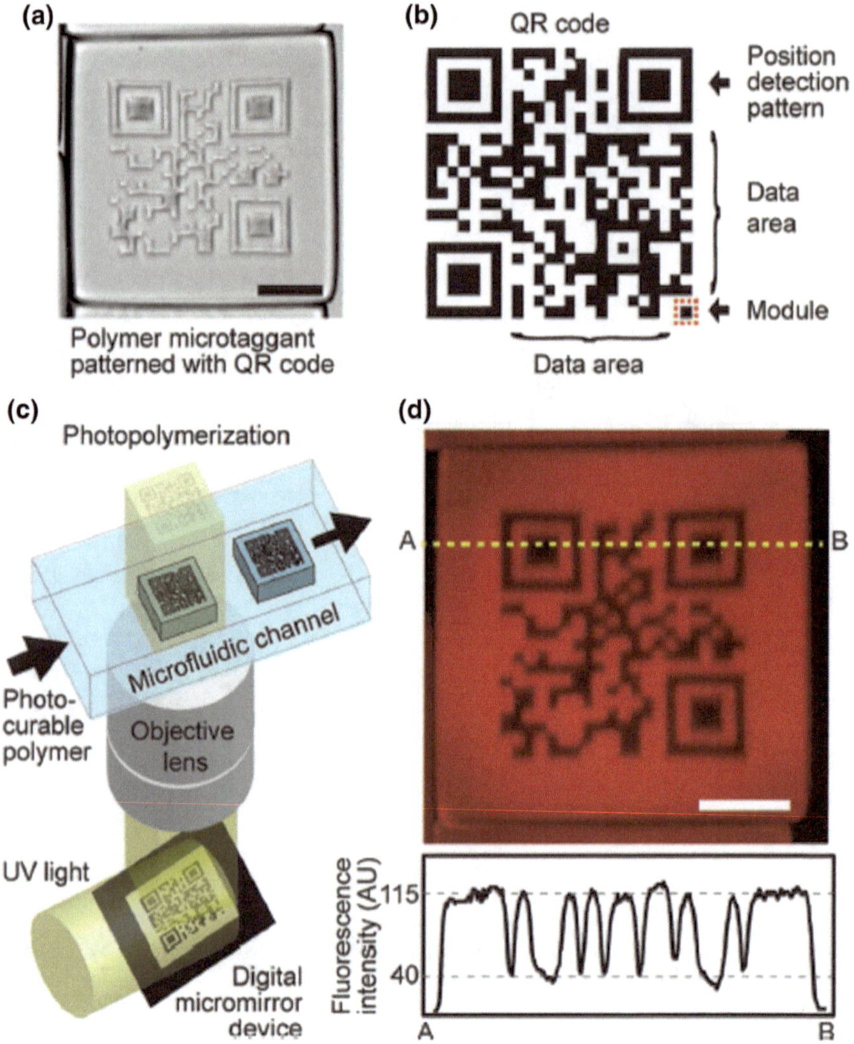

Fig. 5.26 **a** Microscope image of polymer microtaggant patterned with QR Code (scale bar: 200 μm). **b** The structure of QR Code. Three square patterns allow the detection of the edge positions of the code. The module indicates the black and white dots composing QR Code. **c** Fabrication setup of QR-coded polymer microtaggant using an optofluidic maskless lithography system. **d** Fluorescence microscope image of the QR-coded microtaggant and the line profile (**a**–**b**) of the fluorescence intensity (scale bar = 100 μm). Like many other particles with coding capability, these encoded particles can act as a novel method for anti-counterfeiting of drugs (With permission of Wiley)

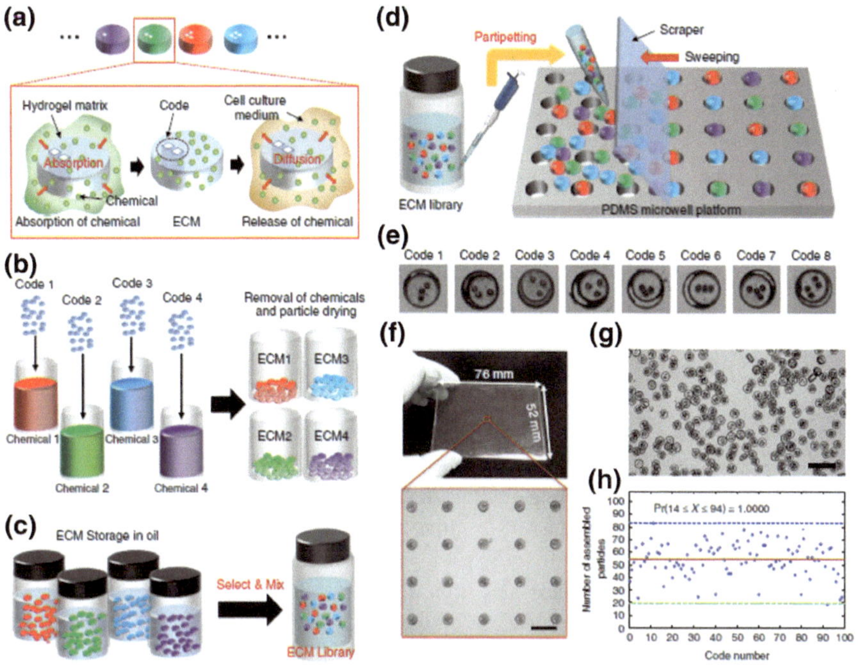

Fig. 5.27 Partipetting. **a** Encoded chemical-laden microparticles (ECMs) loaded with chemical substances and a corresponding 2D graphical barcode. **b** Chemicals are non-specifically absorbed into the ECM hydrogel matrix and then dried. **c** ECMs are stored in mineral oil to prevent cross-contamination. **d** Examples of different ECM 2D graphical barcodes. **e** The partipetting process for filling the microwell particle device. **f** The process for cell-based assays. A microparticle chip is filled via partipetting and then placed face-to-face with a cell microwell device to allow diffusion of chemicals. **g** Proof of concept assay with cell tracker red and green stains. **h** Combinatorial assay with potential for 3 different chemicals to be introduced simultaneously to a cell microwell (With permission of RSC publishing group)

5.3.2 *Selection and Manipulation Using Optics in Biotechnology*

5.3.2.1 Laser can be Used in Microfluidic Systems

One of the merits for using microfluidic system is the use of a small volume to perform important tasks such as cell culturing, particle sorting and separation, and concentration gradient formation. Because of this property of microfluidic system, sometimes manipulation of the samples in microfluidic channels can be useful. Thus, we need special methods to access the samples in fluidic channels of these systems. However, a limit exists for the conventional mechanical approaches in the resolution aspect. One methodological approach that can be a solution for this problem is the laser, because lasers have the ability to make high-resolution energy spot, derived from its special light-generation process. This high-resolution photon energy can be used for the further manipulation of the samples in microfluidic channels.

5.3.2.2 Laser

Laser is an acronym for Light Amplification by Stimulated Emission of Radiation. Stimulated emission differs from typical photon energy generation process, spontaneous emission, in some aspects.

When an atom acquires energy from other energy sources like phonons, electrons in the atom are transferred to a higher energy level. But, the electrons in this energy state are unstable. Therefore, naturally, the electrons would decay to the lower energy level again. This decaying process of the electrons makes radiation of light, which has the same amount of energy as the difference between two energy levels (higher and lower). When excitation energy source is also a light, the incident light and radiated light will have the same frequency. But, they are independent in phase and direction. In other words, two types of light (incident light and radiated light) will have different phase and direction. This is the typical light generation process, called spontaneous emission. However, light emissions from excited atoms can be stimulated. This phenomenon is called stimulated emissions. Stimulated emission differs from spontaneous emission in the aspect of coherence. The picture below represents the stimulated emission (Fig. 5.28).

Suppose that there is an electron existing in the excited energy level, and energy gap between the excited energy level and the grounded energy level amounts to ΔE. If the energy of an incident photon, hv, is same as ΔE, decaying of the electron in excited energy level is stimulated. The stimulated decaying process invokes the

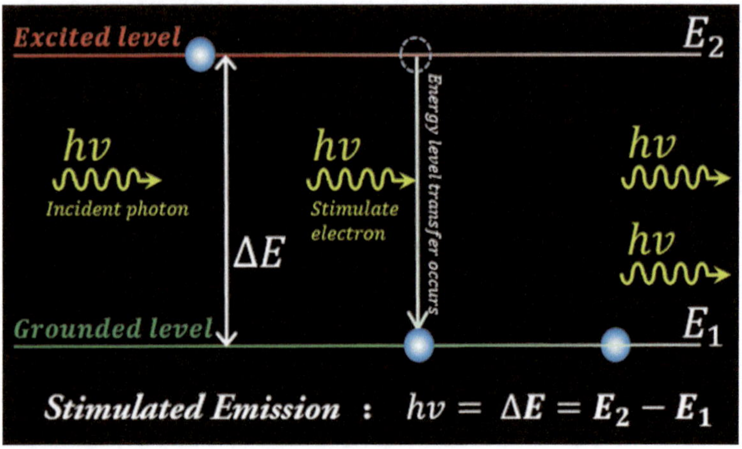

Fig. 5.28 Scheme of stimulated emission process

radiation of photon having the same amount of energy as ΔE. Now, there are two photons, one is the incident photon and the other is radiated photon. This is the stimulated emission. The important thing in this phenomenon is that two photons will have the same frequency, direction and phase.

As you can see from its name, LASER (Light Amplification by Stimulated Emission of Radiation) is the device that amplifies light by using stimulated emission. Because lasers use lights from stimulated emission, the amplified lights in laser have the same frequency, direction, and phase. This coherence in three components gives a laser with two special properties: spatial and temporal coherences. Spatial coherence allows a laser to be focused to a tight spot, enabling applications such as laser cutting and lithography. Temporal coherence allows a laser to emit light with a very narrow spectrum. These two special properties of a laser mean that a laser can focus a light having a specific amount of energy to a very tight spot. This property of a laser makes it useful for the very-small-volume-manipulation in a microfluidic system.

Figure 5.29 represents the structure of a typical laser. A laser consists of 4 main components. Those are gain medium, pumping energy, output coupler, and high reflector. Gain medium is where stimulated emission occurs. The atoms in a gain medium get energy to excite their electrons from a pumping energy, such as electricity or light. Then, from the process of stimulated emission, lights are produced, and the lights generated from stimulated emission are reflected two mirrors, the high reflector and the output coupler. While reflecting between the two mirrors, lights interfere with itself and other lights having other frequency constructively or destructively. As a result, lights having a specific frequency survive and form some number of modes. By repeating this process, lights are amplified. And through the output coupler, amplified lights are emitted. For this purpose, output coupler is transparent.

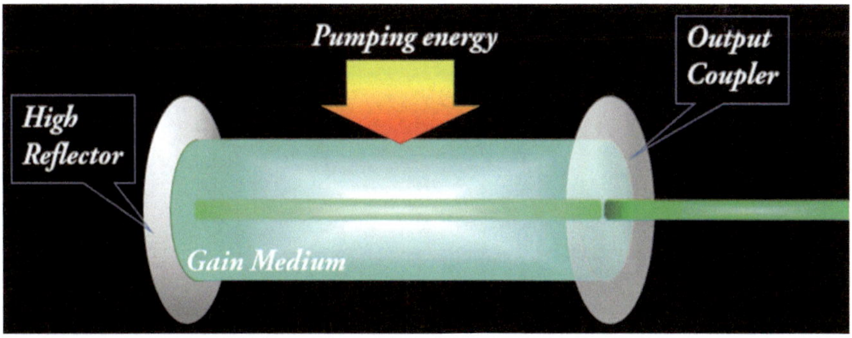

Fig. 5.29 Scheme of a structure of typical laser

5.3.2.3 Opto-Mechanical Selection Using Lasers

Mentioned above, lasers can be used for an energy source to a tight spot [62–65]. This spatially high-resolution photon energy can be used to retrieve specific samples from a microfluidic system. There are some applications that use lasers for selection in microfluidic devices.

Sniper Cloning

When we want to synthesize long DNA, numerous DNA fragments are needed for the bottom-up assembly. These DNA fragments have to be sequence-verified to make the correctly assembled DNAs. However, because of the current limitation of coupling efficiency (<99.5%) in solid-phase synthesis of oligonucleotides, it is very hard to obtain the sequence-verified DNAs by that method. Suppose that we have 0.5% per base, only 1 DNA block out of 100 DNA blocks will have the right sequence, the other 99 DNA blocks will have wrong sequence. This problem concerning error-rate in synthesized DNA blocks makes writing DNAs more difficult. There is a solution by using laser in microfluidic system, called Sniper Cloning [66, 67] (Fig. 5.30). In Sniper Cloning, we first verify the sequence of the DNA blocks by using 454-junior sequencing device. In that platform, we use micro-beads containing DNAs to verify the sequence of DNAs. Beads contain 1

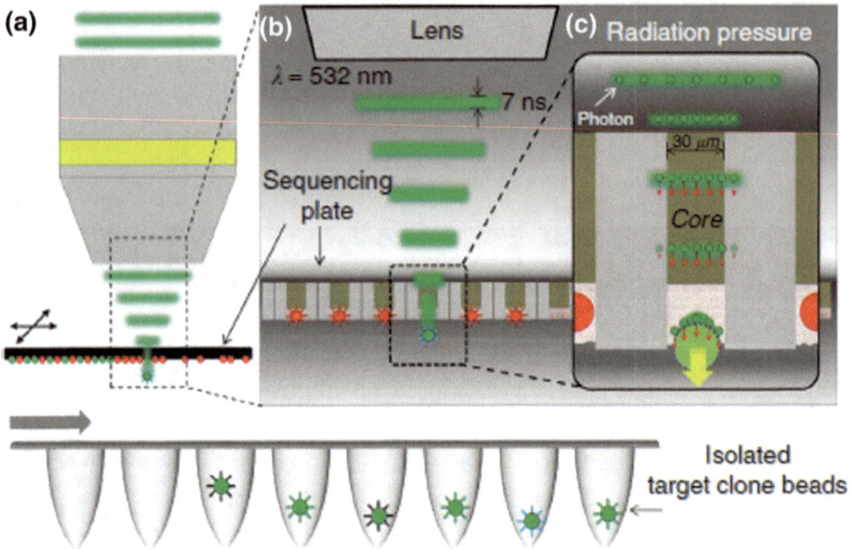

Fig. 5.30 Schematic diagram of the focused pulse laser radiation pressure-driven non-contact target bead sniper system. **a** A motorized stage moves the sequencing plate and locates the target clone bead to the focusing spot of the pulse laser **b** The selectively etched fiber bundle structure of the 454 NGS substrate is well suited for optical releasing. **c** Since the fiber only carries light in the core region and otherwise attenuates light, it also minimizes the horizontal and vertical positioning error (With permission of Springer publishing group)

DNA clone, and 1 bead exists in a well, and this is done by adjusting the size of wells. 454-junior sequencing machine detects the fluorescence from the beads in the sequencing process. After sequencing the blocks, one knows where DNA blocks having the correct sequence are. The remaining task is to retrieve DNA blocks with the correct sequence. Because the sizes of the wells and beads are micro-scale, we need high-resolution retrieval method. The method employed in this platform is using light from the lasers. Mentioned above, the light from lasers can be focused in a very tiny spot, transferring kinetic energy of photons to that spot. Knowing where the correct DNAs are, we move the laser to that spot, and generate pulsed laser beam to the well containing the correct DNA block. The kinetic energy of photons make bead separated from the well, and this bead is dropped to the 96-well plate for further manipulation of the DNA blocks. The figure below represents the main concepts of Sniper Cloning.

Green lines are lights from a laser, having wavelength of 532 nm. This amplified light passes through a lens, and are focused to a tight spot. By adjusting the distance between the lens and the chip that was used for the sequencing of DNAs, diameter of the light can be matched to the well size. Each well has one bead containing one clone of DNA. The wall of wells guides the incident light, then this guided light eventually hits the bead containing correct sequence that we will use for some purposes. This bead containing sequence-verified DNAs falls into one well of a 96-well plate below the sequencing chip. After we purify the beads in the 96-well plate, we can get sequence verified DNAs.

Selective Retrieval in Dynamic Microarray

There is another selective retrieval approach by using lasers. Figure 5.31a explains the main concept of this device [68]. When a particle is moving through the microfluidic channel in this device, there are two possible paths for the particle. Path 1 is to the well that has the size for just one particle and the other is the main stream toward the exit of this device. For some fluid mechanical reasons, the fluidic resistance of path 1 is smaller than the fluidic resistance of path 2. So particles can be trapped in the well easily. After some experiments to the trapped samples, selective retrieval is done by using a laser. When laser is radiated to the cavity beside of the well containing the target particle, fluid in the spot evaporates and forms a bubble. The pressure caused by the formation of the bubble makes the target particle join to the main stream of this fluidic device again. The high resolution of lasers makes this selective retrieval possible. Using other technologies, like encapsulation of the samples, can prevent possible damages from this focused energy of lasers [69].

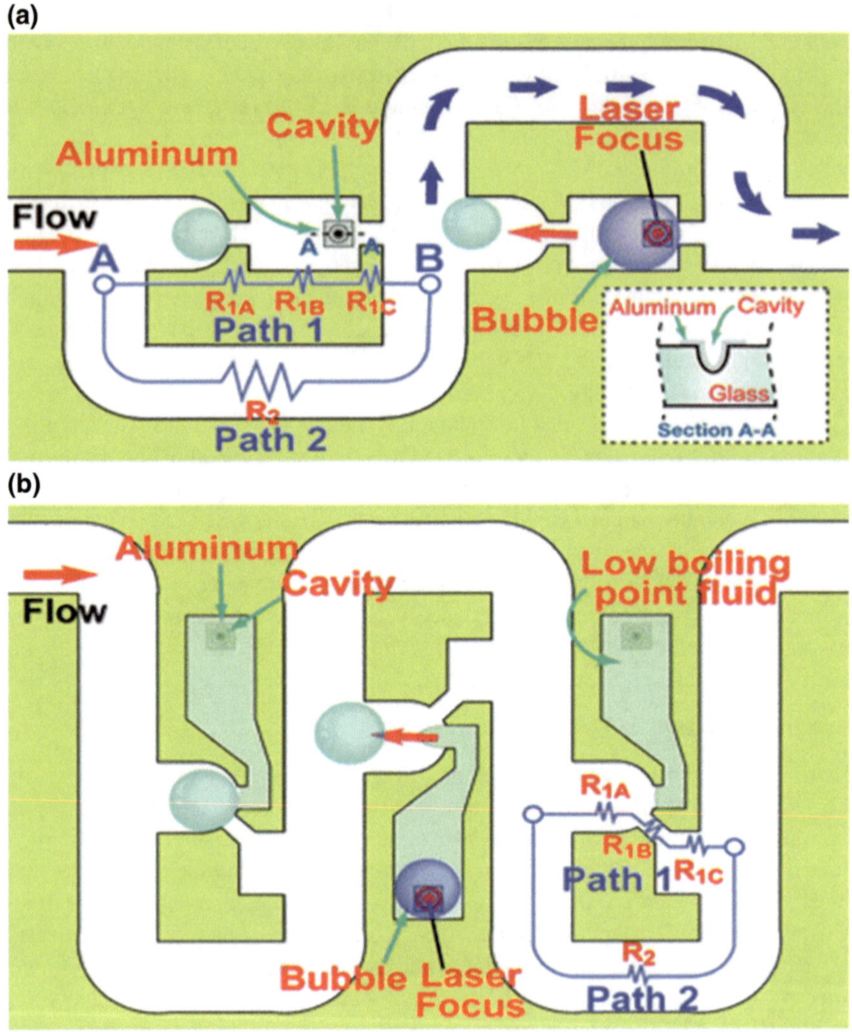

Fig. 5.31 Schematic of **a** device 1 and **b** design 2 for the dynamic microarray system with gentle retrieval mechanism (With the permission from RSC)

5.4 Conclusion

For micro and nano systems for biological applications, the optofluidics has been adopted to address the limitations that the micro or nanofluidics devices have. Imaging, manipulating, and fabricating the structures or the fluids in these systems provide a powerful tool to meet the technical difficulties that the current devices or technologies have. In this section we looked into some examples and principles for the optofluidics that can be applied to the bio MEMS or NEMS systems. While there are some other examples that cannot be categorized into these three subjects, but optofluidics, in many ways, can be developed and converged to other technologies.

References

1. A.R. Hawkins, H. Schmidt, *Handbook of Optofluidics*. CRC Press (2010)
2. X. Fan, I.M. White, Optofluidic microsystems for chemical and biological analysis. Nat. Photonics **5**, 591–597 (2011). https://doi.org/10.1038/nphoton.2011.206
3. D. Erickson, D. Sinton, D. Psaltis, Optofluidics for energy applications. Nat. Photonics **5**, 583–590 (2011). https://doi.org/10.1038/nphoton.2011.209
4. D. Psaltis, S.R. Quake, C. Yang, Developing optofluidic technology through the fusion of microfluidics and optics. Nature **442**, 381–386 (2006). https://doi.org/10.1038/nature05060
5. M. Weibel, W. Caseri, U.W. Suter et al., Preparation of polymer nanocomposites with "ultrahigh" refractive index. Polym. Adv. Technol. **2**, 75–80 (1991). https://doi.org/10.1002/pat.1991.220020204
6. J. Choi, Y.-G. Jung, J. Kim et al., Rapid antibiotic susceptibility testing by tracking single cell growth in a microfluidic agarose channel system. Lab Chip **13**, 280–287 (2013). https://doi.org/10.1039/C2LC41055A
7. H.J. Bae, S. Bae, C. Park et al., Biomimetic microfingerprints for anti-counterfeiting strategies. Adv. Mater. **27**, 2083–2089 (2015). https://doi.org/10.1002/adma.201405483
8. H.J. Bae, S. Bae, J. Yoon et al., Self-organization of maze-like structures via guided wrinkling. Sci. Adv. **3**, e1700071 (2017). https://doi.org/10.1126/sciadv.1700071
9. N.C. Shaner, R.E. Campbell, P.A. Steinbach et al., Improved monomeric red, orange and yellow fluorescent proteins derived from Discosoma sp. red fluorescent protein. Nat. Biotechnol. **22**, 1567–1572 (2004). https://doi.org/10.1038/nbt1037
10. M. Schena, D. Shalon, R.W. Davis, P.O. Brown, Quantitative monitoring of gene expression patterns with a complementary DNA microarray. Science **270**, 467–470 (1995). https://doi.org/10.1126/SCIENCE.270.5235.467
11. M.T. Guo, A. Rotem, J.A. Heyman, D.A. Weitz, Droplet microfluidics for high-throughput biological assays. Lab Chip **12**, 2146 (2012). https://doi.org/10.1039/c2lc21147e
12. H. Song, J.D. Tice, R.F. Ismagilov, A microfluidic system for controlling reaction networks in time. Angew Chemie **115**, 792–796 (2003). https://doi.org/10.1002/ange.200390172
13. J.-C. Baret, O.J. Miller, V. Taly et al., Fluorescence-activated droplet sorting (FADS): efficient microfluidic cell sorting based on enzymatic activity. Lab Chip **9**, 1850 (2009). https://doi.org/10.1039/b902504a
14. J. Eid, A. Fehr, J. Gray et al., Real-time DNA sequencing from single polymerase molecules. Science **323**, 133–138 (2009). https://doi.org/10.1126/science.1162986
15. M.L. Metzker, Sequencing technologies—the next generation. Nat. Rev. Genet. **11**, 31–46 (2010). https://doi.org/10.1038/nrg2626
16. Y. Choi, H. Choi, A.C. Lee, et al., Reconfigurable DNA accordion rack. Angew Chemie Int. Ed. (2018). https://doi.org/10.1002/anie.201709362
17. N.B. Colthup, L.H. Daly, S.E. Wiberley, *Introduction to Infrared and Raman Spectroscopy*. Academic Press (1975)
18. S. Nie, S.R. Emory, Probing single molecules and single nanoparticles by surface-enhanced Raman scattering. Science **275**, 1102–1106 (1997). https://doi.org/10.1126/SCIENCE.275.5303.1102
19. Y. Xia, D.J. Campbell, Plasmons: Why should we care? J. Chem. Educ. **84**, 91 (2007). https://doi.org/10.1021/ed084p91

20. B. Lee, S. Kim, H. Kim, Y. Lim, The use of plasmonics in light beaming and focusing. Prog. Quantum Electron. **34**, 47–87 (2010). https://doi.org/10.1016/J.PQUANTELEC.2009.08.002
21. T. Vo-Dinh, D.L. Stokes, G.D. Griffin et al., Surface-enhanced Raman scattering (SERS) method and instrumentation for genomics and biomedical analysis. J. Raman Spectrosc. **30**, 785–793 (1999). https://doi.org/10.1002/(SICI)1097-4555(199909)30:9%3c785:AID-JRS450%3e3.0.CO;2-6
22. S. Chan, S. Kwon, T.-W. Koo et al., Surface-Enhanced Raman Scattering of Small Molecules from Silver-Coated Silicon Nanopores. Adv. Mater. **15**, 1595–1598 (2003). https://doi.org/10.1002/adma.200305149
23. D.-K. Lim, K.-S. Jeon, J.-H. Hwang et al., Highly uniform and reproducible surface-enhanced Raman scattering from DNA-tailorable nanoparticles with 1-nm interior gap. Nat. Nanotechnol. **6**, 452–460 (2011). https://doi.org/10.1038/nnano.2011.79
24. B. Fazio, C. D'Andrea, A. Foti et al., SERS detection of biomolecules at physiological pH via aggregation of gold nanorods mediated by optical forces and plasmonic heating. Sci. Rep. **6**, 26952 (2016). https://doi.org/10.1038/srep26952
25. R. Phillips, Photopolymerization. J Photochem. **25**, 79–82 (1984). https://doi.org/10.1016/0047-2670(84)85016-9
26. P. Kissel, R. Erni, W.B. Schweizer et al., A two-dimensional polymer prepared by organic synthesis. Nat. Chem. **4**, 287–291 (2012). https://doi.org/10.1038/nchem.1265
27. C.S. Chern, Emulsion polymerization mechanisms and kinetics. Prog. Polym. Sci. **31**, 443–486 (2006). https://doi.org/10.1016/J.PROGPOLYMSCI.2006.02.001
28. E. Andrzejewska, Photopolymerization kinetics of multifunctional monomers. Prog. Polym. Sci. **26**, 605–665 (2001). https://doi.org/10.1016/S0079-6700(01)00004-1
29. G.M. Whitesides, E. Ostuni, S. Takayama et al., Soft lithography in biology and biochemistry. Annu. Rev. Biomed. Eng. **3**, 335–373 (2001). https://doi.org/10.1146/annurev.bioeng.3.1.335
30. J. Zhu, G. Wang, Y. Hao et al., Highly sensitive and spatially resolved polyvinyl alcohol/acrylamide photopolymer for real-time holographic applications. Opt. Express **18**, 18106 (2010). https://doi.org/10.1364/OE.18.018106
31. D.J. Beebe, J.S. Moore, J.M. Bauer et al., Functional hydrogel structures for autonomous flow control inside microfluidic channels. Nature **404**, 588–590 (2000). https://doi.org/10.1038/35007047
32. H. Chun, T.D. Chung, H.C. Kim, Cytometry and velocimetry on a microfluidic chip using polyelectrolytic salt bridges (2005). https://doi.org/10.1021/ac048535o
33. Y.K. Cheung, B.M. Gillette, M. Zhong et al., Direct patterning of composite biocompatible microstructures using microfluidics. Lab Chip **7**, 574 (2007). https://doi.org/10.1039/b700869d
34. D. Dendukuri, D.C. Pregibon, J. Collins et al., Continuous-flow lithography for high-throughput microparticle synthesis. Nat. Mater. **5**, 365–369 (2006). https://doi.org/10.1038/nmat1617
35. D.C. Pregibon, M. Toner, P.S. Doyle, Multifunctional encoded particles for high-throughput biomolecule analysis. Science **315**, 1393–1396 (2007). https://doi.org/10.1126/science.1134929
36. S.W. Song, Y. Jeong, S. Kwon, Photocurable polymer nanocomposites for magnetic, optical, and biological applications. IEEE J. Sel. Top. Quantum Electron. **21**, 324–335 (2015). https://doi.org/10.1109/JSTQE.2014.2381491
37. E.J. Brandrup, E.H. Immergut, E.A. Grulke, et al., *Polymer Handbook Fourth Edition*
38. H.-U. Simmrock, A. Mathy, L. Dominguez et al., Polymers with a high refractive index and low optical dispersion. Angew Chemie Int. Ed. **28**, 1122–1123 (1989). https://doi.org/10.1002/anie.198911221
39. S.E. Chung, W. Park, H. Park et al., Optofluidic maskless lithography system for real-time synthesis of photopolymerized microstructures in microfluidic channels. Appl. Phys. Lett. **91**, 041106 (2007). https://doi.org/10.1063/1.2759988

40. S.E. Chung, S.A. Lee, J. Kim, S. Kwon, Optofluidic encapsulation and manipulation of silicon microchips using image processing based optofluidic maskless lithography and railed microfluidics. Lab Chip **9**, 2845 (2009). https://doi.org/10.1039/b903760h
41. S.E. Chung, W. Park, S. Shin et al., Guided and fluidic self-assembly of microstructures using railed microfluidic channels. Nat. Mater. **7**, 581–587 (2008). https://doi.org/10.1038/nmat2208
42. S.A. Lee, S.E. Chung, W. Park et al., Three-dimensional fabrication of heterogeneous microstructures using soft membrane deformation and optofluidic maskless lithography. Lab Chip **9**, 1670 (2009). https://doi.org/10.1039/b819999j
43. W. Park, S. Han, S. Kwon, Fabrication of membrane-type microvalves in rectangular microfluidic channels via seal photopolymerization. Lab Chip **10**, 2814 (2010). https://doi.org/10.1039/c005173j
44. S.H. Lee, S.-E. Choi, A.J. Heinz et al., Active guidance of 3D microstructures. Small **6**, 2668–2672 (2010). https://doi.org/10.1002/smll.201001248
45. Y. Song, Y. Jeong, T. Kwon et al., Liquid-capped encoded microcapsules for multiplex assays. Lab Chip **17**, 429–437 (2017). https://doi.org/10.1039/C6LC01268J
46. H. Kim, J. Ge, J. Kim et al., Structural colour printing using a magnetically tunable and lithographically fixable photonic crystal. Nat. Photonics **3**, 534–540 (2009). https://doi.org/10.1038/nphoton.2009.141
47. J. Ge, H. Lee, L. He et al., Magnetochromatic Microspheres: Rotating Photonic Crystals. J. Am. Chem. Soc. **131**, 15687–15694 (2009). https://doi.org/10.1021/ja903626h
48. H. Lee, J. Kim, H. Kim et al., Colour-barcoded magnetic microparticles for multiplexed bioassays. Nat. Mater. **9**, 745–749 (2010). https://doi.org/10.1038/nmat2815
49. L. He, Y. Hu, H. Kim et al., Magnetic assembly of nonmagnetic particles into photonic crystal structures. Nano Lett. **10**, 4708–4714 (2010). https://doi.org/10.1021/nl103008v
50. J. Kim, Y. Song, L. He et al., Real-time optofluidic synthesis of magnetochromatic microspheres for reversible structural color patterning. Small **7**, 1163–1168 (2011). https://doi.org/10.1002/smll.201001822
51. S.E. Chung, J. Kim, S.-E. Choi et al., In situ fabrication and actuation of polymer magnetic microstructures. J. Microelectromechanical Syst. **20**, 785–787 (2011). https://doi.org/10.1109/JMEMS.2011.2159093
52. J. Kim, S.E. Chung, S.-E. Choi et al., Programming magnetic anisotropy in polymeric microactuators. Nat. Mater. **10**, 747–752 (2011). https://doi.org/10.1038/nmat3090
53. H. Kim, J. Kim, E.-G. Kim et al., Optofluidic *in situ* maskless lithography of charge selective nanoporous hydrogel for DNA preconcentration. Biomicrofluidics **4**, 043014 (2010). https://doi.org/10.1063/1.3516037
54. L.N. Kim, S.-E. Choi, J. Kim et al., Single exposure fabrication and manipulation of 3D hydrogel cell microcarriers. Lab Chip **11**, 48–51 (2011). https://doi.org/10.1039/C0LC00369G
55. W. Park, S. Han, H. Lee, S. Kwon, Free-floating amphiphilic picoliter droplet carriers for multiplexed liquid loading in a microfluidic channel. Microfluid Nanofluidics **13**, 511–518 (2012). https://doi.org/10.1007/s10404-012-0989-4
56. S.H. Lee, A.J. Heinz, S.-E. Choi et al., Polymer based chemical delivery to multichannel capillary patterned cells. Lab Chip **11**, 605 (2011). https://doi.org/10.1039/c0lc00328j
57. S. Han, H.J. Bae, J. Kim et al., Lithographically encoded polymer microtaggant using high-capacity and error-correctable QR code for anti-counterfeiting of drugs. Adv. Mater. **24**, 5924–5929 (2012). https://doi.org/10.1002/adma.201201486
58. S. Eun Chung, J. Kim, OhD Yoon et al., One-step pipetting and assembly of encoded chemical-laden microparticles for high-throughput multiplexed bioassays. Nat. Commun. **5**, 3468 (2014). https://doi.org/10.1038/ncomms4468
59. S. Song, S. Kim, D. Oh et al., One-Step Generation of a Drug-Releasing Hydrogel Microarray-On-A-Chip for Large-Scale Sequential Drug Combination Screening. Adv. Sci. **6**(3), (2018). https://doi.org/10.1002/advs.201801380
60. D. Lee, A.C. Lee, S. Han et al., Hierarchical shape-by-shape assembly of microparticles for micrometer-scale viral delivery of two different genes. Biomicrofluidics **12**, 031102 (2018). https://doi.org/10.1063/1.5030597

61. D.Y. Oh, H. Na, S.W. Song et al., ELIPatch, a thumbnail-size patch with immunospot array for multiplexed protein detection from human skin surface. Biomicrofluidics **12**, 031101 (2018). https://doi.org/10.1063/1.5032170
62. A.C. Lee, Y. Lee, D. Lee et al., Divide and conquer: A perspective on biochips for single-cell and rare molecule analysis by next-generation sequencing. APL. Bioeng. **3**, 020901 (2019). https://doi.org/10.1063/1.5095962
63. J. Noh, O. Kim, Y. Jung et al., High-throughput retrieval of physical DNA for NGS-identifiable clones in phage display library. mAbs **11**(3), 532–545 (2019). https://doi.org/10.1080/19420862.2019.1571878
64. S. Kim, A.C. Lee, H. Lee et al., PHLI-seq: constructing and visualizing cancer genomic maps in 3D by phenotype-based high-throughput laser-aided isolation and sequencing. Genome. Biol. **19**, 158 (2018). https://doi.org/10.1063/1.5095962
65. O. Kim, D. Lee, A.C. Lee et al., Whole Genome Sequencing of Single Circulating Tumor Cells Isolated by Applying a Pulsed Laser to Cell-Capturing Microstructures. Small. online. (2019). https://doi.org/10.1002/smll.201902607
66. H. Lee, H. Kim, S. Kim et al., A high-throughput optomechanical retrieval method for sequence-verified clonal DNA from the NGS platform. Nat. Commun. **6**, 6073 (2015). https://doi.org/10.1038/ncomms7073
67. H. Yeom, Y. Lee, T. Ryu et al., Barcode-free next-generation sequencing (NGS) error validation for ultra-rare variant detection. Nat. Commun. **10**, 977 (2019). https://doi.org/10.1038/s41467-019-08941-4
68. W.-H. Tan, S. Takeuchi, A trap-and-release integrated microfluidic system for dynamic microarray applications. Proc. Natl. Acad. Sci. USA **104**, 1146–1151 (2007). https://doi.org/10.1073/pnas.0606625104
69. W.-H. Tan, S. Takeuchi, Dynamic microarray system with gentle retrieval mechanism for cell-encapsulating hydrogel beads. Lab Chip **8**, 259–266 (2008). https://doi.org/10.1039/B714573J

Chapter 6
Electricity for Fluidics and Bio-Devices

Hyomin Lee, Sung Jae Kim, Marie Frenea-Robin, Bruno Le Pioufle, Thi Hong Nhung Dinh, Stephane Serfaty and Pierre-Yves Joubert

Abstract This chapter is focused on biomaterial interaction with electromagnetic fields. It gives a wide and deep introduction of biological medium and their electromagnetic properties, which can be used to sense and analyze tissue or cells within lab-on-chip. It consists of three complementary parts. A recall of electrokinetics laws is done and is applied to biological medium (electrolyte with fluid motion) in the first section. The scaling factor towards nano-scale is highlighted considering the ionic concentration distribution. The second section deals with dielectrophoresis forces (DEP), which results from the cell polarization under non-uniform electric field. The dielectric modeling of cells and their DEP response are depicted. Taking benefits of cell dielectric behavior, the applications such as cell sorting, trapping and characterization with electrorotation are presented. The last section introduces the biomaterial behavior at radiofrequency range (from MHz up to several tens of GHz).

H. Lee
Department of Chemical and Biological Engineering, Jeju National University, 102 Jejudaehakro, Jeju-si, Jeju, Republic of Korea

S. J. Kim (✉)
Department of Electrical and Computer Engineering, Seoul National University, Seoul, Republic of Korea
e-mail: gates@snu.ac.kr

M. Frenea-Robin
Laboratoire Ampère, UMR 5005, CNRS, Université de Lyon, Université Lyon 1, 69622 Villeurbanne, France
e-mail: marie.robin@univ-lyon1.fr

B. Le Pioufle
ENS Paris-Saclay, 61 av du Pdt Wilson, 94235 Cachan, France

T. H. N. Dinh · P.-Y. Joubert
C2N/CNRS, Université Paris Sud, 10 Boulevard Thomas Gobert, 91 120 Palaiseau, France

S. Serfaty
SATIE/CNRS, Université Cergy-Pontoise, 95 000 Cergy-Pontoise, France

G. Barbillon et al. (eds.), *Engineering of Micro/Nano Biosystems*, Microtechnology and MEMS, https://doi.org/10.1007/978-981-13-6549-2_6

Applications using thee relaxation times occurring at such high frequency are described to sense or characterize biological medium. Electric Impedance Spectroscopy and Radio-Frequency devices are thus introduced.

Keywords Nano-electrokinetics · Ion concentration polarization · Electrostatics · Dielectrophoresis · Electrorotation · Microwaves · Biodevices · Dielectric properties · Relaxation time

6.1 Introduction

From an electrical engineering point of view, biological materials can be seen as heterogeneous conductive materials. In fact, the human body is mainly composed of liquid medium (water) where charged particles (such as sodium or chloride atoms, proteins) can flow. Under electric field, this ionic composition induces a current flow, which can be monitored. Concerning the biological cell, the lipid bilayer of cellular membrane behaves as an electrical insulator and thus modifies the flow path of the ions in relation with the cellular state, localization and numbers. Furthermore, the water is a dielectric material and its polarization also interacts with the electric field. Thus the human body, organ or cell can be seen as complex electrolyte within which the electric field can be used to sense or act on polarizable particles.

In this chapter, the physics of electrokinetic phenomena will be first presented and applied at nano-scale. Applications will be then depicted concerning handling and sensing cells using dielectrophoresis within lab-on-chip. Results will be then extended towards bioimpedance analysis for biomaterial dielectric properties extraction using large frequency range of the electric field, from kHz up to several GHz.

6.2 Micro- and Nano-Scale Electrokinetics (Hyomin Lee and Sung Jae Kim)

6.2.1 Introduction to Nonlinear Nanoscale Electrokinetics

There has been considerable interest over the past decade in the science and engineering of fluid transport at micrometer and nanometer scales [1–4]. Advances in micro- and nano-electromechanical system (MEMS and NEMS) technology over the past few decades have allowed miniaturized devices to be developed for various micro- and nano-fluidic applications. These miniaturized platforms offer significant advantages compared to conventional bulk analytical instruments: they support precise control of liquids flowing usually under laminar regime, minimize consumption of reagents and samples, favor short reaction times, enable highly parallel and multiplexed analysis, require less power to operate, and potentially have low cost of production [1]. For these advantages, electrokinetic mechanisms have been usually used as the preferred method of transporting fluids in the miniaturized systems because of the ease and low cost of electrode fabrication, excluding moving mechanical parts, and the significant effect of the electric double layer which is the thin polarized layer in the liquid adjacent to the charged surface. Due to these reasons, many seminal works to describe the electrokinetic phenomena in microscale system have been established [1, 5–7].

Nevertheless, electrokinetic phenomena in nanoscale system are not still obvious. In the nanoscale system, the electric double layer would be overlapped so that its equilibrium structure could not be maintained under external driving field [8, 9] called as non-equilibrium state. Conventional theories for microscale system are only applicable to the electric double layer under equilibrium limit. In addition, several factors, usually neglected in microscale system, could play an important role for nanoscale transport phenomena when the system dimension decreased. For example, hydrodynamic slip on solid surface could be occurred [10, 11], while "no slip" condition is usually applied in microscale system. In viewpoints of conventional theories, fluid molecules were attached to solid surface so that they could not move. This is known as 'no-slip condition' on liquid-solid interfaces. However, in actual system, fluid molecules on the interface are not attached to the interface strongly so that they could be movable. This is called as slippage of fluid molecule. Due to this slippage, ionic current could be higher than the value predicted by conventional theories for microscale system. Another example among unique properties of nanofluidic system is 'surface-charge-governed conductance'. Ionic transport in nanofluidic systems have the unique property of surface-charge-governed regime as demonstrated by Stein et al. [12]. In a dilute limit, ionic conductance is independent from the bulk properties of the system such as the electrolyte concentration or the geometrical factor, so that the conductance curve saturates below a specific concentration value, which is determined by surface charge density and called 'surface-charge-governed conductance'. Because the plateau of the conductance curve is only revealed in a nanochannel system, this property has been utilized to demonstrate the validity of the device in the view point of nanofluidic application. Beyond the specific concentration value, the conductance is proportional to the bulk concentration, called 'geometry-governed regime'. These two distinct regimes can be plotted (ionic conductance as a function of bulk concentrations) simultaneously. Since these unique properties were usually neglected in microscale electrokinetic analysis, direct application of the conventional theory could cause non-negligible errors in nanoscale electrokinetic analysis. Thus, aims of recent researches in nanoscale electrokinetics were (1) finding new constraints to describe the unique property of nanoscale, which has never been observed in microfluidic system, (2) establishing the modified model using the new constraints, and (3) predicting the phenomena, which are difficult to observe by experimental manner. To exemplify these concerns, the contents of this book focused on nanoscale electrokinetic transport phenomena inside the nanoscale system (ionic field effect transistor, a.k.a IFET) and nearby the system (ion concentration polarization, a.k.a. ICP) which are the most relevant phenomenon occurs in the biological system, especially inside human body.

At first, governing equations and appropriate boundary conditions for the nanoscale system were revisited from general conservation principle as written in Sect. 6.2.2. In this chapter, fundamental equations such as the Poisson equation, the

Nernst-Planck equations, Stokes equations, and the continuity equation could be found, which are used as governing equations in other chapters.

In Sect. 6.2.4, the model to describe ion concentration polarization (ICP) phenomenon was discussed. The micro-nano hybrid system, which is composed of microchannel reservoir—perm-selective nanoporous membrane—microchannel reservoir and also resembled the cellular structure, has unique phenomenon of as ICP. Through the perm-selective nanoporous media, only counter-ion can be transported across the media while co-ion cannot. Because of the selectivity of transported species, ion concentration is decrease on one side (this is called 'depletion'), while on the other side ion concentration is increase (this is called 'enrichment'). This is called as 'ion concentration polarization'. For various applications such as desalination [13], preconcentration [14], and battery system [15, 16], the concentration distribution inside the depletion and enrichment zone should be the key feature to apply these applications.

Furthermore, we suggested the model to describe the gate modulation in the partially gated ionic field effect transistor. The ionic field effect transistor (IFET) as a nanofluidic device could change the polarity of the nanochannel or nanopore so that the permselectivity, unique property of nanoscale system, which permits only counter-ion pass through the nanoscale system, could be controlled. Using this functionality, switching of macromolecules such as DNA, RNA, and proteins [17] or ionic species such as cation and anion [18, 19] could be possible. In the field of IFET research, the description for the polarity modulation by gate electrode in IFET has been an important issue.

6.2.2 Conservation Principles of Electrokinetic System

Three fundamental conservation principles are needed to describe the transport phenomena in the electrokinetic system. Because the typical electrokinetic system involves a fluid motion u with ionic species which have electric charge, independent variables such as electric potential ψ, concentration of each ionic species cj and hydrodynamic pressure p should be required for the characterization of the system. Each independent variable is related to the conservation principle; (1) the electric potential is related to conservation of electric charge, (2) concentration of ionic species is associated with the conservation of mass, and (3) hydrodynamic pressure and flow field is connected with the conservation of mass and momentum. Therefore, in this chapter, governing equations and boundary conditions for electrokinetic system will be introduced from a viewpoint of conservation principles under the continuum framework which is held down to a length scale greater than 1 nm [10]. Although some other models have been proposed to try to integrate the discrete effect of ion transport or fluid flow in the continuum frameworks with an extra term [20] or by adapting the results of molecular dynamics [21], continuum hypothesis still remain intact in most electrokinetic system.

6.2.2.1 Conservation of Electric Charge

Most electrokinetic systems immersed in an electrolyte solution attain electric charge distribution due to charged surface and ionized species. The electrostatic forces arising from the interactions between them plays essential role in the transport phenomena in the electrokinetic system. In order to describe the electrostatic system, the governing equation and appropriate boundary conditions will be introduced briefly.

The Maxwell's equations are not limited to the electrostatic system. The equations relate the electric field E and the magnetic field H to material properties, the charge density, and conservation principles. The full set of Maxwell's equations in general form is

$$\nabla \cdot \mathbf{D} = \rho_{\mathrm{f}} \tag{6.1}$$

$$\nabla \cdot \mathbf{B} = 0 \tag{6.2}$$

$$\nabla \times \mathbf{E} = -\frac{\partial \mathbf{B}}{\partial t} \tag{6.3}$$

and

$$\nabla \times \mathbf{H} = \frac{\partial \mathbf{D}}{\partial t} + \mathbf{J}_{\mathrm{f}}. \tag{6.4}$$

Here, D is the electric displacement vector, ρ_f is the free charge density, B is the magnetic induction field, E is the electric field, H is the magnetic field, and J_f is the conduction current density associated with the movement of free charge. For a linear and homogeneous dielectric medium, the following constitutive relationship can be written

$$\mathbf{D} = \varepsilon \mathbf{E} \tag{6.5}$$

and

$$\mathbf{B} = \mu \mathbf{H} \tag{6.6}$$

where ε is the dielectric permittivity and μ is the magnetic permeability of the medium. Utilizing Eqs. (6.5) and (6.6), the Maxwell equations can be written for a linear homogeneous dielectric medium as

$$\nabla \cdot \mathbf{E} = \frac{\rho_{\mathrm{f}}}{\varepsilon} \tag{6.7}$$

$$\nabla \cdot \mathbf{H} = 0 \tag{6.8}$$

$$\nabla \times \mathbf{E} = -\mu \frac{\partial \mathbf{H}}{\partial t} \tag{6.9}$$

and

$$\nabla \times \mathbf{H} = \varepsilon \frac{\partial \mathbf{E}}{\partial t} + \mathbf{J}_{\mathrm{f}}. \tag{6.10}$$

It is clear from the above equations that when the electric field varies with time, one needs to consider the magnetic field as well. In this case, the equations governing the electric and magnetic fields are coupled. However, in most of electrokinetic system, the effect of the magnetic field can be negligible or the system is under static case i.e. the absence of time variation so that electric field can be decoupled from the magnetic field. Therefore, the Maxwell's equations for describing the typical electrokinetic system can be reduced in the following form:

$$\nabla \cdot \mathbf{E} = \frac{\rho_{\mathrm{f}}}{\varepsilon} \tag{6.11}$$

$$\nabla \times \mathbf{E} = \mathbf{0}. \tag{6.12}$$

Because the electric field is curl-free, electric field can be represented by gradient of scalar quantity ψ known as electric potential. Hence, the governing equation for conservation of electric charge is

$$\nabla \cdot (\varepsilon \nabla \psi) = -\rho_{\mathrm{e}} \tag{6.13}$$

with the fundamental relationship between the irrotational field and the scalar potential as

$$\mathbf{E} = -\nabla \psi. \tag{6.14}$$

Next, we will briefly introduce the appropriate boundary conditions for electrostatic problems. For electrostatic systems, the Gauss law can be applied to the boundary between two domains to determine the boundary conditions for the normal and tangential components of the electric field. The Gauss law relates the electric field through a closed surface to the enclosed charge and can be denoted as

$$\oint_S \mathbf{E} \cdot \mathbf{n} \mathrm{d}S = \frac{Q}{\varepsilon} \tag{6.15}$$

where S is bounded surface of arbitrary volume V, n is the outward normal to the surface S, and Q is the total charge inside the volume V. Given an interface with a charge per unit area given by σs, separating two domains labeled 1 and 2 with two permittivity of ε_1 and ε_2, respectively, the boundary condition for the normal component of the electric field is given by

$$\varepsilon_1 \mathbf{E}_1 \cdot \mathbf{n}_1 - \varepsilon_2 \mathbf{E}_2 \cdot \mathbf{n}_2 = \sigma_s. \tag{6.16}$$

In the above expression, the direction of n_1 points from domain 2 to domain 1. A line integral over a loop at the boundary gives the following boundary condition for any tangent component of the electric field;

$$\mathbf{E}_1 \cdot \mathbf{t} = \mathbf{E}_2 \cdot \mathbf{t}. \tag{6.17}$$

Equation (6.17) holds for any unit vector tangent to the surface. Boundary conditions (6.16) and (6.17) can be represented in terms of electric potential so that

$$-\varepsilon_1 \frac{\partial \psi_1}{\partial n_1} + \varepsilon_2 \frac{\partial \psi_2}{\partial n_2} = \sigma_s \tag{6.18}$$

and

$$\psi_1 = \psi_2 \tag{6.19}$$

which are hold at the interface between two domain.

There are widely used boundary conditions for microscale or nanoscale electrokinetic system. For a reservoir or electrode surface where the electric potential is applied, the electric potential is specified.

$$\psi|_{\text{reservoir}} = V_0 \tag{6.20}$$

where V_0 is the specified value of the applied potential. Because reservoirs of microdevice are typically considered as an enormous source of ions compared with the microchannels themselves, any voltage drop can be neglected between the electrode and the inlet or outlet to the microchannel. For an insulating wall that has surface charge density σ_s, Eq. (6.18) is simplified because the electric field in the insulating wall (E_2) normal to the surface is approximately zero. Thus the normal boundary condition is a relation between the surface charge density and the potential gradient normal to the surface,

$$\left.\frac{\partial \psi}{\partial n}\right|_{\text{wall}} = -\frac{\sigma_s}{\varepsilon}. \tag{6.21}$$

Typically, the material of the channel is SiO_2, Al_2O_3, or PDMS of which relative permittivity is lower than 10 so that channel wall can be treated as insulating material and Eq. (6.21) is a reasonable approximation. When the relative permittivity of the channel wall cannot be neglected or the electrostatics inside the solid is important, Eqs. (6.18) and (6.19) should be used together. These are summarized in Table in Sect. 6.2.2.4.

6.2.2.2 Conservation of Mass

This section describes the phenomena that lead to flux of charged ionic species in the dilute limit. These fluxes lead to the Nernst-Planck equations. Although certain conditions, i.e. high concentration and high applied voltage, would violate the dilute approximation, we will discuss only the conservation of mass in the dilute limit. The effect of the non-dilute condition will be reconsidered in Sect. 6.2.4 because the effect cannot be negligible in nanometer scale system.

Firstly, we will consider the diffusion flux of the ionic species when concentration gradient is existed. In the dilute limit, Fick's law defines a flux density of species proportional to the gradient of the species concentration and the diffusivity of the species in the solvent as

$$\mathbf{j}_{\mathrm{diff},i} = -D_i \nabla c_i \tag{6.22}$$

where $j_{\mathrm{diff},i}$ is the diffusive species flux density i.e. the amount of species i moving across a surface per unit area due to diffusion, D_i is the diffusivity of species i in the solvent (usually water), and c_i is the concentration of species i. Fick's law is a macroscopic way of representing the summed effect of the random motion of species owing to thermal fluctuations. Fick's law for thermal energy flux caused by a temperature gradient and the Newtonian model for momentum flux induced by a velocity gradient, and the species diffusivity D_i is analogous to the thermal diffusivity $\alpha = k/\rho\ Cp$ and the momentum diffusivity η/ρ.

Second flux is the convective flux of the ionic species when there is an ambient fluid flow. Its form is

$$\mathbf{j}_{\mathrm{conv},i} = c_i \mathbf{u} \tag{6.23}$$

where $j_{\mathrm{conv},i}$ is the convective species flux density i.e. the amount of species i moving across a surface per unit area due to an ambient fluid flow u.

Third flux is the migration flux of the ionic species due to an external force such as electric field in usual electrokinetic system. The migration flux can act differently on different species in general. Let us denote the force acting on one ionic species by F_{ext}, which is applied external force. The ionic species will move under the influence of this external force, and one can write its migration velocity V_i as

$$\mathbf{v}_i = \omega_i \mathbf{F}_{\mathrm{ext}} \tag{6.24}$$

where ω_i is defined as the mobility of the species (velocity per unit applied force). The migration flux can be defined as

$$\mathbf{j}_{\mathrm{mig},i} = c_i \mathbf{v}_i = c_i \omega_i \mathbf{F}_{\mathrm{ext}}. \tag{6.25}$$

At this point, we need to interpret the mobility ω_i, which was introduced in Eq. (6.24). Normally, application of a force to a mass causes it to accelerate so that assigning a constant velocity to the particle as in Eq. (6.24) seems counter-intuitive. However, it should be noted that as soon as a solute particle starts to move under the influence of the external force, it encounters a counteracting drag force due to the other particles (predominantly solvent) surrounding it. Consequently, it is assumed that the solute particle moves at a terminal velocity attained under the combined influence of the external force and the frictional drag force of the surrounding medium. This is a grossly simplified view of the migration velocity and Einstein showed that

$$\omega_i = \frac{D_i}{k_B T} \tag{6.26}$$

where k_B is the Boltzmann constant and T is the absolute temperature of the surrounding. Substituting Eq. (6.26) in Eq. (6.25), one obtains

$$\mathbf{j}_{\mathrm{mig},i} = \frac{D_i}{k_B T} c_i \mathbf{F}_{\mathrm{ext}}. \tag{6.27}$$

As earlier mentioned, Fext is usually the external electric field for electrokinetic system so that Fext can be represented by

$$\mathbf{F}_{\mathrm{ext}} = -z_i e \nabla \psi \tag{6.28}$$

where z_i is the valence of the ionic species and e is the elementary charge. Therefore, the migration flux of the ionic species can be expressed in terms of the independent variables of electrokinetic system.

$$\mathbf{j}_{\mathrm{mig},i} = -\frac{z_i e D_i}{k_B T} c_i \nabla \psi = -\frac{z_i F D_i}{RT} c_i \nabla \psi \tag{6.29}$$

where F is the Faraday constant and the R is the gas constant.

In above representations for ionic flux in terms of each transport mechanism, total ionic flux j_i can be expressed by

$$\mathbf{j}_i = \mathbf{j}_{\mathrm{diff},i} + \mathbf{j}_{\mathrm{mig},i} + \mathbf{j}_{\mathrm{conv},i} = -D_i \nabla c_i - \frac{z_i F D_i}{RT} c_i \nabla \psi + c_i \mathbf{u}. \tag{6.30}$$

Due to the conservative nature of chemical species with no chemical reaction, total ionic flux should be satisfied with

$$\nabla \cdot \mathbf{j}_i = \nabla \cdot \left(-D_i \nabla c_i - \frac{z_i F D_i}{RT} c_i \nabla \psi + c_i \mathbf{u} \right) = 0. \tag{6.31}$$

When there are chemical reactions, Eq. (6.31) becomes

$$\nabla \cdot \mathbf{j}_i = \nabla \cdot \left(-D_i \nabla c_i - \frac{z_i F D_i}{RT} c_i \nabla \psi + c_i \mathbf{u} \right) = R_i \tag{6.32}$$

where R_i is the reaction rate of chemical species i. Equation (6.32) is known as the Nernst-Planck equations which imply the conservation of mass in terms of concentration of species i.

Widely used boundary conditions for Nernst-Planck equations are (1) specified concentration and (2) specified ionic normal flux. Firstly, specified concentration condition is used for reservoir, inlet, or outlet of the system. When the concentration is known at the boundary, condition is

$$c_i|_{\text{reservoir}} = c_0 \tag{6.33}$$

where c_0 is the known concentration at the boundary. Secondly, specified ionic normal flux is chosen in the case of channel wall, chemically reactive surface, or inlet/outlet under a forced transport mechanism. The general form of this type of boundary is

$$\mathbf{n} \cdot \mathbf{j}_i|_{\text{boundary}} = \mathbf{n} \cdot \mathbf{j}_0 \tag{6.34}$$

where j_0 is the known expression of ionic flux though the boundary. When boundary is channel wall, $j_0 = 0$ which means any species cannot penetrate through the boundary. When boundary is chemically reactive surface, j_0 is a function of surface reaction rate. In case of inlet/outlet under a forced transport mechanism, j_0 is fixed to a function of relating transport mechanism. For example, when forced convection is on inlet or outlet, j_0 becomes $c_0\mathbf{u}|_{\text{boundary}}$.

6.2.2.3 Conservation of Momentum

In this section, we discuss the Stokes equation for the conservation of momentum and the continuity equation. In typical electrokinetic system, there are fluid flows induced by ion movement driven by external forces. Thus, fluid motion should be considered and this is governed by the conservation of momentum. The Reynolds number of usual microscale or nanoscale electrokinetic system is low so that viscous forces dominate over inertial forces. As a result, the Navier-Stokes equations can be reduced to the Stokes equations which are linear differential equations so that the Stokes equations are easier to solve than the Navier-Stokes equations.

The flow of the Newtonian fluid system under laminar flow conditions with constant density and viscosity is governed by the momentum conservation equation and is given by the Navier-Stokes equations [22].

$$\rho \frac{\partial \mathbf{u}}{\partial t} + \rho \mathbf{u} \cdot \nabla \mathbf{u} = -\nabla p + \eta \nabla^2 \mathbf{u} + \rho \mathbf{g} + \mathbf{F}_b. \tag{6.35}$$

Equation (6.35) represents the force balance on a fluid element in space. Here ρ is the density of the fluid, t is the time, u is the flow velocity field, p is the pressure, η is the viscosity of the fluid, g is the gravitational acceleration, and F_b is the body force acting on the fluid element in addition to the gravity. In electrokinetic transport processes, one needs to consider the electrical body force acting on the fluid. Further discussion on the electrical body force will be presented later in this section.

Each term of Eq. (6.35) represents forces acting on a unit volume of the fluid. The first term on the left hand side represents the rate of change of momentum at a given location within the flow field. For steady conditions, this term drops out. The second term is due to the fluid inertia and is negligible for low Reynolds number flows [23] and this will be shown in later. The first term of the right hand side represents the contribution of pressure, the second term is due to viscous forces, the third term arises from the body force due to gravity, and the last term arises from other body force which is usually electrical body force in the electrokinetic system. Equation (6.35) can be written for any orthogonal coordinate system [22].

Although the conservation of momentum can be described by the Navier-Stokes equations, it is difficult to solve the equations because of nonlinearity of the inertial term. Thus, we derive the Stokes equations by neglecting the unsteady and inertial terms when the viscous force is dominant. Equation (6.35) is rewritten in non-dimensional form in which the length scale is L_c, the velocity scale is U_c, the pressure scale is $\eta\ U_c/L_c$, and the time scale is L_c/U_c.

$$Re\left(\frac{\partial \tilde{\mathbf{u}}}{\partial \tilde{t}} + \tilde{\mathbf{u}} \cdot \tilde{\nabla} \tilde{\mathbf{u}}\right) = -\tilde{\nabla} \tilde{p} + \tilde{\nabla}^2 \tilde{\mathbf{u}} + \tilde{\mathbf{g}} + \tilde{\mathbf{F}}_b. \tag{6.36}$$

Above expression is non-dimensional form of the Navier-Stokes equations. A tilde denotes dimensionless variable. Re is the Reynolds number, defined by $R_e = \rho U_c L_c/\eta$ of which meaning is the ratio of inertial force and viscous force acting on a fluid element. Considering the microscale or nanoscale electrokinetic system, R_e is usually much smaller than 1 so that R_e can be approximated to 0. Therefore, the left hand side of Eq. (6.36) is negligible. Consequently, following equations are called as Stokes equations with body force terms.

$$\mathbf{0} = -\nabla p + \eta \nabla^2 \mathbf{u} + \rho \mathbf{g} + \mathbf{F}_\text{b}. \tag{6.37}$$

For convenience, the term of gravity force merge into pressure so that modified pressure is usually used rather than original hydrodynamic pressure.

As earlier discussed, the momentum equation given by Eq. (6.37) contains a generic body force term, F_b which can be used to consider any type of force acting on a fluid volume element. In electrokinetic transport processes, this body force arises due to an electric field. The electric body force per unit volume on the fluid, F_E is given by [24]

$$\mathbf{F}_\text{E} = \rho_\text{f} \mathbf{E} - \frac{1}{2} \mathbf{E} \cdot \mathbf{E} \nabla \varepsilon + \frac{1}{2} \nabla \left[\left(\rho \frac{\partial \varepsilon}{\partial \rho} \right)_T \mathbf{E} \cdot \mathbf{E} \right] \tag{6.38}$$

which is derived from the Korteweig-Helmholtz electric for a linear dielectric medium. For the special case of a constant dielectric permittivity and an incompressible fluid, the electrical body force becomes

$$\mathbf{F}_\text{E} = \rho_\text{f} \mathbf{E}. \tag{6.39}$$

Recognizing that the electric field E is related to the electric potential ψ, the body force is represented as

$$\mathbf{F}_\text{E} = -\rho_\text{f} \nabla \psi. \tag{6.40}$$

Introducing the electric force term for the body force in Eq. (6.37), we obtain the following governing equations for conservation of momentum in the electrokinetic system as

$$\mathbf{0} = -\nabla p + \eta \nabla^2 \mathbf{u} - \rho_\text{f} \nabla \psi. \tag{6.41}$$

Additionally, the mass conservation for a fluid is also required to solve the Stokes equations. The general form of the mass conservation for the fluid is given by [24]

$$\frac{\partial \rho}{\partial t} + \nabla \cdot (\rho \mathbf{u}) = 0. \tag{6.42}$$

For the incompressible fluid, the density of the fluid is constant so that

$$\nabla \cdot \mathbf{u} = 0 \tag{6.43}$$

which is called as the continuity equation for the incompressible fluid. Equations (6.41) and (6.43) are the governing equations for usual electrokinetic system.

The Stokes equations and the continuity equation can be solved after setting appropriate boundary conditions corresponding to a given flow situation. The boundary conditions can specify the velocities or pressures at given locations of the flow boundary. Of particular interest with respect to boundary conditions for these equations is the no-slip condition, which is generally applied at the solid boundaries and this condition is

$$\mathbf{u}|_{\text{boundary}} = \mathbf{U}_0 \tag{6.44}$$

where U_0 is the specified velocity. The no-slip condition implies that the fluid in contact with a solid object moves at the same velocity as the object. If the object is stationary such as the wall of microchannel, then the fluid velocity is zero at the surface. Imposition of the no-slip condition usually sets up a velocity gradient in the fluid near the solid-fluid interfaces. Such a gradient gives rise to a viscous stress. The local hydrodynamic stress tensor at a point on the surface on the solid surface is given by

$$\boldsymbol{\sigma}^{\mathrm{H}} = -p\mathbf{I} + \eta\left(\nabla\mathbf{u} + \nabla\mathbf{u}^{\mathrm{T}}\right) \tag{6.45}$$

where σ^H is the hydrodynamic stress tensor, I is the unit tensor, the superscript T means the transpose of the tensor, and the final term on the right hand side represents the viscous stresses. This hydrodynamic stress can be integrated over the surface of the solid object to determine the total hydrodynamic force acting on the object.

6.2.2.4 Coupled Conservation Principles

For the electrokinetic system, independent variables are the electric potential (ψ), ion concentration of i-th species (c_i), pressure (p), and flow field (u) obtained by Eqs. (6.13), (6.32), (6.41), and (6.43) which are derived from conservation principles. Each independent variable is a function of other variables. For example, the electric potential is a function of the ion concentration, the ion concentration is a function of the electric potential and flow field, and flow field is a function of pressure, ion concentration, and electric potential. This situation is called that governing equations are coupled. Thus, governing equations should be solved by coupled manner as similar as simultaneous algebraic equations. Accordingly, coupled governing equations and boundary conditions are summarized in Table 6.1.

Table 6.1 Governing equations and appropriate boundary conditions

Conservative quantity	Name of governing equation	Differential form of governing equation	Boundary conditions
Electric charge	Poisson equation	$\nabla \cdot (\varepsilon \nabla \psi) = -\rho_e$	$\psi\|_{\text{reservoir}} = V_0$ (fixed potential) $\left.\frac{\partial \psi}{\partial n}\right\|_{\text{wall}} = -\frac{\sigma_s}{\varepsilon}$ (surface charge density)
Mass	Nernst-Plank equations	$\nabla \cdot \mathbf{j}_i = \nabla \cdot \left(-D_i \nabla c_i - \frac{z_i F D_i}{RT} c_i \nabla \psi + c_i \mathbf{u}\right) = R_i$	$c_i\|_{\text{reservoir}} = c_0$ (fixed concentration) $\mathbf{n} \cdot \mathbf{j}_i\|_{\text{reservoir}} = \mathbf{n} \cdot \mathbf{j}_0$ (specified normal flux)
Momentum	Stokes equations	$\mathbf{0} = -\nabla p + \eta \nabla^2 \mathbf{u} - \rho_f \nabla \psi$	$\mathbf{u}\|_{\text{boundary}} = \mathbf{U}_0$ (fixed velocity or pressure in alternative condition) $\boldsymbol{\sigma}^{\text{H}} = -p\mathbf{I} + \eta(\nabla \mathbf{u} + \nabla \mathbf{u}^{\text{T}})$ (specified hydrodynamic stress)
Mass of fluid	Continuity equation	$\nabla \cdot \mathbf{u} = 0$	

6.2.3 Microscale Electrokinetics

Electrokinetics has referred the branch of electrodynamics, which treats the laws of the electrical current, but it should contain rigorous consideration of electrolytic current as well. Recent studies have generally defined the electrokinetic phenomenon as the general term associated with the relative motion between two electrically charged phases, i.e. liquid and solid. Staring from pollutant removal using electric field in the soil science area, the fields of application using the electrokinetic phenomenon has been widen various research fields as shown in Fig. 6.1. With aids of splendid advances in micro/nanofabrication and manufacturing technology, the electrokinetics have been applied to precise medicines, drug discovery and separation science in (bio-) medical engineering and complex fluids, colloid science and micro/nanofluidic in physics area. Among such advances, the major phenomena associated with human internal body structure, electroosmosis has been regarded as the major key elements to understand the biological transport phenomena. Other electrokinetic phenomena such as electrophoresis, dielectrophoresis, induced charge electrokinetics, electrowetting, streaming potential, sedimentation potential, etc. were not covered in this book because their engineering applications focused on problems other than phenomena occurred inside biological system.

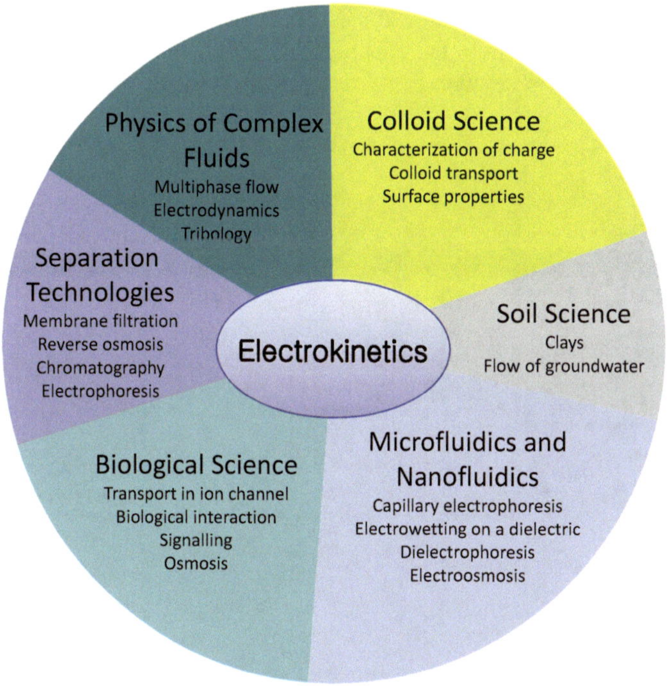

Fig. 6.1 The applications of electrokinetic phenomenon

6.2.3.1 Equilibrium Electroosmosis

Electroosmosis is the motion of liquid induced by an applied electric potential across a porous material, capillary tube, membrane, microchannel, or any other fluid conduit. It is an essential component in separation science, especially capillary electrophoresis. It can occur in natural unfiltered water, human body fluid, buffer solutions and highly concentrated brine as well. Discovered by F. F. Reuss in 1809, water migrated through porous clay diaphragms toward the cathode under the influence of electric field since mineral particles are generally negatively charged. The physics behind electroosmosis is shown in the schematics (Fig. 6.2). Once solid conduit was filled with electrolyte solution, ions in the solution attack the surface so that the surface become to have electrical charges. The charges attract counter-ions to form a thin electrical double layer which has counter-ion rich region, while the center (or bulk) of the conduit remains electro-neutral status. With an external electric potential gradient, the ions inside the electrical double layer migrated toward their counter electrode and the water in the conduit is drawn by the ions due to viscous drag and therefore, flows through the conduit. This situation was governed by Navier-Stokes equations with electrical body forces as

$$\rho\left(\frac{\partial \mathbf{u}}{\partial t} + \mathbf{u}\cdot\nabla\mathbf{u}\right) = -\nabla p + \mu\nabla^2\mathbf{u} + \rho\mathbf{g} + \rho_e\mathbf{E}. \tag{6.46}$$

Since the conduit where the drag forces dominant to induce electroosmosis has under microscale dimension, the equations is simplified in a steady Stokes flow with electrical body force, with neglecting gravity force and pressure gradient as

$$\mu\nabla^2\mathbf{u} = -\rho_e\mathbf{E}. \tag{6.47}$$

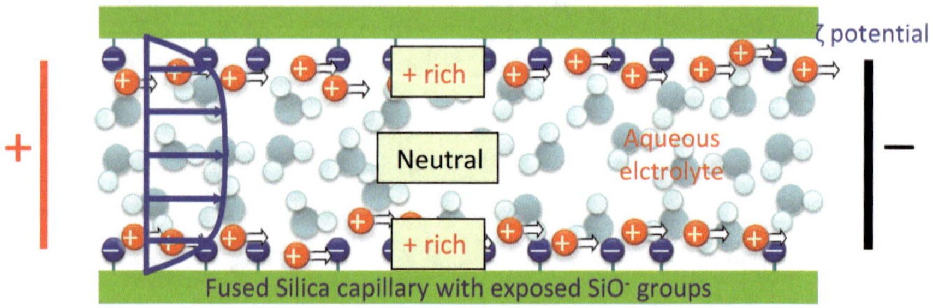

Fig. 6.2 Schematic diagram of electroosmosis

Compared to a pressure driven flow, electroosmosis is surface driven flow so that one can assume thin electrical double layer approximation for a slip velocity known as Helmholtz-Smoluchowski slip velocity defined as

$$\mathbf{u} = -\frac{\varepsilon\zeta}{\mu}\mathbf{E}. \tag{6.48}$$

The Helmholtz-Smoluchowski slip velocity is used as a slip boundary condition to solve the entire flow field for more complicated geometries after solving electrical field. Conventional conditions such as water in silicon microchannel ($\varepsilon = 6.9 \times 10^{-10}$ C^2/Jm, $\zeta = -100$ mV, $\mu_{water} = 10^{-3}$ Ns/m^2 and $|E| = 1$ kV/m) would the slip velocity of ~69 μm/s.

Since the pressure necessary for driving a liquid flow inside a microscale conduit is proportional to the quartic of diameter, electroosmosis is increasingly effective in driving flow as capillary diameter decreases because the flow rate is proportional to the square of diameter in electroosmosis. However, the thin double layer approximation is no longer valid when the electrolyte concentration is extremely diluted or the size of conduit become comparable to the thickness of electrical double layer which has the range of 1–100 nm. The second scenario generally occurs in nanoscale electrokinetics and most of assumption, governing equations and boundary conditions of equilibrium electroosmosis should be modified as non-equilibrium (or the second kind of) electroosmosis.

6.2.4 Nanoscale Electrokinetics

6.2.4.1 Nonlinear Concentration Distribution: Ion Concentration Polarization

Permselective Nanoporous Membrane System

Recently, the ion concentration polarization (ICP) phenomena, which are occurred nearby the nanojunction or nanoporous membrane, have drawn significant attention in a variety of applications such as desalination [13], preconcentration [25], analytical sensors [14], and fuel cells [16]. When the nanochannel is thinner than 100 nm, it has the permselectivity, unique property of nanoscale system, due to the fact that the Debye layer thickness (λ_D) is non-negligible compared with the channel thickness so that the overlap of Debye layer could be happened [24, 26, 27]. Through the permselective nanoporous media, only counter-ion can be transported across the media while co-ion cannot be done. Because of the selectivity of transported species, ion concentration is decrease on one side (this is called 'depletion'), while on the other side ion concentration is increase (this is called 'enrichment') as shown in Fig. 6.3. In other words, the permselectivity is the

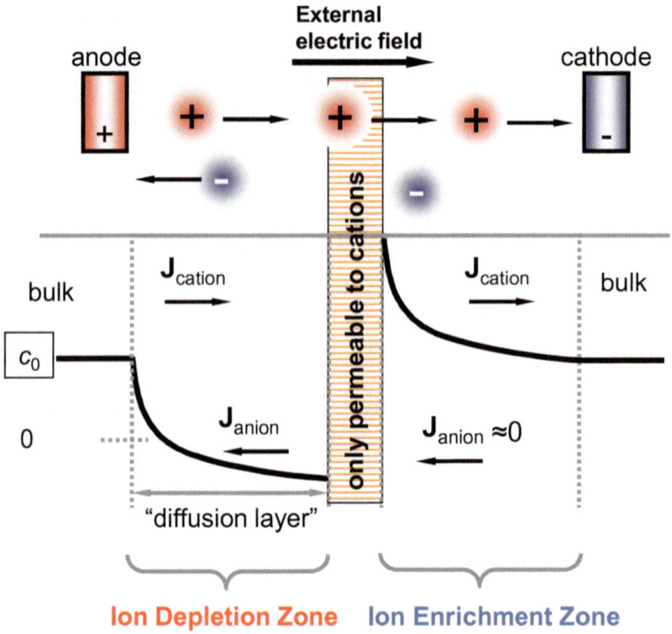

Fig. 6.3 Schematic diagram of ion concentration distribution nearby permselective media

property, which let only counter-ions pass through the permselective media so that ion concentration is depleted on one side while it is enriched on the other side. The phenomena of concentration polarization front and back of permselective media is called as ICP [28, 29]. As long as the permselectivity of the nanoporous media is stand up, ICP is always occurred nearby the nanojunction.

The permselective systems have a unique feature in the voltage against current curves. As shown in Fig. 6.4a, they have a characteristic shape with a region of slow current variation (the plateau) after a region of linearly increased current against the applied voltage. The linearly increased region is called as 'Ohmic' regime and the region of the plateau is called as 'limiting' regime. Beyond the limiting regime, the ionic current is grown once again as the applied voltage is increased, this is called as 'overlimiting' regime. In the limiting and overlimiting regime, ion depletion zone is extended outward to microchannel reservoir, so that the black region is formed nearby nanoporous membrane shown in Fig. 6.4b. Its extended depletion zone has drawn significant attention in scientific and engineering fields [29–32].

In the depletion zone, followings are known as dominant phenomena: (1) ion concentration is depleted below 1/100 dilution against the reservoir concentration [13, 33], (2) the electric field is amplified [34], and (3) strong vortices nearby nanojunction are formed [28]. When the voltage is applied across the cationic permselective nanoporous membrane, cation is permitted to pass through while

(a)

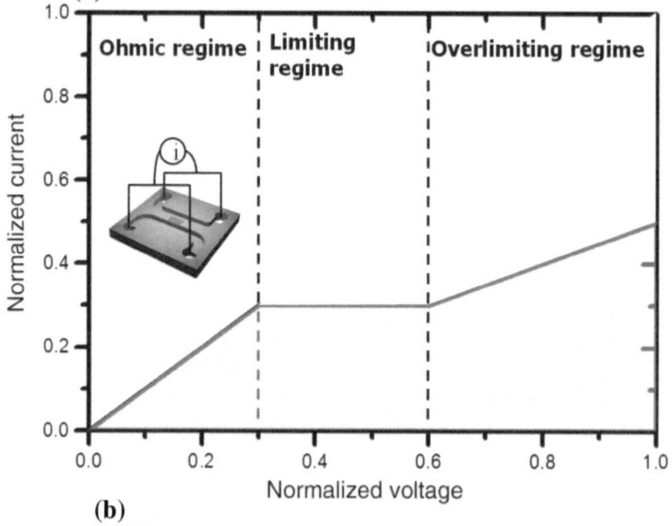

(b)

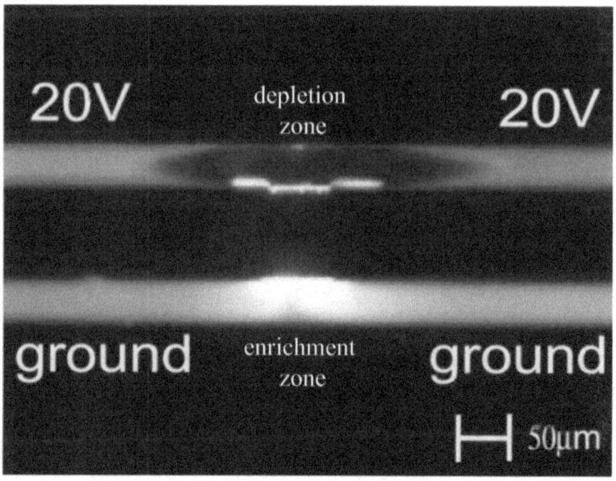

Fig. 6.4 **a** Characteristic curves of ionic current against applied voltage of the Ohmic, limiting, and overlimiting regime. **b** The basic formation of ion-enrichment and ion-depletion in nanofluidic system

anion is not. As a result, in cathodic side in microchannel, ions are abundant while ions are depleted below 1/100 times in anodic side. Due to the depletion of ionic species as charge carriers, the electrical conductivity becomes extremely small. Because the ionic current through the system should be uniform, the electric field is amplified in depletion zone to satisfy the uniform ionic current which was confirmed by the experimental measurement [34]. Additionally, strong vortices are formed in the depletion zone to satisfy the continuity condition [28]. The vortex speed was estimated to be usually over 1000 μm/s, which is at least (10) times higher than that of primary electroosmotic flow under the same electric potential. Its speed is proportional to either the square or cube of the applied voltage so that formation of vortex is nonlinear phenomena. At the steady state, Kim et al. [28] clearly observed the counter-rotating vortices adjacent to the nanojunction. In case of single nanojunction, two strong vortices were observed while four independent vortices were formed in the depletion zone of double nanojunctions.

Models for Ion Concentration Polarization

Classical model (derived by Levich in 1962 [35]) based on Nernst-Planck equations, which consider ion transport only due to diffusion and electromigration, predicts the linear concentration profile inside the ion depletion zone and Ohmic/limiting current behavior as a function of applied voltages. In the case of a cation selective membrane, the ionic currents of cation and anion could be expressed as

$$i_+ = -D_+ F z_+ \frac{dc_+}{dy} - \frac{F^2 z_+^2 D_+ c_+}{RT} \frac{d\psi}{dy} \tag{6.49}$$

and

$$i_- = -D_- F z_- \frac{dc_-}{dy} - \frac{F^2 z_-^2 D_- c_-}{RT} \frac{d\psi}{dy} \tag{6.50}$$

where i_+ is the ionic current by cation transport, i_- is the ionic current by anion transport, D_+ is the diffusivity of cation, D_- is the diffusivity of anion, F is the Faraday constant, z_+ and z_- is the valence of each ionic species, R is the gas constant, T is the absolute temperature, c_+ and c_- is the ion concentration of each ionic species, and ψ is the electric potential, respectively. Equations (6.49) and (6.50) are valid in the domain as shown in Fig. 6.5., where the thickness of ICP layer is b.

In order to analyze the Eqs. (6.49) and (6.50), the condition of electroneutrality is assumed to hold within the fluid; that is,

$$z_+ c_+ + z_- c_- = 0 \tag{6.51}$$

Second assumption is the ideally cation selective membrane. Under second assumption, the anion current must be zero because only cation can pass through the membrane. Its mathematical form is

$$i_-|_{y=b} = 0 \tag{6.52}$$

When there is no anion current source in the fluid, following is hold.

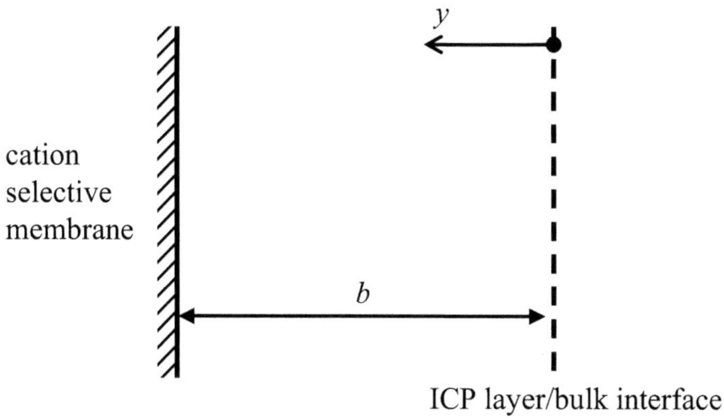

Fig. 6.5 Schematic diagram of ICP layer for the case of cation selective membrane

$$\frac{\mathrm{d}i_-}{\mathrm{d}y} = 0 \tag{6.53}$$

Integrating Eq. (6.53) with respect to y,

$$i_- = C \tag{6.54}$$

where C is the integration constant. Due to Eq. (6.52), i_- should be zero. In next step, the effective concentration is defined by

$$c = \frac{c_+}{\nu_+} = \frac{c_-}{\nu_-} \tag{6.55}$$

where c is the effective concentration, ν_+ and ν_- are the dissociation number. For example of the dissociation number, ν_+ is 1 and ν_- is 2 when the electrolyte is $CuCl_2$. With Eqs. (6.51)–(6.55), (6.49) and (6.50) become

$$i_+ = -D_+ F z_+ \nu_+ \frac{\mathrm{d}c}{\mathrm{d}y} - \frac{F^2 z_+^2 D_+ \nu_+ c}{RT} \frac{\mathrm{d}\psi}{\mathrm{d}y} \tag{6.56}$$

and

$$0 = -D_- F z_- \nu_- \frac{\mathrm{d}c}{\mathrm{d}y} - \frac{F^2 z_-^2 D_- \nu_- c}{RT} \frac{\mathrm{d}\psi}{\mathrm{d}y} \tag{6.57}$$

From the anion current as expressed in Eq. (6.57),

$$\frac{d\psi}{dy} = -\frac{RT}{Fz_-c}\frac{dc}{dy} \tag{6.58}$$

Above expression means that electric migration is same as diffusive transport when the convective transport is negligible. Using Eq. (6.58), the cation current could be reduced in the following form.

$$i_+ = -D_+ F v_- (z_+ - z_-)\frac{dc}{dy} \tag{6.59}$$

in which the condition of electroneutrality as expressed in Eq. (6.51) is applied in the following modified form.

$$z_+ v_+ = z_- v_- \tag{6.60}$$

Since there is a fixed current for a given applied voltage, it follows that c must be linear in distance across the ICP layer. Consider the case of $i_+ > 0$ which corresponds to the formation of the depletion. Due to $(z_+ - z_-) > 0$, (dc/dy) should be negative i.e. the concentration of ions decreases from the ICP layer/bulk interface to the membrane surface. The potential drop is also in the same direction. Integrating Eq. (6.59), concentration profile inside the ICP layer is obtained.

$$c = c_0 - \frac{i_+ y}{D_+ F v_- (z_+ - z_-)} \tag{6.61}$$

where c_0 is the bulk concentration. When $c|y = b = 0$, the current density approaches a limiting value which is called as limiting current.

$$i_L = \frac{D_+ F v_- (z_+ - z_-) c_0}{b} \tag{6.62}$$

The above limiting current was derived by Levich in 1962. Although Levich's model pointed out that the bulk charge makes impossible the existence of steady currents greater than the limiting values, Rubinstein and Shtilman [33] showed that this is not true. They showed the increase of the ionic current beyond the limiting regime (plateau of Fig. 6.4a). This regime is called as overlimiting regime. To describe the transition from limiting regime to overlimiting one, some mechanisms have been proposed: (1) bulk charge effect [33], (2) surface conduction along the microchannel wall [31], (3) electro-convective mixing inside the ICP layer [31], and (4) electroosmotic instability [36]. With these mechanisms, depletion zone could be extended toward the reservoir so that system could enter the overlimiting regime. In next, a modified model, which considered above mechanisms, are briefly introduced.

Rubinstein and Shtilman [33] developed the model which allows one to investigate the role of the bulk charge in developing concentration polarization and to trace the connection with classical theory. As the simplest possible model consider a steady current passing through an ideally permselective membrane immersed in a stirred solution of 1:1 valent electrolyte. They assumed that near the membrane there is an 'unstirred' layer of thickness δ, which does not depend on the magnitude of the voltage V, applied to this layer. The concentration of the electrolyte is assumed to be constant and equal to c_0. The cation concentration within the membrane is supposed to be equal to the fixed charge concentration N. It is assumed that the ions within the unstirred layer are distributed by means of electro-diffusion only. Direct the x-axis normally to the membrane, identifying x = 0 with the outer boundary of the "unstirred" layer. Then the corresponding boundary value problem takes the form

$$\frac{d^2\psi}{dx^2} = -\frac{4\pi F}{\varepsilon}(c_1 - c_2) \tag{6.63}$$

$$\frac{\mathrm{d}}{\mathrm{d}x}\left(\frac{\mathrm{d}c_1}{\mathrm{d}x} + \frac{Fc_1}{RT}\frac{\mathrm{d}\psi}{\mathrm{d}x}\right) = 0 \tag{6.64}$$

$$\frac{\mathrm{d}}{\mathrm{d}x}\left(\frac{\mathrm{d}c_2}{\mathrm{d}x} - \frac{Fc_2}{RT}\frac{\mathrm{d}\psi}{\mathrm{d}x}\right) = 0 \tag{6.65}$$

$$\psi|_{x=0} = 0 \tag{6.66}$$

$$c_1|_{x=0} = c_2|_{x=0} = c_0 \tag{6.67}$$

$$\psi|_{x=\delta} = -V \tag{6.68}$$

$$c_1|_{x=\delta} = N \tag{6.69}$$

$$\left(\frac{\mathrm{d}c_2}{\mathrm{d}x} - \frac{Fc_2}{RT}\frac{\mathrm{d}\psi}{\mathrm{d}x}\right)\Bigg|_{x=\delta} = 0 \tag{6.70}$$

where ε is the dielectric permittivity. The Poisson equation is represented by Eq. (6.63). The Nernst-Planck equations are represented by Eqs. (6.64) and (6.65), which describe the mass transport of charged species under consideration of the diffusion and the electromigration. Because of assumption of unstirred layer, the Stokes equations and the continuity equation for the fluid flow are not solved. Equations from (6.66) to (6.70) are boundary conditions for unstirred ICP layer. Equations (6.66) and (6.67) are the conditions for bulk reservoir. Equations (6.67)–(6.70) are the conditions for the ideally cation selective membrane, so that anion flux through the membrane surface should be zero. Instead of Eq. (6.68), the Donnan equilibrium is often used, which is the following mathematical form.

$$\psi|_{x=\delta} = -V + \frac{RT}{F}\ln\left(\frac{N}{c_0}\right) \tag{6.71}$$

Due to the Poisson equation, their model could describe the effect of the bulk charge to ICP layer in the overlimiting regime. Their results are showed in Fig. 6.4. In their definition, parameter ε is defined by

$$\varepsilon = \frac{\lambda_D^2}{\delta^2} \tag{6.72}$$

Here the dimensionless parameter has the meaning of a square ratio of the Debye length to the thickness of the unstirred layer. The Levich's model of concentration polarization, based on the assumption of local electroneutrality corresponds to the limit (Fig. 6.6).

$$\varepsilon = 0 \tag{6.73}$$

With the parameter of Eq. (6.73), ionic current is saturated in the limiting value as denoted in Fig. 6.4a. Due to the effect of the bulk charge, ionic current is increased with different conductance compared with Ohmic regime. In Fig. 6.4b, the plateau of concentration distribution is extended to the bulk interface, in which the plateau is often called as extended space charge layer.

Dydek et al. [31] revisited the classical problem of diffusion-limited ion transport to a membrane by considering the effects of charged sidewalls. Using simple mathematical models and numerical simulations, they identified three additional mechanisms for the overlimiting regime: surface conduction, electro-convective mixing, and electroosmotic instability. In order to consider the effect of the surface conduction only, they assumed that convection becomes negligible compared to diffusion (small Peclet number limit), of which case is corresponding to long, narrow channels and thin double layers. Under their assumption, the Nernst-Planck equations could be homogenized as follows

$$\frac{dc_+}{dx} + c_+\frac{d\tilde{\psi}}{dx} = -\frac{j}{zFD} \tag{6.74}$$

$$\frac{dc_-}{dx} - c_-\frac{d\tilde{\psi}}{dx} = 0 \tag{6.75}$$

$$c = c_- = c_+ + \frac{\rho_s}{F} \tag{6.76}$$

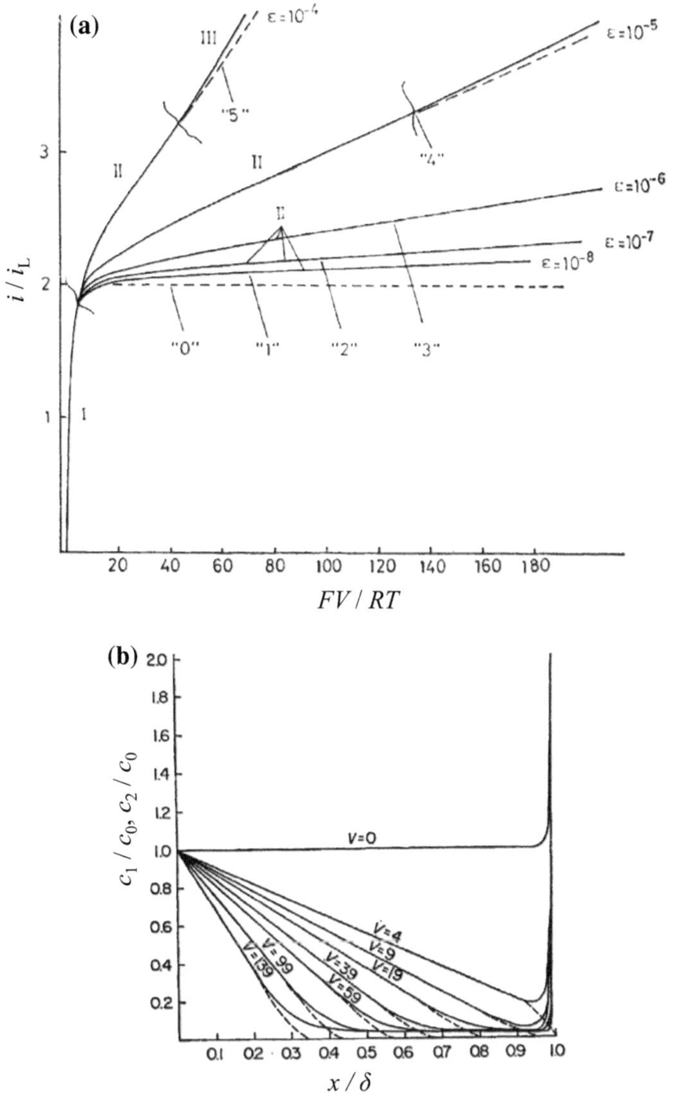

Fig. 6.6 The effect of bulk charge [33]. **a** Calculated voltage against current curves for different values of parameter, ε. **b** Calculated ion concentration profiles at different voltages for $\varepsilon = 10^{-4}$. Solid lines are cation concentration and dashed lines are anion concentration

where c_+ and c_- are the mean concentration of cations and anions, respectively, D is the ionic diffusivity, $\tilde{\psi}$ is dimensionless electric potential scaled by the thermal voltage, RT/zF. Equations (6.74) and (6.75) relate the cation flux to the current density j in steady state for an ideally cation selective membrane. Equation (6.76) means averaged electroneutrality condition including both ionic and fixed surface charge of microchannel wall and ρ_s is defined as

$$\rho_s = \frac{2\sigma_s}{H} \tag{6.77}$$

where σ_s is the surface charge density of microchannel wall and H is the depth of the microchannel. Without the loss of generality, they could be derived analytical solutions of Eqs. (6.74) and (6.75). The current-voltage relation is demonstrated to possess nearly constant conductance in the overlimiting regime. Also, the plateau of the concentration profile is formed as similar as the extended space charge layer of Rubinstein and Shtilman [33]. Additionally, they considered the effect of electro-convective mixing. Under and assumption of dead-end channel, pressure-driven back flow opposes electroosmotic flow and results in two counter-rotating vortices. Concentrated solution flows to the membrane along the sidewalls, and depleted solution returns in the center. To describe the electro-convective mixing, they assumed thin double layers and neutral bulk solution described by the steady 2D Nernst-Planck equations, the Stokes equations, and the continuity equation. These governing equations with appropriate boundary conditions were solved numerically. As a result, numerical solutions to describe the electro-convective mixing were obtained.

These coupled phenomena resembled the ion transportation through cell membrane since the membrane has various-sized nanopore so that effective control of important ions such as Na^+ and K^+ are regulated by this basic transport phenomena. In next, we will introduce more intense mechanism called ionic field effect transistor that is capable of controlling the ion transportation through the nanoporous membrane.

6.2.4.2 Modulated Ion Transportation: Ionic Field Effect Transistor

Ionic Field Effect Transistor: Voltage-Gated Nanoscale System

Recent advances in nano-fabrication methods enable to fabricate rigorous and definite nano-sized structure for various scientific and engineering applications. Nanostructure possesses unique scientific and technological properties that microstructure cannot exhibit. Especially, as decreasing the size of nanostructure below 100 nm, the structure had a perm-selectivity which let only counter-ions can pass through below a critical electrolyte concentration. The perm-selectivity was reported to be depending on the magnitude and polarity of surface charge density and bulk electrolyte concentration. Thus, the active control of the surface charge density at wide range of electrolyte concentration has been drawn significant attentions in both scientific and engineering field [15, 37–41] for manipulating the motion of charged species and this has become one of the important fields in nanofluidics research. The emerging application fields of nanofluidic system were energy harvesting [15, 37], biosensors [38, 39], backflow from shale gas extraction port [40] or desalinations of seawater [41] which enable to create a huge market that never have existed. Those applications were fundamentally originated from controlling the motion of charged species passing through a nanostructure and, therefore, the cost-effective/on-demand/sensitive control has become the most important practical issue of nanofluidic researches.

Various passive types of modulating the motion of a charged species in nanofluidic system were reported such as changing the viscosity of solution in the nanochannel [42], utilizing mechanical friction between DNA and nanopore [43], coating an adhesive material on nanochannel [44] and surface treatment for changing the surface potential [12, 45]. Those platforms employed passive methods which were unable to change the behavior of charged species on-demand, once the devices were fabricated. In contrast, ionic field effect transistor (IFET) can provide an active method which enables to enhance, diminish or even reverse the behavior of charged species in situ by introducing gate potential.

Figure 6.5a shows a schematic of metal-oxide-electrolyte system which is a key-building unit of ionic field effect transistor. When voltage is applied to the metal part called as gate electrode, surface potential on oxide-electrolyte interface is modulated by field effect so that the electric double layer (EDL) inside the electrolyte solution is changed depending on the applied voltage on the gate electrode. This process is how the IFET control the ionic transport in the nanoscale system. The metal-oxide-electrolyte system is integrated on nanochannel wall, then the nanochannel is connected with microchannel. Figure 6.5b shows the schematic of typical ionic field effect transistor.

The first experimental results of an ionic field effect transistor were demonstrated by Gajar and Geis [46] in 1992. The devices had nanochannels about 88 nm in height and 300–900 μm in length. They investigated both the steady-state and transient responses of their ionic FET devices. However, it was found that the response of the IFET to a step voltage in the gate terminal is quite slow. This is mainly due to two reasons. The first is the large geometries of the nanochannel and the second is the existence of processes other than ambipolar diffusion.

In 2004, theoretical modeling of the ionic transport in silica nanotubes revived research interest in IFET devices [45]. Fan et al. [47] experimentally demonstrated field effect modulation of ion transport using gated silica nanotubes shown. Using KCl as the testing solution, they found that as the gate voltage varies from −20 to +20 V, the ionic conductance decreases monotonically from 105 pS down to 45 pS, due to depletion of cations under the applied electric field, which shows a typical p-type transistor behavior. Since then, various IFET have been reported. Using multiple nanopore structures with sub-10 nm diameters and TiO_2 as dielectric material, Nam et al. [18] also showed that the ionic conductivity of the nanopore can be modulated by a gating voltage. A p-type I–V characteristic was also observed, which suggested the major carriers are cations.

Another application is the control of single molecule translocating through the nanoscale system. Since biomolecules usually have multivalent charges, field effect control over the molecule transport in nanochannels can be more effective than monovalent ions. It has been experimentally observed that the fluorescence intensity of 30-base fluorescently labeled single-stranded DNA (ssDNA) in a 1 mM KCl solution could be increased by six times in 2D nanoslits, which is ascribed by the electrostatic interactions between the gating voltage and DNA molecules [48]. It is found that both electroosmotic flow (EOF) and electrostatic interactions arising from the field effect control can effectively regulate the DNA translocation through a nanopore. Recently, Paik et al. [49] demonstrated a gated nanopore structure that is capable of reversibly altering the rate of DNA capture by over three orders of magnitude. They ascribed this extremely large modulation ratio to the counterbalance between the electrophoresis (EP) and the EOF, rather than pure electrostatic interactions between the gating voltage and the DNA molecules. When the gating voltage was low, EOF overwhelmed the EP. Thus DNAs were rejected from entering the pore. In contrast, a high gating voltage reduced the EOF due to the reduced Na+ in the diffuse layer of the nanopore wall so that DNAs were accepted to pass through the nanopore.

To reflect these complex features, widely-used models to describe the electrokinetic phenomena inside the IFET will be discussed in the next section.

Conventional Models for Ionic Field Effect Transistor

In this section, descriptions of the complete set of equations that were used for simulation of electrokinetic transport in IFET are presented. Most models were formulated under Poisson-Nernst-Planck-Stokes coupled governing equations with appropriate boundary conditions shown in Sect. 6.2.2. Gate effects on ionic transport were interpreted as boundary conditions so that the independent differential equations should be needed. Firstly, general governing equations and boundary conditions are rewritten for the metal-oxide-electrolyte system. Secondly, the gate modulated is described by the Poisson-Boltzmann equation and the Laplace equation.

As discussed in Sect. 6.2.4, the Poisson-Nernst-Planck-Stokes (PNPS) equations are used to describe the nanoscale electrokinetic system. The electrostatic potential ψ is governed by the Poisson equation

$$\nabla \cdot (\varepsilon \nabla \psi) = -\rho_{\mathrm{f}} \tag{6.78}$$

where meaning of parameter was discussed in Sect. 6.2.1. In Eq. (6.49), the volume charge density is represented by

$$\rho_{\mathrm{f}} = F \sum_{i} z_i c_i \tag{6.79}$$

c_i in the above equation is provided by the Nernst-Planck equations, which combines the diffusion due to a concentration gradient, the migration due to an electric field, and convection due to a ambient flow field.

$$\nabla \cdot \mathbf{J}_i = \nabla \cdot \left(-D_i \nabla c_i - \frac{z_i F D_i}{RT} c_i \nabla \psi + c_i \mathbf{u} \right) = 0 \tag{6.80}$$

without any chemical reactions. To obtain solutions for flow field, the Stokes equations and the continuity equation should be solved by coupling manner.

$$\mathbf{0} = -\nabla p + \eta \nabla^2 \mathbf{u} - \rho_{\mathrm{f}} \nabla \psi \tag{6.81}$$

$$\nabla \cdot \mathbf{u} = 0 \tag{6.82}$$

With above governing equations from (6.78) to (6.82), appropriate boundary conditions should be determined. The numerical domain for conventional IFET is shown in Fig. 6.5b. Numerical boundaries were divided into four sections: (1) axis of symmetry, (2) inlet/outlet, (3) reservoirs, and (4) nanochannel wall, as denoted in Fig. 6.5b. Dash-dot line represents axis of symmetry in the cylindrical coordinate system, thus this boundary (denoted as '(1)') possessed symmetry conditions. Corresponding mathematical forms about the axis of symmetry are

$$\mathbf{n} \cdot \nabla \psi = 0 \tag{6.83}$$

$$\mathbf{n} \cdot \mathbf{j}_i = 0 \tag{6.84}$$

$$\mathbf{n} \cdot \mathbf{u} = 0 \tag{6.85}$$

In the above conditions, n is the outward normal vector. On the inlet/outlet boundaries of nanoscale system (denoted as '(2)'), electric potential, concentration and pressure were fixed at specific values.

$$\psi = 0(\text{source}) \tag{6.86}$$

$$\psi = V_{\mathrm{D}}(\text{drain}) \tag{6.87}$$

$$c_i = c_0 \tag{6.88}$$

$$p = 0 \tag{6.89}$$

where V_D is the applied voltage on drain electrode and c_0 is the bulk electrolyte concentration. The third boundary type is reservoirs (denoted as '(3)') on which insulating conditions were satisfied.

$$\mathbf{n} \cdot \nabla \psi = 0 \tag{6.90}$$

$$\mathbf{n} \cdot \mathbf{j}_i = 0 \tag{6.91}$$

$$\mathbf{n} \cdot \mathbf{u} = 0 \tag{6.92}$$

Note that these mathematical forms are equal to conditions of the axis of symmetry, but their physical meanings are different. Most importantly, last boundary type is nanochannel wall (denoted as '(4)') on which surface charge density was set up by normal derivative of electric potential and the walls had no-penetration condition and no-slip condition for the Nernst-Planck equations and the Stokes equations, respectively.

$$-\varepsilon \mathbf{n} \cdot \nabla \psi = \sigma_s \tag{6.93}$$

$$\mathbf{n} \cdot \mathbf{j}_i = 0 \tag{6.94}$$

$$\mathbf{u} = \mathbf{0} \tag{6.95}$$

where σ_s is the surface charge density on the nanochannel wall. The detailed expressions of boundary conditions were summarized in Table 6.2 once more.

As earlier mentioned, IFET could be modeled as metal-oxide-electrolyte (MOE) system. In the usual MOE system, the surface charge density modulated by gate voltage is obtained from simple algebraic equations independent of governing equations. The MOE structure depicted in Fig. 6.7a can be naturally incorporated into chip-based fluidic devices to perform electro-fluidic gating. Its geometry is that of a parallel-plate capacitor as denoted in Fig. 6.8. whose metallic gate electrode is separated from the conductive electrolyte by a thin insulating oxide. The electric double layer is modulated by applying a voltage across the capacitor, which generates an electric field normal to the solid-liquid interface. Here we introduce models for the charging behavior of the MOE capacitor. The electric double layer screens electric fields, whether they originate from the chemical charge at the oxide

Table 6.2 Boundary conditions for each equation of the model

Boundary	Poisson Equation	Nernst-Planck Equations	Stokes Equations
Axis of symmetry	[a] $\mathbf{n} \cdot \nabla \psi = 0$	$\mathbf{n} \cdot \mathbf{J}_i = 0$	$\mathbf{n} \cdot \mathbf{u} = 0$
Inlet/outlet	$\psi = 0$ (source) $\psi =$ [b] V_D (drain)	$c_i =$ [d] c_0	$p = 0$
Reservoirs	$\mathbf{n} \cdot \nabla \psi = 0$	$\mathbf{n} \cdot \mathbf{J}_i = 0$	$\mathbf{n} \cdot \mathbf{u} = 0$
Nanochannel wall	$-\varepsilon_f \mathbf{n} \cdot \nabla \psi =$ [c] σ_s	$\mathbf{n} \cdot \mathbf{J}_i = 0$	$\mathbf{u} = \mathbf{0}$

[a] $\mathbf{n}$ is the outward normal vector
[b] V_D is the applied voltage on drain electrode
[c] σ_S is the suface charge density
[d] c_0 is the bulk electrolyte concentration

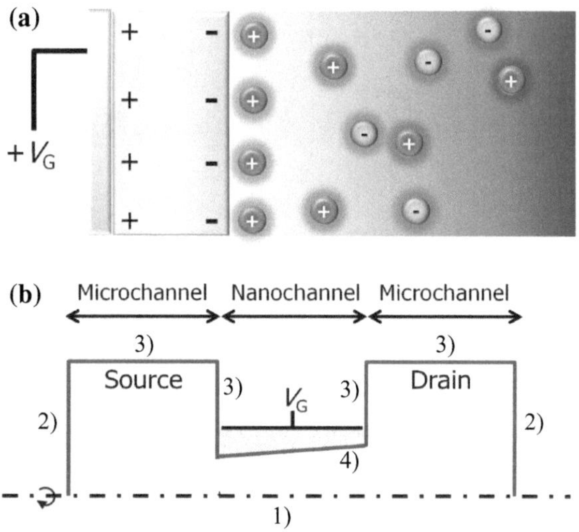

Fig. 6.7 **a** Schematic of metal-oxide-electrolyte system which is a key-building structure of ionic field effect transistor. **b** Schematic of gated nanochannel connected with microchannel reservoir. This structure is called as ionic field effect transistor

surface or from the applied voltage across the capacitor. To treat these two contributions separately, we consider an equivalent circuit model of the MOE capacitor, shown in Fig. 6.8. This simple model allows us to determine the potential and the charge density at every location. It consists of three elements arranged in series: two linear capacitors representing the oxide and the Stern layer and a nonlinear element representing the electric double layer. The potential difference across the oxide layer is $V_G - \psi_s$. The oxide is assumed to have a constant capacitance per unit area, C_{ox}, that accurately describes the dielectric properties of common materials used in micro- and nanofluidic devices such as silicon dioxide (SiO_2), aluminum dioxide (Al_2O_3), and poly-dimethylsiloxane (PDMS) [50–52]. The capacitive charge density induced at the surface of the insulator, σ_{ox}, is given by

$$\sigma_{ox} = C_{ox}(V_G - \psi_s) \tag{6.96}$$

Within the basic Stern model, the potential drops linearly across the solid-liquid interface by an amount $\psi_s - \zeta$. The relationship between this potential drop and the charge density screened by the double layer, σ_{DL}, is given by

$$\sigma_{DL} = C_{Stern}(\psi_s - \zeta) \tag{6.97}$$

where C_{Stern} is the phenomenological Stern capacitance per unit area. C_{Stern} reflects the structure and dielectric properties of the solid-liquid interface. The basic Stern model has been widely applied to model the charging of the double layer and is well supported by experimental evidence [53–55]. When the surface reactions between functional groups on the oxide layer and ionic species in the electrolyte solution is considered, the net charge density on the surface of the oxide layer must equal the

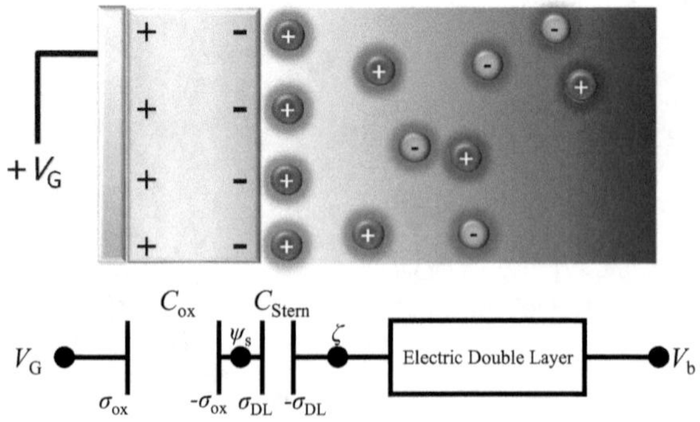

Fig. 6.8 Equivalent circuit model of the MOE capacitor. The oxide layer are modeled as capacitors of specific capacitance C_{ox} and C_{Stern}, respectively

chemical charge density from the ionized surface groups charge density from the ionized surface groups, σ_{chem}; therefore,

$$\sigma_{chem} = \sigma_{DL} - \sigma_{ox} \tag{6.98}$$

However, in most cases, the Stern layer and chemistry of the oxide/electrolyte interface are neglected [56, 57] so that σ_{chem} is zero, and then remaining surface charge densities become

$$\sigma_{DL} = \frac{2\varepsilon\kappa RT}{F}\sinh\left(\frac{F}{2RT}(\zeta - V_b)\right) \tag{6.99}$$

and

$$\sigma_{ox} = C_{ox}(V_G - \zeta) \tag{6.100}$$

Therefore, gate modulated zeta potential can be determined by Eqs. (6.21)–(6.23) when the Stern layer and the surface reaction are neglected.

6.2.5 Concluding Remarks

Over the past decade, microfluidic and nanofluidic applications have drawn to significant attentions in science and engineering fields. Although the conventional theory well-described the microscale system, additional constraints to describe the nanoscale system should be necessary. However, these additional constraints could not be found by only experimental manner because observation inside the nanoscale system was impossible. In this chapter, we introduced microscale electrokinetics which started from the basic conservation laws and then, focused on two nanoscale electrokinetic phenomena occurs near (or through) a permselective nanoporous system; (1) ion concentration polarization and (2) ionic field effect transistor. They are closely related to the unique ion transportation for direct ion separation or selective ion transportation inside human body or cell-cell environments. Conclusively, due to the difficulties of direct observation inside these nanoscale system, the theoretical formulation with nonlinear constraints would be one of the possible manners to investigate the nanoscale system.

6.3 Electrostatics for Biodevices: Dielectrophoresis (Marie Frenea-Robin, Bruno Le Pioufle)

In this section, the effect of dielectrophoretic forces on cells is examined. Dielectrophoresis arises from the interaction between a polarized particle and a non-uniform electric field in which the particle is immersed. In the case where a stationary electric field is applied, living cells polarize and move towards field minima or maxima, depending on their polarization contrast with the medium. This contrast is characterized by the Clausius-Mossotti factor f_{CM}, which is a function of the complex permittivities of both cell and medium. This principle, referred to as conventional dielectrophoresis (cDEP) is commonly used for cell sorting or cell trapping, as will be illustrated with some examples found in the literature. Propagative electric fields are also often used to apply dielectrophoretic forces, which corresponds to travelling wave dielectrophoresis (tw-DEP) in the case of linear propagation, or electrorotation (E ROT), when a rotational electric field is considered. Applications where these two other forms of dielectrophoresis are implemented will also be reported in this chapter. Lastly, examples regarding DEP of biomolecules will be briefly reviewed.

6.3.1 Introduction, Basics of the Dielectrophoresis Phenomenon

This section introduces the basics of dielectrophoresis, from the interaction between an electric field and a polarized particle, up to the generalized formula of the DEP force.

6.3.1.1 Dielectrophoretic Force Induced on a Spherical Particle

A spherical particle (radius R, complex permittivity ε_p^*) get polarized once exposed to an external electric field $\vec{E}$. The particle behaves as an electrostatic dipole $\vec{m}$ which is a function of the sphere radius R and of the complex permittivities of both particle and medium (respectively ε_p^* and ε_m^*, that appear in the proportional dependency to the Clausius Mossotti factor $f_{CM} = \frac{\varepsilon_p^* - \varepsilon_m^*}{\varepsilon_p^* + 2\varepsilon_m^*}$):

$$\vec{m} = 4\pi\varepsilon_m f_{CM}\vec{E} \tag{6.101}$$

where the complex permittivity ε^* defined in Eq. 6.102 depends on the permittivity ε and conductivity of the particle and medium:

$$\varepsilon^* = \varepsilon - i\frac{\sigma}{\omega} \tag{6.102}$$

and i is the complex number ($i^2 = -1$).

This electrostatic dipole interacts with the external electric field, and experiences the dielectrophoretic force:

$$\overrightarrow{F_{DEP}} = \left(\vec{m}.\vec{\nabla}\right)\vec{E} \tag{6.103}$$

where $\vec{\nabla}$ is the space derivative vector $\vec{\nabla} = \left(\frac{\partial}{\partial x}\frac{\partial}{\partial y}\frac{\partial}{\partial z}\right)^t$

In the case where the electric field varies sinusoidally with the time:

$$\vec{E} = E_{xo}\cos(\omega t + \varphi_x)\vec{x} + E_{yo}\cos(\omega t + \varphi_y)\vec{y} + E_{zo}\cos(\omega t + \varphi_z)\vec{z} \tag{6.104}$$

ω being the angular frequency of the applied field, and $\vec{x}$, $\vec{y}$, $\vec{z}$ are the unitary vectors of the Euclidean space. $\varphi_x, \varphi_y, \varphi_z, E_{xo}, E_{yo}, E_{zo}$ are respectively the phase shifts and field modulus on the $\vec{x}$, $\vec{y}$, $\vec{z}$ directions.

The dielectrophoresis force then expresses as follows:

$$\overrightarrow{F_{dep}} = 2\pi\varepsilon_m R_{cell}^3 \Re(f_{CM})\overrightarrow{\nabla E^2} + 2\pi\varepsilon_m R_{cell}^3 \Im(f_{CM})\left(E_{xo}^2\overrightarrow{\nabla\varphi_x} + E_{yo}^2\overrightarrow{\nabla\varphi_y} + E_{zo}^2\overrightarrow{\nabla\varphi_z}\right) \tag{6.105}$$

6.3.1.2 Case of Cells

In the case of cells, the dielectrophoretic force direction and amplitude thus depend on the Clausius-Mossotti factor f_{CM} which is a function of the difference of complex permittivities between the cell and the medium [58]:

$$f_{CM} = \frac{\varepsilon_{cell}^* - \varepsilon_m^*}{\varepsilon_{cell}^* + 2\varepsilon_m^*} \tag{6.106}$$

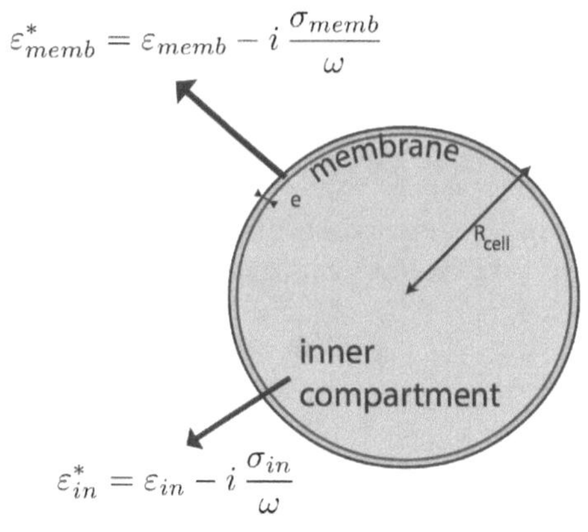

Fig. 6.9 Single shell model of the cell. The cell is composed of an averaged intra-cellular compartment (defined by its permittivity ε_{in}, its conductivity σ_{in} and radius R_{cell}), surrounded by the cell membrane (permittivity ε_{memb}, conductivity σ_{memb} and thickness *e*)

The complex permittivity of a spherical shaped cell is commonly simplified to an approximated value ε^*_{cell} obtained thanks to the Single shell model [59]. This model describes the cell as an averaged inner compartment (which dielectric properties are σ_{in} and ε_{in}), surrounded by an outer cytoplasmic membrane (dielectric properties σ_{memb} and ε_{memb}, radius R_{cell}, thickness e, see Fig. 6.9):

$$\varepsilon^*_{cell} = \varepsilon^*_{memb}\left[\frac{\left(\frac{R_{cell}}{R_{cell}-e}\right)^3 + 2\left(\frac{\varepsilon^*_{in}-\varepsilon^*_{memb}}{\varepsilon^*_{in}+2\varepsilon^*_{memb}}\right)}{\left(\frac{R_{cell}}{R_{cell}-e}\right)^3 - \left(\frac{\varepsilon^*_{in}-\varepsilon^*_{memb}}{\varepsilon^*_{in}+2\varepsilon^*_{memb}}\right)}\right] \tag{6.107}$$

6.3.2 *Using Dielectrophoresis to Induce Translational Motion*

6.3.2.1 Handling Cells Using a Stationary Electric Field

The phenomenon described in this section, arising in the presence of a stationary field, is sometimes referred to as 'conventional dielectrophoresis' (cDEP). In electric fields with constant phase, Eq. 6.105 giving the time-averaged force acting on a cell simplifies to:

$$\overrightarrow{F_{dep}} = 2\pi\varepsilon_m R^3_{cell}\Re(f_{CM})\overrightarrow{\nabla E^2_{RMS}} \tag{6.108}$$

This equation indicates that the DEP force intensity is cell-volume dependent and that the force cancels if the field is uniform ($\overrightarrow{\nabla E^2_{RMS}} = 0$, where E_{RMS} denotes the root-mean-square electric field). The DEP force varies with the square of the applied field magnitude, and its direction is therefore not affected by the field polarity. It is due to the fact that a field reversal results in a reversed induced dipole moment. It also implies that both direct current (DC) and alternating current (AC) can be used. Dielectrophoresis should therefore not be confused with *electrophoresis*, which refers to the motion of charged particles in DC electric fields.

It can be seen from Eq. 6.108 that the direction of the force depends on the sign of the real part of the f_{CM} factor. The case where $\Re(f_{CM}) > 0$ corresponds to the cell being more polarisable than its surrounding medium, resulting in an induced dipole moment aligned with the applied field. As a consequence, the cell is translated towards regions of highest electric field magnitude under the effect of *positive* DEP (pDEP, see Fig. 6.10a). Conversely, the case where $\Re(f_{CM}) < 0$ corresponds to the cell being less polarisable than its immersion medium, resulting in an induced dipole moment directed against the applied field. The cell is then directed towards areas of low electric field strength, which is referred to as *negative* DEP (nDEP, see Fig. 6.10b).

For a spherical particle, the real part of the f_{CM} factor is comprised between −0.5 and 1. This term therefore not only determines the direction of the force, but it also affects its intensity. As can be seen from Fig. 6.11, the dielectrophoretic behaviour of a cell can be adjusted by tuning the field frequency. It should be noted that the conductivity and permittivity of the cell themselves vary with the frequency, due to the heterogeneous structure of biological particles (Fig. 6.12). In the low frequency range (below a few tens of kHz), the DEP force depends on the respective conductivities of the cell and its immersion medium. The cell membrane blocks current

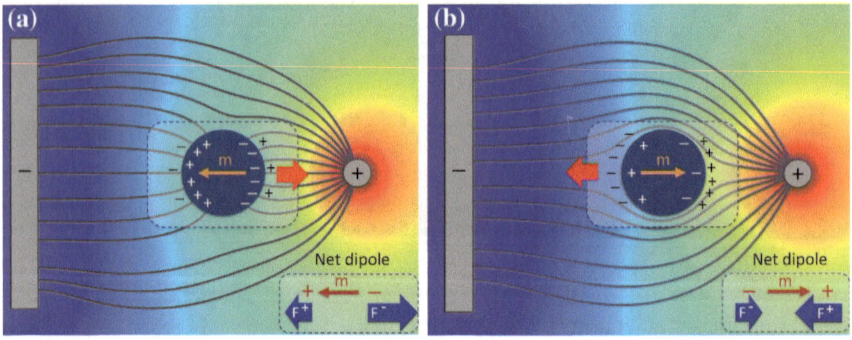

Fig. 6.10 Illustration of conventional DEP principle. **a**. pDEP arises when the cell is more polarisable than its immersion medium. Due to the non-uniform character of the field, a different electric force is induced on the two poles of the net effective dipole. This results in a translation of the cell towards the region of high electric field strength. **b**. nDEP corresponds to the case where the induced dipole opposes the applied field. Due to the force imbalance, the cell is now pushed down the gradient, from higher to lower field intensities. It should be noted that the direction of motion remains unchanged when the electrode polarity is reversed

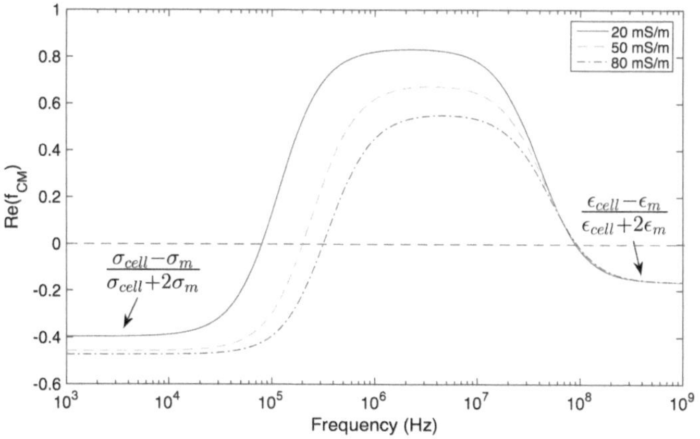

Fig. 6.11 Plot of the DEP response of a Jurkat cell (human T lymphocyte) for three different values of the medium conductivity. The shape of the curves is typical for a living cell with intact membrane. The following values were introduced in Eq. 6.107 [60]: $\sigma_{memb} = 3 \times 10^{-6}\,\mathrm{S \cdot m^{-1}}$; $\varepsilon_{memb} = 6 * \varepsilon_0\,\mathrm{F \cdot m^{-1}}$; $\sigma_{in} = 0.4\,\mathrm{S \cdot m^{-1}}$; $\varepsilon_{in} = 45 * \varepsilon_0\,\mathrm{F \cdot m^{-1}}$; $R_{cell} = 5\,\mu\mathrm{m}$; $e = 5\,\mathrm{nm}$

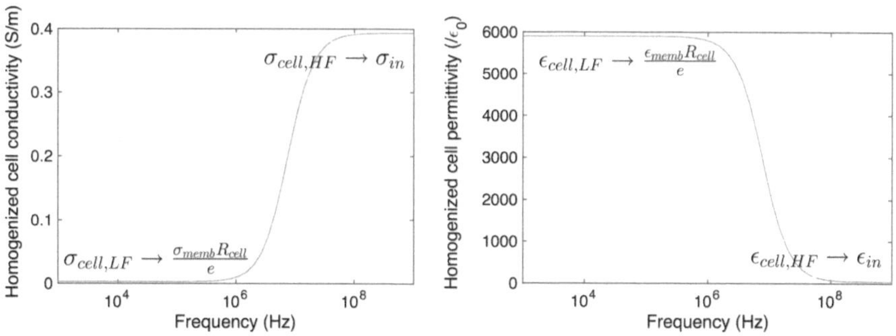

Fig. 6.12 Evolution of cell effective conductivity and permittivity with respect to frequency (from Eq. 6.107)

flow, which leads to the cell being less polarizable than the medium (i.e. nDEP force). With increasing frequency, the high membrane resistance is progressively bypassed by the membrane capacitance, and the magnitude and direction of the force depend on the interplay between the dielectric properties of the membrane and cytoplasm. At frequencies higher than a few tens of MHz, the DEP response is dominated by the respective permittivities of cell interior and surrounding medium.

As can be seen on Fig. 6.11. There are two frequencies at which the direction of the DEP force exerted on the cell reverses, commonly referred to as "crossover frequencies".

The lower crossover frequency value, corresponding to a transition from nDEP to pDEP, can be shifted by modifying the surrounding medium conductivity. Hence this parameter plays an important role on the cell dielectrophoretic behaviour. Low conductive media (σ_m being a few tens of mS $\cdot$ m^{-1}) are currently used in DEP experiments because in more conductive physiological media ($\sigma_m \approx 1$ S $\cdot$ m^{-1}) observation of pDEP is only possible at very high frequencies and Joule heating effects can become a concern. The low conductivity buffer used must be formulated to maintain pH and osmolarity within the physiological range.

As previously mentioned, the force intensity is directly proportional to the gradient of the squared electric field strength. If the electrode gap is scaled down by a factor K, the DEP force gets scaled up by a factor of K^3. Hence significant forces can be generated by applying a few volts to miniaturized electrodes. The electric field distribution is determined by the choice of the electrode geometry. The spatial non-uniformities required for the DEP effect is commonly generated by 2D co-planar electrodes of interdigitated, castellated or polynomial design [61]. Yet, a variety of alternative electrode structures have also been developed to allow exploitation of interesting 3D effects, such as "top-bottom" electrodes [62] or sidewall conducting PDMS electrodes in microfluidic devices [63, 64]. Insulating structures in a microchannel with remote electrodes can also be exploited to shape the electric field [65], following the principle of IDEP (insulator-based dielectrophoresis). Demierre et al. proposed the concept of "liquid electrodes" combining planar electrodes with IDEP technique [66]. This configuration exploits metal electrodes fabricated on the bottom of dead-end chambers placed on the side of a microfluidic channel and acting similarly to thick electrodes embedded in the channel walls. Detailed reviews of microfabrication technologies and microelectrode geometries employed in DEP applications can be found in the literature [67, 68].

Dielectrophoresis is a technique currently employed for cell handling in microfluidic systems. As a contactless and label-free manipulation method, it provides a non invasive tool to perform various tasks such as sample preparation (concentration, filtering, enrichment), cell sorting or cell isolation prior to further detection or measure of response to various stimuli. DEP has been used for instance in immunosensors to enhance the capture and detection of pathogenic bacteria by bringing the cells to the sensor site while increasing the contact between the target and the immobilized antibodies, prior to sandwich ELISA detection of captured cells [69]. Hamada and coworkers proposed similarly to detect bacteria by first capturing cells onto microelectrodes using pDEP and then measuring the variation of electrode impedance due to the presence of trapped bacteria [70].

DEP traps were also designed to perform parallel single-cell assays. Monitoring of many individual cells at the same time involves their patterning in array format, which can be achieved using quadrupole electrode units [71–73]. This electrode arrangement offers the advantage of providing strong confinement under nDEP, as cells less polarizable than the surrounding suspending medium will be directed away from electrode edges towards the center of the interelectrode gap [61]. Silicon Biosystems has developed and brought to market a chip called DEPArray^{TM} for massively parallel handling of cells with single cell resolution. Programmable electrodes are used to generate DEP "cages" within which single cells may be trapped, levitated and manipulated individually. A DEP cage corresponds to a field minimum created above a microelectrode by applying a sinusoidal voltage to the associated electrode and to the conductive Indium Tin Oxide (ITO) lid, and a counterphase sinusoidal voltage to the surrounding electrodes. The dynamic traps can be reconfigured dynamically under software control, so as to move the cells above the device plane towards the desired target location.

DEP-driven cell assembly also constitutes a promising route to engineer biomaterials that mimic the structure of native tissues and develop cell-culture platforms reproducing in vivo cellular microenvironment. Ho et al. have demonstrated rapid heterogeneous patterning of liver cells using concentric radiating tip electrode array. The obtained alternate radial pattern of hepatocytes and endothelial cells mimicked the lobular morphology of real liver tissue [74]. Such bottom-up approach may provide better control over the relative spatial organization of cells compared to more traditional top-down approaches based on scaffolds. It also offers the advantage to promote cell-cell interactions. Menad et al. have recently shown that cell aggregates of regular size and shape could be fabricated using nDEP. Those cell assemblies could be maintained by cell-cell adhesion after the electric field removal and indications of cell activity into aggregates were observed after overnight incubation [75]. This may offer a new approach to produce 3D tumor spheroid models for in vitro therapeutic screening.

One of the main applications of dielectrophoresis is the selective capture of target cells from a mixed population, which can be achieved without the need for labeling. Since dielectric properties of cells are dependent upon various parameters including morphological factors such as cell volume and membrane area, different cell types will exhibit different crossover frequencies in a given medium.

The simplest separation approach consists in dividing a cell mixture into two distinct sub-populations, by choosing an operating frequency at which one fraction of cells is submitted to nDEP and the other one to pDEP. Physical separation of the two sub-populations is then achieved by superimposing a liquid flow above the electrode array. Based on this principle, Becker et al. have shown that human breast cancer cells could be separated from blood using the castellated electrode design described on Fig. 6.13a [76].

Doh and Cho presented a continuous cell separation chip using three planar electrodes placed at the bottom of a microfluidic channel. Cells with pDEP affinity moved towards the chamber walls while cells with nDEP affinity remained in the central streamline. The laminar nature of the flow enabled the collection of cell fractions through different outlets [77]. Bisceglia et al. developed a method and a microfluidic device for the extraction and concentration of pathogens from a suspension of blood cells, based on the large tolerance of micro-organisms towards osmotic shocks. A hypotonic medium selectively alters blood cells and changes their dielectric properties while keeping most micro-organisms functional. Following this approach, micro-organisms were trapped using pDEP, while altered

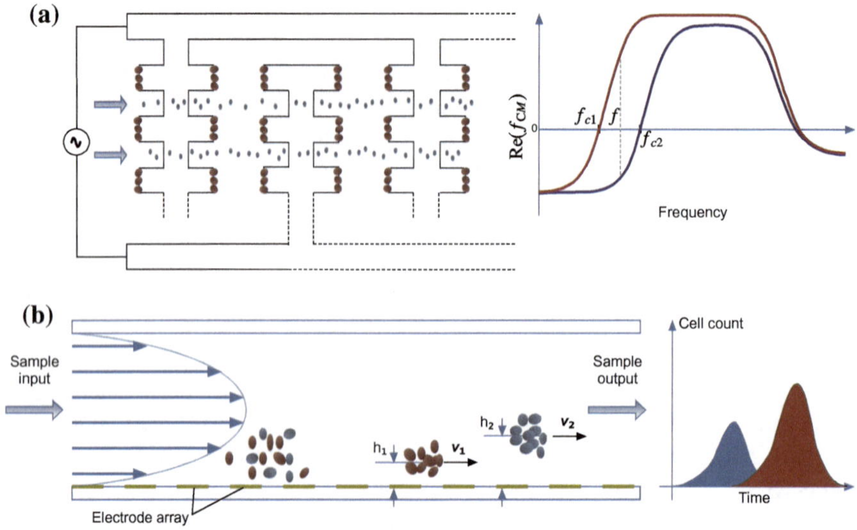

Fig. 6.13 **a**. Illustration of DEP cell separation for a mixture containing two cell subpopulations. The field frequency f is chosen so as to fall between the DEP crossover frequencies of the two cell types f_{c1} and f_{c2}. Red cells exposed to positive DEP are trapped at the electrode tips, while blue cells experiencing nDEP are focused into bands and carried away by the fluid stream. **b**. Side view of a DEP-FFF separation chamber. Cells of type 1 and 2 submitted to nDEP are levitated at two different heights h_1 and h_2 above the electrode array placed at the chamber floor. Since they are closest to the middle of the channel height, blue cells move at a faster velocity $(v_1 > v_2)$ and reach the device output first

blood cells submitted to negative dielectrophoresis were driven by the flow towards the device outlet [78].

Cell separation can also be achieved using dielectrophoresis field flow fractionation (DEP-FFF), a chromatographic approach in which cells are positioned at different heights within the separation chamber using a combination of dielectrophoretic and sedimentation forces. Due to the parabolic flow profile, different cell subpopulations travel at different velocities according to their distance from the chamber floor and therefore emerge from the outlet at different times (Fig. 6.13b). Provided the chamber length is long enough, this method allows to fractionate cells with subtle differences. It also enables to discriminate more than two cell classes. A continuous flow microfluidic processing chamber was recently proposed by Gascoyne and coworkers to achieve a throughput as high as 10^6 cells • min^{-1} and isolate circulating tumor cells from 10 mL clinical blood specimens in less than 1 h [79]. The ApoStream® technology developed by ApoCell Inc is based on this approach.

6.3.2.2 Handling Cells Using a Propagative Electric Field

The second term of Eq. 6.105 is related to the twDEP force acting on a spherical particle. This term exists only in non-uniform electric fields with spatially dependent phase and if the imaginary component of the Clausius Mossoti factor is not null. A travelling field can be generated using sequentially phase-shifted AC voltages applied to an array of finger electrodes (see Fig. 6.14). The resulting translational force propels the particle along the electrodes, with or against the field direction, depending on whether the imaginary component of the Clausius Mossoti factor is negative or positive, respectively. In practice the operating conditions (field frequency, medium conductivity) are chosen so as to meet the following requirements: (i) cells must be repelled from the electrode array by nDEP ($\Re(f_{CM})<0$), rather than trapped on the electrodes by pDEP and (ii) $\Im(f_{CM}) \neq 0$.

twDEP is also employed in spiral cell concentrators: using parallel spiral electrodes excited with phase-shifted signals enables to produce a force acting along the radial direction and to collect cells of interest at the centre of the device. This principle was exploited by Gascoyne and coworkers for detecting malaria in blood samples by DEP. Application of four phase signals to the spiral electrode array caused normal erythrocytes to be trapped at the electrode edges by pDEP while parasitised cells were levitated and carried towards the centre of the spiral by the travelling field [80]. Van Den Driesche et al. developed a continuous cell separation device comprising a finger electrode array placed at the bottom of the microfluidic channel so as to produce a travelling electric field perpendicular to the pressure driven flow. Jurkat cells and Lactobacillus casei bacteria present in the input sample responded differently to the transverse twDEP force. At a given frequency, only Jurkat cells were deflected from their initial flow path, which enabled to collect them at the opposite side of the separation chamber [81].

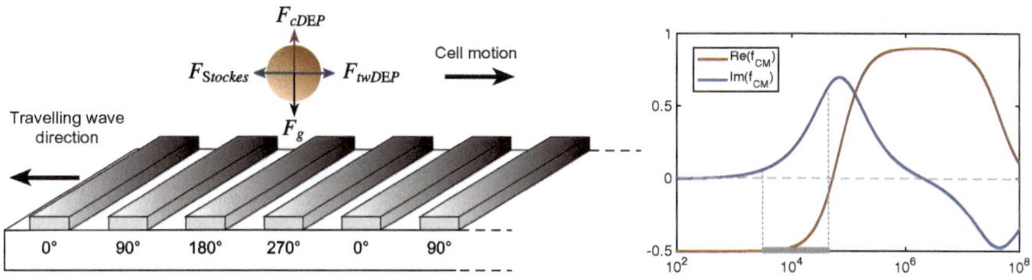

Fig. 6.14 Illustration of twDEP principle. In the frequency range delimited by the grey band (see DEP cell spectra on the right), cells are levitated by nDEP above the electrode array ($\Re(f_{CM})<0$) to a height at which F_{cDEP} and the sedimentation force F_g balance. Due to the F_{twDEP} force component, cells are also translated along the electrodes against the travelling field direction ($\Im(f_{CM}) > 0$), at a velocity limited by Stokes drag

6.3.3 Using Rotating Fields to Induce Electrorotation

This section describes how the cell dielectric parameters can be estimated through its electromechanical response when immersed in a rotary propagative electric field. In an electrorotation experiment, the angular frequency of the electric field is varied in order to produce a cell mechanical velocity spectrum. The resulting spectrum is highly informative on the cell components properties.

6.3.3.1 Extraction of the Cell Dielectric Parameters from an Electrorotation Experiment—Principle

Once immersed within a rotational electric field, the cell experiences a torque, which direction $\vec{z}$ is normal to the electrode plan. This phenomenon, named electrorotation, is due to a phase lag between the field and induced dipole moment.

$$\overrightarrow{\Gamma_{dep}} = \vec{m} \otimes \vec{E} = -4\pi R^3 \varepsilon_m \Im(f_{CM}) E^2 \vec{z} \tag{6.109}$$

The cell consequently rotates, with a rotational velocity Ω_{cell} that depends on the angular frequency of the electric field. The steady-state electrorotation rate corresponds to the speed at which the torque and the hydrodynamic friction balance. A spectrum of the rotational velocity can thus be measured:

$$\Omega_{cell}(\omega) = \frac{\varepsilon_0 \varepsilon_m E^2}{2\eta} \Im(f_{CM}) \tag{6.110}$$

where η represents the dynamic viscosity of the medium.

The mechanical velocity is dependent on the cell dielectric parameters (complex permittivities of the membrane and of the cell content) that can thus be estimated. To do so, the distance between the measured spectrum of the rotational velocity of the cell $\Omega_{cell}(\omega)$ and the theoretical one, which is a function of the complex permittivities of its different domains (Eqs. 6.110, 6.106 and 6.107 in the case of a cell modeled by the single shell), is minimized thanks to least square algorithm [82].

6.3.3.2 Using Octupole Electrodes to Levitate the Cell

When conducting an electrorotation experiment, a common strategy consists in first trapping the cell using a four-electrode set, by applying a stationary electric field that induces cDEP as described previously, and secondly superposing the rotating electric field [59].

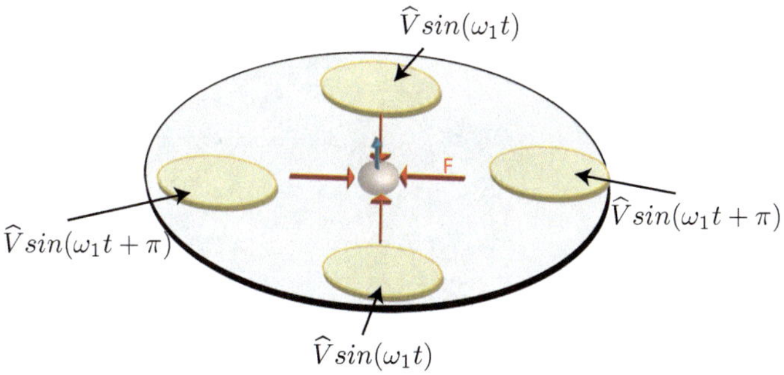

Fig. 6.15 Single cell trapped by negative dielectrophoresis (red arrows) at the center of a four electrode set, being powered by alternate polarities of sinusoidal voltages. The cell experiences a levitation force (blue arrow) at the center of the electrode set, due to higher order of the force

For the trapping step, the four electrodes are powered with opposite phase sine waves (angular frequency ω_1, amplitude $\hat{V}$, see Fig. 6.15). The resulting force field pushes the cell towards the center of the electrode set due to nDEP.

The cell also experiences a levitation force, that cannot be described by the Eq. 6.105 as the electric field is null for any altitude along the line at the center of the electrode set. Indeed Eq. 6.105 remains true as long as the electric field is homogenous all over the volume of the handled dielectric particle, that can be considered in that case as a dipole. In the case where the cell is centered by the four-electrode set, it experiences higher moments of the force [83, 84] and it is finally pushed not only where the electric field is minimal but also where its gradient is minimal (if the second order is considered). The second order approximation of the dielectrophoretic force along the z-axis can be expressed as a function of the first order and second order potential energies U_1 and U_2:

$$F_{depz} = -\frac{\partial}{\partial z}(U_1 + U_2) \tag{6.111}$$

where

$$\begin{aligned} U_1 &= 2\pi\varepsilon_m R_{cell}^3 f_{CM} E^2 \text{ and} \\ U_2 &= \frac{2}{3}\pi\varepsilon_m R_{cell}^5 f_{CM2}\left(\left(\frac{\partial^2\Phi_e}{\partial x^2}\right)^2 + \left(\frac{\partial^2\Phi_e}{\partial y^2}\right)^2 + \left(\frac{\partial^2\Phi_e}{\partial z^2}\right)^2\right) \\ &+ \frac{4}{3}\pi\varepsilon_m R_{cell}^5 f_{CM2}\left(\left(\frac{\partial^2\Phi_e}{\partial y\partial z}\right)^2 + \left(\frac{\partial^2\Phi_e}{\partial z\partial x}\right)^2 + \left(\frac{\partial^2\Phi_e}{\partial x\partial y}\right)^2\right) \end{aligned} \tag{6.112}$$

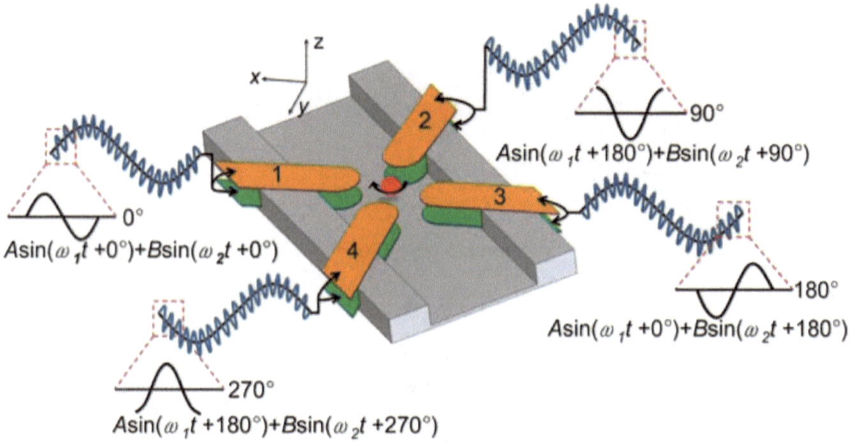

Fig. 6.16 Single cell captured, levitated and submitted to electrorotation experiment with an octupolar electrode set. The cell is centered and levitated midway between the two layers of electrodes, thanks to the application of a stationary field (voltage amplitude A and angular frequency ω_1) to which the rotational field is superposed (voltage amplitude B and angular frequency ω_2, 90° phase shift). From [86], reproduced with the permission of Royal Society of Chemistry

Φ_e being the applied external potential $\vec{E} = -\vec{\nabla}\Phi_e$ and f_{CM2} the second order generalized Clausius-Mossotti factor:

$$f_{CM2} = \frac{\varepsilon_{cell}^* - \varepsilon_m^*}{2\varepsilon_{cell}^* + 3\varepsilon_m^*} \tag{6.113}$$

The cell stabilizes at an altitude where the levitation force is balanced with the gravitational force [85].

The z positioning of the cell above the plan of the device surface is better controlled using octopolar electrodes [86]. In that case, the two superposed quadrupolar set of electrodes repulse the cell along the z axis, the equilibrium position being slightly below midway between the two planes, due to the gravitational force. Such a 3D device structure is more complicated to fabricate than the simple coplanar device, but ensures an easier control of the cell altitude, getting rid of any friction between the cell and the device ceiling or floor (see Fig. 6.16).

6.3.3.3 Some Application Examples of Electrorotation Experiments

The dielectric parameters of various biological components, including either single cells or cell assemblies could be determined using electrorotation experiments. For example, the viability of unicellular protozoan pathogens was assessed using that method, for the disinfection of water [87]. For more fundamental purposes, the dielectric properties (conductivity and permittivity) of the different components (outer wall, cytoplasm) of procaryotic cells were investigated in [88]. The dielectric parameters of eucaryotic cells were monitored using the same method, to determine the evolution of the conductance and permittivity of their cytoplasmic membrane and inner compartment, once submitted to permeabilizing pulsed electric field [82]. The same protocole was used to investigate the structural changes in spheroids in [89].

The biocomponent dielectric parameters were estimated by minimizing the distance between the measured spectrum of the rotational velocity of the cell $\Omega_{cell}(\omega)$ and the theoretical one, using least square criteria. The dielectric properties (σ and ε of the outer membrane, periplasmic space, inner membrane, and cytoplasmic compartment) of bacteria were assessed by this method using a spectrum range for ω up to 1 GHz, on the basis of a three-shell model [88]. In the same way, the dielectric properties of the external wall and inner space of ellipsoid shaped yeasts were determined in [90] as well as the dielectric behaviour of various shape algae [91].

6.3.4 Dielectrophoresis of Biomolecules

Biological applications of DEP are not limited to cells and DEP has also proved to be useful for the concentration, separation and detection of biological entities as small as proteins, DNA and viruses. However, while the DEP behavior of cells is well understood, the polarization mechanism of biomolecules is not fully elucidated. In the case of DNA, the underlying mechanism inducing DEP motion is mainly attributed to counterion response to AC electric fields. Negatively charged phosphate groups along the DNA backbone attract positive counterions from the surrounding ionic solution, which migrate in response to the applied field (Fig. 6.17).

The resulting shift in the counterion cloud results in an induced dipole moment which is assumed to be responsible for DNA DEP behavior [92]. Washizu and coworkers pioneered the concept of DNA DEP more than twenty years ago by demonstrating the stretching of DNA molecules under a field strength of 10^6 V/m and a frequency of 1 MHz. The size (number of base pairs) of the stretched DNA molecule could thereby be determined and target DNA fragments could be cut using a UV beam [93]. This group also demonstrated that synthesized oligonucleotides having 22 bases could be separated from λ-DNA (having 48.5 kbp) using DEP-FFF. The larger molecules showed higher polarisation and were retained onto the electrodes placed at the bottom of the separation chamber, while the smaller oligonucleotides could flow more easily through the channel [94]. Tuukkanen et al. also demonstrated trapping of DNA molecules as small as 27 base pairs, which shows that DNA manipulation can be achieved over a broad range of sizes [95]. It should be noted that real-time observation of DNA motion is typically achieved using fluorescence microscopy and therefore requires DNA staining.

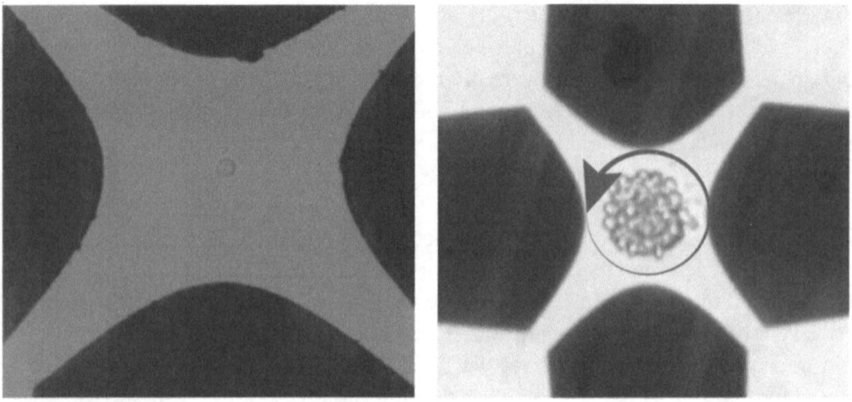

Fig. 6.17 Capture and electrorotation experiment of biocomponents for the estimation of their dielectric parameters. Left figure: B16F10 cell (murine melanoma cell) motioned by a set of four polynomial electrodes (inter-electrode distance 70 μm). Right figure: a spheroid (U87MG cell line) is positioned in a larger set of electrodes (inter-electrode distance 150 μm) for electrorotation experiment. From [89], reproduced with the permission of Springer

DEP is also an interesting tool for protein concentration, which is often a prerequisite in bioanalytical applications, since proteins, unlike DNA, cannot be amplified. Most proteins are around 1–10 nm in size. Given the fast decay of F_{DEP} with the cube of the particle radius and the influence of Brownian motion on small objects, large electric field gradients are therefore required to enable their manipulation (∇E^2 as high as 10^{15}–10^{16} $V^2 \cdot m^{-3}$). The different devices reported for protein DEP were recently reviewed by Nakano and Ros [96]. Clarke et al. increased the concentration of immunoglobulin G proteins by a factor of 300 at the tip of a capillary glass of inner diameter 100–150 nm using electrodeless dielectrophoresis. An electric field of 10^6 $V \cdot m^{-1}$ was generated in the tip of the nanopipette on application of a 1 voltage between the electrodes in the pipette (ground electrode) and bath solution (working electrode). The tip acts as a constriction used to create a high-field gradient with a local maximum, allowing protein trapping by pDEP [97]. While this study was aimed at concentrating proteins in the bulk, it is also possible to immobilize proteins on surface using DEP without the need for any chemical modification. Otto et al. have achieved dielectrophoretic immobilisation of antibodies on a regular array of cylindrical sub-microelectrodes and could demonstrate that the antibodies capability to bind to target proteins was preserved after DEP application [98].

6.3.5 Conclusion

In this section, we have presented the basic principles of three AC electokinetic phenomena, namely dielectrophoresis, twDEP and electrorotation. A summary Fig. 6.18 schematically depicts some device geometries discussed above. While the given examples were mainly focused on cell handling applications, there are many other applications of DEP to nano biosystems, such as controlled assembly of carbon nanotubes for the development of biosensors [100] or specific capture of

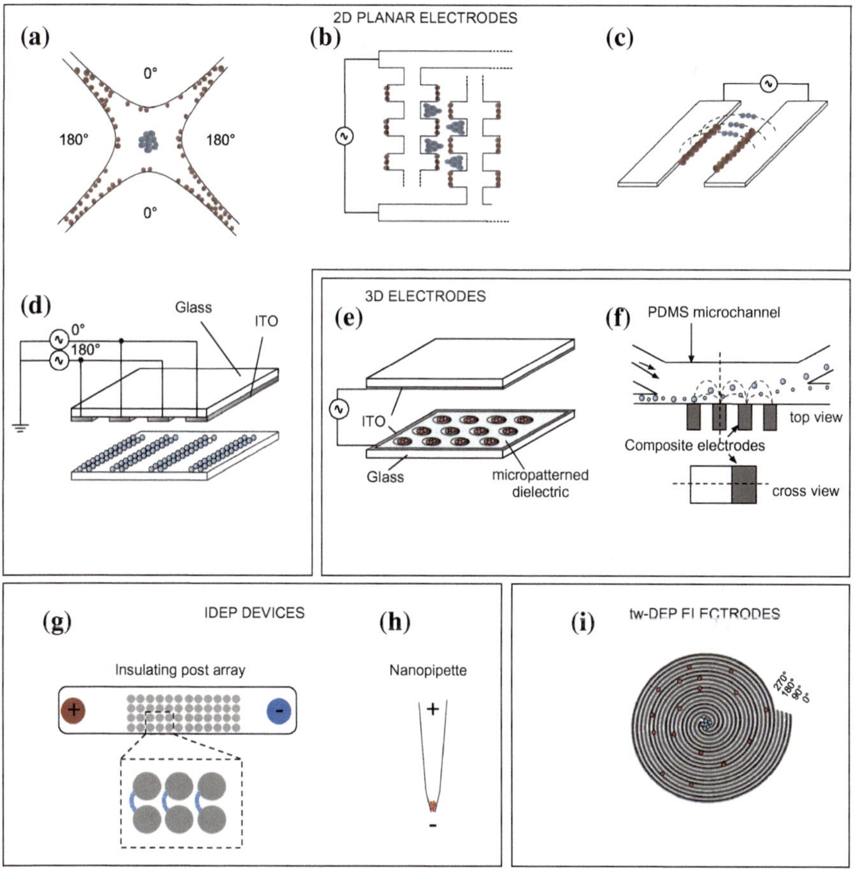

Fig. 6.18 Bioparticles trapped by nDEP and pDEP are represented in blue and red, respectively. **2D planar electrodes**: **a** Quadrupolar electrodes of polynomial design. **b** Interdigitated array (IDA) of rectangular electrodes. **c** Parallel electrodes than can also be arranged as a large IDA as in [99]. **d** **3D electrodes**: **e** Top bottom electrodes (adapted from [62]). **f** Sidewall electrodes [63]. **IDEP electrodes**: **g** Circular insulating post array used to repel proteins from the constrictions where the electric field is maximum [65]. **h** Nanopipette used to trap proteins using pDEP [97]. **tw-DEP** electrodes: **i** Spiral electrodes used to separate cells using a propagative electric field [80]

proteins using beads coated with antibodies [101], for instance. Dielectrophoresis can also be combined with other force fields such as acoustophoresis or magnetophoresis, so as to develop hybrid systems allowing multitarget and multicriteria cell separation [102]. It should also be noted that depending on the operating conditions (frequency and intensity of the electric field), other electric field-induced forces may arise during a DEP experiment, which requires a proper understanding of other AC electrokinetic phenomena involving liquid actuation such as AC electroosmosis and electrothermal effects (see Sect. 6.2). While these phenomena may induce perturbations in some cases, they can also be exploited in conjonction with DEP to provide a large panel of manipulation approaches [103, 104].

6.4 RF and Microwaves for Fluidics and Biodevices (Thi Hong Nhung Dinh, Stephane Serfaty, Pierre-Yves Joubert)

6.4.1 Introduction

Due to their structure and composition, organic materials have free and bound electrically charged entities. When excited by an external electromagnetic field, drifts and displacements of those charges are provoked, inducing conduction and polarization currents within the material. The dielectric properties of the material are a measure of these induced microscopic conduction and polarization phenomena, which constitute the response of the material to an external electromagnetic field.

Since the dielectric properties are determined by and informative about the nature, the composition and the state of the organic material, they are considered as valuable characterization parameters in many scientific and engineering fields involving organic material, such as agriculture, food industry [105–107] and healthcare [108]. In agrifood engineering, the dielectric properties are used e.g. to monitor the quality of food products [107, 109, 110], or to evaluate their state of integrity during maturation [111] or storage [112, 113]. They are also used to predict and control the effects of electro-thermal treatments such as sterilization or pasteurization on food products [105, 114, 115]. In the healthcare domain, the dielectric properties provide relevant information about the physiological state of biological cells and tissues. They are a way of monitoring the state and changes occurring in biological material, which can for example lead to the elaboration of medical diagnosis of tissue pathologies such as inflammation, burns, tumors, etc. [116–122]. Indeed, natural evolutions of organic material such as maturation or ageing, material modifications due to processing such as heating, freezing, or thawing…, or cell functional and/or morphological alterations appearing e.g. with pathologies, result in either structural alterations or changes in the complex biochemical content of the material. In all cases, these alterations induce significant modifications of the dielectric properties of the material. For example, cell membranes altered by electroporation feature modified dielectric properties, which can be read out through the measurement of the changes of the global cell bio-impedance in a 20 MHz frequency range [123]; in the same way, the dielectric properties of tissues with tumors are reported as exhibiting more than 70% changes relatively to those of sound tissues at around 100 MHz [124].

As a result, the dielectric properties constitute relevant indicators either for the monitoring of organic material changes, or for the detection, the characterization and the monitoring of pathologies. Due to the complex structure of organic material which includes macromolecules of various sizes and shapes, the dielectric properties are strongly frequency dependant. For these reasons, a wide range of dielectric characterization methods have been developed, involving electromagnetic fields operating especially in the radiofrequency (RF) and microwave (MW) bandwidths, i.e. including frequencies ranging from a few kHz to tens of GHz. This chapter aims at presenting the basics of the interactions taking place between electromagnetic fields and organic material, and at presenting typical dielectric sensing and characterization devices developed for various scales of organic material investigations. The chapter is organized as follows. Firstly, basics on the dielectric properties of organic material will be recalled, with emphasis on biomaterial such as cells and tissues. Secondly, examples of dielectric characterization methods and devices used for healthcare applications and operating from cell up to tissue scales will be presented.

6.4.2 Basics on the Dielectric Properties of Organic Material

6.4.2.1 Overview of Organic Material Structure and Composition

Biomaterials are highly heterogeneous materials which can be considered at various dimension scales, at described in Fig. 6.19. They are constituted of complex liquids including organic and mineral elements, macromolecules, water-insoluble matter such as fibers or tendons. The organization of these constitutive elements allows intra and extra cellular structures to be formed. At meso or macroscopic scales these structures appear to be homogeneous, and feature highly variable structural properties which, on a mechanical standpoint, range from liquid to solid state materials [125].

At a tissue scale, these structures can be considered as mainly immerged in an aqueous liquid. The resulting complex fluids include organic electrolytes and bio-molecules which interact with one another, and feature properties close to those of aqueous solutions, due to their molecular asymmetry, their electrically polarized nature and their hydrogen and molecular bonding. The particular arrangement and biochemical composition of each tissue provide specific functionalities,

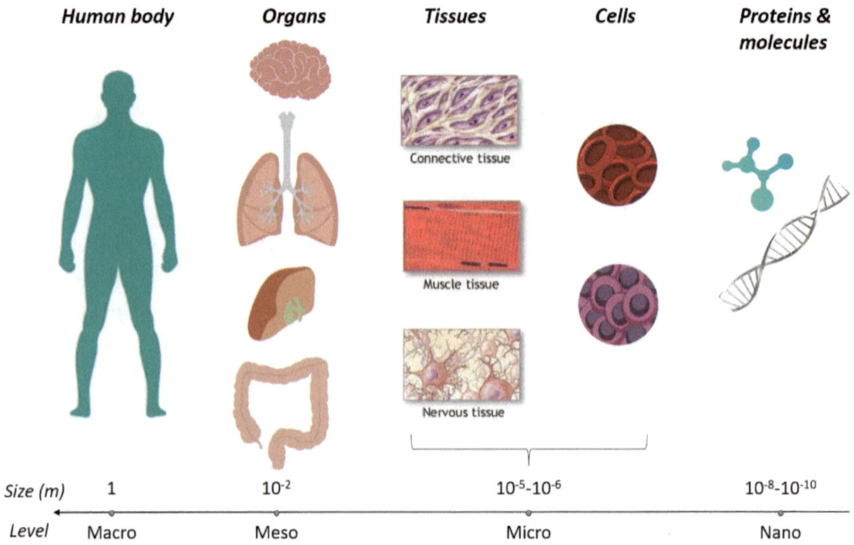

Fig. 6.19 Schematic representation of the dimensions scales in the human body

the biological and physical roles of which are fundamental within the whole biological system. Biological tissues may fall into two main families: (i) the connective tissues, which are constituted of entangled cells included in a kind of matrix featuring various stiffness properties, ranging from fluid like material (e.g. blood), to gel like material (e.g. adipose tissue) and solid state materials (e.g. bones); and (ii) the epithelial tissues featuring joined cells, such as skin.

At a microscopic scale, cells strongly contribute to the macroscopic evolution of the biological system. To do so, they possess specific functionalities, such as autoreplication, environmental protection, acquisition of nutritive elements, communication, degradation and renewing of intrinsic molecules, production of energy, etc. Despite their small size (1–100 μm), cells generally contains numerous components and feature structural properties of complex fluids [126].

The inner part of the cell includes a nucleus, a cytoskeleton made of a network of microtubules, some intermediary filaments, organelles of different size and shapes, proteins, all immerged in an intracellular solution (cytosol). The cell is covered by a phospholipid double layer membrane reinforced by proteins molecules. These proteins are formed with specific three dimensional structured polymers including hydrophobic sites, which depend on the affinity of hydrophilic polar groups present in the solution surrounding the cell. They are accompanied by large mounds of organic material. In aqueous solutions, these macromolecules behave like electrically polarized molecules featuring permanent or induced bipolar electrical moments of various magnitudes, depending on the size and shape of the molecules. Carbohydrates are also present at the surface of membranes, which are involved in cell communications, and also responsible for the secretion of lubricating gels in some tissues such as cartilage, tendons or ligaments. In aqueous solutions they modify the dispersion of water according to their concentration.

Finally, a large amount of electrolytes such as sodium, calcium, magnesium and chlorides play important roles in cell functioning and biological interactions. Numerous vital processes depend on the equilibrium between inner and outer cell electrolytes concentrations. The cellular membrane are globally impermeable to passive ions exchanges, but allow physiologically controlled unidirectional or bidirectional going through of ions. At that scale of operation, a loss of that subtle equilibrium e.g. due to a microbial infection may result in a chain reaction with possible severe consequences such as irreversible structural changes in the tissue. Such biological dysfunctions should be detected and characterized at an early stage so that an appropriate therapy can be proposed.

The involved nano/micro-scale biochemical content and structural changes result in the changes of the dielectric properties of the organic material, which may be observed at larger (micro to meso) scales of observations, providing appropriate dielectric sensing device can be developed.

6.4.2.2 Dielectric Properties of Biomaterial

Electrolytes, macromolecules, proteins, organic and mineral components which are present in cells and tissues are electrically active since they feature free electrical charges and bound electrical charge distributions, the displacement of which being induced by an electrical field applied. These effects are macroscopically described by the complex permittivity ε of the biological material (considered as homogeneous and generally isotropic) and which reads [127, 128]:

$$\varepsilon = \varepsilon' - j\varepsilon'' = \varepsilon_0 \varepsilon_r - j\frac{\sigma}{\omega} \tag{6.114}$$

In this complex expression, $j = \sqrt{-1}$ and ω is the angular frequency. The real part $\varepsilon' = \varepsilon_0 \varepsilon_r$ accounts for the permittivity of the material (where ε_0 is the permittivity of vacuum and ε_r is the dielectric constant of the material, also designated as the relative permittivity). The permittivity measures the induced polarization of the material by units of electrical field, i.e. the extent to which the bound charge distributions can be distorted and polarized under the influence of the field, and the consequent energy stored within the material. For organic material, these charge distributions are mainly associated with double layer membrane proteins, with the surface of solvated macromolecules or with other polar molecules [129].

The imaginary part ε'' of the complex permittivity ε is relative to the dielectric losses associated with the displacement and conduction currents induced within the material by the applied field. In biological material, conduction currents are mainly due to the ionic free charges moving through the tissue under the influence of the field. These losses are proportional to the effective conductivity σ (Eq. 6.115) the material, expressed in S/m.

Each electrically active entity within the material structure will exhibit its own characteristic response to the applied field, according to its own nature, shape and size. As a result, the macroscopic terms ε' and ε'' (or ε_r and σ) exhibit a frequency dependant behavior. Besides, the proportion of the dielectric losses to the stored energy is given by the loss tangent so that:

$$\tan\delta = \frac{\varepsilon''}{\varepsilon'} = \frac{\sigma}{\varepsilon_0\varepsilon_r\omega} \tag{6.115}$$

In order to sense structural modifications of biological material through their macroscopic dielectric properties changes, it is relevant to chose the frequency of investigation at which the changes of ε' and ε'' are both significant, i.e. in the same order of magnitude. This condition is reached for:

$$\tan\delta = 1 \tag{6.116}$$

As a result, for a given material, the ideal sensing angular frequency ω_n would be so that:

$$\omega_n = \sigma/\varepsilon_0\varepsilon_r \tag{6.117}$$

Around this characteristic angular frequency ω_n, the frequency response of the material to an electrical stimulus features a Debye-type form [130] which, in terms of relative permittivity, reads:

$$\varepsilon_r = \varepsilon_{r\infty} + \frac{\varepsilon_{rs} - \varepsilon_{r\infty}}{1 + j\frac{\omega}{\omega_n}} \tag{6.118}$$

where ω_n is the characteristic angular frequency relative to the so-called relaxation time$\tau = 1/\omega_n$, $\varepsilon_{r\infty}$ is the relative permittivity at an angular frequency sufficiently high compared to ω_n and ε_{rs} is the static relative permittivity. A similar form can be written for the conductivity.

6.4.2.3 Relaxation Times Relevant for the Sensing and Characterization of Organic Material

Because of the complexity of the organic material structure, dielectric properties changes due to material alterations may occur at various scales. The difficulty for developing devices able to sense these alterations lies in the choice of the most relevant investigation frequency, since several relaxation times may be distinguished for biological tissues. Indeed, due to multiple interactions occurring at multiple scales, four relaxation times are experimentally observed in the 1 Hz–100 GHz [130–133]. They are modeled by means of the Cole-Cole-Davidson weighed complex sum form [131, 132, 134]:

$$\varepsilon = \varepsilon_\infty - j\frac{\sigma_s}{\omega} + \sum_n \frac{\Delta\varepsilon_n}{1 + \left(j\frac{\omega}{\omega_n}\right)^{1-\alpha_n}} \tag{6.119}$$

where $\Delta\varepsilon_n$ is the permittivity change ($\Delta\varepsilon_n = \varepsilon_n - \varepsilon_\infty$), ε_n is the static permittivity for relaxation n, α_n is the a frequency dispersion coefficient relative to relaxation n. This model can be adjusted to fit with experimental data and is generally limited to a 3-pole simplified model in a 3 decade frequency bandwidth [30, 32, 33]. It may also be relevant to consider the imaginary and real parts dependancy so that the permittivity frequency response writes [135, 136]:

$$\varepsilon = \varepsilon_\infty - j\frac{\sigma_s}{\omega} + \frac{\Delta\varepsilon}{\left(\left(j\frac{\omega}{\omega_n}\right)^{\alpha} + \left(j\frac{\omega}{\omega_n}\right)^{1-\beta}\right)^{\gamma}} \tag{6.120}$$

where α, β, γ are real coefficients taking values in the [0, 1] range. The adjustment of these coefficients allows models rather close to observed dispersion data to be elaborated in a large frequency bandwidth.

In order to sense material structural alterations, it is necessary to determine the relevant frequency range. Published experimental studies on biological tissues point out that a main relaxation time arises around a few tens of MHz for typical tissues, depending on the actual aqueous content of the tissue [137, 138] (Fig. 6.20).

High water content biological tissue exhibit three main dispersion areas commonly called α, β and γ dispersions, as depicted in Fig. 6.20. In the low frequencies (below 10 kHz), the α-dispersion is characterized by a rather large dispersion on the relative permittivity, which may be partly attributed to the diffusion of electrolytes. Other complexes mechanisms have also been reported in the literature [139]. Several interactions occurring next to the cell membrane surfaces are also involved,

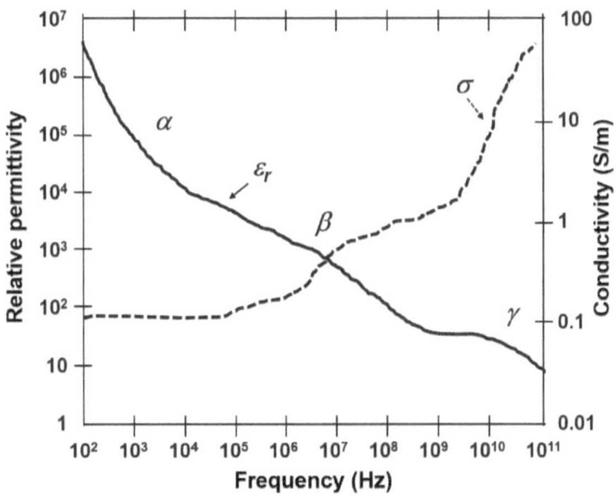

Fig. 6.20 Frequency dependence of the dielectric parameters of typical high water content biological tissues [140]

such as hydrophilic—hydrophobic interactions in the protein double layer membrane, ionic canals, etc. Due to the size and shape of the involved macromolecules structures, several dispersions might be superimposed. The ionic equilibrium between extra and intra cellular media which are separated from around 6 nm, can also significantly modify the dispersion in that area.

The β-dispersion, is considered as extending along a 4 decade frequency bandwidth typically centered around 10 MHz (Fig. 6.20). It is due for one part to the polarization of various macromolecules within the intra cellular media, but mainly to the interfacial polarization of the cell membranes. The latter acts as a barrier for passive ion transport, the efficiency of which depending on the frequency of the applied electrical field. These interfacial polarization phenomena have been theoretically and experimentally studied [141]. Some simplified models build with double layered spherical systems included in ionic liquids foresee the presence of a number of dispersion directly related to the number of interfaces [142, 143]. Besides, it was experimentally established that damaged cell membranes modify the β-dispersion. Hence, numerous biomedical applications are based on the monitoring of the β-dispersion which is related to physio-pathological state changes of biological material, induced by cell morphology or cell physiology modifications. Finally, the γ dispersion is related to the polarization of water molecules and therefore occurs in a higher frequency range (>1 GHz). The detection of pathologies involving water content modifications such as enhanced vascularization due to the presence of tumors can be detected through the monitoring of the γ-dispersion of the dielectric properties of tissues.

6.4.3 Dielectric Characterization Techniques

Pathological alterations implying micro-scale or nanoscale structural modifications of cells inner material and cell interfaces should be detectable through the monitoring of the complex dielectric properties of tissues at a mesoscopic scale of investigation. To that respect, the use of electromagnetic waves corresponding to the β-dispersion, seems particularly relevant to monitor tissue changes (i) due to cell structural modifications in dense areas through the changes of the polarizability of the material, or (ii) to the modifications of dissolved ionic components in soft material through the modification of the conductivity. Higher frequencies of investigation should help to downsize the scale of investigation, so that higher order dispersions and refined cell model could be considered.

For that reason, numerous dielectric characterization techniques have been developed in a wide frequency bandwidth ranging from a few kHz up to tens of GHz. Most developed devices operate at a meso-scopic scale to carry out tissue level investigations. However, micro-scale dielectric sensing devices involving microfluidic environment are more and more developed so as to investigate down to the biological cell level, and constitute high accuracy label-free biosensors. In this section, the main dielectric characterizations techniques are recalled and some examples of organic material dielectric sensing applications involving meso-scale (tissue-level) or micro scale (cell-level) characterizations, are presented.

6.4.3.1 Electrical Impedance Spectroscopy

The electrical impedance spectroscopy (EIS) is a measurement technique allowing the bio-impedance of organic material to be sensed over a frequency bandwidth which, in practice, generally operates from a few kHz up to 100 MHz. The basic principle of the measurement is to inject a small frequency-adjustable alternating current within the material, and to measure the voltage drop appearing between two (or more) electrodes, which are either deposited on or inserted within the material.

The bio-impedance of the material is determined using ohm's law, and gives information which is related to the local dielectric properties of the investigated material. This technique is widely used and is proven efficient in various medical applications such as tissue pathology diagnostic or monitoring, e.g. for cell [123], tumor [145] or wound healing [146] characterizations. In its simplest form, EIS can be implemented using a two-electrode scheme, as presented in Fig. 6.21a. Two electrodes are placed in contact with the material and connected to the impedance

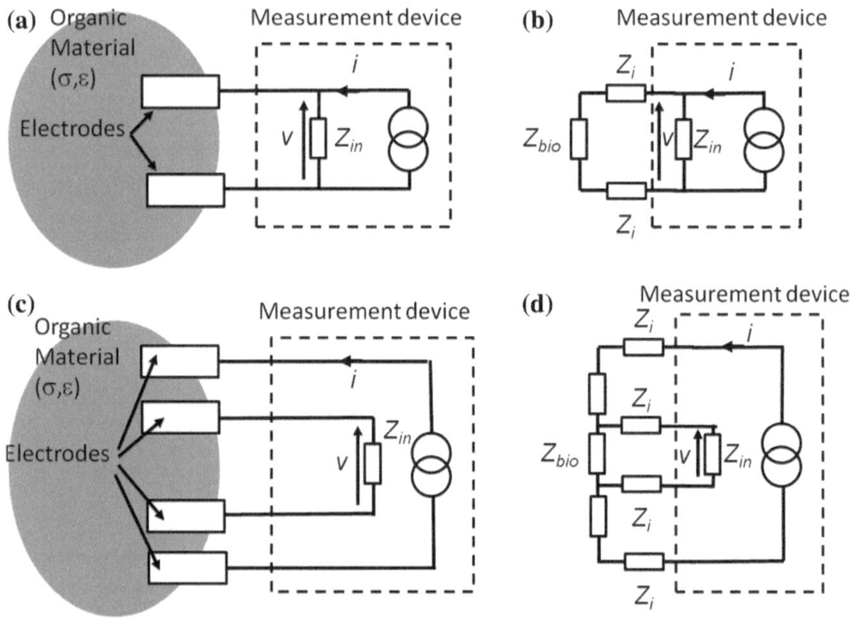

Fig. 6.21 Bio-impedance measurement schemes for Electrical Impedance spectrosctopy; **a**. Two-electrode measurement scheme. **b**. Electrical equivalent circuit model of Two-electrode measurement scheme. **c**. Four-electrode measurement scheme—Electrical equivalent circuit model of four-electrode measurement scheme. Inspired by [144]

measurement device. The electrical equivalent circuit is presented in Fig. 6.21b, in which Z_{bio} is the bio-impedance to be measured, $\underline{Z_i}$ are interface impedances, Z_{in} is the measurement device input impedance, supposed to be "infinitely high", so that:

$$\frac{v}{i} = 2Z_i + Z_{bio} \tag{6.121}$$

The injected current i being known, the voltage drop v being measured, it is possible to estimate Z_{bio} providing Z_i is small enough or previously determined using a calibration procedure. A highly demonstrative experiment aiming at illustrating the relevance of the EIS to characterize organic material was proposed in [147] as a experimental training for undergraduate students. The experiment is based on set up presented in Fig. 6.21a. It consists in inserting a pair of pin electrodes into a potato used as an easy to handle biomaterial, and in measuring the resulting bio-impedance at the end of the electrode pair in the 100 Hz–20 MHz bandwidth. In this frequency range, it is mostly the ionic conduction that affects the bio-impedance. Here, Z_i is considered as negligible and a v/i Bode plot enables Z_{bio} to be visualized over the whole frequency range. The observed electrical phenomenon can be modeled as depicted in Fig. 6.22: an elementary equivalent circuit is established which includes (i) a resistance R_{ext}, accounting for the electrical conductivity of the extracellular matter, (ii) a resistance R_{in} accounting for the intracellular conductivity and (iii) a capacitor C_{ext} modeling the cell membrane ion-barrier.

This model is quite coarse, however it gives a rather good image of the macroscopic electrical behavior of tissues, e.g. for human skin [148]. It can be refined introducing a so-called constant phase element which is a fractional power capacitor modeling the discrepancies between cells within the tissue [149]. The resulting bio-impedance electrical model then reads:

$$Z_{bio} = R_\infty + \frac{R_\infty - R_0}{1 + (j\omega\tau)^\alpha} \tag{6.122}$$

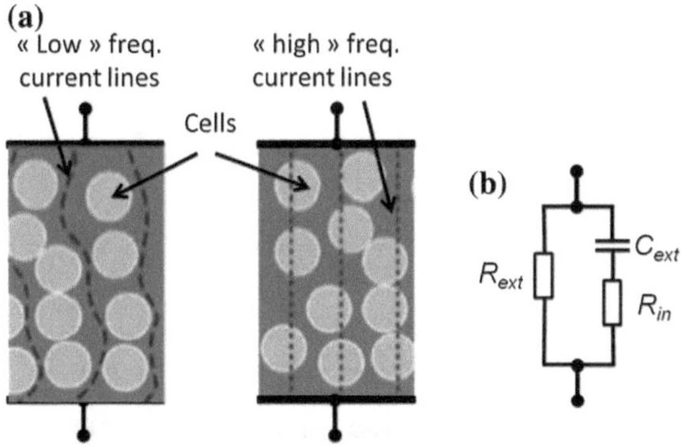

Fig. 6.22 Conductivity dispersion (**a**) and equivalent electrical circuit model (**b**) of the bio-impedance of an elementary organic material (potatoe), [147]. Typical values issued from the fit with experiment data are: $R_{ext} \approx 8$ kΩ, $R_{in} \approx 300$ Ω, $C_{ext} \approx 1$ nF [147]

where R_{-} is the high frequency resistance, R_0 is the static resistance, is the relaxation time, and α is a parameter taking values in the [0 1] interval.

Here, the electrical circuit model of Fig. 6.22b. provides an acceptable model that fits with the measured potato EIS data. In the low frequencies (f < 10 kHz), $Z_{bio} = R_{ext}$ since ionic current lines are limited to the extracellular matter (C_{ext} acts as an open circuit). For high frequencies (f > 10 MHz) $Z_{bio} = R_{ext} // R_{in}$ since ionic current flows through both the extra and intra cellular matter (C_{ext} acts as a short circuit). In this frequency range, the resistance of the material is reduced ($R_{in} \ll R_{ext}$) which corresponds to an increased conductivity. For midrange frequencies (around 200 kHz), a transient electrical behaviour is observed. It corresponds to a β-type dispersion in which the cell membrane barrier progressively opens to the ion flow, and in which consequently the conductivity increases (as presented in Fig. 6.20. In order to put into light the change of the material electrical properties with the alteration of the biomaterial, Français et al. propose to provoke an electroporation of the cell membrane [147] and to measure the resulting bio-impedance changes. The electroporation consists in applying electrical field pulse sequence to the organic material (e.g. a sequence of 10 rectangular pulses of 100 µs duration separated from 1 s, and featuring a 500 V/cm intensity between the electrodes), so as to actively induce the formation of pores within the cell membranes. After electroporation, the measured EIS data are significantly changed and may be modeled by new electrical element values. In particular, the R_{ext} value is reduced by a 20–30% rate, resulting from the increase of the ionic conductivity of the material resulting from the provoked poration of the cell membranes. Based on the approach, the EIS difference between electropored and untreated tissues (lung, heart, …) of small animals was successfully highlighted in a 10 MHz bandwidth, showing 50–80% magnitude changes at 200 kHz [150]. Also, normal and malignant mice breast tissues were successfully diagnosed using EIS in a 100 Hz–1 MHz bandwidth [151].

The two-electrode measurement scheme might be found inaccurate when the interface impedances Z_i cannot be ignored or previously measured. To that end, a four-electrode scheme depicted in Fig. 6.21b. can be preferred. In this scheme, the bio-impedance Z_{bio} is directly given by the ratio v/i, whatever the values of Z_i, providing the measurement device feature an high input impedance Z_{in} and providing appropriate placement of the four electrode set (Fig. 6.22). Lukaski et al. evaluated the efficiency of the four-electrode bio-impedance measurement to non-invasively and reliably monitor the healing of cutaneous wounds [146]. Surface electrodes were placed at each side of a wound (19 mm in length, 17 mm in width, and 2 mm in depth) as depicted in (Fig. 6.23) and the monitoring of the real part (R), the imaginary part (X_c) and the phase angle (PA) of the bio-impedance was carried out over the healing period (tens of days) at a 50 kHz frequency. In this work, the real part R was found to be inversely proportional to the amount of extracellular fluid and directly proportional to the fibrin clot and epithelialization, thus correlated to the successful wound healing. Similarly, X_c was found to be proportional to the cell mass, through the polarizability of the cell membrane. As a result, increases in X_c indicate epidermal proliferation whereas X_c decreases with

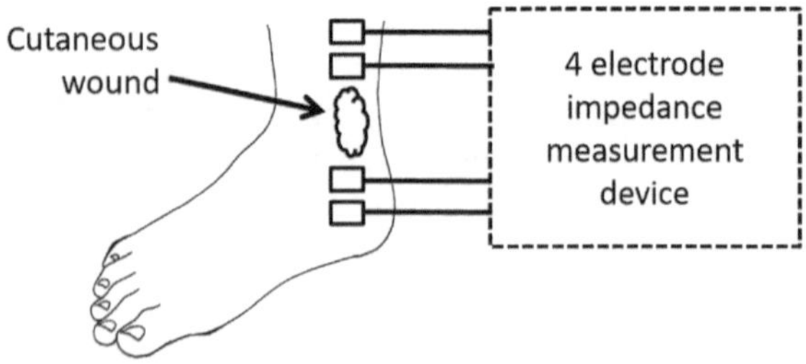

Fig. 6.23 Four-electrode EIS of a cutaneous wound, inspired by [146]

infection and cell loss. Healthy membranes cause a delay in the transit of voltage and current; thus, the greater the *PA*, the healthier the cell membranes, and decreases in *PA* reflect impaired membrane function. For uncomplicated wound healing, the observed general pattern is a continuous increase in R, X_c, and *PA* that is consistent with decreased extracellular fluid, increased cell mass, vitality and epithelialization.

The EIS approach can be extended to electrodes arrays, in order to provide spatially resolved bio-impedance characterizations. To this end, a two dimensional (2D) bio-impedance measurement device was proposed in [145] to locate and characterize pathological epithelial tissues in a 0.1 Hz–100 kHz bandwidth. The device uses 64 units of four electrode sets (256 electrodes), providing a 2D bio-impedance image of the investigated tissue with a 400 μm × 400 μm resolution. The device was successfully implemented on a pig tissues (skin, tong) to detect and characterize wound extends. Based on this approach, 2D or even 3D EIS imaging techniques were developed for breast tumor detection [119]. Even if contact issues, resolution issues, and electrode distribution issues were observed, preliminary clinical examinations showed promising results fully or partially consistent with X-ray mammography and biopsy [152]. More recently, a flexible micro-electrode array has been proposed in [153] for the early detection of pressure ulcers in vivo. The array is fabricated by means of an inkjet printed technology and features a hexagonal configuration of 55 equally spaced (≈2 mm apart) gold electrodes printed on a 35 μm polyethylene naphthalate flexible substrate, and electrically connected to the tissue via highly conductive hydrogel bumps printed above the electrodes. The array is placed on a wound in vivo and the electrical impedance is collected in a 100 Hz–1 MHz bandwidth for each pair of neighboring electrodes. Maps of impedance magnitude, phase angle and post-processed damage threshold are constructed based on the location of each electrode measurement pair. The array enables the non-invasive assessment of pressure induced tissue damage on a rat model, with sufficient sensitivity and robustness to carry out the early detection of mild and reversible wounds that are not visually observable. Even if actual clinical test have to be performed, this device arise promising applications to prevent irreversible tissue damage. The examples presented above illustrate macroscopic implementations of EIS, which give macroscopic electrical information related to microscopic phenomenon. A reduced scale of investigation may be obtained using lab on chips in a microfluidic environment, where cell EIS might be carried out.

For example, Fernandez et al. [154] proposed a flexible label free sensor aiming at assessing the viability and biomass of cells, through the measurement of their bio-impedance. To do so, yeast cells are grown, collected and suspended in a dionized water solution. The cell suspension flows within microfluidic channels equipped with parallel electrodes placed at each side of the channel. The electrical impedance viewed from the electrodes is modeled with an electrical circuit constituted of electrode stray resistances and capacitances, of the resistance of the extracellular solution flowing in the channel, of the intracellular cytoplasmic resistance R_{cyt} and of the cell membrane capacitance (C_m). In order to solely sense R_{cyt} and C_m that are related to the cell dielectric properties, a differential microfluidic device featuring two channels equipped with electrode pairs separated by different distances is build up. The differential structure of the device enables to determine the permittivity of the cell from the evaluation of R_{cyt} and C_m, using a Cole-Cole model of the cell permittivity in a 0.3–50 MHz bandwidth. Experimental results show that viable cells exhibit an electric impedance spectrum which is different in amplitude and in relaxation time from heat treated cells, which are altered in their shape and electrical parameters (R_{cyt} and C_m) due to membrane degradation. In the same way, Bürgel et al. [123] proposed an on-chip single cell microfluidic integrated device allowing for the electroporation and the EIS of single cells in suspension (Fig. 6.24). Paired electrodes are used to induce the cell electroporation and to carry out the EIS analysis. Here, the electroporation is achieved by exposing the cell to an electric field of 2 kV/cm at 50 kHz in a single passage. The passage of single cells is detected and impedance measurements are performed at 8 frequencies distributed in the 20 kHz–20 MHz range, constituting the EIS. After baseline removal (subtraction of the EIS of cell-free solutions) and electrical modeling of the channel/cell impedances, the effect and dynamics of the electroporation of HeLa cells were clearly demonstrated through on-chip EIS analysis.

For practical reasons, EIS techniques are generally limited to a 100 MHz bandwidth above which the simple electrical equivalent R-C modeling of the sensing system no longer applies [155]. However higher investigation frequencies may be necessary to reach more thorough characterizations involving higher

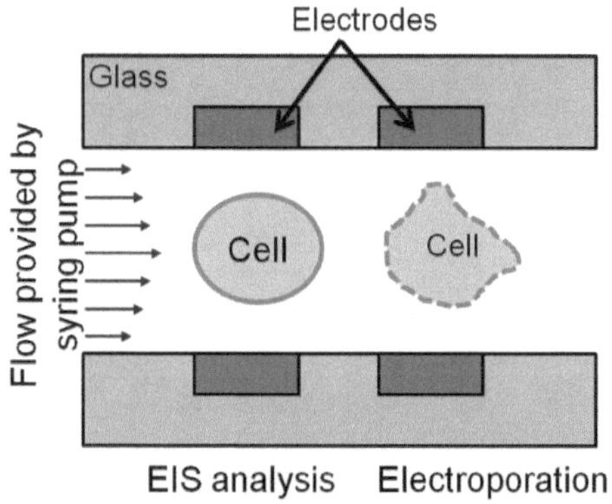

Fig. 6.24 Single cell EIS and electroporation microfluidic device. Schematic presentation, as described in [123]

frequency dispersions or reduced investigation scales, so that refined material models such as layered tissues, multi-shell cell models, etc. can be elaborated. To do so, waveguide based dielectric sensing devices must be considered.

6.4.3.2 Waveguide Based Dielectric Sensing

Designing high frequency sensing devices requires the use of waveguide based sensors. Indeed, high frequency excitation signals generate electromagnetic wavelengths that might reduce down to the same dimensional sizes as the device, so that quasi static assumptions no longer apply. The design of waveguides based sensing devices enables electromagnetic wave propagations to be controlled, so that appropriate sensor model might be developed. Numerous waveguides based techniques have been developed for dielectric characterizations since the 1940s [156]. They fall into two main categories: the transmission measurement techniques, and the reflectometric measurement techniques.

In transmission measurement methods, a two-port waveguide is implemented (Fig. 6.25). The incident wave is injected in port-1 and transmitted to port-2. The organic sample is inserted in or placed next to the waveguide, so that the transmitted electromagnetic wave is altered by the dielectric properties of the sample. A vector network analyzer (VNA) is used to determine the complex transmission coefficient S_{21} defined by the ratio between the transmitted wave (measured in the port-1 reference plane) and the incident wave (measured in the port-2 reference plane). In the same way, the reflectometric measurement method uses a waveguide device including or placed next to the sample under test (Fig. 6.25). The characterization of the sample consists in determining the complex reflection coefficient S_{11} defined as the ratio between the reflected and injected waves in the port-1

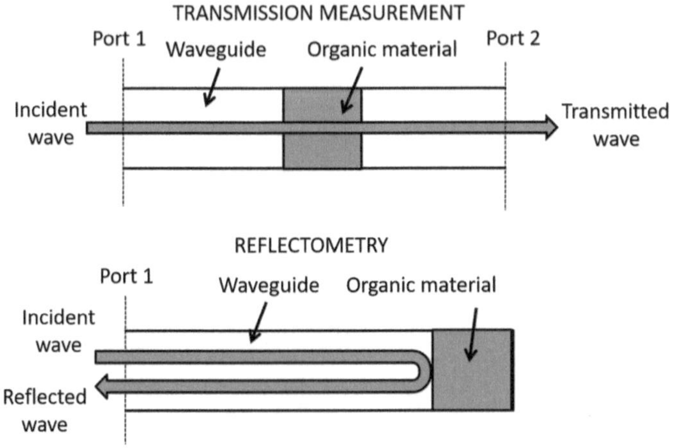

Fig. 6.25 Transmission and reflectrometric waveguide based measurement methods

reference plane, by means of a VNA. Reflectrometric methods based on the use of coaxial probes are the most popular techniques and have been widely used for the dielectric characterization of organic material, thanks to their easy and non-invasive implementation features. Coaxial probes are generally made of a open-ended coaxial transmission line, possibly terminated by a dielectric bead, and set directly in contact with the considered material, enabling ex vivo as well as in vivo characterizations. The accuracy of the method requires (i) a good contact between the probe and the material, which is generally reached with organic materials in the form of liquids, gels or malleable solids, and (ii) some calibration process using reference liquids [156] such as di-ionized water, ethanol, and methanol solutions [157].

Gabriel et al. [158] elaborated a database of the dielectric properties of a wide range of excised animal tissue in a 20−37 °C temperature range, and over a 20 GHz bandwidth by means of coaxial probes, and a four dispersion Cole-Cole model associated to a static ionic conductivity term (Eq. 6.6) have been used to reduce the complexity of various human and animal tissues (brain, fat, breast, skin, bone, liver, etc.) [138]. A 3 mm diameter precision open-ended coaxial probe was used in [120] to assess the in vivo complex permittivity of normal and malignant human liver tissues, converting the measured reflexion coefficient into the tissue complex permittivity using the de-embedding and rational function models described in [157]. In the context of the enhancement of controlled electrical intra-cardiac chirurgical operations, Brusson et al. [118] used a precision coaxial probe in the 0.8–2.6 GHz range to correlate the depth of electrically induced heart tissue burns (and quantified by microscopic observations) with the measured complex reflection of the implemented probe, in the 0–6 mm depth range. Caduff et al. proposed a waveguide based multisensory approach to carry out non invasive glucose monitoring [159]. Indeed, it has been established that blood glucose variations induce changes in the skin and underlying tissues (SAUT) dielectric properties. In order to sense these changes, a set dielectric sensors were designed to investigate the SAUT in the 1–200 kHz low frequency range, the 0.1–100 MHz middle frequency range and a 1–2 GHz microwave range. The kHz range sensors feature interdigited EIS-like electrodes, the MHz range sensors are constituted of 3 reflectometric capacitive sensors of various sizes, and the GHz range sensors are constituted of 2 coplanar waveguide transmission lines. The sensors were designed so as to investigate the SAUT at various skin depths, allowing glucose induced dielectric changes to be sensed within the underlying tissues, or non-glucose related changes e.g. due to skin-sweat or temperature changes, to be sensed in the upper skin layers. All dielectric sensors are integrated within the same sensory plateform, set on a bandage like device placed around the arm, together with conventional sweat, temperature and optical diffuse reflectance sensors. Statistical clinical studies highlighted the correlation of the developed non invasive multisensory data-fused glucose monitoring approach, with observed blood glucose concentrations [159].

As illustrated above, reflectometric dielectric sensors are relevant for non-invasive ex vivo or in vivo implementations on tissues. For practical reasons transmission methods are less suitable for this kind of characterizations since the considered sample should be inserted into the waveguide [156]. However waveguide transmission sensing techniques find wide applications in microfluidic characterization devices, where micro-device designs allow micro-strip transmission lines to be judiciously combined with microfluidic channels. For example, Leroy et al. [160] proposed a microfuidic biosensor designed for the dielectric spectroscopy of micro-scale particules flowing in a microchannel. The sensor is based on a discontinuated coplanar microstrip transmission line, constituted of 15 μm thick gold electroplated electrodes deposited on a fused silicate substrate. The microstrip line discontinuity is designed to form a capacitive sensor in a detection micro-area, above which a SU8 microchannel are patterned and covered with a PDMS top (Fig. 6.26a). Dielectric measurements are performed on 12 μm diameter polystyrene beads up to 8 GHz, through the measurement of the transmission coefficient S_{21} of the transmission line carried out by a VNA. Measurements clearly demonstrate the ability of the device to sense the presence of single particule suspended in a di-ionized water solution, with a 1fF capacitance contrast. Similar experiments conducted with a 5 GHz resonant implementation allows 10 MHz resonance shift to be observed with single beads, from which adequate post processing computations enable the permittivity of the bead to be estimated. Finally, trials carried out on cancerous colorectal cells showed promising results, although dispersions in the size of the cells make the post processed estimation of the cell permittivity less accurate than that obtained with calibrated beads. Grenier et al. proposed a biosensor based on the use of a coplanar transmission line with a fluidic channel integrated on the top surface (Fig. 6.26b) [161]. The device is implemented as a two-port transmission sensor operated in the 0.4–35 GHz, and used to quantify the concentration of human umbilical vein endothelial cells suspended in a sucrose and fetal bovine water-based solution. The complex permittivity of the biological solution was estimated from the measured S_{21} parameter through a dedicated de-embedding technique and modeling of the measurements using numerical computations a Cole-Cole modeling. A 100% change in the relative permittivity was observed between the cell-free solution and the 100% cell concentration fluid at 20 GHz, validating the relevance of the developed device as a non invasive label-free biosensor allowing cell quantification and counting to be carried out.

A last example is provided by the sensor developed in [162], which aim at distinguishing the inner content of cells in order to discriminate cancer cell according to their evolution stage, by means of waveguide transmission based microfluidic biosensors. The developed sensor is a resonating device constituted of a meandered inductor and two inter-digitated comb capacitors set in parallel and implemented in a microwave coplanar waveguide, integrated with a SU8 patterned chamber on the top (Fig. 6.26c). The device acts as a stop-band transmission line for which the S_{21} coefficient shows a minimum value at the resonance frequency of the formed LC resonator. Cells are deposited in the chamber using a droplet technique, and are trapped within the inter-digitated comb capacitor fingers (Fig. 6.26c). The presence of cells locally modifies the inter-digitated comb capacitor capacitance, and hence the resonance frequency of the whole resonator. For example, the presence of two cells in the chamber induces a shift of the S_{21}

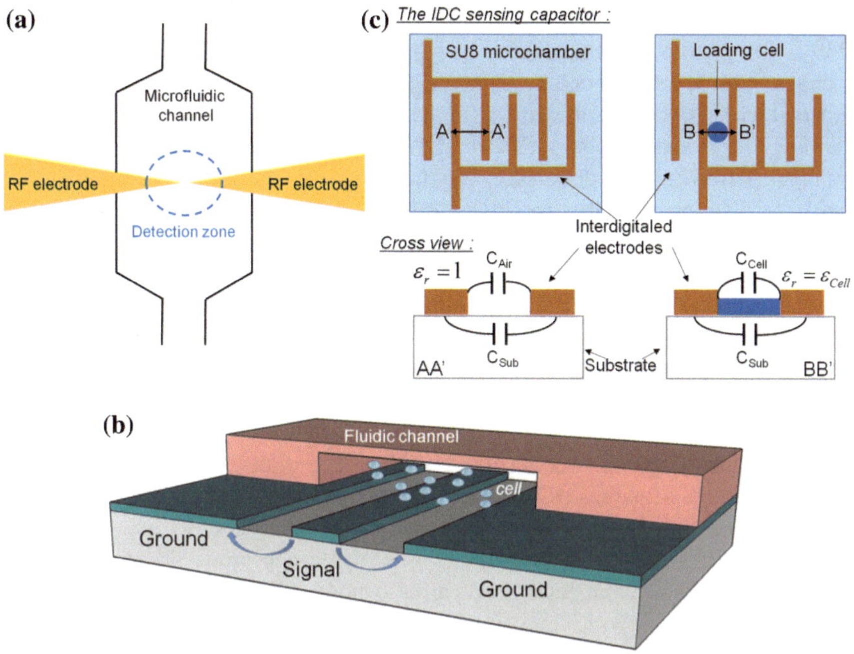

Fig. 6.26 Transmission line based dielectric sensors. **a**. Sensing capacitance inserted in a transmission line in microfluidic channel, from [160], **b**. Schematic view of RF interaction between a coplanar waveguide and a biological solution, which is localized in a microfluidic channel, from [161], **c**. Inter-digitated comb capacitors (IDC) set in a microwave coplanar waveguide, from [162]

resonance frequency of 89 MHz, for a 7 GHz cell-free sensor resonance. The study carried out on various cell lines allowed a clear correlation to be established between cell lines—especially between cells with low and high aggressiveness levels—and the electromagnetic signature constructed from measurements provided by the implemented biosensor and Cole-Cole modeling.

6.4.3.3 Contact Free Dielectric Sensing

EIS and waveguide based dielectric sensing methods use probes or sensors requiring an electrical contact with the organic material under test. These techniques are proved relevant however induce some possible contact issues including bio-compatibility, loss of accuracy, loss of non-invasivity or comfort in the case of in vivo or wearable devices [156, 163]. For that reasons, free space or non-contact sensing techniques have been developed. In the low frequency spectrum (kHz up to tens of MHz), inductive methods generally referred as magnetic impedance techniques (MIT), use a non-contact magnetic coupling induced between an inductive antenna and the organic material to electrically characterize the material under test [119]. However, in practice, these techniques are mostly sensitive to conductivity changes occurring in the organic material, and are hardly sensitive to the dielectric constant [164, 165]. In the upper range of the RF and MW frequencies (1–100 GHz), free space techniques using MW antennas and RADAR imaging based approaches have been developed for non-contact and non-invasive dielectric characterizations especially for breast cancer detections [122, 166–168]. More recently, integrated systems operating in the 1–10 GHz band based on time domain multi-static RADAR techniques involving antenna arrays have been developed and previous clinical validations have been obtained for the efficient detection and localization of beast tumors [169, 170]. These systems allow advanced 2D dielectric image reconstructions to be achieved, and offer highly promising perspectives in terms of large scale development of low cost and easy to use breast cancer diagnosis devices.

Microwave bandwidth contact-free dielectric sensors often operate in the γ-dispersion area of tissues (Fig. 6.21), where the polarization of water molecules is mainly responsible for the dielectric behaviour of the tissue. Frequencies higher than the GHz are therefore relevant to investigate cancerous tissues since they feature increased vascularization compared to normal tissues. However GHz sensors are hardly sensitive to the conductivity of organic materials, although tumors and other tissue pathologies are known to features changes of both the real and imaginary part of the complex permittivity [116]. For that reasons, midrange RF frequency inductive techniques operating in the tens to hundreds of MHz bandwidth and related to the β-dispersion, have been developed for the non-contact assessment of the complex permittivity of organic materials.

For example, the technique developed in [171] is based on the use flat multi-turn split-conductor transmission-line resonator (MTLR) [172] distantly interacting as a transmit and receive inductive sensor with the considered organic material. The MTLR is monitored by a distant control coil fed with a low intensity RF current. When electromagnetically excited by the coil, the resonator generates an amplified RF magnetic field and induces the circulation of eddy currents and displacement currents within the organic material which depend on its dielectric properties. By reciprocity these currents modify the generated magnetic field, and hence, the impedance of the RF resonator. As a result, the measurement of the resonator impedance gives access to the dielectric properties of the investigated material. It is actually carried out through the evaluation of the impedance changes induced at the ends of the control coil (Z_{mes}) electromagnetically coupled to the resonator, by means of a VNA equipped with an impedance measurement kit. The whole set up is modeled by an equivalent electrical model as depicted in Fig. 6.27. [171]. In this electrical model, the R_c and L_c stand for the control coil, R_1, L_1, C_1 are the equivalent electrical lump elements of the MTLR, and R_i, L_i are the additional lumped elements induced by the conductivity and the permittivity of the material loading the MTLR, respectively [173]. Experimental feasibility of organic material non-contact characterization was realized for the in-line monitoring of milk fermentation during yogurt formation in [174]. To do so, milk solutions acidified with various concentrations of glucono-δ-lactone (GDL) were considered. The solutions were incubated at 37 °C so that micelles of casein aggregate under the combined action of the GDL and temperature, and so that the jellification process starts. Since the acidified milk solution features typical dielectric properties of $\sigma \approx 1.1$ S/m and $\varepsilon_r \approx 65$, the MTLR was designed so that the resonance frequency is around 300 MHz (according to Eq. 6.117). As a result, the used MTLR is a planar 19 mm large resonator constituted of two rolled-up 1 mm width transmission lines constituted of 35 μm-thick copper tracks deposited by photolithography on each side of a 250 μm-thick low-loss dielectric substrate (CuFlon). The designed MTLR features a practical resonance frequency $f_0 = 306$ MHz. The non-contact dielectric monitoring system was implemented during the jellification process of milk solutions featuring GDL concentrations of 13–26 g/l of GDL, and the dielectric parameters where extracted from the measurement of Z_{mes} using the electrical equivalent model (Fig. 6.27) and a calibration procedure implemented using a calibrated NaCl solution [175]. The resulting time evolution of both dielectric parameters is presented in Fig. 6.28. The observed conductivity and relative permittivity behaviour and kinetics are induced by the structural changes of the organic material during jellification, and are consistent with those observed with contact techniques [176]. Such non-contact approach offers interesting perspectives for the design of wearable sensors such bandage-like devices allowing the non-contact monitoring of wound healing or glucose content.

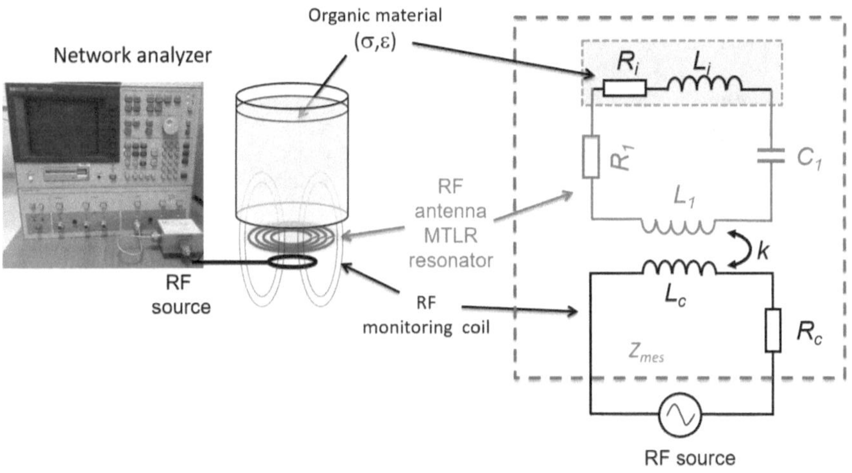

Fig. 6.27 Experiment set up and associated equivalent electrical modeling for non-contact dielectric characterization of organic material

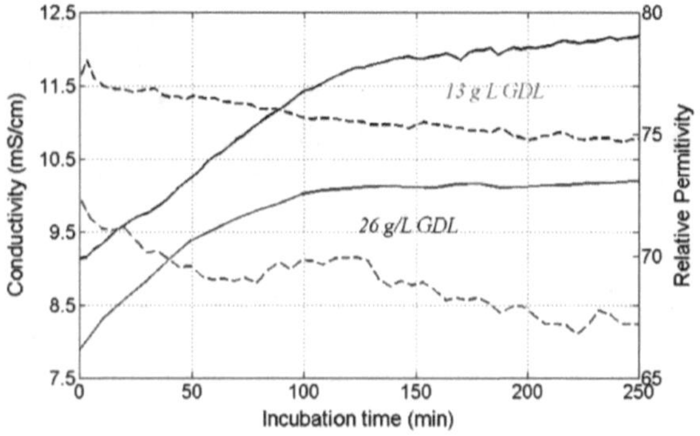

Fig. 6.28 Evolution of the conductivity and relative permittivity versus incubation time for two milk samples acidified with different doses of glucono-δ-lactone (GDL)

6.4.4 Concluding Remarks

In this chapter, basics on the dielectric properties of organic material have been recalled and some examples of widely used dielectric characterization techniques have been presented. Finally, examples of dielectric sensing devices operating either at a tissue level or at cell–level of investigations have been presented.

These examples aimed at illustrating the relevance of dielectric characterizations to detect and monitor tissue pathologies in an accurate, non-invasive, non toxic and easily implemented manner. Thanks to the progress made in micro-fabrication processes as well as in sensing technology including data processing and communication, dielectric sensing devices are particularly good candidates be integrated in cheap and efficient point-of-care testing devices, as well as well as wearable sensing plateforms including microfluidic lab-on-chip biosensors or state-of-health physiological sensors.

Acknowledgements This Sect. 6.4 has benefited from the financial support of the LabeX LaSIPS (ANR-10-LABX-0040-LaSIPS) managed by the French National Research Agency under the "In-vestissements d'avenir" program (nANR-11-IDEX-0003-02). The authors acknowledge project supports from ENS Cachan-Universite Paris Saclay, Institut d'Alembert, Ecole Centrale de Lyon and CNRS.

References

1. D. Mark, S. Haeberle, G. Roth, F. von Stetten, R. Zengerle, Microfluidic lab-on-a-chip platforms: requirements, characteristics and applications. Chem. Soc. Rev. **39**(3), 1153–1182 (2010)
2. A. van den Berg, H.G. Craighead, P. Yang, From microfluidic applications to nanofluidic phenomena. Chem. Soc. Rev. **39**(3), 899–900 (2010)
3. W. Sparreboom, A. van den Berg, J.C.T. Eijkel, Principles and applications of nanofluidic transport. Nat Nano **4**(11), 713–720 (2009)
4. H. Daiguji, Ion transport in nanofluidic channels. Chem. Soc. Rev. **39**(3), 901–911 (2010)

5. T.M. Squires, M.Z. Bazant, Induced-charge electro-osmosis. J. Fluid Mech. **509**, 217–252 (2004)
6. J.Y. Kim, B.J. Yoon, Electrophoretic motion of a slightly deformed sphere with a nonuniform zeta potential distribution. J. Colloid Interface Sci. **251**(2), 318–330 (2002)
7. P.R.C. Gascoyne, J. Vykoukal, Particle separation by dielectrophoresis. Electrophoresis **23** (13), 1973–1983 (2002)
8. R. Yan, W. Liang, R. Fan, P. Yang, Nanofluidic diodes based on nanotube heterojunctions, Nano Lett. (2009)
9. L.-J. Cheng, L.J. Guo, Ionic current rectification, breakdown, and switching in heterogeneous oxide nanofluidic devices. ACS Nano **3**(3), 575–584 (2009)
10. L. Bocquet, E. Charlaix, Nanofluidics, from bulk to interfaces. Chem. Soc. Rev. **39**(3), 1073–1095 (2010)
11. U. Vermesh, J.W. Choi, O. Vermesh, R. Fan, J. Nagarah, J.R. Heath, Fast nonlinear ion transport via field-induced hydrodynamic slip in sub-20-nm hydrophilic nanofluidic transistors. Nano Lett. **9**(4), 1315–1319 (2009)
12. D. Stein, M. Kruithof, C. Dekker, Surface-charge-governed ion transport in nanofluidic channels. Phys. Rev. Lett. **93**(3), 35901 (2004)
13. S.J. Kim, S.H. Ko, K.H. Kang, J. Han, Direct seawater desalination by ion concentration polarization. Nat Nano **5**(4), 297–301 (2010)
14. Y.-C. Wang, A.L. Stevens, J. Han, Million-fold preconcentration of proteins and peptides by nanofluidic filter. Anal. Chem. **77**(14), 4293–4299 (2005)
15. H. Daiguji, P. Yang, A.J. Szeri, A. Majumdar, Electrochemomechanical energy conversion in nanofluidic channels. Nano Lett. **4**(12), 2315–2321 (2004)
16. E. Kjeang, N. Djilali, D. Sinton, Microfluidic fuel cells: a review. J. Power Sources **186**(2), 353–369 (2009)
17. R. Karnik, K. Castelino, A. Majumdar, Field-effect control of protein transport in a nanofluidic transistor circuit. Appl. Phys. Lett. **88**(12) (2006)
18. S.-W. Nam, M.J. Rooks, K.-B. Kim, S.M. Rossnagel, Ionic field effect transistors with sub-10 nm multiple nanopores. Nano Lett. **9**(5), 2044–2048 (2009)
19. S.-W. Nam, M.-H. Lee, S.-H. Lee, D.-J. Lee, S.M. Rossnagel, K.-B. Kim, Sub-10-nm nanochannels by self-sealing and self-limiting atomic layer deposition. Nano Lett. **10**(9), 3324–3329 (2010)
20. M.Z. Bazant, M.S. Kilic, B.D. Storey, A. Ajdari, Towards an understanding of induced-charge electrokinetics at large applied voltages in concentrated solutions. Adv. Colloid Interface Sci. **152**(1–2), 48–88 (2009)
21. M. Haan, J.F. Gwan, A. Baumgaertner, Correlated movements of ions and water in a nanochannel. Mol. Simul. **35**(1–2), 13–23 (2009)
22. R.B. Bird, W.E. Stewart, E.N. Lightfoot, *Transport Phenomena* (Wiley, 2007)
23. J. Happel, H. Brenner, *Low Reynolds Number Hydrodynamics: With Special Applications to Particulate Media* (Springer, Netherlands, 1983)
24. W.B. Russel, D.A. Saville, W.R. Schowalter, *Colloidal Dispersions* (Cambridge University Press, 1992)
25. S.H. Ko, S.J. Kim, L.F. Cheow, L.D. Li, K.H. Kang, J. Han, Massively parallel concentration device for multiplexed immunoassays. Lab Chip **11**(7), 1351–1358 (2011)
26. A. Plecis, R.B. Schoch, P. Renaud, Ionic transport phenomena in nanofluidics: experimental and theoretical study of the exclusion-enrichment effect on a chip. Nano Lett. **5**(6), 1147–1155 (2005)
27. B.S. Reto, H. Jongyoon, R. Philippe, Transport phenomena in nanofluidics. Rev. Mod. Phys. **80**(3), 839 (2008)
28. S.J. Kim, Y.-C. Wang, J.H. Lee, H. Jang, J. Han, Concentration polarization and nonlinear electrokinetic flow near a nanofluidic channel. Phys. Rev. Lett. **99**(4), 44501 (2007)
29. S.J. Kim, Y.-A. Song, J. Han, Nanofluidic concentration devices for biomolecules utilizing ion concentration polarization: theory, fabrication, and applications. Chem. Soc. Rev. **39**(3), 912–922 (2010)

30. A. Mani, T.A. Zangle, J.G. Santiago, On the propagation of concentration polarization from microchannel − nanochannel interfaces part I: analytical model and characteristic analysis. Langmuir **25**(6), 3898–3908 (2009)
31. E.V. Dydek, B. Zaltzman, I. Rubinstein, D.S. Deng, A. Mani, M.Z. Bazant, Overlimiting current in a microchannel. Phys. Rev. Lett. **107**(11), 118301 (2011)
32. P. Kim, S.J. Kim, J. Han, K.Y. Suh, Stabilization of ion concentration polarization using a heterogeneous nanoporous junction. Nano Lett. **10**(1), 16–23 (2009)
33. I. Rubinstein, L. Shtilman, Voltage against current curves of cation exchange membranes. J. Chem. Soc. Faraday Trans. 2 Mol. Chem. Phys. **75**, 231–246 (1979)
34. S.J. Kim, L.D. Li, J. Han, Amplified electrokinetic response by concentration polarization near nanofluidic channel. Langmuir **25**(13), 7759–7765 (2009)
35. R.F. Probstein, *Physicochemical Hydrodynamics: An Introduction* (Wiley, 2005)
36. C.L. Druzgalski, M.B. Andersen, A. Mani, Direct numerical simulation of electroconvective instability and hydrodynamic chaos near an ion-selective surface. Phys. Fluids **25**(11) (2013)
37. W. Guo, L. Cao, J. Xia, F.-Q. Nie, W. Ma, J. Xue, Y. Song, D. Zhu, Y. Wang, L. Jiang, Energy harvesting with single-ion-selective nanopores: a concentration-gradient-driven nanofluidic power source. Adv. Funct. Mater. **20**(8), 1339–1344 (2010)
38. M. Wanunu, W. Morrison, Y. Rabin, A.Y. Grosberg, A. Meller, Electrostatic focusing of unlabelled DNA into nanoscale pores using a salt gradient. Nat Nano **5**(2), 160–165 (2010)
39. Y. Kim, K.S. Kim, K.L. Kounovsky, R. Chang, G.Y. Jung, J.J. dePablo, K. Jo, D.C. Schwartz, Nanochannel confinement: DNA stretch approaching full contour length. Lab Chip **11**(10), 1721–1729 (2011)
40. T.C. Kinnaman, The economic impact of shale gas extraction: a review of existing studies. Ecol. Econ. **70**(7), 1243–1249 (2011)
41. S.J. Kim, S.H. Ko, K.H. Kang, J. Han, Direct seawater desalination by ion concentration polarization. Nat Nano **8**(8), 609 (2013)
42. D. Fologea, J. Uplinger, B. Thomas, D.S. McNabb, J. Li, Slowing DNA translocation in a solid-state nanopore. Nano Lett. **5**(9), 1734–1737 (2005)
43. M. Utkur, C. Jeffrey, D. Valentin, A. Aleksei, T. Gregory, Slowing the translocation of double-stranded DNA using a nanopore smaller than the double helix. Nanotechnology **21** (39), 395501 (2010)
44. M. Wanunu, A. Meller, Chemically modified solid-state nanopores. Nano Lett. **7**(6), 1580–1585 (2007)
45. H. Daiguji, P. Yang, A. Majumdar, Ion transport in nanofluidic channels. Nano Lett. **4**(1), 137–142 (2004)
46. S.A. Gajar, M.W. Geis, An ionic liquid-channel field-effect transistor. J. Electrochem. Soc. **139**(10), 2833–2840 (1992)
47. R. Fan, M. Yue, R. Karnik, A. Majumdar, P. Yang, Polarity switching and transient responses in single nanotube nanofluidic transistors. Phys. Rev. Lett. **95**(8), 86607 (2005)
48. R. Karnik, R. Fan, M. Yue, D. Li, P. Yang, A. Majumdar, Electrostatic control of ions and molecules in nanofluidic transistors. Nano Lett. **5**(5), 943–948 (2005)
49. K.-H. Paik, Y. Liu, V. Tabard-Cossa, M.J. Waugh, D.E. Huber, J. Provine, R.T. Howe, R.W. Dutton, R.W. Davis, Control of DNA capture by nanofluidic transistors. ACS Nano **6**(8), 6767–6775 (2012)
50. S.M. Sze, *Physics of Semiconductor Devices* (Wiley, 1981)
51. M.J. McNutt, C.T. Sah, Determination of the MOS oxide capacitance. J. Appl. Phys. **46**(9), 3909–3913 (1975)
52. I. Klammer, A. Buchenauer, H. Fassbender, R. Schlierf, G. Dura, W. Mokwa, U. Schnakenberg, Numerical analysis and characterization of bionic valves for microfluidic PDMS-based systems. J. Micromechanics Microengineering **17**(7), S122 (2007)
53. S.H. Behrens, D.G. Grier, The charge of glass and silica surfaces. J. Chem. Phys. **115**(14), 6716–6721 (2001)

54. T. Hiemstra, W.H. Van Riemsdijk, G.H. Bolt, Multisite proton adsorption modeling at the solid/solution interface of (hydr)oxides: a new approach: I. Model description and evaluation of intrinsic reaction constants. J. Colloid Interface Sci. **133**(1), 91–104 (1989)
55. F.H.J. van der Heyden, D. Stein, C. Dekker, Streaming currents in a single nanofluidic channel. Phys. Rev. Lett. **95**(11), 116104 (2005)
56. N. Hu, Y. Ai, S. Qian, Field effect control of electrokinetic transport in micro/nanofluidics. Sens. Actuators B Chem. **161**(1), 1150–1167 (2012)
57. L.-H. Yeh, S. Xue, S.W. Joo, S. Qian, J.-P. Hsu, Field effect control of surface charge property and electroosmotic flow in nanofluidics. J. Phys. Chem. C **116**(6), 4209–4216 (2012)
58. T.B. Jones, *Electromechanics of Particles* (Cambridge University Press, 2005)
59. Y. Huang, R. Holzel, R. Pethig, X.-B. Wang, Differences in the AC electrodynamics of viable and non-viable yeast cells determined through combined dielectrophoresis and electrorotation studies. Phys. Med. Biol. **37**(7), 1499 (1992)
60. C. Reichle, T. Schnelle, T. Müller, T. Leya, G. Fuhr, A new microsystem for automated electrorotation measurements using laser tweezers. Biochim. Biophys. Acta **1459**(1), 218–229 (2000)
61. Y. Huang, R. Pethig, Electrode design for negative dielectrophoresis. Meas. Sci. Technol. **2** (12), 1142 (1991)
62. D.R. Albrecht, G.H. Underhill, T.B. Wassermann, R.L. Sah, S.N. Bhatia, Probing the role of multicellular organization in three-dimensional microenvironments. Nat. Methods **3**(5), 369–375 (2006)
63. N. Lewpiriyawong, C. Yang, Y.C. Lam, Continuous sorting and separation of microparticles by size using AC dielectrophoresis in a PDMS microfluidic device with 3-D conducting PDMS composite electrodes. Electrophoresis **31**(15), 2622–2631 (2010)
64. A.-L. Deman, M. Brun, M. Quatresous, J.-F. Chateaux, M. Frenea-Robin, N. Haddour, V. Semet, R. Ferrigno, Characterization of C-PDMS electrodes for electrokinetic applications in microfluidic systems. J. micromechanics microengineering **21**(9), 95013 (2011)
65. B.H. Lapizco-Encinas, B.A. Simmons, E.B. Cummings, Y. Fintschenko, Insulator-based dielectrophoresis for the selective concentration and separation of live bacteria in water. Electrophoresis **25**(10–11), 1695–1704 (2004)
66. N. Demierre, T. Braschler, P. Linderholm, U. Seger, H. van Lintel, P. Renaud, Characterization and optimization of liquid electrodes for lateral dielectrophoresis. Lab Chip **7**(3), 355–365 (2007)
67. M. Li, W.H. Li, J. Zhang, G. Alici, W. Wen, A review of microfabrication techniques and dielectrophoretic microdevices for particle manipulation and separation. J. Phys. D. Appl. Phys. **47**(6), 63001 (2014)
68. R. Martinez-Duarte, Microfabrication technologies in dielectrophoresis applications—a review. Electrophoresis **33**(21), 3110–3132 (2012)
69. L. Yang, Dielectrophoresis assisted immuno-capture and detection of foodborne pathogenic bacteria in biochips. Talanta **80**(2), 551–558 (2009)
70. R. Hamada, H. Takayama, Y. Shonishi, L. Mao, M. Nakano, J. Suehiro, A rapid bacteria detection technique utilizing impedance measurement combined with positive and negative dielectrophoresis. Sens. Actuators B Chem. **181**, 439–445 (2013)
71. J. Voldman, M.L. Gray, M. Toner, M.A. Schmidt, A microfabrication-based dynamic array cytometer. Anal. Chem. **74**(16), 3984–3990 (2002)
72. M. Frenea, S.P. Faure, B. Le Pioufle, P. Coquet, H. Fujita, Positioning living cells on a high-density electrode array by negative dielectrophoresis. Mater. Sci. Eng. C **23**(5), 597–603 (2003)
73. X. Guo, R. Zhu, Controllably moving individual living cell in an array by modulating signal phase difference based on dielectrophoresis. Biosens. Bioelectron. **68**, 529–535 (2015)
74. C.-T. Ho, R.-Z. Lin, W.-Y. Chang, H.-Y. Chang, C.-H. Liu, Rapid heterogeneous liver-cell on-chip patterning via the enhanced field-induced dielectrophoresis trap. Lab Chip **6**(6), 724–734 (2006)

75. S. Menad, L. Franqueville, N. Haddour, F. Buret, M. Frenea-Robin, nDEP-driven cell patterning and bottom-up construction of cell aggregates using a new bioelectronic chip. Acta Biomater. **17**, 107–114 (2015)
76. F.F. Becker, X.B. Wang, Y. Huang, R. Pethig, J. Vykoukal, P.R. Gascoyne, Separation of human breast cancer cells from blood by differential dielectric affinity. Proc. Natl. Acad. Sci. U. S. A. **92**(3), 860–864 (1995)
77. I. Doh, Y.-H. Cho, A continuous cell separation chip using hydrodynamic dielectrophoresis (DEP) process. Sens. Actuators A Phys. **121**(1), 59–65 (2005)
78. E. Bisceglia, M. Cubizolles, F. Mallard, F. Vinet, O. Francais, B. Le Pioufle, Micro-organism extraction from biological samples using DEP forces enhanced by osmotic shock. Lab Chip **13**(5), 901–909 (2013)
79. S. Shim, K. Stemke-Hale, A.M. Tsimberidou, J. Noshari, T.E. Anderson, P.R.C. Gascoyne, Antibody-independent isolation of circulating tumor cells by continuous-flow dielectrophoresis. Biomicrofluidics **7**(1), 11807 (2013)
80. P. Gascoyne, C. Mahidol, M. Ruchirawat, J. Satayavivad, P. Watcharasit, F.F. Becker, Microsample preparation by dielectrophoresis: isolation of malaria. Lab Chip **2**(2), 70–75 (2002)
81. S. Van Den Driesche, V. Rao, D. Puchberger-Enengl, W. Witarski, M.J. Vellekoop, Continuous cell from cell separation by traveling wave dielectrophoresis. Sens. Actuators B Chem. **170**, 207–214 (2012)
82. C.I. Trainito, O. Français, B. Le Pioufle, Monitoring the permeabilization of a single cell in a microfluidic device, through the estimation of its dielectric properties based on combined dielectrophoresis and electrorotation in situ experiments. Electrophoresis **36**(9–10), 1115–1122 (2015)
83. M. Washizu, T.B. Jones, Multipolar dielectrophoretic force calculation. J. Electrostat. **33**(2), 187–198 (1994)
84. T.B. Jones, Basic theory of dielectrophoresis and electrorotation. IEEE Eng. Med. Biol. Mag. **22**(6), 33–42 (2003)
85. L.F. Hartley, K.V.I.S. Kaler, R. Paul, Quadrupole levitation of microscopic dielectric particles. J. Electrostat. **46**(4), 233–246 (1999)
86. S.-I. Han, Y.-D. Joo, K.-H. Han, An electrorotation technique for measuring the dielectric properties of cells with simultaneous use of negative quadrupolar dielectrophoresis and electrorotation. Analyst **138**(5), 1529–1537 (2013)
87. C. Dalton, A.D. Goater, J. Drysdale, R. Pethig, Parasite viability by electrorotation. Colloids Surf. A Physicochem. Eng. Asp. **195**(1), 263–268 (2001)
88. R. Hölzel, Non-invasive determination of bacterial single cell properties by electrorotation. Biochim. Biophys. Acta (BBA)-Molecular Cell Res. **1450**(1), 53–60 (1999)
89. C.I. Trainito, E. Bayart, F. Subra, O. Français, B. Le Pioufle, The Electrorotation as a tool to monitor the dielectric properties of spheroid during the permeabilization. J. Membr. Biol., 1–8 (2016)
90. M. Kriegmaier, M. Zimmermann, K. Wolf, U. Zimmermann, V.L. Sukhorukov, Dielectric spectroscopy of Schizosaccharomyces pombe using electrorotation and electroorientation. Biochim. Biophys. Acta (BBA)-General Subj. **1568**(2), 135–146 (2001)
91. T. Müller, T. Schnelle, G. Fuhr, Dielectric single cell spectra in snow algae. Polar Biol. **20**(5), 303–310 (1998)
92. D. Bakewell, N. Vergara-Irigaray, D. Holmes, Dielectrophoresis of biomolecules. JSM Nanotechnol. Nanomedicine **1**(1003), 1–14 (2013)
93. M. Washizu, O. Kurosawa, Electrostatic manipulation of DNA in microfabricated structures. IEEE Trans. Ind. Appl. **26**(6), 1165–1172 (1990)
94. T. Kawabata, M. Washizu, Dielectrophoretic detection of molecular bindings. Ind. Appl. IEEE Trans. **37**(6), 1625–1633 (2001)
95. S. Tuukkanen, A. Kuzyk, J.J. Toppari, H. Häkkinen, V.P. Hytönen, E. Niskanen, M. Rinkiö, P. Törmä, Trapping of 27 bp-8 kbp DNA and immobilization of thiol-modified DNA using dielectrophoresis. Nanotechnology **18**(29), 295204 (2007)

96. A. Nakano, A. Ros, Protein dielectrophoresis: advances, challenges, and applications. Electrophoresis **34**(7), 1085–1096 (2013)
97. R.W. Clarke, S.S. White, D. Zhou, L. Ying, D. Klenerman, Trapping of proteins under physiological conditions in a nanopipette. Angew. Chemie Int. Ed. **44**(24), 3747–3750 (2005)
98. S. Otto, U. Kaletta, F.F. Bier, C. Wenger, R. Hölzel, Dielectrophoretic immobilisation of antibodies on microelectrode arrays. Lab Chip **14**(5), 998–1004 (2014)
99. M. Suzuki, T. Yasukawa, H. Shiku, T. Matsue, Negative dielectrophoretic patterning with colloidal particles and encapsulation into a hydrogel. Langmuir **23**(7), 4088–4094 (2007)
100. N. Liu, X. Cai, Y. Lei, Q. Zhang, M.B. Chan-Park, C. Li, W. Chen, A. Mulchandani, Single-walled carbon nanotube based real-time organophosphate detector. Electroanalysis **19**(5), 616–619 (2007)
101. M. Javanmard, S. Emaminejad, C. Gupta, J. Provine, R.W. Davis, R.T. Howe, Depletion of cells and abundant proteins from biological samples by enhanced dielectrophoresis. Sens. Actuators B Chem. **193**, 918–924 (2014)
102. U. Kim, H.T. Soh, Simultaneous sorting of multiple bacterial targets using integrated dielectrophoretic-magnetic activated cell sorter. Lab Chip **9**(16), 2313–2318 (2009)
103. I.-F. Cheng, T.-Y. Chen, W.-C. Chao, Increasing local density and purity of molecules/bacteria on a sensing surface from diluted blood using 3D hybrid electrokinetics. Biomicrofluidics **10**(3) (2016)
104. J. Oh, R. Hart, J. Capurro, H.M. Noh, Comprehensive analysis of particle motion under non-uniform AC electric fields in a microchannel. Lab Chip **9**(1), 62–78 (2009)
105. M.E. Sosa-Morales, L. Valerio-Junco, A. López-Malo, H.S. García, Dielectric properties of foods: reported data in the 21st century and their potential applications. LWT-Food Sci. Technol. **43**(8), 1169–1179 (2010)
106. S. Ryyniinen, The electromagnetic properties of food materials : a review of the basic principles **26**, 409–429 (1995)
107. M.T. Jilani, M. Zaka, A.M. Khan, M.T. Khan, S.M. Ali, A brief review of measuring techniques for characterization of dielectric materials. Int. J. Inf. Technol. Electr. Eng. **1**(1), 1–5 (2012)
108. C. Gabriel, S. Gabriel, E. Corthout, The dielectric properties of biological tissues: {I. Literature} survey. Phys. Med. Biol. **41**(11), 2231 (1996)
109. H. Lizhi, K. Toyoda, I. Ihara, Discrimination of olive oil adulterated with vegetable oils using dielectric spectroscopy. J. Food Eng. **96**(2), 167–171 (2010)
110. L. Ragni, P. Gradari, A. Berardinelli, A. Giunchi, A. Guarnieri, Predicting quality parameters of shell eggs using a simple technique based on the dielectric properties. Biosyst. Eng. **94**(2), 255–262 (2006)
111. W. Guo, X. Zhu, S.O. Nelson, R. Yue, H. Liu, Y. Liu, Maturity effects on dielectric properties of apples from 10 to 4500 MHz. LWT-Food Sci. Technol. **44**(1), 224–230 (2011)
112. K. Sacilik, A. Colak, Determination of dielectric properties of corn seeds from 1 to 100 MHz. Powder Technol. **203**(2), 365–370 (2010)
113. L. Ragni, A. Al-Shami, G. Mikhaylenko, J. Tang, Dielectric characterization of hen eggs during storage. J. Food Eng. **82**(4), 450–459 (2007)
114. M.S. Venkatesh, G.S.V. Raghavan, An overview of microwave processing and dielectric properties of agri-food materials. Biosyst. Eng. **88**(1), 1–18 (2004)
115. Y. Wang, T.D. Wig, J. Tang, L.M. Hallberg, Dielectric properties of foods relevant to RF and microwave pasteurization and sterilization. J. Food Eng. **57**(3), 257–268 (2003)
116. D. Haemmerich, D.J. Schutt, A.W. Wright, J.G. Webster, D.M. Mahvi, Electrical conductivity measurement of excised human metastatic liver tumours before and after thermal ablation. Physiol. Meas. **30**(5), 459–466 (2009)
117. K. Heileman, J. Daoud, M. Tabrizian, Dielectric spectroscopy as a viable biosensing tool for cell and tissue characterization and analysis. Biosens. Bioelectron. **49**, 348–359 (2013)
118. M. Brusson, J. Rossignol, S. Binczak, G. Laurent, B. de Fonseca, assessment of burn depths on organs by microwave. Procedia Eng. **87**, 308–311 (2014)

119. Y. Zou, Z. Guo, A review of electrical impedance techniques for breast cancer detection. Med. Eng. Phys. **25**(2), 79–90 (2003)
120. A.P. O'Rourke, M. Lazebnik, J.M. Bertram, M.C. Converse, S.C. Hagness, J.G. Webster, D. M. Mahvi, Dielectric properties of human normal, malignant and cirrhotic liver tissue: in vivo and ex vivo measurements from 0.5 to 20 GHz using a precision open-ended coaxial probe. Phys. Med. Biol. **52**(15), 4707–4719 (2007)
121. E. Marzec, K. Wachal, The electrical properties of leg skin in normal individuals and in patients with ischemia. Bioelectrochemistry Bioenerg. **49**(1), 73–75 (1999)
122. S. Ha, M.J. Hamamura, W.W. Roeck, J. Hugg, D.J. Wagenaar, D. Meier, B.E. Patt, O. Nalcioglu, Feasibility study of a unilateral RF array coil for MR-scintimammography. Phys. Med. Biol. **56**(21), 6809–6822 (2011)
123. S.C. Bürgel, C. Escobedo, N. Haandbæk, A. Hierlemann, Sensors and actuators B: chemical on-chip electroporation and impedance spectroscopy of single-cells. Sens. Actuators B. Chem. **210**, 82–90 (2015)
124. J. Cameron, Physical properties of tissue. A comprehensive reference book, edited by F.A. Duck. Med. Phys. **18**(4), 834 (1991)
125. M.A. Meyers, P.-Y. Chen, A.Y.-M. Lin, Y. Seki, Biological materials: structure and mechanical properties. Prog. Mater. Sci. **53**(1), 1–206 (2008)
126. G. Bao, S. Suresh, Cell and molecular mechanics of biological materials. Nat. Mater. **2**(11), 715–725 (2003)
127. L. Bernard, N. Burais, L. Nicolas, Numerical formulations to compute induced electromagnetic in the human body, in *2006 12th Biennial IEEE Conference on Electromagnetic Field Computation*, pp. 487–487
128. A. Vander Vorst, A. Rosen, Y. Kotsuka, *RF/Microwave Interaction with Biological Tissues* (Wiley, Hoboken, NJ, USA, 2005)
129. R. Pethig, D.B. Kells, The passive electrical properties of biological systems: their significance in physiology, biophysics and biotechnology **32**(8), 933–970 (1987)
130. R.M. Hill, L.A. Dissado, Debye and non-Debye relaxation. J. Phys. C Solid State Phys. **18** (19), 3829–3836 (2000)
131. S. Grimnes, O.G. Martinsen, *Bioimpedance & Bioelectricity Basics*, vol. XXXIII, no. 2 (2008)
132. V. Raicu, Dielectric dispersion of biological matter: model combining Debye-type and 'universal' responses. Phys. Rev. E. Stat. Phys. Plasmas. Fluids. Relat. Interdiscip. Topics **60** (4), 4677–4680 (1999)
133. O.G. Martinsen, S. Grimnes, H.P. Schwan, O. Rikshospitalet, Interface phenomena and dielectric properties of biological tissue. Encycl. Surf. Colloid Sci. (7), 2643–2652 (2002)
134. P.M. Buff, M.B. Steer, G. Lazzi, Cole-Cole dispersion models for aqueous gelatin-syrup dielectric composites. IEEE Trans. Geosci. Remote Sens. **44**(2), 351–355 (2006)
135. V. Raicu, N. Kitagawa, A. Irimajiri, A quantitative approach to the dielectric properties of the skin. Phys. (College. Park. Md). **45**(2), 1–4 (2000)
136. R.F. Harrington, *Time-Harmonic Electromagnetic Fields*, 2nd edn. (Institute of Electrical & Electronics Enginee, 2001)
137. D. Ireland, A. Abbosh, Modeling human head at microwave frequencies using optimized debye models and FDTD method. IEEE Trans. Antennas Propag. **61**(4), 2352–2355 (2013)
138. S. Gabriel, R.W. Lau, C. Gabriel, The dielectric properties of biological tissues: II. Measurements in the frequency range 10 Hz–20 GHz. Phys. Med. Biol. **41**(11), 2251–2269 (1996)
139. S. Gabriel, R.W. Lau, C. Gabriel, The dielectric properties of biological tissues: III. Parametric models for the dielectric spectrum of tissues. Phys. Med. Biol. **41**(11), 2271–2293 (1996)
140. D. Miklavcic, N. Pavselj, F.X. Hart, Electric properties of tissues, in *Wiley Encyclopedia of Biomedical Engineering* (Wiley, 2006), pp. 1–12
141. A.V. Korjenevsky, Maxwell-Wagner relaxation in electrical imaging. Physiol. Meas. **26**(2), S101-10 (2005)

142. T. Hanai, T. Imakita, N. Koizumi, Analysis of dielectric relaxations of w/o emulsions in the light of theories of interfacial polarization. Colloid Polym. Sci. **260**(11), 1029–1034 (1982)
143. A. Irimajiri, T. Suzaki, K. Asami, T. Hanai, Dielectric modeling of biological cells: models and algorithm. Bull. Inst. Chem. Res. Kyoto **69**(4), 421–438 (1991)
144. L. Nicolas, L. Pichon, Laurent Bernard Caractérisation électrique des tissus biologiques et calcul des phénomènes induits dans le corps humain par des champs électromagnétiques de fréquence inférieure au GHz (2007)
145. C.T. Ching, J. Chen, Sensors and actuators B: chemical a non-invasive, bioimpedance-based 2-dimensional imaging system for detection and localization of pathological epithelial tissues. Sens. Actuators B Chem. **206**, 319–326 (2015)
146. H.C. Lukaski, M. Moore, Bioelectrical impedance assessment of wound healing **6**(1), 209–212 (2012)
147. O. Français, B. Le Pioufle, C. Trainito, Etude et mise en oeuvre d'un microsystème fluidique pour la caractérisation diélectrique de cellules biologiques par électrorotation, in *Symposium de Génie Electrique*
148. U. Pliquett, M.R. Prausnitz, Electrical impedance spectroscopy for rapid and noninvasive analysis of skin electroporation, in *Electrochemotherapy, Electrogenetherapy, and Transdermal Drug Delivery: Electrically Mediated Delivery of Molecules to Cells*, ed. by M.J. Jaroszeski, R. Heller, R. Gilbert (Humana Press, Totowa, NJ, 2000), pp. 377–406
149. K.S. Cole, R.H. Cole, Dispersion and absorption in dielectrics I. Alternating current characteristics. J. Chem. Phys. **9**(4), 341 (1941)
150. D.A. Dean, T. Ramanathan, D. Machado, R. Sundarajan, Electrical impedance spectroscopy stydy of biological tissues. J. Electrostat. **66**, 165–177 (2008)
151. S.M. Hesabgar, A. Sadeghi-naini, G. Czarnota, A. Samani, Dielectric properties of the normal and malignant breast tissues in xenograft mice at low frequencies (100 Hz–1 MHz). Measurement **105**, 56–65 (2017)
152. V. Cherepenin, A. Karpov, A. Korjenevsky, V. Kornienko, A 3D electrical impedance tomography (EIT) system **22**, 9–18 (2001)
153. S.L. Swisher, M.C. Lin, A. Liao, E.J. Leeflang, Y. Khan, F.J. Pavinatto, K. Mann, A. Naujokas, D. Young, S. Roy, M.R. Harrison, A.C. Arias, V. Subramanian, M.M. Maharbiz, Impedance sensing device enables early detection of pressure ulcers in vivo. Nat Commun. **6** (Mar 2015)
154. R.E. Fernandez, E. Lebiga, A. Koklu, A.C. Sabuncu, A. Beskok, Flexible bioimpedance sensor for label-free detection of cell viability and biomass flexible bioimpedance sensor for label-free detection of cell viability and biomass (Sep 2015)
155. S. Huclova, D. Erni, J. Fröhlich, Modelling and validation of dielectric properties of human skin in the MHz region focusing on skin layer morphology and material composition. J. Phys. D. Appl. Phys. **45**(2), 025301 (2011)
156. A.P. Gregory, R.N. Clarke, A review of RF and microwave techniques for dielectric measurements on polar liquids. IEEE Trans. Dielectr. Electr. Insul. **13**(4), 727–743 (2006)
157. D. Popovic, L. McCartney, C. Beasley, M. Lazebnik, M. Okoniewski, S.C. Hagness, J.H. Booske, Precision open-ended coaxial probes for in vivo and ex vivo dielectric spectroscopy of biological tissues at microwave frequencies. IEEE Trans. Microw. Theory Tech. **53**(5), 1713–1721 (2005)
158. C. Gabriel, C. Gabriel, S. Gabriel, S. Gabriel, E. Corthout, E. Corthout, The dielectric properties of biological tissues: I. Literature survey. Phys. Med. Biol. **41**(11), 2231–2249 (1996)
159. A. Caduff, M. Mueller, A. Megej, F. Dewarrat, R.E. Suri, J. Klisic, M. Donath, P. Zakharov, D. Schaub, W.A. Stahel, M.S. Talary, Characteristics of a multisensor system for non invasive glucose monitoring with external validation and prospective evaluation. Biosens. Bioelectron. **26**(9), 3794–3800 (2011)

160. J. Leroy, C. Dalmay, A. Landoulsi, F. Hjeij, C. Mélin, B. Bessette, C. Bounaix Morand Du Puch, S. Giraud, C. Lautrette, S. Battu, F. Lalloué, M.O. Jauberteau, A. Bessaudou, P. Blondy, A. Pothier, Microfluidic biosensors for microwave dielectric spectroscopy. Sens. Actuators, A Phys. **229**, 172–181 (2015)
161. K. Grenier, D. Dubuc, P. Poleni, M. Kumemura, Integrated broadband microwave and microfluidic sensor dedicated to bioengineering **57**(12), 3246–3253 (2009)
162. L.Y. Zhang, C. Bounaix Morand Du Puch, C. Dalmay, A. Lacroix, A. Landoulsi, J. Leroy, C. Mélin, F. Lalloué, S. Battu, C. Lautrette, S. Giraud, A. Bessaudou, P. Blondy, M.O. Jauberteau, A. Pothier, Discrimination of colorectal cancer cell lines using microwave biosensors. Sens. Actuators A Phys. **216**, 405–416 (2014)
163. L.-F. Chen, C.K. Ong, C.P. Neo, V.V. Varadan, V.K. Varadan, *Microwave Electronics: Measurement and Materials Characterization* (2004)
164. J. Heller, J.R. Feldkamp, Auto-tuned induction coil conductivity sensor for in-vivo human tissue measurements. Meas. Sci. Rev. **9**(6), 162–168 (2009)
165. A. Barai, S. Watson, H. Griffiths, R. Patz, Magnetic induction spectroscopy: non-contact measurement of the electrical conductivity spectra of biological samples. Meas. Sci. Technol. **23**(8), 085501 (2012)
166. S.C. Hagness, A. Taflove, J.E. Bridges, Two-dimensional FDTD analysis of a pulsed microwave confocal system for breast cancer detection: fixed-focus and antenna-array sensors. IEEE Trans. Biomed. Eng. **45**(12), 1470–1474 (1998)
167. X. Li, S.C. Hagness, B.D. Van Veen, D. Van Der Weide, Space-time beamforming for breast cancer detection B, 2–5 (2003)
168. E.G. Fear, J.M. Sill, Preliminary investigations of tissue sensing adaptive radar for breast tumor detection, in *Engineering in Medicine and Biology Society, 2003. Proceedings of the 25th Annual International Conference of the IEEE*, vol. 4, pp. 3787–3790 (2003)
169. H. Song, S. Sasada, T. Kadoya, M. Okada, K. Arihiro, X. Xiao, T. Kikkawa, Detectability of breast tumor by a hand-held impulse-radar detector: performance evaluation and pilot clinical study. Sci. Rep **7**(1) (2017)
170. H. Bahramiabarghouei, S. Member, E. Porter, S. Member, A. Santorelli, S. Member, B. Gosselin, M. Popovi, L.A. Rusch, Flexible 16 antenna array for microwave breast cancer detection **62**(10), 2516–2525 (2015)
171. G. Masilamany, P.-Y. Joubert, S. Serfaty, B. Roucaries, Y. Le Diraison, Radiofrequency inductive probe for non-contact dielectric characterizations of organic medium. Electron. Lett. **50**(7) (2014)
172. S. Serfaty, N. Haziza, L. Darrasse, S. Kan, Multi-turn split-conductor transmission-line resonators. Magn. Reson. Med. **38**(4), 687–689 (1997)
173. T.H.N. Dinh, M. Wang, S. Serfaty, D. Placko, P.-Y. Joubert, Non contact estimation of the dielectric properties of organic material using an inductive RF sensor and a multifrequency approach, in *Proceedings of the IEEE Conference on Cosmetic Measurements and Testing, June 6–7, Cergy Pontoise, France*, no. 1, pp. 1–4 (2017)
174. T.H.N. Dinh, S. Serfaty, P.-Y. Joubert, Radiofrequency MTLR sensor for the non-contact dielectric monitoring of milk solutions during fermentation process, in *Proceedings of the 18th International Symposium on Applied Electromagnetics and Mechanics (ISEM) 2017*, pp. 1–2 (2017)
175. T.H.N. Dinh, M. Wang, S. Serfaty, P.Y. Joubert, Contactless radio frequency monitoring of dielectric properties of egg white during gelation. IEEE Trans. Magn. **53**(4), 1–7 (2017)
176. T.H.N. Dinh, E. Martincic, S. Serfaty, P.-Y. Joubert, Monitoring of yogurt formation using a contactless RF dielectric sensor, in *IEEE SENSORS 2016*, pp. 1–3 (2016)

Chapter 7
Magnetophoresis in Bio-Devices

Anne-Laure Deman and Damien Le Roy

Abstract Associating magnetism to microfluidics is a powerful approach to address challenges in biomedical applications. Indeed, due to the versatility of this approach, it can be exploited in applications as diverse as blood fractionation and Circulating Tumor Cell separation and detection. The separation and manipulation of sub-mm particles, such as magnetically labelled biological cells, magnetic micro- and nano-particles, are achieved thanks to magnetophoresis forces arising from magnetic flux gradients, engineered inside of microfluidics devices. In this chapter, basic concepts to understand physical phenomena, to design and optimize magnetic bio-micro-devices are presented. Finally, few examples of such devices are given to illustrate the potential of this approach.

Keywords Magnetism theory · Magnetophoresis · Magnetic field gradient · Micromagnet · Microcoil · Micro-concentrators · Ferromagnetic micro-patterns · Superparamagnetic particles · Separation and manipulation · Trapping · Immuno-assays · Magnetic labelling

7.1 Introduction

In microdevices, magnetism can be used to realize valves, to achieve pumping functions, or the manipulation of micro- or nano-objects inside microfluidic channels. The objects responding to the magnetic field could be either particles or cells, using extrinsic or native magnetic properties. *Magnetophoresis*, which refers

A.-L. Deman (✉)
Institut des Nanotechnologies de Lyon (INL), CNRS, UMR 5270,
Université de Lyon 1, Université de Lyon, Villeurbanne 69622, France
e-mail: anne-laure.deman-him@univ-lyon1.fr

D. Le Roy
Campus LyonTech La Doua, Université Lyon 1 et Institut Lumière Matière,
10, Rue Ada Byron, Villeurbanne 69622, France
e-mail: damien.le-roy@univ-lyon1.fr

G. Barbillon et al. (eds.), *Engineering of Micro/Nano Biosystems*,
Microtechnology and MEMS, https://doi.org/10.1007/978-981-13-6549-2_7

to the motion of an object in a non-uniform magnetic field, permits to address numerous biomedical issues. Indeed, micro-and nano-particles can be functionalized by multiple coating, proteins, or molecules offering multifunctionality, and magnetic labelling of cells can be obtained using specific cell biomarkers. By mastering microfluidic designs, magnetic field gradients can be patterned to optimize targeting and controllability. Moreover, magnetic interactions offer contactless manipulation, thus preserving biological entities, and do not depend, or in a negligible way, on the pH or the conductivity of the medium, and can be designed to operate over a wide range of temperature. In addition, congestion surrounding microfluidic device is limited, as the magnetic approach can be implemented with permanent magnets, and so do not necessarily require external source of energy. In this chapter, we will first underline some magnetism principles that serve to describe magnetophoresis. We will then highlight all forces involved in microfluidic devices. In a third part, we will focus on the specificities that come with the implementation of magnetophoresis in microdevices, in particular the fabrication and integration of magnetic micropatterns. We will finish this chapter with some examples of bio-analysis performed using magnetophoretic devices.

7.2 Magnetism Principles for Magnetophoresis

To start addressing the question of *magnetophoresis*, which refers to the motion of an object in a non-uniform magnetic field, we will briefly recall some magnetostatic principles and apply them to the use of magnetic forces in microsystems. We will place the discussion in the context of an interaction between three elements: (1) a point-like magnetic dipole that is the object to manipulate, (2) a source of local variation of magnetic field that is the remote and (3) and an external magnetic field to assist the remote control. We focus our attention on static and quasi-static magnetic interactions, thus considering that the magnetic moment of the target object is aligned with the magnetic field it is submitted to.

7.2.1 Origin of the Magnetophoretic Force: A Magnetic Particle in an External Magnetic Field

The magnetic particle is described here by its magnetic moment $\vec{m}$. According to Ampère, a magnetic moment is equivalent to a tiny current loop. If the circulating current is I (in A), and if $\vec{S}$ is the oriented surface, then:

$$\vec{m} = I\vec{S} \tag{7.1}$$

provided that the current flows in a plane. The direction of $\vec{m}$ is given by the right-hand corkscrew rule.

To discuss the interaction between a magnetic moment and an external magnetic field, it is then convenient to describe the magnetic moment as its equivalent current loop and to consider the Laplace force $\overrightarrow{F_L}$ acting on it when submitted to a magnetic induction $\vec{B}$. The expression of $\overrightarrow{F_L}$ acting on a current loop of length ℓ, with circulating I and submitted to a magnetic induction $\vec{B}$ (in free space: $\vec{B} = \mu_0 \vec{H}$, and more generally one writes: $\vec{B} = \mu_r \mu_0 \vec{H}$, μ_r the relative dimensionless permeability of the medium), is:

$$\overrightarrow{F_L} = I \oint \overrightarrow{d\ell} \times \vec{B} \quad (7.2)$$

By calculating the moment of the Laplace force, one can show that a magnetic field $\vec{H}$, creates a torque $\vec{\Gamma}$ on a magnetic moment $\vec{m}$, expressed as:

$$\vec{\Gamma} = \vec{m} \times \vec{B} \quad (7.3)$$

This first result shows that when submitted to an external field, a magnetic moment will tend to align parallel to the field.

The '*potential energy*' of a magnetic moment in a field, also known as the *Zeeman energy*, E_Z, that is, apart from a constant:

$$E_Z = -\vec{m} \cdot \vec{B} \quad (7.4)$$

When a magnetic moment is in a uniform external field, there is a torque but no translational force. In contrast, if the external field is not uniform, the potential energy will depend on the position and this will lead to a net force that can be expressed as:

$$\vec{F} = -\vec{\nabla} \mathrm{E} = \vec{\nabla}\left(\vec{m} \cdot \vec{B}\right) = \mu_r \mu_0 \vec{\nabla}\left(\vec{m} \cdot \vec{H}\right) \quad (7.5)$$

This develops in:

$$\vec{F} = \mu_r \mu_0 \left(\left(\vec{m} \cdot \vec{\nabla}\right)\vec{H} + \left(\vec{H} \cdot \vec{\nabla}\right)\vec{m} + \vec{m} \times \left(\vec{\nabla} \times \vec{H}\right) + \vec{H} \times \left(\vec{\nabla} \times \vec{m}\right)\right) \quad (7.6)$$

As there is no induced current in the particles, the third and fourth terms vanish (from Ampere's law) and assuming that the particle as a constant moment (which is

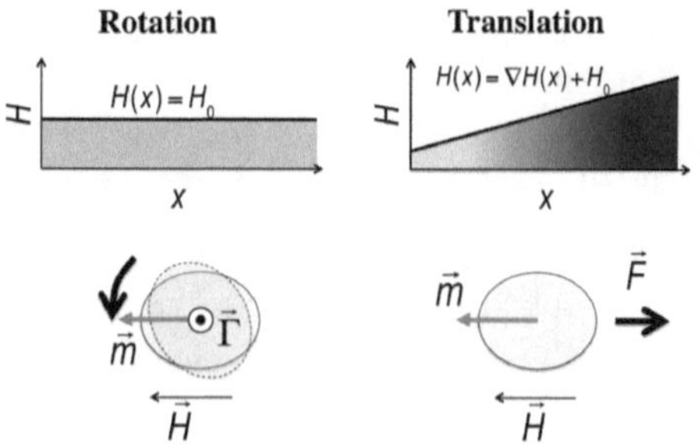

Fig. 7.1 Action of a magnetic field on a magnetic particle

the case for superparamagnetic particles), the second term can be neglected, which reduces (7.6) to:

$$\overrightarrow{F} = \mu_0(\overrightarrow{m} \cdot \overrightarrow{\nabla})\overrightarrow{H} \tag{7.7}$$

This expression of the translational force shows that (i) it is non-zero only in non-uniform magnetic field, and (ii) if the magnetic moment is parallel (/antiparallel) to the magnetic field then it will be attracted to the region of maximum (/minimum) field, as illustrated in Fig. 7.1.

7.2.2 Role of the Magnetic Nature of the Object to Manipulate

Magnetophoresis is implemented in microfluidic systems to manipulate, sort, and trap magnetic micro- or nano-objects, mainly. These objects are mainly: magnetic micro- or nano-particles [1–3], magnetically labeled cells [4–6] or natively paramagnetic cells such as deoxygenated Red Blood Cells (RBC) [7–9]. The magnetic susceptibility establishes the relationship between H and the magnetization M (the volume fraction of m), as $M = \chi H$. Based on this, micro- and nano-objects can be divided into four different categories: diamagnetic, paramagnetic, ferromagnetic, or superparamagnetic substances. The magnetic response of those materials to an applied field is schematized in Fig. 7.2.

7.2.2.1 Diamagnets

A diamagnet is a substance composed of atoms that do not have any net magnetic moment. This is usually the case of atoms with closed-shell electronic structures, either monoatomic rare gases, model ionic solids like NaCl, or covalent solids like diamond, silicon or germanium. Superconductors are perfect diamagnets, with a magnetic susceptibility of −1. When subjected to an applied magnetic field, the diamagnetic substance develops a negative moment, proportional to the field.

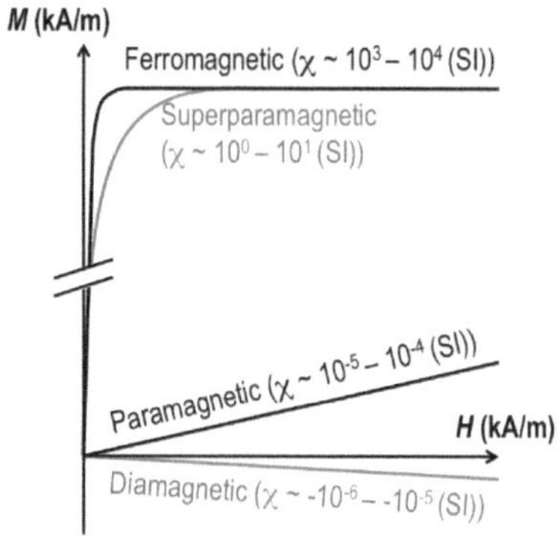

Fig. 7.2 Magnetic response of substances to a magnetic field as a function of their magnetic nature

The theory of diamagnetism, initiated by Paul Langevin in early 20th century, considers that the applied field reduces the effective current of the electron orbits, which results in a magnetic moment that is opposite to the applied field.

In diamagnetic substances, the susceptibility does not vary with temperature (values of dimensionless volumetric magnetic susceptibility χ (SI) of various diamagnetic substances, fluids and solids, are reported in Table 7.1). Note that a superconductor is a perfect diamagnet with a susceptibility of -1. According to the above discussion, the experienced magnetic force thus scales with $-\nabla H^2$. Diamagnetic substances are therefore repelled from the regions of maximum magnetic field.

In general, organic matter is diamagnetic, and animal bodies have therefore an effective diamagnetic behavior. If the repelling force compensates for gravity, levitation can happen, as demonstrated in the experiment of A. Geim and M. Berry who made a frog fly over a superconducting coil. They received for it the Ig Nobel price in 2000.

Table 7.1 Dimensionless susceptibility χ of some substances at room temperature (Unit: 10^{-6} [SI]) [8–12]

Substance	χ (unit 10^{-6} [SI])
Water	−9.03
Bovine Serum Albumin (BSA)	−10.38
Blood plasma	−7.7
Ethanol	−7.28
Acetone	−5.78
WBC	−0.13
RBC in air (in H_2O)—oxyHb	−9.21 (−0.18)
RBC in air (in H_2O)—deoxyHb	−5.70 (3.33)
RBC in air (in H_2O)—met	−5.25 (3.78)
Cu	−22
Bi	−166
HOPG (//)	−450

Although diamagnets are generally considered insensitive to magnetic fields gradients as their intrinsic magnetic susceptibility are the smallest of all materials, significant response can be obtained by adjusting the magnetic susceptibility contrast between the diamagnetic object and its surrounding medium. For instance, in microfluidic devices, the medium can be doped with paramagnetic substances.

7.2.2.2 Paramagnets

In the same report on the theory of diamagnetism published in 1905, Paul Langevin addressed the dependence of paramagnet susceptibilities on the temperature (T) in $1/T$, measured by Pierre Curie in the late 19th century. Its generalization led to the so-called Curie-Weiss equation that establishes the relation between the magnetic susceptibility, χ, and the temperature as:

$$\chi = \frac{C'}{T - \theta} \tag{7.8}$$

with C' a constant, and θ a parameter that depends on the material.

In his simple description, Langevin assumed that the atoms in the material exhibit a net magnetic moment, which results from the fact that all spins and orbital moments of an atom do not cancel out. The atomic magnetic moments, hold by the unpaired electrons, do not interact with each other. In the absence of external magnetic field, the moments point in random directions, globally compensating for each other. Therefore, there is no net magnetization. When a magnetic field is applied, the moments tend to align with the field, but the thermal fluctuation impedes the alignment. Therefore, paramagnet present relatively small magnetic susceptibility that decays with the increase of temperature. From this description, if μ_{at} represents the atomic moment (here we assume they are all the same), the magnetization M of the paramagnet is given by the so-called Langevin function that is:

$$M = M_0\left[coth\left(\frac{\mu_0\mu_{at}H}{k_BT}\right) - \frac{k_BT}{\mu_0\mu_{at}H}\right] \tag{7.9}$$

where M_0 is the maximum magnetization that the paramagnet can reach (all the magnetic moments are aligned with the applied field).

M_0 is the product of the volume density of atoms n (in m^{-3}) and the atomic moment $\mu_{at}(M_0 = n\mu_{at})$, H the applied field, k_B the Boltzmann constant.

When expressed as a series, the Langevin function first term leads to:

$$M = M_0\frac{\mu_0\mu_{at}H}{3k_BT} \tag{7.10}$$

Table 7.2 Dimensionless susceptibility χ of some substances at room temperature (Unit: 10^{-6} [SI]) [13, 14]

Substance	χ (Unit: 10^{-6} [SI])
Al	20
Ca	22
Mg	12
Pd	805
Gd	476300

So the magnetic susceptibility $\chi = \frac{M}{H}$ is:

$$\chi = M_0 \frac{\mu_0 \mu_{at}}{3k_B T} \tag{7.11}$$

Table 7.2 gives the magnetic susceptibility of some paramagnets.

Considering that μ_{at} is only few μ_B (μ_B the Bohr magneton, 9.274 10^{-24} A·m^2), this result shows that a huge value of magnetic field, of the order of 10^5–10^6 kA·m^{-1} would be needed to reach M_0, which is possible experimentally.

7.2.2.3 Ferromagnets and Ferrimagnets

To describe ferromagnets and ferrimagnets, Weiss introduced in 1907 the notion of *molecular field,* which accounts for interactions between all individual atomic moments, and is proportional to the magnetization. This field does not exist in reality but it is a convenient way to approximate the effect of interactions between atomic moments, which is described in quantum mechanics by the Heisenberg Hamiltonian. In the case of ferromagnets, two adjacent magnetic moments are coupled parallel while in ferrimagnets there exists two sublattices holding different magnetic moments and coupled antiparallel. The strength of the interaction between two adjacent moments is characterized by the exchange stiffness A. Values of A for some ferromagnetic substances are reported in Table 7.3. The characteristic of ferromagnets (and ferrimagnets) is its spontaneous magnetization M_S, which represents the alignment of magnetic moments located on a crystal lattice. In the absence of magnetic field, the moments tend to align on preferential directions. The associated magnetic energy landscape that determines the orientation of the magnetization in the absence of any applied magnetic field is the magnetic anisotropy energy. Above a critical temperature that is named the Curie point, the magnetic order is abruptly reduced and the substance becomes a paramagnet. The Curie temperature can be as high as 1400 K (cobalt). It scales with the abovementioned exchange stiffness A. Apart from the spontaneous magnetization, ferromagnets are characterized by their structure in magnetic domains, already suggested by Weiss as an explanation for the absence of any remanent magnetization in a large fraction of ferromagnetic substances like iron. The domain structure results from a compromise between the exchange interactions between adjacent atomic moments that tends to

Table 7.3 Characteristic parameters of some ferromagnets [13, 15, 16]

Material	M_S (kA/m)	K_1 (kJ/m^3)	A (pJ/m)
Fe	1710	48	21
Co	1440	410	31
Ni	488	−5	8
$Ni_{80}Fe_{20}$	840	0.15	10
Fe_3O_4 (magnetite)	480	−13	7
γ-Fe_2O_3 (maghemite)	400	4.6	(10)
CoPt	810	4900	10
$SmCo_5$	860	17200	12
$Nd_2Fe_{14}B$	1280	4900	8

align them collinearly, the minimization of a self-energy term, also referred as magnetostatic energy, due to the dipole field that is created by the magnetization, and the anisotropy. Inside a ferromagnetic (or a ferrimagnetic) substance, the dipole field is named demagnetizing field, as it points in the opposite direction with respect to the magnetization.

In magnetically-hard compounds, the magnetic anisotropy energy is large enough to dominates over the magnetostatic energy term and the magnetic domains can remain significantly uncompensated when the applied field is removed. The net magnetization when the field is brought to zero is called the remanent magnetization (M_R). The applied field needed to bring the net magnetization to zero is called the coercive field. In contrast, the magnetic domains are compensated in magnetically-soft substances in which the anisotropy term is relatively small. Figure 7.3 shows characteristic magnetization curves for hard and soft ferromagnets. M_R and H_C are the remanent magnetization and the coercive field, respectively.

In most cases, the anisotropy originates from the crystalline structure. A high degree of symmetry, like in cubic lattices, leads to magnetic softness while a lower degree of symmetry, like hexagonal and tetragonal lattices leads to magnetic

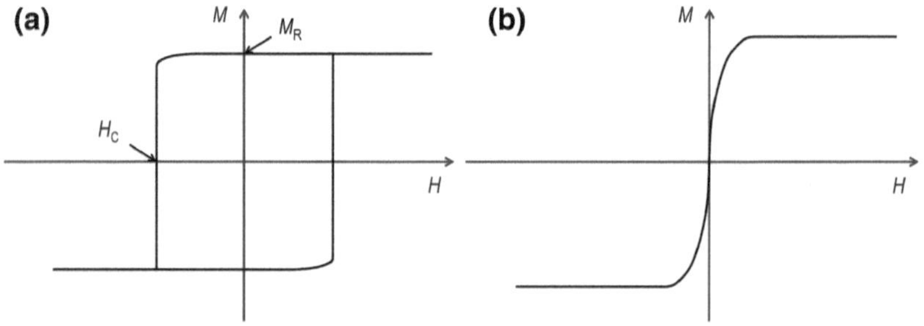

Fig. 7.3 Characteristic magnetization curves for hard **a** and soft **b** ferromagnets

hardness. The uniaxial term of the anisotropy energy density can be written as $K_1 \sin^2\alpha$ with α the angle between the magnetization and its easy axis and K_1 the uniaxial anisotropy constant (in J/m^3) (see Table 7.3).

The magnetic response of soft ferromagnets to an applied field depends on its shape, through its demagnetizing tensor. In general, the demagnetizing field is not uniform within the volume of the ferromagnetic substance, except in ellipsoid shapes. Approximating the shape to an ellipsoid permits to predict the magnetic response to an applied field as the susceptibility is given by the simple following expression:

$$\chi = \frac{\chi_0}{1 + N\chi_0} \tag{7.12}$$

where N is a shape-dependent parameter called the demagnetizing factor (with values between 0 and 1), χ_0 the intrinsic magnetic susceptibility (related to the relative dimensionless magnetic permeability of the material, μ_r, by $\chi_0 = \mu_r - 1$).

The way to estimate the demagnetizing factor is discussed in the following (Sect. 7.3.3) in the context of the integration of magnetic flux concentrators.

In the case of objects of spherical shape, like for magnetic particles, the demagnetizing factor is equal to 1/3.

7.2.2.4 Superparamagnetism

The magnetic structure of nano-sized ferromagnets is usually single domain. Below a certain critical size, the energy cost of creating a domain wall is higher than the cost of magnetostatic energy. A well-established picture considers that all the magnetic moments of the object rotate coherently, and so add up to form giant magnetic moments, called macrospin. This model, known as the Stoner-Wohlfarth model, was proposed by two theoretical physicists, Edmund Clifton Stoner and Erich Peter Wohlfarth, in 1948. Note that if this model can well describe the magnetization reversal in the smallest size, it is often observed that before the magnetic structure breaks into multi-domain, the system loses the coherent rotation leading to other mechanism like curling reversal. It is also often observed that the anisotropy has a dominant uniaxial symmetry term, which can be originated from slightly elongated shape. Thus, a simple but commonly encountered model used to describe nano-sized ferromagnet is a macrospin with its magnetic anisotropy energy having two minima at 180° from each other. This is represented in Fig. 7.4.

The anisotropy energy barrier separating two minima can be expressed as the product of an effective anisotropy constant K_{eff} (in J/m^3) and the volume of the object V. In the bulk, the anisotropy constant K_1 varies over a wide range, it is of the order of 1–10 kJ/m^3 for relatively soft ferromagnets and 10^3–10^4 kJ/m^3 for hard ferromagnets. When a field is applied, the energy barrier becomes asymmetric, one of the two energy wells is the favored compared to the other and thus the energy

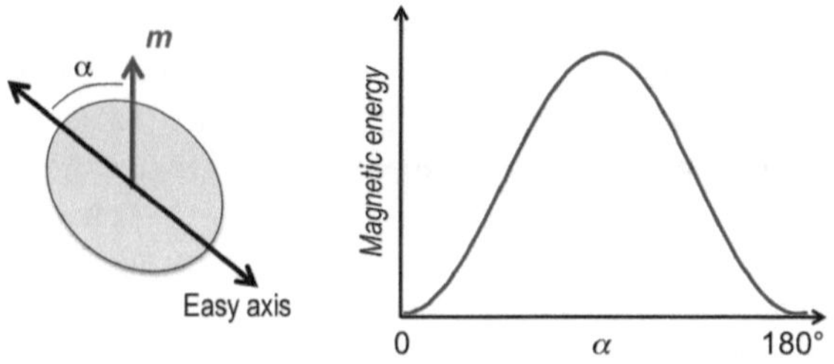

Fig. 7.4 Magnetic anisotropy energy profile for a macrospin with uniaxial anisotropy, as a function of the angle that the macrospin makes with the easy axis, denoted α

barrier is reduced. Néel proposed that the macrospin would flip its orientation at a time that follows an Arrhenius law:

$$\tau = \tau_0 exp\left(\frac{K_{eff} V}{k_B T}\right) \tag{7.13}$$

$1/\tau_0$ is the attempt frequency, typically of the order of 1 GHz, which is the ferromagnetic resonance frequency in the demagnetizing field. There is a continuous variation of the relaxation time with temperature. Assuming a measurement time of about 100 s in magnetometers, a commonly used figure of merit is the so called blocking temperature, that corresponds to the temperature for which: $K_{\mathrm{eff}} V = 25\ \mathrm{k_B} T$. Figure 7.5 displays the temperature dependence of the relaxation times for Fe nanoparticles.

Table 7.4 gives the uniaxial term of the anisotropy for benchmark ferromagnets and some critical sizes: D_{coh} is the maximum particle diameter for coherent rotation and D_{sp} is the maximum diameter for «stable» magnetization direction.

l_{ex} reflects the compromise between the exchange energy that align two adjacent magnetic moments and the magnetostatic energy.

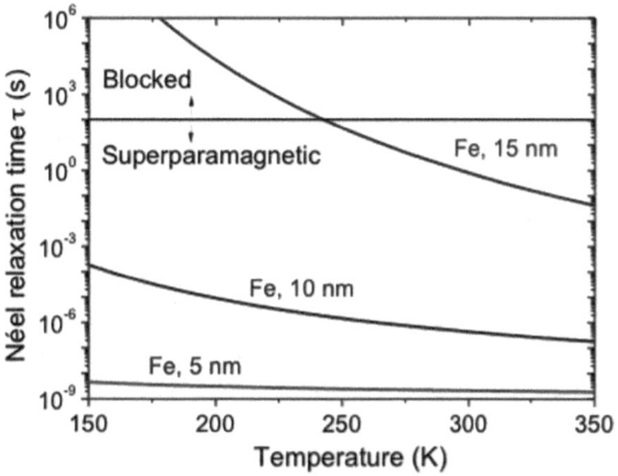

Fig. 7.5 Temperature dependence of the Néel relaxation time of Fe nanoparticles with diameters of 5, 10, and 15 nm

Table 7.4 Characteristic parameters and lengths in some ferromagnets [13]

Material	M_S (kA/m)	K_1 (kJ/m^3)	l_{ex} (nm) $l_{ex} = \sqrt{\frac{A}{\mu_0 M_S^2}}$	D_{coh} (nm) $D_{coh} = 2\sqrt{24} l_{ex}$	D_{sp} (nm) $D_{sp} \approx \sqrt{\frac{50 k_B T}{K_1}}$
Fe	1710	48	2.4	24	16
Co	1440	410	3.4	34	8
Ni	488	−5	5.1	50	34
Fe_3O_4	480	−13	4.9	48	26
CoPt	810	4900	3.5	34	3.4
$SmCo_5$	860	17200	3.6	36	2.2
$Nd_2Fe_{14}B$	1280	4900	1.9	19.4	3.4

Far from the blocking state (i.e. for measuring time greater than the Néel relaxation time), an ensemble of non-interacting macrospins in the superparamagnetic state responds to an applied field as a paramagnet with giant elementary moments, typically in the order of 10^2 to 10^4 μ_B, while it is of the order of few μ_B in paramagnet. As a consequence, there is no remanence magnetization and the magnetization curve is similar to the one reported in Fig. 7.3b. Therefore, one can describe the system magnetization with a Langevin function of the form:

$$M = M_0 \left[\coth\left(\frac{M_S V_P H}{k_B T} \right) - \frac{k_B T}{M_S V_P H} \right] \tag{7.14}$$

where V_P is the volume of the particle, M_S its spontaneous magnetization and M_0 the magnetization of the particle ensemble.

At relative low field, the magnetization of the ensemble varies linearly with the applied field, and the susceptibility is:

$$\chi = \frac{\mu_0 n (M_S V_P)^2}{3 k_B T} \tag{7.15}$$

where n is the volume density of particles.

As compared to paramagnetic substances, the field needed to reach saturation is relatively low, of the order of 10^1–10^2 kA/m. When the applied field is greater than this critical value, the magnetization of the particle saturates and the susceptibility can then be expressed as:

$$\chi = \frac{n M_S V_P}{H} \tag{7.16}$$

Most of the objects that are manipulated in microfluidic devices are superparamagnetic. They can be either individual nanoparticles or composites particles that are composed of nanoparticles embedded in a non-magnetic matrix.

7.3 Magnetic Micro- or Nano-Object Transport in Magnetophoretic System

In this section, we describe forces experienced by micro- or nano-object in microfluidic channel, during magnetophoresis experiments. For the sake of readability, we will describe forces in presence, one by one, considering a model magnetic particle (i.e. homogeneous and isotropic shape and magnetic properties).

Magnetic particles are submitted to various forces when flowing in a micro-channel [17–19]: (1) magnetophoretic force due to magnetic field gradients $\overrightarrow{F_{\mathrm{m}}}$, (2) fluidic drag force $\overrightarrow{F_{\mathbf{d}}}$, (3) gravitational force $\overrightarrow{F_{\mathrm{g}}}$, (4) buoyance forces $\overrightarrow{F_{\mathrm{b}}}$, (5) thermal kinetic energy (Brownian motion), and finally (6) forces resulting from inter-particle interactions, interactions between particles and micro-channel walls and interactions between particles and fluid. Analyzing the forces balance is required to anticipate particles motion in micro-channels in order to design sorting and trapping magnetic functions.

7.3.1 *Magnetophoretic Force*

Considering a particle of volume V_p and magnetization M_p, the magnetophoretic force in free space, can be expressed as follow based on Eq. (7.7):

$$\overrightarrow{F_{\mathrm{m}}} = \mu_0\left(\vec{m}\cdot\vec{\nabla}\right)\vec{H} = \mu_0 V_p\left(\vec{M}_p\cdot\vec{\nabla}\right)\vec{H} \tag{7.17}$$

The particle magnetization is expressed as:

$$\vec{M}_p = f(H)\vec{H} \tag{7.18}$$

where $f(H)$ is a function of which expression depends on whether the particle magnetization is saturated or not, i.e. on the magnitude of the magnetic field.

At low magnetic field, the particle magnetization is not saturated and the magnetization varies linearly with the applied field and so $f(H) = \chi_m$, χ_m being the measured magnetic susceptibility.

As mentioned in the section introducing ferromagnets, the internal magnetic field in magnetic objects, $\overrightarrow{H_{\mathrm{int}}}$, differs from the external field, $\vec{H}$, by a quantity that scales with the magnetization. Thus, at a given position in the volume: $\overrightarrow{H_{\mathrm{int}}} = \vec{H} + \overrightarrow{H_{\mathrm{d}}}$, with $\overrightarrow{H_{\mathrm{d}}}$ the so-called demagnetizing field, that is opposed to the magnetization that creates it, as schematized in Fig. 7.6. Therefore, one should distinguish the intrinsic susceptibility $\chi_0(\vec{M}_p = \chi_0\overrightarrow{H_{\mathrm{int}}})$ from the measured susceptibility, $\chi_m(\vec{M}_p = \chi_m\vec{H})$.

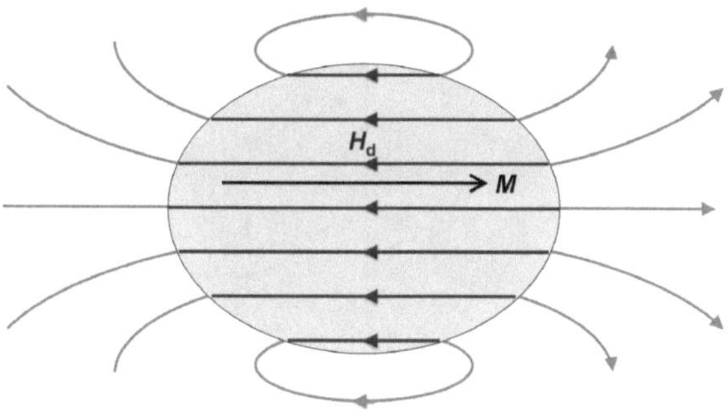

Fig. 7.6 Schematics of the field created inside and outside a uniformly magnetized ellipsoid object. It is generally referred inside as the demagnetizing field, and outside as the dipolar field

As a result, the magnetization can be expressed as:

$$\vec{M}_P = \chi_m \vec{H} = \frac{\chi_0}{1 + N\chi_0} \vec{H} \tag{7.19}$$

For spherical particles where $N = 1/3$, Eq. (7.19) becomes:

$$\vec{M}_P = \chi_m \vec{H} = \frac{3\chi_0}{3 + \chi_0} \vec{H} \tag{7.20}$$

In a microfluidic device, the micro-particle is immersed in a fluid of magnetic susceptibility χ_f, and magnetic permeability μ_f. As described by Furlani et al. [20] the expression of the force on a magnetized particle is then:

$$\vec{F}_m = \mu_f V_p \frac{3(\chi_0 - \chi_f)}{(\chi_0 - \chi_f) + 3(1 + \chi_f)} \left(\vec{H} \cdot \vec{\nabla}\right) \vec{H} \tag{7.21}$$

For large magnetic fields, the particle magnetization is saturated (all atomic moments being aligned along the magnetic field) and $\vec{M}_p = \vec{M}_{sp}$, $\left|\vec{M}_{sp}\right| = M_{sp}$ being the magnetization at saturation.

As a consequence, depending on the magnetic field, when $\mu_f \approx \mu_0$ ($|\chi_f| \ll 1$), $\overrightarrow{F_m}$ can be expressed as follow:

$$\vec{F}_m = \begin{cases} \mu_0 V_p \frac{3(\chi_0 - \chi_f)}{(\chi_0 - \chi_f) + 3} \left(\vec{H} \cdot \vec{\nabla}\right) \vec{H}, & H < \left(\frac{(\chi_0 - \chi_f) + 3}{3(\chi_0 - \chi_f)}\right) M_{sp}, \\ \mu_0 V_p \left(\vec{M}_{sp} \cdot \vec{\nabla}\right) \vec{H}, & H \geq \left(\frac{(\chi_0 - \chi_f) + 3}{3(\chi_0 - \chi_f)}\right) M_{sp}, \end{cases} \tag{7.22}$$

For micro- or- nano-particles, depending on their magnetic properties, the expression of $\overrightarrow{F_m}$ becomes:

$$\chi_0 << 1, \vec{F}_m = \begin{cases} \mu_0 V_p (\chi_0 - \chi_f) \left(\vec{H} \cdot \vec{\nabla} \right) \vec{H}, & H < \frac{M_{sp}}{\chi_0 - \chi_f}, \\ \mu_0 V_p \left(\vec{M}_{sp} \cdot \vec{\nabla} \right) \vec{H}, & H \geq \frac{M_{sp}}{\chi_0 - \chi_f}, \end{cases} \tag{7.23}$$

$$\chi_0 > > 1, \vec{F}_m = \begin{cases} \mu_0 V_p 3 \left(\vec{H} \cdot \vec{\nabla} \right) \vec{H}, & H < \frac{M_{sp}}{3}, \\ \mu_0 V_p \left(\vec{M}_{sp} \cdot \vec{\nabla} \right) \vec{H}, & H \geq \frac{M_{sp}}{3}, \end{cases} \tag{7.24}$$

Equation (7.23) shows that the particle will be subjected either to a positive force in the case of $\chi_0 > \chi_f$ or to a negative force, if $\chi_f > \chi_0$, and then either attracted or repelled from maximum of magnetic field gradient [21]. These motions refer to positive and negative magnetophoresis [22].

To summarize, when magnetophoresis is implemented in microfluidic devices, particles magnetization can be saturated or not based on its own magnetic properties and on the magnitude of the applied magnetic field. As a result, the magnetophoretic force can be either proportional to $\left(\vec{H} \cdot \vec{\nabla} \right) \vec{H}$ or to $\vec{\nabla} \vec{H}$, as reported in Table 7.5.

The size of the object, its magnetic properties, as well as the magnetic field and its gradient are key parameters to improve magnetophoretic forces. In microfluidic devices, magnetophoretic forces reported in literature ranges from few pN to several nN, depending on the objects and magnetic system implemented (see Sect. 7.3).

7.3.2 *Fluidic Drag Force*

Viscous drag force acts on particle in the opposite direction to their motion. In low Reynolds fluid flow conditions, the drag force is expressed using Stokes' law, as follow [1]:

$$\vec{F}_d = 6\pi\eta R_p (\vec{v}_f - \vec{v}_p) f_D \tag{7.25}$$

where R_p is the radius of the particle, $\vec{v_p}$ its the velocity, η and $\vec{v_f}$ respectively the viscosity of the external medium and the velocity of the fluid. f_D is the drag

Table 7.5 Expression of magnetophoretic force depending on the magnitude of the magnetic field and on the magnetic susceptibility of the particle $(\Delta\chi = (\chi_0 - \chi_f))$

Non-saturated particles (low magnetic field)		Saturated particles (large magnetic field)
$\chi_0 \ll 1$	$\chi_0 \gg 1$	
$\vec{F_m} = \mu_0 V_p \Delta\chi \left(\vec{H} \cdot \vec{\nabla} \right) \vec{H}$	$\vec{F_m} = \mu_0 V_p 3 \left(\vec{H} \cdot \vec{\nabla} \right) \vec{H}$	$\vec{F_m} = \mu_0 V_p \left(\vec{M}_{sp} \cdot \vec{\nabla} \right) \vec{H}$

coefficient of the particle that takes into account the influence of a solid wall in the vicinity of the particle, z being the distance of the particle to the wall [1, 19].

$$f_D = \left[1 - \frac{9}{16}\left(\frac{R_p}{R_p + z}\right) + \frac{1}{8}\left(\frac{R_p}{R_p + z}\right)^3 - \frac{45}{256}\left(\frac{R_p}{R_p + z}\right)^4 - \frac{1}{16}\left(\frac{R_p}{R_p + z}\right)^5\right]^{-1} \tag{7.26}$$

f_D varies between 1, for particle distant from the wall to 3 for particle in contact with micro-channel wall ($z = 0$). One can notice that for applications with particle moving through the channel, $f_{D,}$ varies and $\overrightarrow{F_d}$ can change during particle displacement [23]. This coefficient is frequently ignored for experiments in which the particle radius is relatively small compared to the dimensions of the channel. In case of cell labeled with magnetic particle, the apparent radius can be estimated according to the size of magnetic particles relative to that of the cells.

In most microfluidic applications, the fluid flow profile is not uniform but varies along the channel section (see laminar flow in Chap. 2). However, particles diameter being usually smaller than dimensions of microfluidic channel dimensions, the fluid velocity is considered relatively constant across the particle [23]. The drag force, given by Eq. (7.25), is then estimated at a time t, with particle velocity at t and fluid flow velocity at the position of the particle at t. Most of the publications related to magnetophoretic functions in microfluidic devices assume that the fluid has an average velocity in the entire channel section. See Chap. 2 to take into account v_f profile depending on the position of the particle in that channel for various section shape i.e. rectangular, triangular, cylindrical and aspect ratio.

Drag force is typically in the order of few to few tens of pN in microfluidic devices. For example, $F_d \approx 8.4$ pN for 1 μm diameter particle flowing at 1 mm/s in a fluid with a viscosity of 0.89 mPa·s.

7.3.3 *Gravitational and Buoyance Forces*

The gravitational force and the associated buoyance forces are expressed as [19]:

$$\overrightarrow{F_g} + \overrightarrow{F_b} = -V_p(\rho_p - \rho_f)\vec{g} \tag{7.27}$$

with ρ_p and ρ_f, the density of the particle and the solution, respectively, and $\vec{g}$ the acceleration due to gravity.

Gravitational force and buoyance forces are neglected for sub-micrometer or nanoscale particles [19, 24]. Indeed, they are much lower than magnetic forces, as illustrated here: 1 μm diameter Fe_3O_4 particle $(\rho_p = 5000\ \mathrm{kg/m^3})$ flowing in water $(\rho_f = 1000\ \mathrm{kg/m^3})$, experiences $F_g = 2.56\,10^{-2}$ pN and $F_b = 0.5110^{-2}$ pN, several order of magnitude below applied magnetic forces. For larger particles, typically $R_p > 5\,\mu\text{m}$, these forces should be considered.

7.3.4 Other Forces

Other forces (interactions of the particles with their environment: other particles, channel walls, fluid) contribute to the overall trajectory of magnetic particles in a magnetophoretic microfluidic device [25]. Particle/micro-channel wall interactions result from electrostatic and electrodynamic (van der Waals) forces experienced by particles in solution. Indeed, particle and micro-channel walls in contact with electrolytic solution can present a surface charge that induces a double-layer at their surface and their overlapping induces electrostatic interactions [19]. Van der Waals force, on its hand, originates from attractive electromagnetic interaction between electrical dipole and/or induced dipoles [19]. Both forces can generate unwanted particle sticking to the micro-channel walls. This can be avoided by modifying the pH and the ionic strength of the solution or by coating micro-channel wall or particle surfaces with proteins (Bovine Serum Albumin, BSA for example) [1]. These forces quickly decrease as the distance between surfaces increases and are negligible at distances greater than tens of nanometers. Particle/particle interactions can be electrostatic and magnetic. Electrostatics ones, generated by the electric double-layers, are repulsive forces whereas magnetic interactions between particles can lead to the creation of particle clusters that possess their own dynamics [1]. These inter-particle effects as well as particle/fluid interactions are usually ignored for particle suspension at low volume concentration [25, 26]. If these interactions are considered, they give rise to a complex model solved numerically.

7.3.5 Particle Transport Models

All the forementioned forces have an effect on particle transport and two models are used to predict particle trajectories depending on whether Brownian motion is neglected or not. Particles suspended in a fluid undergo random collisions with fluid molecules, generating a random movement of particles, the Brownian motion. Particle diffusion due to Brownian motion is neglected for particles having diameter greater than tens of nanometer. Gerber and co-workers [26] defined a criterion to estimate particle diameter, D_p, below which Brownian motion influences particle displacement:

$$\left|\vec{F}\right| \cdot D_p \leq k_B \cdot T \tag{7.28}$$

$\left|\vec{F}\right|$ being the magnitude of the total force acting on the particle. For instance, Gerber et al. estimated a critical particle diameter of 40 nm for the capture of Fe_3O_4 particles in water. For particles with a diameter below or equal to D_p, motion of individual particle is predicted using drift-diffusion analysis, whereas for particle with diameter larger than D_p, Brownian motion is neglected and classical Newtonian physics is employed to foresee particle trajectory.

In the first case, because of thermal agitation and diffusion, particle transport cannot be precisely monitored and a statistic approach must be used to predict their trajectories. Particle transport is modeled using a drift-diffusion equation for the particle volume concentration c:

$$\frac{\partial c}{\partial t} = \vec{\nabla} \cdot \vec{J} = 0 \tag{7.29}$$

where $\vec{J} = \vec{J}_{\mathrm{D}} + \vec{J}_{\mathrm{F}}$ is the total flux of particles, which includes a contribution $\vec{J}_{\mathrm{D}} = -D\vec{\nabla}c$ due to diffusion, and a contribution $\vec{J}_{\mathrm{F}} = c\vec{U}$ due to the drift of particle under the influence of applied forces. D is the diffusion coefficient with $D = \gamma kT$, where γ is the mobility of a particle. $\vec{U}$ is the drift velocity, with $\vec{U} = \gamma\vec{F}$, where $\vec{F} = \overrightarrow{F_{\mathrm{m}}} + \overrightarrow{F_{\mathrm{d}}} + \overrightarrow{F_{\mathrm{g}}} + \overrightarrow{F_{\mathrm{b}}} + \cdots$, is the total force acting on particle. Equation 7.29 can be written as follow:

$$\frac{\partial c}{\partial t} = D\vec{\nabla}^2 c - \gamma c\vec{\nabla} \cdot \vec{F} \tag{7.30}$$

In most applications, Brownian motion is neglected and this model is rarely used in literature. For details to solve Drift-diffusion transport see initial work of Gerber [27], Fletcher [28], and more recently wok of Furlani [29].

The second model is the most commonly used in literature, as micro- and nano-objects that are mainly manipulated in magnetophoretic microsystems have a diameter larger than D_p. This model uses Newton's second law, to predict particle trajectory, as expressed in Eq. (7.31):

$$m_p \frac{d\vec{v}_p}{dt} = \sum \text{forces} = \vec{F}_m + \vec{F}_d + \vec{F}_g + \vec{F}_b + \cdots \tag{7.31}$$

where m_p is the mass of the particle.

For sub-micrometer sized particles, the initial term, $m_p \frac{d\vec{v}_p}{dt}$, is often ignored due to their small mass [30, 31]. Thereby, for the inertial term to be considered particle with a mass of 1 pg should have acceleration above 100 m/s^2, which is unusually large for microfluidic applications. Considering expression of the different forces, Eq. (7.31) can be expressed as follow:

$$m_p \frac{d\vec{v}_p}{dt} = \mu_0 \left(\vec{m} \cdot \vec{\nabla} \right) \vec{H} + 6\pi\eta R_p (\vec{v}_f - \vec{v}_p) f_D - V_p (\rho_p - \rho_f) \vec{g} \quad (7.32)$$

The balance of forces can be tuned by modifying particle properties $(R_p, M_s, \rho_p, \chi_p)$, fluid properties (η, ρ_f, χ_f) and the magnetic source (B).

A schematic reporting the main forces experienced by particles in magnetophoretic microsystems is reported in Fig. 7.7a (direction of forces being arbitrary). Figure 7.7b also reports order of magnitude of forces applied on particle according to its size, in typical microfluidic device operating conditions. For magnetophoretic force, we considered that magnetic particles are saturated, and two cases were calculated: the case of Fe_3O_4 particles of which maximum size reach in general few tens of nanometers, and the case of composite particles, composed of a polymer matrix and Fe_3O_4 nanoparticles, of which size can reach tens of micrometers. Based on the choices made, it results that magnetophoretic force is mainly in competition with the drag force. In most publications one considers that the particle magnetization is not saturated, and so the balance of forces can be expressed as follows:

$$\mu_0 V_p \Delta\chi \left(\vec{H} \cdot \vec{\nabla} \right) \vec{H} = 6\pi\eta R_p (\vec{v}_f - \vec{v}_p) f_D \quad (7.33)$$

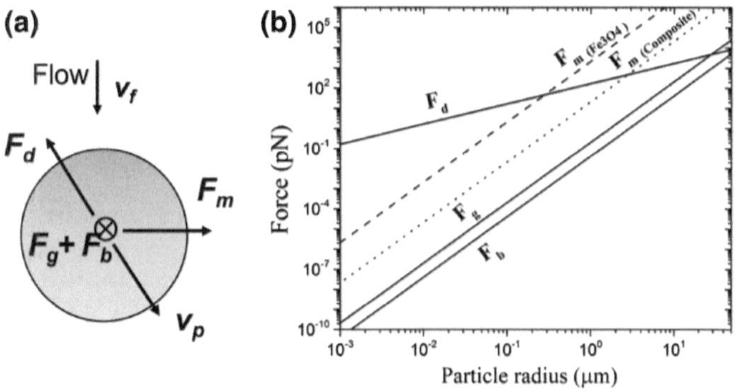

Fig. 7.7 **a** Schematic representation of forces experienced by a particle inside magnetophoretic device (arbitrary direction of forces) **b** Order of magnitude of forces as functions of particle radius for following settled particles parameters and fluidic and magnetic conditions: particle density of $\rho_p = 5000\ kg/m^3$ (Fe_3O_4) flowing in water ($\rho_f = 1000\ kg/m^3$, $\eta = 0.89$ mPa.s), with a velocity of 1 mm/s and submitted to a magnetic field gradient of $\nabla B = 10^3\ T.m^{-1}$. For calculus, we considered that two type of magnetic particles that are saturated, Fe_3O_4 particles with magnetization at saturation $M_s = 510\ kA{\cdot}m^{-1}$ (F_m, dash), and composite particles (1 vol.% of Fe_3O_4 nanoparticles in a polymer matrix) with $M_s = 5.1\ kA{\cdot}m^{-1}$ (F_m, dot)

7.4 Implementation of Magnetophoresis in Microsystems

We will focus in this section on the integration of magnetic flux sources in microsystems, which will be used as the remote control.

7.4.1 *Sources of Magnetic Field and Magnetic Field Gradient at the Micrometer Scale*

To start with, it is worth taking a look at an estimate of the magnitude of the magnetophoretic force. A strong centimeter sized magnet, like the strongest NdFeB-based commercial ones, generates a magnetic field gradient of the order of 10 T/m, and a magnetic field of few 0.1 T close to its surface. Therefore, the force per volume of target object (with magnetization of 100 kA·m^{-1}) will be of the order of 10^{-21} N/nm^3. For particles with a size of 10 nm, this leads to a force in the range of 1 aN (1 attoNewton = 10^{-18} N), a value comparable to the gravity force acting on them. In order to efficiently manipulate nanosized object, it is then of first importance to generate strong magnetic field gradients, and the way to do so is to scale down the magnetic flux source to the micrometer scale.

7.4.1.1 A Variety of Magnetic Field Sources

A magnetic field can be produced by a permanent magnet or an electrical current passing through a coil. These sources of magnetic stray field can be combined with magnetic concentrators that will focus the flux lines due to their large magnetic susceptibility, producing field gradient in their surroundings. Figure 7.8 illustrates the three main approaches that are used to create magnetic field gradients in microsystems.

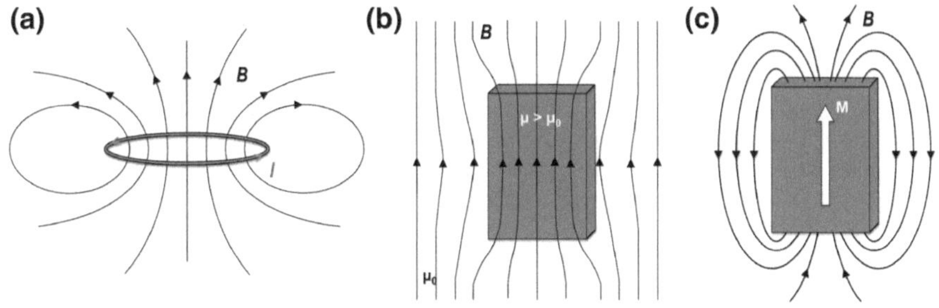

Fig. 7.8 The three reported solutions to generate magnetic field gradients at the micrometer scale: **a** micro-coils made of conducting materials [32], **b** micro-concentrators of magnetic flux made of magnetically-soft materials [33] and permanent micro-magnets made of magnetically-hard materials [34]

7.4.1.2 Downscaling a Magnetic Field Source

Considering a uniformly magnetized object, the field that emanates from it depends on its magnetization and its shape. The magnetization is a bulk property, it does not depend on the size of the object: a large magnet and a small magnet with the same magnetization are capable of creating similar magnetic field around them. On the other hand, the distance on which the magnetic stray field decays scales with the size of the object. Therefore, there is a remarkable advantage to reduce the size of magnetic flux sources as it increases the gradient of the produced magnetic stray field. Figure 7.9 shows the pattern of magnetic stray field around a magnet with a square section of side a, magnetized upwards.

In other words, reducing the size of a magnet by a factor k multiplies the maximum field gradient by k. For micrometer size magnets, the stray field gradient is in the range of 10^3–10^6 T/m. For the same target particle as considered in the introduction of this section, the force that was 1 aN with a centimeter sized magnet, is increased to the range of 1–10^3 fN (1 fN = 10^{-15} N), a value well greater than the gravity force.

Interestingly, downscaling coils is also favorable as the admissible current density can be increased while reducing the size of the conductor. Indeed, Joule heating scales with the volume of the conductor whereas cooling losses through heat flow is proportional to the surface [35].

7.4.2 *Micro-Coils*

Micro-coils are tiny wires of electrical conductor. The building block is a current loop, as illustrated in Fig. 7.10, which can be added up in a planar spiral [36, 37], a 3D solenoid [38], so as to increase the produced flux intensity. Complying with the

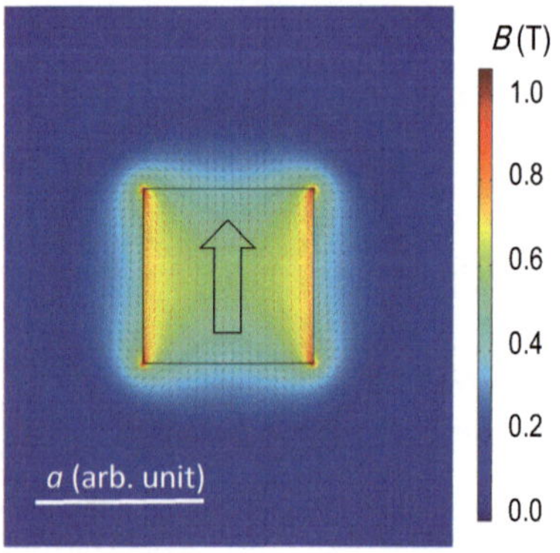

Fig. 7.9 Simulated stray field B produced by a magnet with a square section of side a, with a remanent induction of 1.17 T (comparable to the one of NdFeB magnets). The hollow arrow indicates the direction of the magnetization

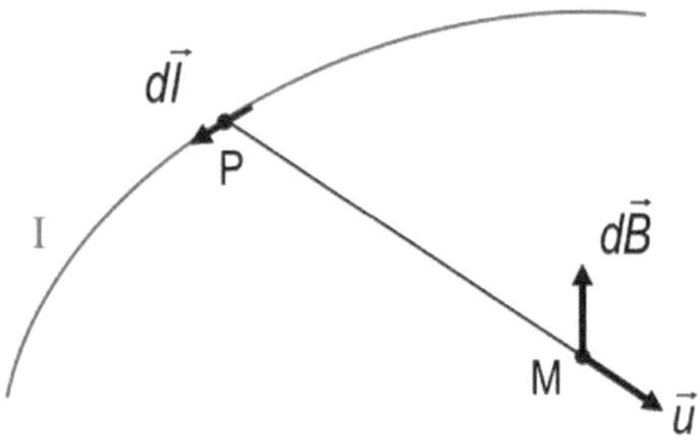

Fig. 7.10 Schematics of the magnetic induction B induced by an electric conductor

framework of this chapter, we will only consider micro-coils as micro-sources of magnetic flux, which differs from their first application as probe heads in magnetic resonance devices [39].

The magnetic induction B created at a point M of the space by an electric circuit element positioned at a point P is given by the Biot-Savart law (7.34):

$$d\vec{B}_M = \frac{\mu_0}{4\pi}\frac{Id\vec{l} \times \vec{u}}{r^2} \tag{7.34}$$

$\vec{u}$ is the unit vector in the plane of the circuit and perpendicular to the element $\vec{dl}$, r the distance from the current element (P) to the considered point (M) and I the current circulating in the element.

It comes that the magnetic induction B generated in the center of a planar coil with N_S windings of radius r is given by:

$$B = \frac{\mu_0 I}{2r} N_S \tag{7.35}$$

The magnitude of the induction increases with the current, the number and the density of windings. However electric current flowing in a resistive wire inevitably produces heat that is called Joule, ohmic or resistive heating. This limits the current intensity but also the number of windings as the resistance of the coil scales with the length of the conductor. The maximum amplitude of magnetic induction reached with micro-coils is in the range of 1 to 10 mT. It could be enhanced by adding a magnetic core at the center that would concentrate the flux, as for bulk electromagnets.

When the micro-coil design integrates several turns, independently to the selected geometry, classical approach combining electrodeposition and photolithography leads to relatively tedious processes, as compared to the fabrication of other magnetic flux micro-sources. It is interesting to note that this could be overcome by exploiting the complex and fine structures in nature, as demonstrated by Kamata et al., who prepared micro-coils using helical microalgae as a biotemplate [40].

Due to relatively low performances in static conditions with respect to the other options, micro-coils are rarely used in this framework but have a great potential for actuators operating at high frequency.

Moreover, the possibility to alternate the produced flux at a time scale below the relaxation times of the superparamagnetic particles could be a way to exploit the bi-directionality of the magnetic force and thus open a new route of development.

7.4.3 Magnetic Flux Micro-sources from Ferromagnets

Ferromagnets are divided into uniformly magnetized regions, called Weiss domains, separated by domain walls. The driving force of domain formation is the minimization of magnetostatic energy, which is the energy stored in the magnetic stray field emanating from the ferromagnet. Schematics in Fig. 7.11 illustrate how closure domain structure (Fig. 7.11b) cancels out the magnetic stray field. The closure domain structure is reached by cooling the ferromagnet across its Curie temperature. The focus of this section is the use of ferromagnets as magnetic field and magnetic field gradient sources and therefore we will consider them in the magnetized configuration (Fig. 7.11a). As previously discussed in Sect. 7.2.1, we will distinguish two types of ferromagnets, the so-called hard magnetic materials that can remain magnetized in the absence of applied field and the so-called soft magnetic materials of which the domain structure systematically falls in a configuration like in Fig. 7.11b at zero field. The former type can be used to prepare permanent micromagnets while the latter serves for micro-concentrators of magnetic flux.

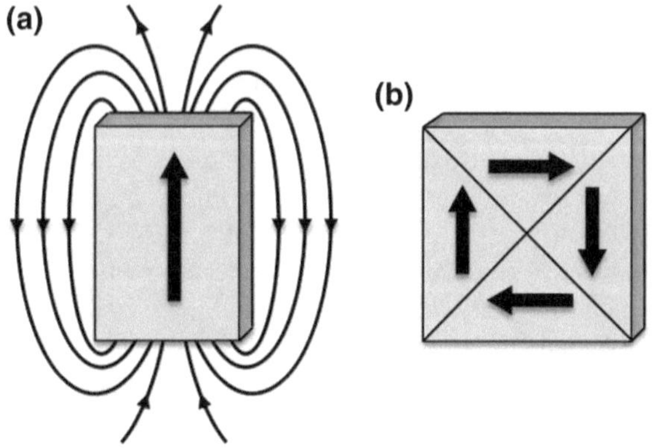

Fig. 7.11 Magnetic domain structure in ferromagnets and magnetic stray field (red lines) in two limit cases: **a** a single domain and **b** a closure domain structure

7.4.3.1 Micromagnets

Permanent magnets define objects that produce magnetic stray field in their environment, in the absence of external excitation, such as a circulating current or a magnetic field. This particularity implies that the predominant constituting phase retains a fraction of its magnetization, which is called the remanent magnetization. An important ingredient is the magnetic anisotropy. The large majority of produced permanent magnets nowadays get their hard properties from a large magneto-crystalline anisotropy. When looking back at the evolution of permanent magnet performances over the last century, the main achievements came with the discovery of new hard magnetic compounds. The main breakthrough came in the 1950's with the discovery of ferrimagnetic hexagonal ferrites, of which the high anisotropy permits to manufacture them in any shape and so opening new possibilities for device designs.

The magnetic performances of magnets are generally assessed by a figure of merit called the energy product, denoted $(BH)_{max}$, which scales the amount of energy stored in the stray field. It corresponds to a working point on the induction curve B(H) that maximizes the area under the curve in its second quadrant (see Fig. 7.12). A high $(BH)_{max}$ reflects a high enough resistance to demagnetization and a large magnetization. Beyond a certain value of coercive field H_C, the maximum energy product is only a function of M. In fact a hard magnetic phase is characterized by a maximum theoretical energy product that is:

$$(BH)_{MAX} = \frac{1}{4}\mu_0 M_S^2 \tag{7.36}$$

Table 7.6 gives characteristic parameters for some hard magnetic substances. When using magnets as magnetic flux sources, one exploits its stray field, which is

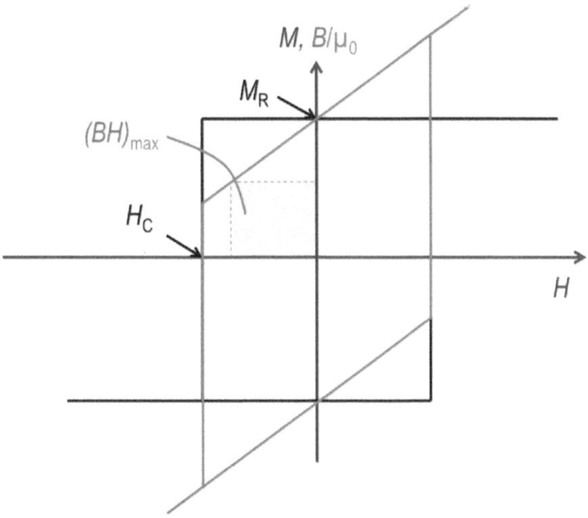

Fig. 7.12 Characteristic magnetization and magnetic induction curves of a permanent magnet

Table 7.6 Characteristic parameters of some hard magnetic phases [41, 42]

Material	M_S (kA/m)	Theoretical $(BH)_{MAX}$ (kJ·m^{-3})
$SrFe_{12}O_{19}$–$BaFe_{12}O_{19}$	200–380	12–45
Alnico 5	1110	(in practice: ≤ 50)
hcp–Co	1370	–
Fe_3C	1190	–
$L1_0$–CoPt	810	200
$L1_0$–FePt	1140	406
$SmCo_5$	860	231
$Nd_2Fe_{14}B$	1280	512
$B8_1$–MnBi	580	106
$L1_0$–MnAl	600	113

the field created outside its volume. In general, the stray field is calculated using either the Amperian approach, in which the magnetization is replaced by an equivalent distribution of current density, or the Coulombian approach, in which the magnetization is replaced by an equivalent distribution of magnetic charges. The two approaches are illustrated in Fig. 7.13 for a uniformly magnetized cubic magnet.

The calculation of magnetic stray field is general performed by finite element method, though analytical solutions for simple shapes can be found in literature [43].

Fabrication of Bulk Magnets

Bulk magnets are generally prepared by powder metallurgical processes, which can be briefly summarized as follows: (1) a mixture of the raw elements is first molten in furnaces and cooled down to form cast magnets, (2) the obtained material is then either ball milled or jet milled to obtain fine powders, with particle sizes ranging from few micrometers up to hundreds of micrometers, (3) this powder is then pressed into molds and heated to be sintered. The as obtained «sintered magnets»

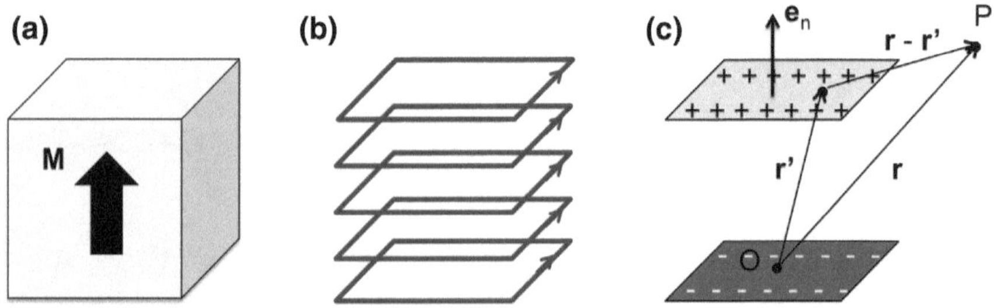

Fig. 7.13 Equivalent approaches for the calculation of the magnetic stray field emanating from a uniformly magnetized magnet with a square section **a** fields produced by an equivalent distribution of currents **b** and an equivalent distribution of charges **c**

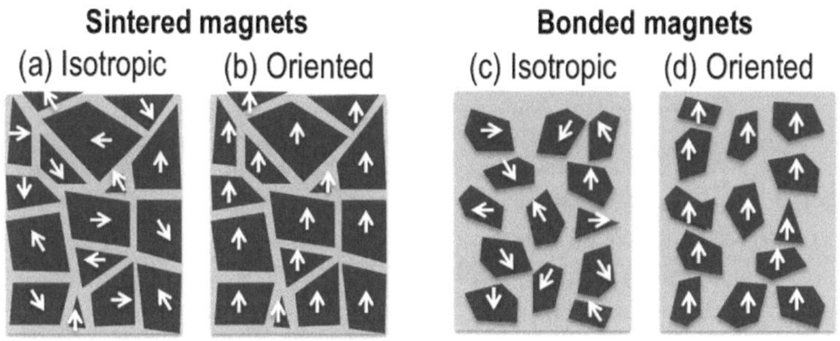

Fig. 7.14 Schematics of the microstructure of bulk-manufactured magnet

are relatively dense (density greater than 90%) but brittle. A solution to obtain more mechanically robust magnets is to bond the fine powder with a polymer, thus facilitating magnets shaping, for example by injection moulding, which opens up new possibilities for device designers. In addition, these so called «bonded magnets» show higher resistance to corrosion. However, they present less magnetization than their sintered magnets counterparts because of the relatively large volume fraction of non-magnetic binder (the volume fraction of the magnetic powder is typically 60–80%). Typical binders used are epoxy resin, polyamides or nitrile rubbers. Schematics of Fig. 7.14 show the different microstructures of magnets.

Classification of Permanent Magnets

One can put magnets in four categories, according to the constituting elements and structure: hexagonal ferrites, alnicos, metal alloys and rare-earth intermetallics. They all have their advantages and weaknesses. Hard hexagonal ferrites have the general formula $MO–6(Fe_2O_3)$, with M being mostly Sr or Ba. They are produced nowadays in large quantity, in the range of 10^6 tons/year. They have limited performances but present the great advantages of being cheap and excellent resistance to corrosion. Alnicos are nanostructured materials consisting of Fe–Co needles in a non magnetic Al-Ni matrix, with traces of Cu and Ti. They are obtained by spinodal decomposition and the anisotropic nanostructure is obtained by means of thermal process under magnetic field. They present good thermal stability, relatively high magnetization but suffer from relatively low resistance to demagnetization and therefore are generally found in rod shapes. Hard metal alloys are usually made by direct casting of 3*d* transition metal elements taken among Mn, Fe, Co, Ni with other metallic elements, of various types like Ga, Al, Pd, Pt, Bi. They have in common a non-cubic structure. Binary $L1_0$ alloys like FePt and CoPt present impressive hard magnetic properties and good resistance to corrosion but suffer from the high cost of Pt. Rare-earth intermetallic magnets, like NdFeB–based magnet, show the highest magnetic performances and are relatively low cost. Their main weakness is their poor resistance to corrosion, which requires coating them in order to limit degradation over time. For use in microfluidic systems intended for biology, the operating temperature window is rather restricted to a narrow range, typically 10–40 °C, where all magnets described above can be used with relatively steady performances.

Integration of Permanent Micromagnets in Microsystems

The integration of permanent magnets in microsystems is challenging mainly because hard magnetic properties are highly sensitive to both chemical composition (alloy composition, structure, chemical order) and microstructure (grain size, grain boundary phases, defects). This usually implies some preparation constraints like thermal treatments, or sometimes processing steps under magnetic field. In addition, once prepared it is necessary to apply a strong external magnetic field so that to get a maximum remanent magnetization.

Micro-Magnets Fabrication Routes

Machining bulk high performance sintered magnets to obtain sub-millimeter sized magnets, for example by spark cut, is restricted to relatively coarse features as inevitable degradation of the surface will be detrimental for the magnetic hardness. In addition, machining or deforming bulk magnets is not suitable for preparing micro-pattern arrays.

Instead, two main approaches were developed to fabricate such micro-magnet arrays: patterning of films or powder positioning.

Several film deposition and patterning methods can be used, including sputtering [44], pulsed-laser deposition [45], evaporation [46], or electrochemical deposition [47]. Electro-deposition has been found as an efficient way to prepare thick films of rare earth-free magnetic films. Pulsed laser deposition technique is usually restricted to relatively low deposition rate and relatively small deposition areas, and is difficult to scale up. Sputtering, like electrochemical deposition, can be adapted for deposition on large areas and with relatively high deposition rates of 10–40 μm/h [48].

Films can be topographically patterned depositing them on thick resist masks and then performing lift-off [49], or else onto pre-etched substrates (predominantly silicon) using deep reactive ion etching [50]. For hard magnetic compounds that are less sensitive to corrosion (rare-earth free magnetic films), an additional option is to pattern the films, post deposition, through wet etching or focused ion beam. The as-obtained arrays of permanent micro-magnets are heated to form the hard magnetic phase, either during deposition, or with post-deposition annealing, and then magnetized in the unidirectional field of a superconducting coil to reach the full remanence of the micro-magnets.

Local change of the structure or of the chemical order can be another route to obtain patterned hard magnetic films at the micrometer scale. As an example, Okuda et al. induced local magnetic hardness in NdFeB film that were initially poorly crystallized and so magnetically soft, by employing pulsed annealing through a mask [51].

Other approaches were developed to create such multipolar structures in continuous magnet films, magnetized locally in opposite directions. To reverse the magnetization in specific zones, one can use electrical pulses in conducting wires positioned in contact to the film [52, 53], or else using the so-called thermo-magnetic patterning [54], which consists in locally reversing the magnetization of a uniformly magnetized film by heating micrometer-sized areas with a laser pulse through a mask, under moderate magnetic field. The process is schematized in Fig. 7.15. The advantages of this method are that it creates multipolar configuration that is favorable for magnetic stray field strength and the absence of topography patterns facilitates the integration in devices. The main disadvantage is the rather limited reported depth (less than 2 μm) of the magnetization reversal. Other means to reverse locally the magnetization of hard films have been proposed since, using soft ferromagnetic masks that locally concentrates magnetic flux in the opposite direction with respect to the initial film magnetization direction [55].

Film-based techniques offer the great advantages of reproducibility, and fine controls over geometries and microstructuration. However, they suffer from a tendency of peeling off from the substrate due to mechanical stress building-up in thick deposited films. Also, their fabrication processes are relatively expensive, slow and tedious. As the hard magnetic phases are usually obtained after heat treatment, the thermal expansion coefficients of the film and the substrate may set an upper limit on the film thickness.

Methods to transfer well-defined arrays of film-based structures in polymer matrix were developed in order to ease the integration in polymer-based microsystems [56–59].

In contrast to film-based methods, powder-based methods employ particles used in bulk-manufactured magnet industry as building blocks, which have already desired hard magnetic properties. Bonded micro-magnet arrays are generally obtained by filling cavities present in resist masks or pre-etched silicon substrates with a mixture of magnetic powder and a polymer binder. Among the most ubiquitous techniques of fabrication, one can cite replica molding, squeegee coating or screen printing. Other techniques like inkjet printing do not use physical master. Note that generally the powder-based approaches integrate a process step with an applied magnetic field in order to align the particles and confer an overall anisotropy to the bonded micro-patterns. In order to go down in sizes, it is also possible

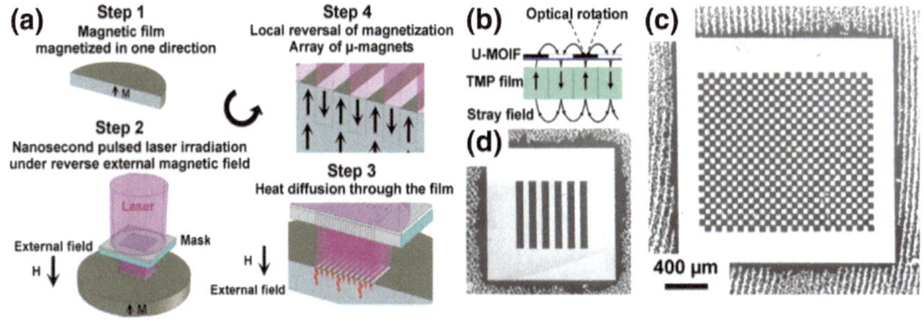

Fig. 7.15 Thermo-Magnetic Patterning process developed by Dumas-Bouchiat et al. Reprinted from [54], with the permission of AIP Publishing

to use magnetic molds to precisely position magnetic particles on a surface and then transfer the formed pattern arrays in a medium like a polymer [60].

Many efforts were done to develop high performance micro-magnets, in various contexts and geometries. The performances are assessed by local characterization of the stray field, and its gradient using magneto-optical indicator films (MOIF) [59], Hall micro-probes [56, 61] or measurements of forces exerted on "colloidal tips" in a magnetic force microscopy set up.

In the context of microfluidic systems integrating magnetophoretic forces, the micro-magnets offer the great advantage of producing strong stray fields without any external excitation, which is beneficial for compact and low-power consumption devices. The main disadvantage of micro-magnets is that the generated force pattern cannot be easily modulated in real-time. This can be an issue to consider when releasing the trapped object is desired. Using magnetically-soft ferromagnets can be a way to overcome this limitation.

7.4.3.2 Micro-Concentrators of Magnetic Flux

The "micro-magnet" term is often abusively used to refer magnetically-soft micro-patterns, although a magnet is characterized by its ability to retain a fraction of its magnetization when the field is removed. Here we make the distinction between micro-magnets and magnetically-soft micro-patterns that only become magnetized in the presence of an external magnetic field. They are used to concentrate an external magnetic flux that can be delivered either by an electric circuit or a permanent magnet. In the absence of external field, the soft magnetic object structure collapses in multi-domains and there is no stray field. Micro-concentrators are characterized by a large change of their magnetization state when submitted to a relatively low external magnetic field. The performance of the flux guide is related to the intrinsic magnetic permeability μ of the constituting material. The higher the magnetic permeability, the more concentrated the magnetic flux is. In addition to the permeability, an important parameter is the maximum flux the material can concentrate, which is determined by the spontaneous magnetization M_S. Figure 7.16 shows the magnetic flux inside and nearby a soft magnetic material that is magnetized in a field of 0.2 T created by bulk magnets.

Table 7.7 presents some values of magnetization M_S and magnetic relative permeability μ_r of some commonly encountered magnetically-soft materials at room temperature. One can find several classes of magnetically-soft materials: low-carbon mild steels, Fe-Ni alloys, Co-Fe alloys, soft ferrites. They generally result from a compromise between high saturation magnetization and the addition of non-magnetic elements to increase their magnetic softness (among which Al, Si, C, Mo, Cu). Unlike hard magnetic compounds, additional elements are intended to render the alloy amorphous or reduce chemical ordering. These elements can also help to increase electrical resistance so that to improve high frequency performances. In the bulk state, grain-orientation, lamination, annealing under magnetic

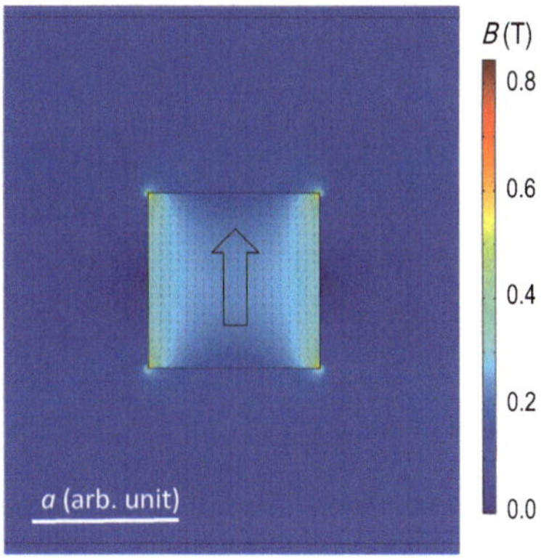

Fig. 7.16 Simulated stray field B produced by a cubic magnetic flux guide of side a, in the gap between two permanent magnets

Table 7.7 Characteristic parameters of some magnetically-soft materials [13, 62, 63]

Material	M_S (kA/m)	μ_r
Soft Fe	1710	300–5,000
Hypernik ($Ni_{50}Fe_{50}$)	1270	6,000–40,000
Metglas 2826 ($Ni_{50}Fe_{50}$-based)	700	50,000–800,000
Metglas 2714 (Co-based)	450	80,000–1,000,000
Permalloy ($Ni_{80}Fe_{20}$)	830	8,000–100,000
Supermalloy ($Ni_{80}Fe_{15}Mo_5$)	700	100,000–300,000
Mumetal ($Ni_{77}Fe_{16}Cu_5Mo_2$)	520	20,000–100,000
Ni	480	100–600
Mn-Zn ferrites	290–400	500–10,000
Ni-Zn ferrites	200–330	10–1,000

field can be used to induce anisotropy and thus improve magnetic permeability in a desired direction.

Dependence of the Magnetic Response on the Patterns Shape

When one wants to integrate micro-sources of magnetic field gradient in microsystems, it is of first importance to consider the shape contribution to the apparent susceptibility (M/H_{ext}), which will predominate in most cases.

In general the demagnetizing field H_d is not uniform within the volume of the object and the use of finite element modeling to describe the magnetic response to an external field is necessary. The demagnetizing field is uniform within ellipsoid shapes (of axes $2a$, $2b$ and $2c$, see Fig. 7.17) and it is thus possible to get a rapid estimate of the demagnetizing field by considering the closest ellipsoid shape to the given object. Table 7.8 gives the analytical expressions of the demagnetizing factors for revolution ellipsoids. The apparent susceptibility is then simply related to

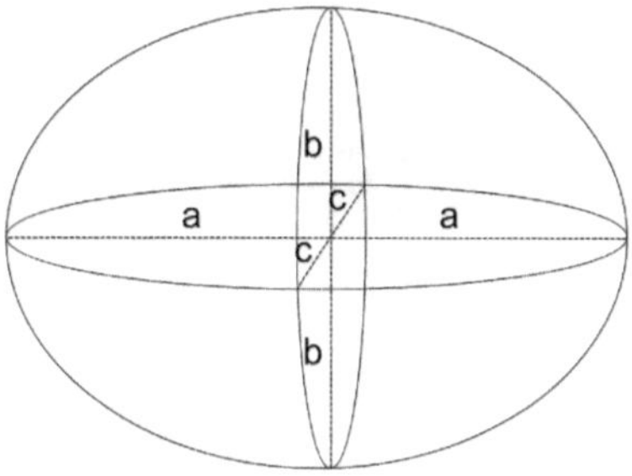

Fig. 7.17 General ellipsoid

Table 7.8 Demagnetizing factors for revolution ellipsoids [64]

Ellipsoid shape	Demagnetizing factors
Prolate (cigar shape), with $s = c/a$ $a = b < c$	$N_c = \frac{1}{s^2-1}\left[\frac{s}{\sqrt{s^2-1}} ln\left(s + \sqrt{s^2-1}\right) - 1\right]$ $N_a = N_b = \frac{1-N_c}{2}$
Sphere $a = b = c$	$N_a = N_b = N_c = \frac{1}{3}$
Oblate $a < b = c$	$N_c = N_b = \frac{1}{2(s^2-1)}\left[\frac{s^2}{\sqrt{s^2-1}} arcsin\left(\frac{\sqrt{s^2-1}}{s}\right) - 1\right]$ $N_a = 1 - 2N_c$

the intrinsic value and the demagnetizing factors along the three orthogonal axes satisfy:

$$N_a + N_b + N_c = 1 \tag{7.37}$$

Fabrication of soft micro-patterns In magnetophoretic microsystems, magnetic flux micro-concentrators are certainly the most used, and this can be partly explained by their ease of micro-fabrication compared to hard magnetic phases that require thermal treatment. Poor cristallinity and low degree of chemical ordering, which is generally obtained in the as prepared state, is not detrimental, and even can be favorable for magnetic softness. This is especially true when we restrict the use to static conditions (variation of the applied field in the Hz range), which is the general case for microfluidic systems integrating magnetophoretic functions.

Figure 7.18 shows some experimental realizations. Apart from multipolar micro-patterning, all the aforementioned micro-fabrication techniques described in the section devoted to micro-magnet arrays can be applied to micro-concentrators. Film-based methods, where films were micro-patterned to prepare batches of well-defined and fully dense micro-concentrators, are largely employed. Films are prepared by electro-deposition, sputtering, evaporation or pulsed laser deposition [65]. As for micro-magnets, thicknesses ranging from 1 to 100 μm are desired and thus high rate deposition techniques are suitable. Micro-patterning can then be achieved by different methods, including lift-off using photoresist masks [66] or by depositing the films on pre-etched substrates [56].

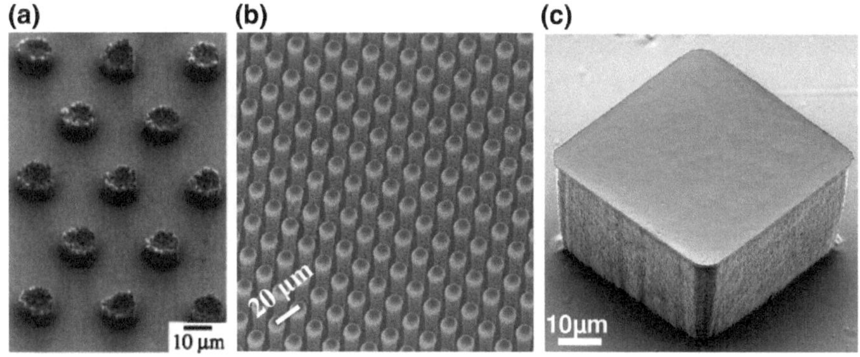

Fig. 7.18 Micro-concentrators prepared by film micro-patterning: Ni structures made by lift-off (reprinted from [66], with the permission of AIP Publishing) **a** FeCo structures made by deposition on pre-etched Si substrates [56] **b** and iPDMS pillar made by soft lithography **c** (reprinted from [67], with the permission of AIP Publishing)

However, film-based methods suffer from high processing cost, and also are limited to suitable substrates using sometimes a buffer layer to ensure adhesion.

In contrast, powder compaction and positioning offer the great advantage of low cost processes. Like for permanent micro-magnets, soft powder-loaded polymer composites can be micro-patterned using the same methods as described for composites with hard magnetic powder, including replica molding [67], or ink printing [68, 69]. However, low compaction inevitably leads to additional demagnetizing field effects at the grain scale, which can be detrimental for magnetic susceptibility and so magnetic flux guiding performances. High aspect ratio structures can be obtained by organizing the soft particles in chains within the polymer matrix [70].

Micro-concentrators of magnetic flux constitute an appealing solution as the reachable magnetic forces are comparable to the ones obtained with permanent micro-magnets but they offer in addition the possibility to modulate the force intensity in real time by varying the external flux. Their fabrication and their integration in microsystems are relatively easier than for micro-magnets as the exploited magnetic properties are less sensitive to micro-fabrication processes.

7.5 Magnetophoretic Functions Dedicated to Bio-Analysis

Implementation of magnetophoretic functions in microsystems permit to address numerous biological, medical, chemical and environmental applications. They are part of the enthusiasm of recent years for micro total analysis systems, μTAS. Employed either to control micro-object motion in channels, or involved in the sensing process, magnetic bio-device field of research is very active, and many researches were published those past years. Objects handled in magnetophoretic devices are magnetic micro- and nano-particles (MMPs and MNPs), magnetically labeled cells and cells presenting intrinsic magnetic properties. Currently, magnetophoretic implementation for biomedical applications is realized following three main strategies, using: (i) external permanent magnet, (ii) integrated permanent magnets and (iii) integrated micro-concentrators of magnetic flux. As explained in Sect. 7.3.1, miniaturization of magnetic material down to micron-scale permits to obtain high magnetic field gradient separator (HGMS) which can be required when small particles or particles with low magnetization are employed in microfluidic systems.

In this section, we will describe these three different approaches by highlighting the type of manipulated micro-object. The first paragraph is focused on magnetic particles and the second paragraph deals with magnetic bio-devices.

7.5.1 Magnetic Micro- and Nano-Particles Used in Microsystems

Magnetic particles (MPs) research fields are multidisciplinary. They require inputs from chemistry, biology, physics, medicine, engineering, whether it concerns their magnetic properties, synthesis, surface functionalization, magnetophoretic manipulation or applications. MNPs possess high surface-to-volume ratio, that is advantageous for detection process, and also magnetic properties that allow their manipulation in the sample as their transport for subsequent sensing functions. They are largely used in biomedical applications such as drug targeting, contrast agent for MRI, specific cell labelling and separation, diagnostic and hyperthermia as reported by those reviews [71–75]. They are also largely exploited in environmental applications mainly to perform pollutant removal or detection [76–79]. In microfluidic devices, manipulation of MPs addresses biomedical applications [4, 80], mainly for cell isolation, immunoassay and DNA extraction, but also environmental [81, 82] applications, for toxin or environmental reagent detection or chemical catalysis [83].

Particles must meet some general requirements to be compatible with biomedical analysis, i.e. biocompatibility, biodegradability (for in vivo applications), stability in various medium, narrow size distribution and regular shape that largely impact measurement reproducibility. They may also be superparamagnetic, in order to exhibit large magnetization in presence of applied magnetic field to achieve large magnetophoretic forces; and no coercivity, i.e. no remanent magnetization in absence of magnetic field, to allow a switch off of the magnetophoretic force and a better dispersibility in solution.

Magnetic particles, also called magnetic beads, of different sizes are employed in micro-fluidic devices: nano-particles (10–100 nm), sub-micrometer particles (0.1–1 μm) and micrometer particles (1–50 μm). Iron oxide particles, magnetite and maghemite particles, are the most commonly used. We can also find particles made of pure ferromagnetic metals, Fe, Ni, or Co, alloys such as Permalloy, $CoPt_3$ or FePt and oxide ferrite such as M–Fe_2O_4 (M being, Mg, Zn, Mn, Ni, Co,...). Different particles properties (magnetization at saturation, maximum diameter for nanoparticles to be superparamagnetic, and critical diameter for particle to switch from single to multi-domains) are given in Table 7.9 for various materials [84–86].

MNP preparation involves several steps including particle synthesis, coating or encapsulation and functionalization. MNP synthesis is a very active field of research in literature, and many reviews were published those past years [87–89]. Different synthesis pathways are reported: (i) chemical methods such as co-precipitation and thermal decomposition which are the most commonly used; (ii) physical methods, such as gas-phase deposition and electron beam lithography, which struggle to control particle size down to the nanometer scale; and (iii) more recently, microbial method that exploits the ability of bacteria such as *Thermoanaebacter* species and *Shewanella* species to synthetize Fe_3O_4 NP under anaerobic conditions by the reduction of Fe(III). Currently, co-precipitation is preferred for its simplicity and important yield. If MNP shape uniformity is required as well as a narrow particle size distribution, thermal decomposition is favored. Both methods permit to obtain magnetic nanoparticles with typical size ranging from 2 to 50 nm.

Once particles are synthetized, a protection layer, impenetrable, can prevent MNP from oxidation or erosion and can also reduce metal degradation and related toxicity (Co, Ni, Mn). This results in MNPs with a core-shell structure, a magnetic core, coated by a shell, isolating the core from the environment. Such MNP surface coating may also improve the colloidal and physical stability of the particles. Coating strategies can be basically divided into two approaches: coating with organic shell (polymers and surfactants) and coating with inorganic materials such as carbon, precious metals (e.g. gold) or oxides (e.g. SiO_2). Another approach, which differs from core-shell structure, consists in embedding dispersed MNPs in a dense polymer (e.g. polystyrene) or silica matrix to form composite. Such composite

Table 7.9 Saturation magnetization is given at T = 0 K, the first order anisotropy constant at room temperature. d_{sp} is the calculated maximum diameter for spherical superparamagnetic nanoparticle, D_{crit} is the critical diameter at room temperature which particle switch to multi-domain particle. Adopted from [86–89]

Material	M_S (kA·m^{-1})	K (kJ·m^{-3})	d_{sp} (nm)	D_{crit} (nm)
Fe	1710	48	17	14
Co	1440	410	8	70
Ni	488	−5	32	55
Fe_3O_4	480	−11.13	26	128
γ-Fe_2O_3	400	4.6	–	166
$CoFe_2O_4$	475	180	10	–
$MnFe_2O_4$	560	−2.8	41	–
$NiFe_2O_4$	300	−5.1	33	–

particles can have different structure depending on synthesis process: (i) MNPS can be dispersed in a continuous matrix, (ii) they can be dispersed on the coating of larger particles, like a shell of MNPs, (iii) they can form agglomerates of individual particles that are connected through their protective shells [90]. This approach permits to obtain micron-size particles, typically 0.1–50 μm, that contain more magnetic materials than primary magnetic nano-particle, and permit to reach higher magnetic forces. They are thus largely used in microfluidic systems. One can notice that larger magnetophoretic forces are exerted on larger particles that contain more magnetic material, giving rise on more efficient magnetic functions, but in that case, gravitational forces are also increased and thus may be considered in forces balance.

Finally, particles can be functionalized with specific molecules such as nucleic acids, peptides or proteins to provide bio-functionality in order to perform cell isolation, immunoassay and DNA extraction. Proteins can bind/adsorb to hydrophobic surfaces such as the one of polymer-coated particles. Strong binding between the particle surface and the proteins can also be obtained via specific group at the surface beads such as hydroxyl (–OH), carboxyl (–COOH) and amines ($–NH_2$) which via an activating agent bind to $–NH_2$ or –SH groups on the proteins. Bio-functionalization can also be processed via specific and strong complementary recognition interaction such as antigen-antibody or streptavidin-biotin. Various types of functionalized magnetic particles are commercially available [5]. Lists of cancer biomarkers, ligands for viruses and for proteins of biopharmaceutical interest can be found in these articles [5, 84].

As a conclusion, ideal properties of MNPs for bio-devices are:

- superparamagnetic behavior,
- spherical shape and narrow size distribution,
- physico-chemical robustness,
- high binding capacities,
- low non-specific binding,
- minimal cell perturbations, in case of cell labeling, as describe hereafter.

7.5.2 Magnetic Bio-Devices

Magnetic bio-devices are employed for two main types of applications: the detection of bio-molecular markers using magnetic beads, and the manipulation of cells. For both cases, similar methods are employed to implement actives functions (using permanent magnets or micro-concentrators of magnetic flux). The strategies developed to manipulate magnetic target objects are either trapping or deviation towards a dedicated area in the device or towards specific outlet. We will describe now different magnetic bio-devices, which are in a first part dedicated to bio-molecular marker detection, and in a second part, dedicated to cell manipulation.

7.5.2.1 Magnetic Bio-Devices for Bio-Molecular Marker Detection Using Magnetic Micro- or Nano-Particles

Manipulation of MNPs or MMPs in bio-devices are mainly dedicated to the detection of biomarkers such as nucleic acids and proteins. Their detection in blood or serum permit avoiding invasive methods, i.e. biopsy, and their detection at low concentrations allows early diagnosis.

In the case of nucleic acid detection, the sample preparation requires two primary steps, cell separation and lysis, followed by nucleic acid extraction and purification. Cell sorting can be performed using magnetophoresis approach as discussed hereafter. Once cells have been lysed, a purification step can be realized with different microfluidic functions based on filtration, electrostatic interaction, silica-based surface affinity and functionalized magnetic microparticles [91–93]. Steps following DNA extraction such as mixing of reagents, washing particles containing DNA, incubating them in an elution buffer to detach DNA, and proceeding to PCR (polymerase chain reaction) for subsequent quantification can also be performed in microfluidic device [92, 94, 95, 99]. Bio-devices integrating several steps for DNA quantification using digital microfluidic [97, 98] or in-channel approaches [92–94, 96] are reported in literature. Digital microfluidic (DMF) principle is to control fluid as discrete and individual droplets using an array of independent electrodes on a microchip. DNA extraction using DMF present the advantage of high flexibility allowing MNP separation, mixing with reagent, and washing, while limiting reagent consumption [100]. For example, Hung et al. [98] presented a microchip for genomic DNA extraction from whole blood using DMF. The magnetic bead collection and washing procedures are shown in Fig. 7.19A. After beads and the remnant supernatant merged with Wash Buffer 1, an external permanent magnet created a magnetophoretic force on magnetic beads that were collected toward the magnet. By applying a voltage on the opposite side of the droplet, it was split into two droplets, magnetic beads being collected in one droplet. The washing procedure was pursued with another buffer from another droplet. Based on in-channel approach, Strohmeier et al. [93], proposed an innovative method for magnetic beads transport in multiple microfluidic chambers containing reagents necessary for DNA purification. The strategy was based on the coupling of a centrifugal microfluidic LabDisk with a stationary permanent magnet. As shown in Fig. 7.19B, the microfluidic structure is composed of microfluidic chambers, isoradially arranged on a central microfluidic LabDisk. Chambers were first loaded with liquids. Transportation of magnetic beads in successive chambers was achieved by incremental rotation of the LabDisk with respect to the non-rotative permanent magnet. Based on this preliminary work, they developed a fully automated centrifugal-microfluidic LabDisk system, with pre-stored reagents, that performs DNA extraction and PCR for the detection of a panel of bacterial pathogens [99]. The Lab-disk design and the associated Lab-Disk player for point-of care processing are presented in Fig. 7.19C.

DNA extraction can also be achieved using micro-concentrators of magnetic flux. For example, Lou et al. [101] proposed a device for micro-magnetic selection

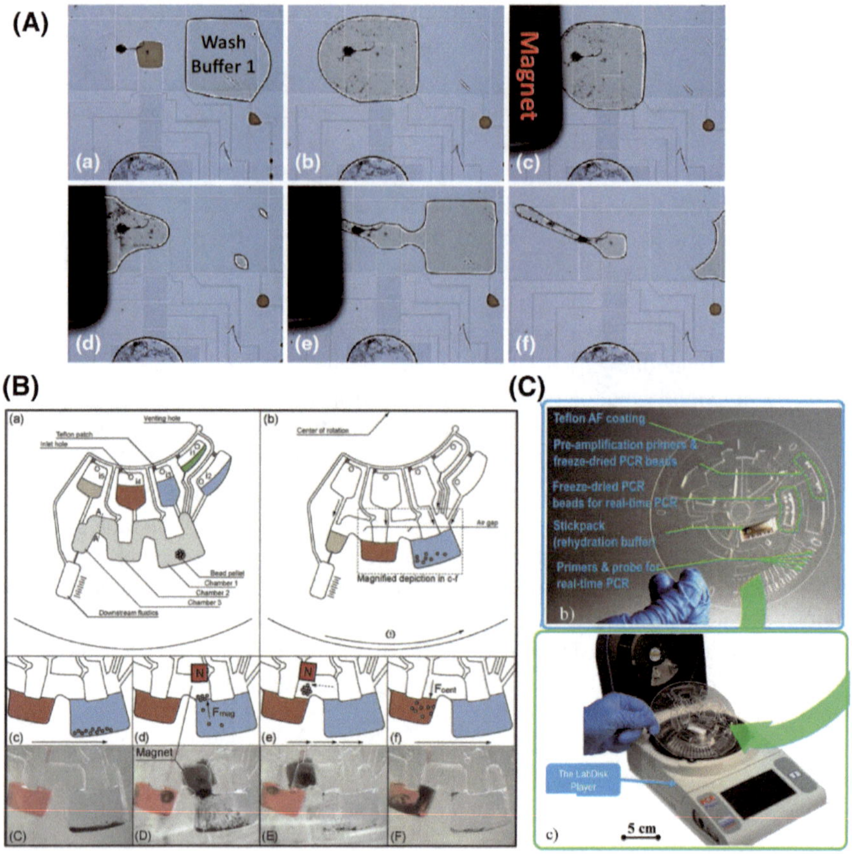

Fig. 7.19 **A** Magnetic beads collection and washing procedure using DMF (Reproduced from [98], copyright (2015), with permission of Springer); **B** Schematic representation of a centrifugal LabDisk microfluidic structure, and of different steps with associated pictures of the device: **c** beads are centrifuged in chamber 1, **d** LabDisk stopped in a defined position relative to the magnet, beads are attracted by the magnet and move across the air gap between chamber 1 and 2, **e** the Labdisk is rotated of 0.5° while the stationary magnet holds the beads, **f** the LabDisk is accelerated and beads are centrifuged into chamber 2 (Reproduced from [93], with permission of The Royal Society of Chemistry.); **C** Top, LabDisk photo, down, photo of the portable LabDisk-Player for processing the LabDisk at the point-of-care (Reproduced from [99], with permission of The Royal Society of Chemistry)

of aptamers, nucleic acid molecules, in microfluidic channels using micro-fabricated nickel strips. As shown in Fig. 7.20, the device comprises 3 inlets: two were used to introduce in the channel the sample made of magnetic beads bound to the target aptamers and unbound oligonucleotides, and the central inlet was used to introduce a buffer solution. Because of laminar flow, sample and buffer streams did not mix. By positioning an external magnet under the device, Ni strips focused magnetic field lines and created around them regions of high magnetic field gradient. Forces generated on magnetic beads deviated them from their trajectory towards the center of the channel, whereas, unbound aptamers were directed into the waste outlet. As a consequence, beads with bonded target nucleotides were exited solely by the product outlet.

In the case of proteins, numerous magnetic bio-devices were developed. Different proteins are presents in blood, with concentrations ranging from pg·mL^{-1} to mg·mL^{-1}. Low blood sample volume used in microfluidic systems requires techniques with high sensitivity, as no method exists for direct amplification of proteins, as PCR for DNA. Achieving low limit of detection (LOD), is also interesting for detection of toxin and environmental agent in serum, water or food samples. Tekin and Gijs [102] published a review on ultra-sensitive protein detection in microfluidic systems. Immunoassays are mainly used to detect a target protein via the specific recognition between a target antigen (Ag) and an antibody (Ab). Different immunoassay technics are employed, the most commonly used being the sandwich immunoassay.

Magnetic-immunoassay is currently performed as follow in micro-device: Ab-coated magnetic beads are transported, via magnetic field, towards the region in the micro-channel where Ag are localized. After Ab-Ag immunocomplex formation, magnetic beads can be conveyed to the area of detection, which is also a crucial step to reach low LOD. The last step is the Ag detection, which can be performed either without involving magnetic beads [102] i.e. by fluorescence, electrochemistry, (electro)chemiluminescence or mass spectrometry, or by using magnetic beads as labels [102], i.e. by isomagnetophoresis [103], or by monitoring the coverage of a surface by magnetic beads. Microchips using external magnet are

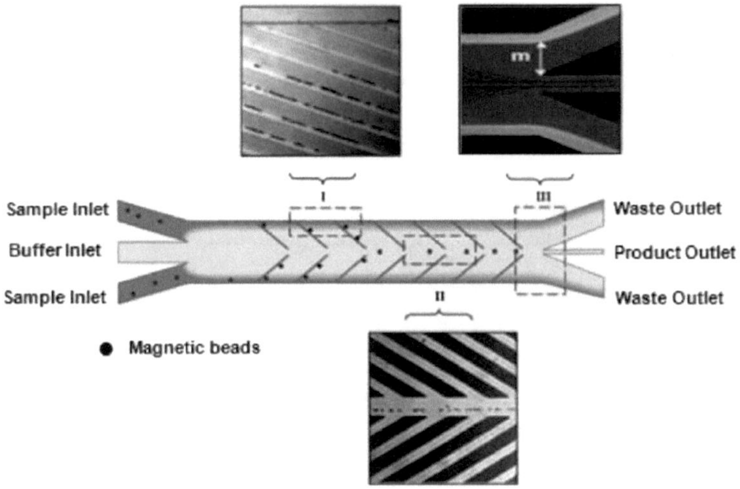

Fig. 7.20 Schematic of the flow pattern and of magnetic beads deflection within the micro-channel. Optical micrographs along the channel demonstrate the separation process (Reproduced from [101], with permission of PNAS)

reported in literature. Indeed, Tekin et al. [104] proposed a system using a magnetic bead surface coverage assay (Fig. 7.21A). First large magnetic beads functionalized with primary Ab specifically captured antigen present in a sample using active microfluidic mixing. In a second step, these beads were exposed to a surface patterned with fixed smaller magnetic beads coated with secondary Ab. A permanent magnet was positioned underneath the device. The generated magnetic field gradient forced larger magnetic beads towards the patterned substrate and biomolecular recognition permitted to trap larger beads on smaller ones. Note that the attractive magnetic interaction between the two types of beads improved Ag-Ab immunocomplex formation. Non-specific adsorption of large magnetic beads was limited by exploiting viscous drag force in the channel after magnet removal. Quantification of antigen concentration was performed by counting the number of bounded large beads. Their immunoassay protocol allowed the detection of (TNF-α) with a LOD of 1 fg·mL^{-1}.

Transport of magnetic beads in micro-channel using external magnet was also reported by Sasso et al. [105]. They developed a system in which magnetic beads were pulled from a reagent stream to the next one in the presence of an external magnet. After this transfer, the solution that carries beads was removed via a waste outlet, and beads flowed into the incubation spiral where Ag-Ab recognition occurred. Phurimsak et al. [106] developed a continuous flow reactor that consists in a microfluidic chamber with co-flowing reagent streams dedicated to C-creatine protein (CRP) detection via a sandwich immunoassay and fluorescent label. Functionalized magnetic particles were deflected across the reaction chamber under the influence of magnetic field gradient due to the presence of an external magnet. They crossed alternating streams of reagents and washing buffers as shown in Fig. 7.21B. The strategy of using ferromagnetic micro-concentrator field lines was also demonstrated for the detection of proteins. The concentrators were obtained by patterning thin or thick microstructures made of nickel or permalloy [68, 107–109]. Yu et al. [108] realized a microfluidic device integrating a nickel array that was able to trap magnetic beads in eight parallel channels as shown in Fig. 7.21C. The obtained self-assembled Ab-magnetic bead patterns allowed to simultaneously detecting eight different cancer biomarkers in the eight parallel branches of the device. A more anecdotal method that consists in packing micrometric magnetic beads in microfluidic chamber [110] was also published.

7.5.2.2 Magnetic Bio-Devices for Cell Manipulation

Two approaches are developed to manipulate cells using magnetophoretic forces: label-free approach and the use of magnetic labels on cells.

Label-Free Cell Manipulation

Two strategies are reported to manipulate label-free cells in magnetic bio-devices: manipulation of diamagnetic cells using ferromagnetic fluids and manipulation of cells possessing intrinsic magnetic properties. Manipulation of diamagnetic cells

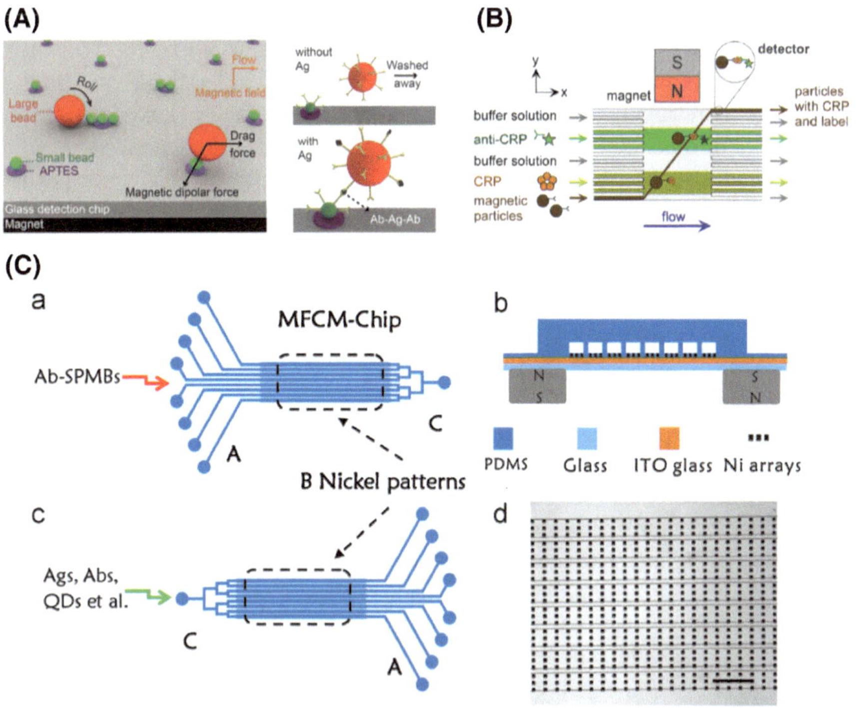

Fig. 7.21 **A** Detection area covered with a small (1.0 mm) bead pattern. Large beads loaded with Ab-Ag immunocomplexes at their surface are transported in this area and roll on the pattern under a magnetic field until they can bind specifically to the small beads by the formation of Ab-Ag-Ab immunocomplexes (Reproduced from [104], with permission of The Royal Society of Chemistry). **B** Principle of the multilaminar flow platform for a CRP sandwich assay: functionalized magnetic particles move through alternating streams of reagents and washing buffers via an external magnet (Reproduced with permission from [106], copyright (2014) American Chemical Society) **C** Schematic of the microfluidic device developed to perform detection of cancer biomarkers using self-assembled magnetic bead patterns. In d, optical image on the nickel pattern array channels (scale bare is 500 μm) (Reproduced from [108], copyright (2013), with permission of Elsevier)

remains challenging, in particular regarding ferrofluids biocompatibility. The reader can find information on this approach in these reviews [4, 111, 112]. In contrast, cells with intrinsic magnetic properties (deoxygenated red blood cells (RBC), and malaria infected RBCs) can be manipulated in bio-devices using standard aqueous buffers.

Red blood cells (RBC) have a paramagnetic behavior when deoxygenated (in veins) and a diamagnetic behavior when oxygenated (in arteries). In RBC vertebrates, hemoglobin is an iron-containing molecule that carries oxygen from the respiratory organs (lungs or gills) to the rest of the body. Oxygenated hemoglobin is diamagnetic due to the presence of paired electrons on its Fe atoms. Deoxyhemoglobin is the form of hemoglobin without the bound oxygen to the Fe atoms, thus containing four or five unpaired electrons. As a consequence, deoxy-hemoglobin presents a paramagnetic behavior, deoxygenated RBCs have a relative magnetic susceptibility of $\Delta\chi_{deox} = \chi_{deoxRBC} - \chi_{water} = 3.310^{-6}$ in water. Although this value is relatively small, it allows exerting sufficient magnetic forces on deox-RBCs in order to trap or separate them from other cells in total blood. This was illustrated by Melville et al. [113] in 1975 and since then, numerous works were published with improvement in the efficiency of RBCs recovery [7–9, 114–118]. For example, Jung et al. [116], developed a microfluidic system with a separation function consisting of six-stages (Fig. 7.22a). This approach allowed separating RBCs from human whole blood sample with a throughput up to 50 μL/h with an efficiency of 86% for RBCs. Each separation stage is a micro-concentrator of magnetic flux of thick ferromagnetic nickel microstructures that allow obtaining magnetic field gradient over the entire channel thickness. A schematic of the device with picture of separation at different stage is shown in Fig. 7.22a.

Besides general blood separation, magnetophoretic forces were used for separation of malaria infected RBCs (iRBCs) in blood. Human malaria is caused by four types of parasites, among which *Plasmodium falciparum* being the most fatal. To prosper in RBCs, malaria parasite needs to clear out hemoglobin which is toxic to them. They convert it in an insoluble crystal known as hemozoin. The ion Fe^{3+} in hemozoin has a stronger paramagnetic character than iron in hemoglobin (Fe^{2+}), and thus increase the infected RBCs magnetic susceptibility relative to water $(\Delta\chi = 1.810^{-6})$ compared to oxygenated RBC $(\Delta\chi = -0.1810^{-6})$, while remaining lower than deox RBCs [119, 120]. During infection, *P. falciparum* parasite is present at different maturation stages in RBCs, late one being the more paramagnetic. Several works were reported in literature to perform separation of

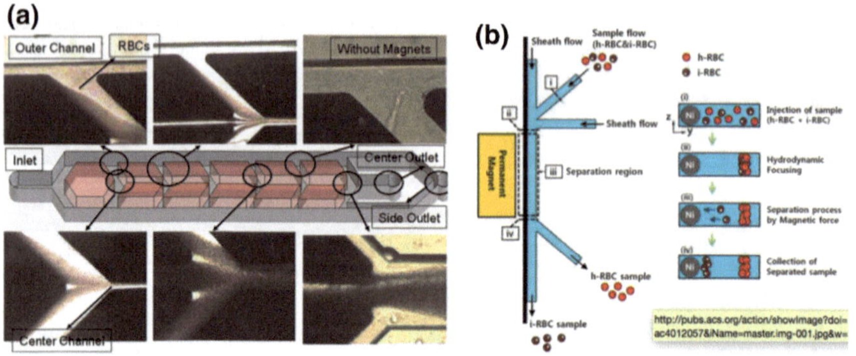

Fig. 7.22 **a** The RBCs are attracted towards ferromagnetic structures through the center channel, while white blood cells WBCs and other rare cells travel along the outer channel. The remaining RBCs in the outer channel are attracted and separated again subsequent separation stages. Micrographs of the blood cell separation at each stage are reported (Reproduced from [116], copyright (2010), with permission of Springer). **b** Schematic diagram of i-RBC separation using the paramagnetic characteristics of hemozoin in i-RBCs (Reproduced with permission from [123], copyright (2013) American Chemical Society)

iRBCs [121–125]. Nam [123] presented a system dedicated to magnetic separation of iRBCs at various stages of maturation (Fig. 7.22b). Their micro-channel possesses three inlets and two outlets. RBCs and i-RBCs mixture was injected in the central inlet, and sheath fluids were injected from the two side inlets. Their role was to focus RBCs at an optimized distance of the nickel wire. RBCs flowed in the channels, and infected ones were attracted toward the nickel wire due to magnetophoretic forces, whereas healthy ones kept flowing along the focus streamline. Using this device, the obtained recovery rate was 99% for late stage i-RBCs et 73% for early ring stage i-RBCs. Notice that, malaria diagnosis in magnetic bio-devices can also be achieved via detection of parasite biomarkers (proteins) present in whole blood, serum and plasma, using functionalized magnetic beads.

Magnetically Labeled Cells

Cells that do not possess intrinsic magnetic properties can be magnetically labeled via two approaches: (i) by internalization into the cytoplasm of nanoparticles (by endositosis [126], phagositosis [127] or transfection [128]) and (ii) by attachment of nanoparticles on their surface. The first approach, in which MNP cross the cell membrane and enter the cytoplasm, is mainly dedicated to enable MRI, drug targeted delivery or to study intracellular process. For the second approach, [129] attachment of NP on the cell membrane can be performed through specific ligand-receptor interaction, such as Ag-Ab, specific lectins or other sugar binding protein, avidin and streptavidin, for example. Other technics can be used to magnetically modify cells. For example by coating MNP on cell membrane using electrostatic interactions, binding of ferritin and magnetoferritin, entrapment of cells into biocompatible magnetic polymers and gels… [129]. Note, that it is generally accepted that attached beads do not modify cells viability, except on sensitive cell population such as stem and progenitor cells [14]. Depending on the labeling approach, the expression of the magnetophoretic force can be adjusted. Either the magnetic cell iron content can be estimated using magnetophoretic method [130], or in the particular case of immunomagnetically labeled cells, the magnetic force can be slightly adapted to depict the tagging of a single cell with numerous smaller magnetic particles [131], as expressed in Eq. (7.38)

$$\overrightarrow{F_{\mathrm{m}}} = \mu_0 V_p \beta j \vartheta \lambda \left(\overrightarrow{M}_p \cdot \overrightarrow{\nabla}\right) \overrightarrow{H} \tag{7.38}$$

β being the number of nanoparticles conjugated to the targeted Ab, j the number of Ag binding sites per cells, ϑ the fraction of Ag on the cell surface bound by targeted Ab and, λ the valence of the Ab binding, i.e. the number of Ag that binds an anti-body, and V_p, the volume of a MNP.

This expression can be modified to take into account binding of secondary Ab to sites present on the primary Ab [132]. The product $\beta j\vartheta\lambda$, is equivalent to the number of bounded magnetic particles on cells, N, $\overrightarrow{F_{\rm m}}$ can therefore be written as [65]:

$$\overrightarrow{F_{\rm m}} = N\overrightarrow{F_{\rm P}} \tag{7.39}$$

$\overrightarrow{F_{\rm P}}$ being the force exerted on a single nano-particle.

Sajay et al. [133] proposed the following expression for the magnetic force, considering that magnetic particles are saturated:

$$\overrightarrow{F_{\rm m}} = \mu_0 V_p \Delta\chi_c \left(\vec{\rm H}\cdot\vec{\nabla}\right)\vec{\rm H} \tag{7.40}$$

$\Delta\chi_c$ being the effective magnetic susceptibility of the cell-particle complex:

$$\Delta\chi_c = N\frac{R_p^3}{R_c^3}\Delta\chi_p \tag{7.41}$$

R_p and R_c being respectively the radius of the nano-particle and the cell, $\Delta\chi_p$, the effective susceptibility of the nano-particle.

One can notice that, the radius of the cell-particle complex must be considered for other forces, such as the drag and the gravitational forces described in Sect. 7.2. The cell-particle complex radius can be approximated by:

$$R = \left(R_c^3 + R_p^3\right)^{\frac{1}{3}} \tag{7.42}$$

Magnetophoretic manipulation of labelled cells in microfluidic devices are mainly focused on separation of white blood cells (WBCs) [133, 134] from blood sample, bacteria from serum or blood sample [135] and rare circulating cells from blood sample [4, 14, 84, 136]. In particular, enrichment of a target cell population and subsequent culture is an important step in the fields of molecular genetics, proteomics, biology and medicine. Regarding biomedical field, many applications require sorting of rare cells that are defined as cells representing less than 1% of the total number of cells. Among examples, we can cite circulating tumor cells (CTCs), circulating endothelial cells (CEC), endothelial progenitor cells (EPCs). Isolation of rare cells represent a big challenge for diagnostic and monitoring of diseases of strong societal interest such as cancer, cardiovascular disease and prenatal diagnostic. Labeling based on specific attachment of MNPs on the cell membrane, in particular immunomagnetic approach in which a micro- or nano-particles bind to cells through Ab-Ag recognition, suits particularly well this application. This approach is the most widely used in microfluidic systems. At the macroscale, magnetic activated cell sorting (MACS) systems consist in attracting labeled cells on a tube wall using an external magnetic field whereas unlabeled cells are eluted. Then target cells are released once the external magnetic field is removed. The U.S. Food and Drug Administration (FDA) approved the Veridex system, CellSearch^{TM}. This apparatus is a commercial system based on MACS method used to detect CTCs and currently employed for clinical diagnosis of breast, colorectal and lung cancers. However, this system works only for EpCAM surface marker expression, limiting its capture efficiency. Downscaling MACS allow portability (point-of care diagnosis), and improved control of magnetic and hydrodynamic forces exerted on target cells in order to achieve better separation efficiency. μMACS is widely employed for CTC detection in microfluidic systems [65, 84, 136].

Three parameters allow assessing the efficiency of cell separation: (i) the purity, i.e. the number of target cells among separated cells, (ii) the recovery, i.e. the number of separated cells versus the total number of cells or versus the number of target cells in the original suspension, and (iii) the viability of cells, i.e. subsequent capacity of separated cells to be considered as alive with living cell assay or cell culture. Two approaches are reported to address these challenges, positive enrichment, i.e. capturing labeled CTCs and eluting blood cells, or negative selection, i.e. capture labeled blood cells and eluting CTCs. In the first case, recovery and purity rates are reported to be superior. However, this approach presents some limitations. Because of the heterogeneous nature of target cells, they do not all express the same Ags, and some target cells with unknown surface biomarkers may be excluded from separation. In addition, the removing of MNPs from cell surface can remain challenging, and may damage cells. Indeed, the detachment process can involve saturated protein solutions, enzymes able to cleave beads form the cells, temperature induced method, which are either not efficient for all cells or may diminish cell viability.

μMacs implementing a single external permanent magnet are reported [137–142]. Their interest is to release trapped particles when removing the magnet. However, low magnetophoretic forces are reached due to magnet size and the large distance between the magnetic field gradient and the target objects in the channel. For instance, Kang et al. [139] and Wang et al. [142] proposed two different systems both based on the entrapment of labeled CTCs in a flow where both CTCs and blood cells are present. In presence of the magnetic field gradient, labeled cells were deviated towards specific compartments, and could then be released by removing the magnet. Oskumur et al. [138], combined size based filtration assay, inertial focusing and immuno-magnetic assay for efficient capture of both EpCAM positive and EpCAM negative cancer cells. In a first step, separation of WBCs and CTCs from RBCs and platelets was achieved using deterministic lateral displacement. In a second step, alignment of cells in the micro-channel was obtained using inertial focusing, and finally magnetically labeled cells, either CTCs, or WBCs, were deflected into a collection channel using magnetophoretic forces. The inertial focusing strategy, permitted to obtain the alignment of cells in a near single line, in order to prevent cellular collisions during magnetophoresis and to precisely control their position in the main channel in order to deflect them using minimal magnetic force.

Microsystems integrating an array of millimeter size permanent magnets are also reported [143, 144]. Multiple magnetic field gradients can then be achieved directly in the channel while still allowing the release of trapped particles by removing magnets. For instance, Lee et al. [143] developed a μ-MixMACS chip for CTC isolation using negative selection. It consisted in a mixing module for effective binding between WBCs and MNPs coated with CD45 Ab and a magnetic-activated cell sorting module to capture WBCs inside the channel. This resulted in the depletion of magnetically labeled WBCs. In contrast, CTCs kept flowing in the channel and were recovered in the outlet. As schematically showed in Fig. 7.23a, the bio-device was composed of three superposed channels: a micromixer, an incubation chamber and an enrichment channel. The magnet array comprised 21 NdFeB pieces. They were arrayed in a laser-cut plastic cartridge with 2 mm gap between adjacent magnets with alternating polarity. The authors compared the separation efficiency obtained in their device with a commercially available kit, EasySep™ Human CD45 Depletion Kit. They reported better CTCs recovery rate in their microsystem.

Integrated arrays of permanent micron-size magnets were also reported for cell manipulation [44, 145]. They presented the advantage to achieve higher forces thanks to magnet down-scaled down to cell size. However, releasing of trapped cells remains challenging so this approach is more adapted for in channel observation or treatment of cells.

Micro-concentrators of magnetic flux were also largely employed for labeled cells sorting in bio-device, and various methods are reported in literature [135, 146–148]. Ferromagnetic thin film approach is reported in Esmaeilsabzali et al. [146] publication. They proposed a system for detection and isolation of prostate cancer cells from blood. This system featured large trapping surface consisting in V-shaped Permalloy ferromagnetic micro-traps as shown in Fig. 7.23b.

The sample continuously flowed through the channels and labeled cells were deviated towards magnetic microtraps. In order to display magnetophoretic force in the whole channel thickness, thick approaches were also reported. For instance, Malik et al. [135], presented a 3D polymer –based microchip for immuno-magnetic capture and release of *Listeria monocytogenes*. Their device was made of cyclo olefin copolymer (COC) and contained an array of 3D micro-pillars (diameter of 20 μm; height of 67 μm) coated with a layer of nickel. A picture of the bio-device highlighting the integrated pillars, and a schematic of principle are shown in Fig. 7.23c. The fabrication process involving hot-embossing and electroless deposition is compatible with mass-production. In addition, they designed their system in a way that the magnetic field is applied perpendicular to the flow in order to facilitate the release of particles when the magnetic field is removed. It comprised a wide capture chamber, containing the pillars, to process large volumes. Faivre et al. [67] developed an elegant approach based in composite polymer for labeled cell manipulation. They proposed a trapping device using thick composite iron-PDMS (I-PDMS) microstructures to generate large magnetic field gradient. Saliba et al. [148] developed an original method, consisting in an array of column of self-assembled bio-functionnalized superparamagnetic particles, on which target

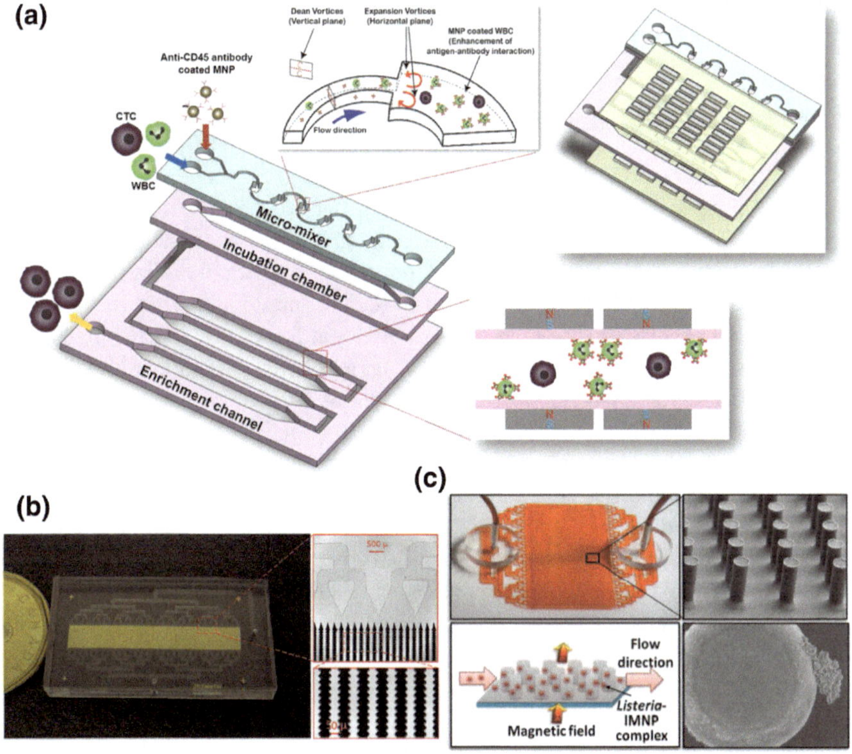

Fig. 7.23 **a** Schematic of a μ-MixMACS chip for one-step CTC isolation using the negative depletion approach. It is composed of three parts: a microfluidic mixer to favor binding between CD45 conjugated MNPs and WBCs, an incubation chamber that permit stable MNP conjugation to the WBCs, and MACS function dedicated to capture the labeled WBCs and elute CTCs through the outlet (Reproduced from [143], copyright (2017), with permission of Elsevier). **b** Photograph of the microchip integrating patterned Permalloy strips. The resulting V-shaped magnetic microtraps seen under microscope are shown (Reproduced from [146], copyright (2016), with permission of Springer). **c** Photograph of 3D polymer-based chip and of the integrated magnetic pillars, and schematic of principle of the device (Reproduced from [135], with permission of The Royal Society of Chemistry)

cells are trapped through Ag-Ab recognition. The substrate was patterned by microcontact printing in order to locally deposit water-based ferrofluid onto glass. These ferromagnetic dots acted as micro-concentrator of magnetic flux in order to assemble magnetic particles in columns. Then, these magnetic particles were functionalized and, via Ag-Ab interaction, cells were trapped on the columns.

7.6 Conclusion

Associating magnetism to microfluidics permit to address numerous challenges for biomedical applications. This combination permit precise manipulation of functionalized micro-particles or biological entities such as magnetically labeled cells or natively paramagnetic RBCs. Indeed, micro- and nano-objects exposed to a non–uniform magnetic field undergo a translational magnetic force referred as magnetophoresis. This force depends on the magnetic nature of the target object and can be greatly increased by integrating micro-sources of magnetic flux directly inside of microfluidic channel. Thus, the magnetic force can predominate, in this environment, over other forces such as drag or gravitational forces. The precise balance of these forces allows describing the motion of target objects and to manipulate them. The implementation of magnetophoretic functions in bio-device are mainly performed using permanent micro-magnet or micro-concentrators of magnetic flux. Both options can lead to comparable maximum force intensities for any given target objects. Permanent micro-magnets do not require any exterior flux to operate, which permits to develop devices with incomparable compactness, but in turns are bound to remain active, with fixed force strength. In contrast, the micro-concentrators of magnetic flux offer easy and real time modulation of the generated force, allowing releasing the trapped object when required. Concerning bio-analysis, magnetic particles present the advantage of versatility in their functionalization. They can thus serve to develop various detection routes, like for nucleic acid and proteins, or for labeling diamagnetic cells, which was notably exploited for blood fractionation, CTCs trapping and sorting. As they do not set restrictions on the target objects, magnetophoretic forces can be easily associated to other forces, dielectrophoresis for instance [149], to improve their separation efficiency.

References

1. M.A.M. Gijs, F. Lacharme, U. Lehmann, Microfluidic applications of magnetic particles for biological analysis and catalysis. Chem. Rev. **110**, 1518–1563 (2010)
2. N. Pamme, On-chip bioanalysis with magnetic particles. Curr. Opin. Chem. Biol. **16**, 436–443 (2012)
3. N. Pamme, J.C.T. Eijkel, A. Manz, On-chip free-flow magnetophoresis: Separation and detection of mixtures of magnetic particles in continuous flow. J. Magn. Magn. Mater. **307**, 237–244 (2006)
4. M. Hejazian, W. Li, N.T. Nguyen, Lab on a chip for continuous-flow magnetic cell separation. Lab Chip (2015). https://doi.org/10.1039/c4lc01422g
5. L. Borlido, A.M. Azevedo, A.C.A. Roque, M.R. Aires-Barros, Magnetic separations in biotechnology. Biotech. Adv. **13**, 1374–1385 (2013)
6. T.P. Forbe, S.P. Forry, Microfluidic magnetophoretic separations of immunomagnetically labeled rare mammalian cells. Lab Chip **12**, 1471 (2012)
7. J. Jung, H.K. Han, Lateral-driven continuous magnetophoretic separation of blood cells. Appl. Phys. Lett. **93**, 223902 (2008)

8. M. Zborowski, G.R. Ostera, L.R. Moore, S. Milliron, J.J. Chalmers, Red blood cell magnetophoresis. Biophys. J. **84**, 2638–2645 (2003)
9. B.Y. Qu, Z.Y. Wu, F. Fang, Z.M. Bai, D.Z. Yang, S.K. Xu, A glass microfluidic chip for continuous blood cell sorting by a magnetic gradient without labeling. Anal. Bioanal. Chem. **392**, 1317–1324 (2008)
10. J. Luo, X. He, D. Andre' d'Avignon, J.J.H. Ackerman, D.A. Yablonskiy, Protein-induced water 1H MR frequency shifts: contributions from magnetic susceptibility and exchange effects. J. Magn. Reson. **202**(1), 102 (2010)
11. M.D. Graham, P.R. Selvin, Separation of lanthanide-binding cells. IEEE T. Magn. **18**(6), 1523 (1982)
12. A.D. Buckingham, T. Schaefer, W.G. Schneider, Solvent effects in nuclear magnetic resonance spectra. J. Chem. Phys. **32**, 1227 (1960)
13. J.M.D. Coey, *Magnetism and Magnetic Materials* (Cambridge University Press, Cambridge, 2010)
14. B.D. Plouffe, S.K. Murthy, L.H. Lewis, Fundamentals and application of magnetic particles in cell isolation and enrichment: a review. Rep. Prog. Phys. **78**(1), 016601 (2015)
15. P. Talagala, P.S. Fodor, D. Haddad, R. Naik, L.E. Wenger, P.P. Vaishnava, V.M. Naik, Determination of magnetic exchange stiffness and surface anisotropy constants in epitaxial $Ni_{1-x}Co_x$(001) films. Phys. Rev. B **66**, 144426 (2002)
16. W. Wu, X.H. Xiao, S.F. Zhang, T.C. Peng, J. Zhou, F. Ren, C.Z. Jiang, Synthesis and magnetic properties of maghemite (gamma-Fe_2O_3) short-nanotubes. Nanoscale Res. Lett. **5** (9), 1474–1479 (2010)
17. E.P. Furlani, Magnetic biotransport: analysis and applications. Materials **3**, 2412–2446 (2010)
18. R. Wirix-Speetjens, W. Fyen, K. Xu, J. De Boeck, G. Borghs, A force study of on-chip magnetic particle transport based on tapered conductors. IEEE T. Magn. **41**, 4128 (2005)
19. X. Han, Y. Feng, Q. Cao, L. Li, Three-dimensional analysis and enhancement of continuous magnetic separation of particles in microfluidics. Microfluid. Nanofluid. (2016). https://doi.org/10.1007/s10404-014-1516-6
20. E.P. Furlani, K.C. Ng, Analytical model of magnetic nanoparticle transport and capture in the microvasculature. Phys. Rev. E **73**, 061919 (2006)
21. Q. Cao, X. Han, L. Li, Configurations and control of magnetic fields for manipulating magnetic particles in microfluidic applications: magnet systems and manipulation mechanisms. Lab Chip (2014). https://doi.org/10.1039/c4lc00367e
22. W. Zhao, R. Cheng, J.R. Miller, L. Mao, Label-free microfluidic manipulation of particles and cells in magnetic liquids. Adv. Funct. Mater. (2016). https://doi.org/10.1002/adfm.201504178
23. R. Zhou, Q. Yang, F. Bai, J.A. Werner, H. Shi, Y. Ma, C. Wang, Microfluid. Nanofluid. **20**, 110 (2016)
24. A. Sinha, R. Ganguly, I.K. Puri, Magnetic separation from superparamagnetic particle suspensions. J. Magn. Magn. Mater. **321**, 2251–2256 (2009)
25. R. Gerber, Magnetic filtration of ultra-fine particles. IEEE T. Magn. **20**, 1159 (1984)
26. R. Gerber, M. Takayasu, F.J. Friedlaender, Generalization of HGMS theory: the capture of ultra-fine particles. IEEE T. Magn. **19**, 2115 (1983)
27. M. Takayasu, R. Gerber, F.J. Friedlaender, Magnetic separation of submicron particles. IEEE T. Magn. **19**, 2112 (1983)
28. D. Fletcher, Fine particle high gradient magnetic entrapment. IEEE T. Magn. **27**, 3655 (1991)
29. E.P. Furlani, K.C. Ng, Nanoscale magnetic biotransport with application to magnetofection. Phys. Rev. E **77**, 061914 (2008)
30. A. Sinha, R. Ganguly, A.K. De, I.K. Puri, Single magnetic particle dynamics in a microchannel. Phys. Fluids **19**, 117102 (2007)

31. K. Nandy, S. Chaudhuri, R. Ganguly, I.K. Puri, Analytical model for the magnetophoretic capture of magnetic microspheres in microfluidic devices. J. Magn. Magn. Mater. **320**, 1398–1405 (2008)
32. R. Fulcrand, A. Bancaud, C. Escriba, Q. He, S. Charlot, A. Boukabache, A.M. Gué, On chip magnetic actuator for batch-mode dynamic manipulation of magnetic particles in compact lab-on-chip. Sens. Actuat. B Chem. **160**(1), 1520–1528 (2011)
33. P.T. Tseng, D. Di Carlo, J.W. Judy, Rapid and dynamic intracellular patterning of cell-internalized magnetic fluorescent nanoparticles. Nano Lett. **9**(8), 3053–3059 (2009)
34. L. Zanini, N.M. Dempsey, D. Givord, G. Reyne, F. Dumas-Bouchiat, Autonomous micro-magnet based systems for highly efficient magnetic separation. Appl. Phys. Lett. **99**, 232504 (2011)
35. N.M. Dempsey, in *Nanoscale Magnetic Materials and Applications*, ed. by J.P. Liu, E. Fullerton, O. Gutfleisch, D.J. Sellmyer (Springer, New York, 2009), pp. 661–680
36. A. Beyzavi, N.T. Nguyen, Modeling and optimization of planar microcoils. J. Micromech. Microeng. **18**, 095018 (2008)
37. R. Fulcrand, D. Jugieu, C. Escriba, A. Bancaud, D. Bourrier, A. Boukabache, A.M. Gué, Development of a flexible microfluidic system integrating magnetic micro-actuators for trapping biological species. J. Micromech. Microeng. **19**, 105019 (2009)
38. K. Yamauchi, J.W.G. Janssen, A.P.M. Kentgens, Implementing solenoid microcoils for wide-line solid-state NMR. J. Magn. Reson. **167**, 87–96 (2004)
39. A. Haase, F. Odoj, M. Von Kienlin, J. Warnking, F. Fidler, A. Weisser, M. Nittka, E. Rommel, T. Lanz, B. Kalusche, M. Griswold, NMR probeheads for in vivo applications. Concept Magn. Reson. A **12**(6), 361–388 (2000)
40. K. Kamata, Z. Piao, S. Suzuki, T. Fujimori, W. Tajiri, K. Nagai, T. Iyoda, A. Yamada, T. Hayakawa, M. Ishiwara, S. Horaguchi, A. Belay, T. Tanaka, K. Takano, M. Hangyo, Spirulina-templated metal microcoils with controlled helical structures for THz electromagnetic responses. Sci. Rep. **4**, 4919 (2014)
41. J.M.D. Coey, New permanent magnets; manganese compounds. J. Phys. Condens. Mat. **26**, 064211 (2014)
42. J.M.D. Coey, Hard magnetic materials: a perspective. IEEE T. Magn. **47**, 12 (2011)
43. R. Engel-Hebert, T. Hesjedal, Calculation of the magnetic stray field of a uniaxial magnetic domain. J. Appl. Phys. **97**, 074504 (2005)
44. V. Zlabotskii, A. Dejneka, S. Kubinova, D. Le Roy, F. Dumas-Bouchiat, D. Givord, N.M. Dempsey, E. Sykova, Life on magnets: stem cell networking on micro-magnet arrays. PLoS ONE **8**(8), e70416 (2013)
45. M. Nakano, K. Yamaguchi, T. Yanai, F. Yamashita, H. Fukunaga, Change in the direction of anisotropy in PLD-fabricated Sm-Co thick film magnets. J. Appl. Phys. **105**, 7 (2009)
46. F. Ruiyi, F. Qingqing, Z. Sheng, P. Chubing, D. Daosheng, New permanent magnetic MnBiDy alloy films. J. Appl. Phys. **76**, 10 (1994)
47. F.M. Rhen, G. Hinds, C. O'Reilly, J.M.D. Coey, Electrodeposited FePt films. IEEE T. Magn. **39**, 5 (2003)
48. N.M. Dempsey, A. Walther, F. May, D. Givord, K. Khlopkov, O. Gutfleisch, High performance hard magnetic NdFeB thick films for integration into micro-electro-mechanical systems. Appl. Phys. Lett. **90**, 092509 (2007)
49. C. Delcour, G. Bugnicourt, N.M. Dempsey, F. Dumas-Bouchiat, C. Villard, Combined magnetic and chemical patterning for neural architectures. J. Phys. D Appl. Phys. **47**, 425403 (2014)
50. A. Walther, C. Marcoux, B. Desloges, R. Grechishkin, D. Givord, N.M. Dempsey, Micro-patterning of NdFeB and SmCo magnet films for integration into micro-electro-mechanical-systems. J. Magn. Magn. Mater. **321**, 590–594 (2008)
51. T. Okuda, A. Sugimura, O. Eryu, L.K.E.B. Serrona, N. Adachi, I. Sakamoto, A. Nakanishi, Nd-Fe-B thin films with perpendicular magnetic anisotropy and high coercivity prepared by pulsed laser annealing. Jpn. J. Appl. Phys. **42**, 6859–6864 (2003)

52. J. Topfer, B. Pawlowski, H. Beer, K. Plotner, P. Hofmann, J. Herrfurth, Multi-pole magnetization of NdFeB magnets for magnetic micro-actuators and its characterization with a magnetic field mapping device. J. Magn. Magn. Mater. **270**, 124–129 (2004)
53. J. Topfer, V. Christoph, Multi-pole magnetization of NdFeB sintered magnets and thick films for magnetic micro-actuators. Sens. Actuat. A Phys. **113**, 257–263 (2004)
54. F. Dumas-Bouchiat, L. Zanini, M. Kustov, N.M. Dempsey, R. Grechishkin, K. Hasselbach, J.C. Orlianges, C. Champeaux, A. Catherinot, D. Givord, Thermomagnetically patterned micromagnets. Appl. Phys. Lett. **96**, 102511 (2010)
55. O.D. Oniku, P.V. Ryiz, A. Garraud, D.P. Arnold, Imprinting of fine-scale magnetic patterns in electroplated hard magnetic films using magnetic foil masks. J. Appl. Phys. **115**, 17A718 (2014)
56. D. Le Roy, G. Shaw, R. Haettel, K. Hasselbach, F. Dumas-Bouchiat, D. Givord, N.M. Dempsey, Fabrication and characterization of polymer membranes with integrated arrays of high performance micro-magnets. Mater. Today Commun. **6**, 50–55 (2016)
57. M. Pallapa, J.T.W. Yeow, A review of hybrid techniques for the fabrication of hard magnetic microactuators based on bonded magnetic powder. Smart Mater. Struct. **24**, 025007 (2015)
58. O.D. Oniku, B.J. Bowers, S.B. Shetye, N. Wang, D.P. Arnold, Permanent magnet microstructures using dry-pressed magnetic powders. J. Micromech. Microeng. **23**, 075027 (2013)
59. N.M. Dempsey, D. Le Roy, H. Marelli-Mathevon, G. Shaw, A. Dias, R.G.B. Kramer, M. Le Viet Cuong Kustov, L. Zanini, C. Villard, K. Hasselbach, C. Tomba, F. Dumas-Bouchiat, Micro-magnetic imprinting of high gradient magnetic flux sources. Appl. Phys. Lett. **104** (26), 262401 (2014)
60. C. Velez, I. Torres-Diaz, L. Maldonado-Camargo, C. Rinaldi, D. Arnold, Magnetic assembly and cross-linking of nanoparticles for releasable magnetic microstructures. ACS Nano **9**(10), 10165–10172 (2015)
61. G. Shaw, R.G.B. Kramer, N.M. Demspey, K. Hasselbach, A scanning hall probe microscope for high resolution, large area, variable height magnetic field imaging. Rev. Sci. Instrum. **87** (11), 113702 (2016)
62. B.D. Cullity, C.D. Graham, *Introduction to Magnetic Materials*, 2nd edn. (Wiley-IEEE press, Piscataway, 2008)
63. O. Gutfleisch, M.A. Willard, E. Bruck, C.H. Chen, S.G. Sankar, J.P. Liu, Magnetic materials and devices for the 21st century: stronger, lighter, and more energy efficient. Adv. Mater. **23**, 821–842 (2011)
64. J.A. Osborn, Demagnetizing factors of the general ellipsoid. Phys. Rev. **67**, 11 (1945)
65. P. Chen, Y.Y. Huang, K. Hoshino, X. Zhang, Multiscale immunomagnetic enrichment of circulating tumor cells: from tubes to microchips. Lab Chip **14**, 446 (2014)
66. T. Deng, M. Prentiss, G.M. Whitesides, Fabrication of magnetic microfiltration systems using soft lithography. Appl. Phys. Lett. **80**(3), 461 (2002)
67. F. Faivre, R. Gelszinnis, J. Degouttes, N. Terrier, C. Riviere, R. Ferrigno, A.L. Deman, Magnetophoretic manipulation in microsystem using carbonyl iron-polydimethylsiloxane microstructures. Biomicrofluidics **8**, 054103 (2014)
68. M. Abonnenc, A.L. Gassner, J. Morandini, J. Josserand, H.H. Girault, Magnetic track array for efficient bead capture in microchannels. Anal. Bioanal. Chem. **395**, 747–757 (2009)
69. P. Chen, Y.Y. Huang, G. Bhave, K. Hoshino, X. Zhang, Inkjet-Print micromagnet array on glass slides for immunomagnetic enrichment of circulating tumor cells. Ann. Biomed. Eng. **44**(5), 1710–1720 (2016)
70. D. Le Roy, D. Dhungana, L. Ourry, M. Faivre, R. Ferrigno, A. Tamion, V. Dupuis, V. Salles, A.L. Deman, Anisotropic ferromagnetic polymer: a first step for their implementation in microfluidic systems. AIP Adv. **6**(5), 056604 (2016)
71. C.C. Berry, A.S.G. Curtis, Functionalisation of magnetic nanoparticles for applications in biomedicine. J. Phys. D Appl. Phys. **36**, R198 (2003)
72. Q.A. Pankhurst, J. Connolly, S.K. Jones, J. Dobson, Applications of magnetic nanoparticles in biomedicine. J. Phys. D Appl. Phys. **36**, R167 (2003)

73. A. Singh, S.K. Sahoo, Magnetic nanoparticles: a novel platform for cancer theranostics. Drug Discov. Today **19**(4), 474–481 (2014)
74. E. Agostinelli, F. Vianello, G. Magliulo, T. Thomas, T.J. Thomas, Nanoparticle strategies for cancer therapeutics: nucleic acids, polyamines, bovine serum amine oxidase and iron oxide nanoparticles. Int. J. Oncol. **46**(5), 16 (2015)
75. J.K. Oh, J.M. Park, Iron oxide-based superparamagnetic polymeric nanomaterials: design, preparation, and biomedical application. Prog. Polym. Sci. **36**, 168–189 (2011)
76. S.C.N. Tang, I.M.C. Lo, Magnetic nanoparticles: essential factors for sustainable environmental applications. Water Res. **47**, 2613–2632 (2013)
77. R.B. Dominguez, G.A. Alonso, R. Muños, A. Hayat, J.L. Marty, Design of a novel magnetic particles based electrochemical biosensor for organophosphate insecticide detection in flow injection analysis. Sens. Actuat. B Chem. **208**, 491 (2015)
78. X. Qu, P.J.J. Alvarez, Q. Li, Applications of nanotechnology in water and wastewater treatment. Water Res. **47**, 3931–3946 (2013)
79. M. Wierucka, M. Biziuk, Application of magnetic nanoparticles for magnetic solid-phase extraction in preparing biological, environmental and food samples. TRAC-Trend. Anal. Chem. **59**, 50–58 (2014)
80. T. Jamshaid, R.T.T. Neto, M. Eissa, N. Zine, M.H. Kunita, A.E. El-Salhi, A. Elaissari, Magnetic particles: from preparation to lab-on-a-chip, biosensors, microsystems and microfluidics applications. TRAC-Trend. Anal. Chem. **79**, 344–362 (2016)
81. J.C. Jokerst, J.M. Emory, C.S. Henry, Advances in microfluidics for environmental analysis. Analyst **137**, 24 (2012)
82. Q. Ramadan, M.A.M. Gijs, Microfluidic applications of functionalized magnetic particles for environmental analysis: focus on waterborne pathogen detection. Microfluid. Nanofluid. **13**, 529–542 (2012)
83. J.H. Kang, J.K. Park, Magnetophoretic continuous purification of single-walled carbon nanotubes from catalytic impurities in a microfluidic device. Small **3**(10), 1784–1791 (2007)
84. Y. Zhu, K. Kekalo, C.N. Dong, Y.Y. Huang, F. Schubitidze, K.E. Griwold, I. Baker, J.X. J. Zhang, Magnetic-nanoparticle-based immunoassays-on-chip: materials synthesis, surface functionalization, and cancer cell screening. Adv. Funct. Mater. (2016). https://doi.org/10.1002/adfm.201504176
85. D.L. Leslie-Pelecky, R.D. Rieke, Magnetic properties of nanostructured materials. Chem. Mater. **8**, 1770–1783 (1996)
86. H. Lee, T.H. Shin, J. Cheon, R. Weissleder, Recent developments in magnetic diagnostic systems. Chem. Rev. **115**, 10690–10724 (2015)
87. A.H. Lu, E.L. Salabas, F. Schüth, Magnetic nanoparticles: synthesis, protection, functionalization, and application. Angew. Chem. Int. Ed. **46**, 1222–1244 (2007)
88. L.H. Reddy, J.L. Arias, J. Nicolas, P. Couvreur, Magnetic nanoparticles: design and characterization, toxicity and biocompatibility, pharmaceutical and biomedical applications. Chem. Rev. **112**, 5818–5878 (2012)
89. S. Laurent, D. Forge, M. Port, A. Roch, C. Robic, L. Vander Elst, R.N. Muller, Magnetic iron oxide nanoparticles: synthesis, stabilization, vectorization, physicochemical characterizations, and biological applications. Chem. Rev. **108**, 2064–2110 (2008)
90. D. Horak, M. Babic, H. Mackova, M.J. Benes, Preparation and properties of magnetic nano-and microsized particles for biological and environmental separations. J. Sep. Sci. **30**, 1751–1772 (2007)
91. J. Kim, M. Johnson, P. Hill, B.K. Gale, Microfluidic sample preparation: cell lysis and nucleic acid purification. Integr. Biol. **1**, 574–586 (2009)
92. P. Liu, X. Li, S.A. Greenspoon, J.R. Scherer, R.A. Mathies, Integrated DNA purification, PCR, sample cleanup, and capillary electrophoresis microchip for forensic human identification. Lab Chip **11**, 1041 (2011)

93. O. Strohmeier, A. Emperle, G. Roth, D. Mark, R. Zengerle, F. von Stetten, Centrifugal gas-phase transition magnetophoresis (GTM) – a generic method for automation of magnetic bead based assays on the centrifugal microfluidic platform and application to DNA purification. Lab Chip **13**, 146–155 (2013)
94. S. Park, Y. Zhang, S. Lin, T.H. Wang, S. Yang, Advances in microfluidic PCR for point-of-care infectious disease diagnostics. Biotechnol. Adv. **29**, 830–839 (2011)
95. Y. Zhang, H.-R. Jiang, A review on continuous-flow microfluidic PCR in droplets: advances, challenges and future. Anal. Chim. Acta (2016). https://doi.org/10.1016/j.aca.2016.02.006
96. T. Kokalj, E. Pérez-Tuiz, J. Lammertyn, Building bio-assays with magnetic particles on a digital microfluidic platform. New Biotechnol. (2015). https://doi.org/10.1016/j.nbt.2015.03.007
97. M. Kühnemund, D. Witters, M. Nilsson, J. Lammertyn, Circle-to-circle amplification on a digital microfluidic chip for amplified single molecule detection. Lab Chip (2014). https://doi.org/10.1039/c4lc00348a
98. P.Y. Hung, P.S. Jiang, E.S. Lee, S.K. Fan, Y.W. Lu, Genomic DNA extraction from whole blood using a digital microfluidic (DMF) platform with magnetic beads. Mycrosyst. Technol. (2015). https://doi.org/10.1007/s00542-015-2512-9
99. G. Czilwik, T. Messinger, O. Strohmeier, S. Wadle, F. von Stetten, N. Paust, G. Roth, R. Zengerle, P. Saarinen, J. Nittymäki, K. McAllister, O. Sheils, J.O. Leary, D. Mark, Rapid molecular infectious disease diagnostics on a fully automated centrifugal-microfluidic LabDisk system using highly sensitive nested PCR with integrated sample preparation. Lab Chip **15**, 3749–3759 (2015)
100. N. Vergauwe, S. Vermeir, J.B. Wacker, F. Ceyssens, M. Cornaglia, R. Puers, M.A.M. Gijs, J. Lammertyn, D. Witters, A highly efficient extraction protocol for magnetic particles on a digital microfluidic chip. Sens. Actuat. B Chem. **196**, 282–291 (2014)
101. X. Lou, J. Qian, Y. Xiao, A.E. Gerdon, E.T. Lagally, P. Atzberger, T.M. Tarasow, A. J. Heeger, H.T. Soh, Micromagnetic selection of aptamers in microfluidic channels. P. Natl. Acad. Sci. USA **106**, 2989–2994 (2009)
102. H.C. Tekin, M.A.M. Gijs, Ultrasensitive protein detection: a case for microfluidic magnetic bead-based assays. Lab Chip (2013). https://doi.org/10.1039/c3lc50477h
103. Y.K. Hahn, J.K. Park, Versatile immunoassays based on isomagnetophoresis. Lab Chip **11**, 2045 (2011)
104. H.C. Tekin, M. Cornaglia, M.A.M. Gijs, Attomolar protein detection using a magnetic bead surface coverage assay. Lab Chip **13**, 1053 (2013)
105. L.A. Sasso, I.H. Johnston, M. Zheng, R.K. Gupte, A. Undar, J.D. Zahn, Automated microfluidic processing platform for multiplexed magnetic bead immunoassays. Microfluid. Nanofluid. **13**, 603–612 (2012)
106. C. Phurimsak, M.D. Tarn, S.A. Peyman, J. Greenman, N. Pamme, On-chip determination of C-reactive protein using magnetic particles in continuous flow. Anal. Chem. **86**, 10552–10559 (2014)
107. J.D. Adams, U. Kim, H.T. Soh, Multitarget magnetic activated cell sorter. P. Natl. Acad. Sci. USA **105**, 18165–18170 (2008)
108. X. Yu, H.S. Xia, A.D. Sun, Y. Lin, K. Wang, J. Yu, H. Tang, D.W. Pang, Z.L. Zhang, On-chip dual detection of cancer biomarkers directly in serum based on self-assembled magnetic bead patterns and quantum dots. Biosens. Bioelectron. **41**, 129–136 (2013)
109. E. Mirowski, J. Moreland, S.E. Russek, M.J. Donahue, Manipulation of magnetic particles by patterned arrays of magnetic spin-valve traps. Appl. Phys. Lett. **84**, 1786 (2007)
110. B. Teste, F. Mallogi, A.L. Gassner, T. Georgelin, J.M. Siaugue, A. Varenne, H. Gurault, S. Descroix, Magnetic core shell nanoparticles trapping in a microdevice generating high magnetic gradient. Lab Chip **11**, 833 (2011)
111. P. Kauffmann, A. Ith, D. O'Brien, V. Gaude, F. Boué, S. Combe, F. Bruckert, B. Schaack, N.M. Dempsey, V. Haguetb, G. Reyne, Diamagnetically trapped arrays of living cells above micromagnets. Lab Chip (2011). https://doi.org/10.1039/c1lc20232d

112. R.J. Yang, H.H. Hou, Y.N. Wang, L.M. Fu, Micro-magnetofluidics in microfluidic systems: a review. Sens. Actuat. B Chem. **224**, 1–15 (2006)
113. D. Melville, F. Paul, S. Roath, Direct magnetic separation of red cells from whole blood. Nature **255**, 706 (1975)
114. K.H. Han, A.B. Frazier, Continuous magnetophoretic separation of blood cells in microdevice format. J. Appl. Phys. **96**, 5797–5802 (2004)
115. K.H. Han, A.B. Frazier, Paramagnetic capture mode magnetophoretic microseparator for high efficiency blood cell separations. Lab Chip **6**, 265–273 (2006)
116. J. Jung, Y. Choi, K.H. Han, A.B. Frazier, Six-stage cascade paramagnetic mode magnetophoretic separation system for human blood samples. Biomed. Microdevices **12**, 637–645 (2010)
117. J.D. Chen, D. Chen, T. Yuan, Y. Xie, X. Chen, A microfluidic chip for direct and rapid trapping of white blood cells from whole blood. Biomicrofluidics **7**, 034106 (2013)
118. L.R. Moore, F. Nehl, J. Dorn, J.J. Chalmers, M. Zborowski, Open gradient magnetic red blood cell sorter evaluation on model cell mixtures. IEEE T. Magn. **49**, 309–315 (2013)
119. S. Hackett, J. Hamzah, T.M.E. Davis, T.G. St Pierre, Magnetic susceptibility of iron in malaria-infected red blood cells. Biochim. Biophys. Acta **1792**, 93–99 (2009)
120. L.R. Moore, H. Fujioka, P.S. Williams, J.J. Chalmers, B. Grimberg, P.A. Zimmerman, M. Zborowski, Hemoglobin degradation in malaria-infected erythrocytes determined from live cell magnetophoresis. FASEB J. **20**, 747–749 (2006)
121. C. Ribaut, A. Berry, S. Chevalley, K. Reybier, I. Molais, D. Parzy, F. Nepveu, F. Benoit-Vical, A. Valentin, Concentration and purification by magnetic separation of the erythrocytic stages of all human plasmodium species. Malaria J. **7**, 45 (2008)
122. S.C. Bhakdi, A. Ottinger, S. Somsri, P. Statongno, P. Pannadaporn, P. Chimma, P. Malasit, K. Pattanapanyasat, H.P.H. Neumann, Optimized high gradient magnetic separation for isolation of plasmodium-infected red blood cells. Malaria J. **9**, 38 (2010)
123. J. Nam, H. Huang, H. Lim, C. Lim, S. Shin, Magnetic separation of malaria-infected red blood cells in various developmental stages. Anal. Chem. **85**, 7316–7323 (2013)
124. W.T. Wu, A.B. Martin, A. Gandini, N. Aubry, M. Massoudi, J.F. Antaki, Design of microfluidic channels for magnetic separation of malaria–infected red blood cells. Microfluid. Nanofluid. **20**, 41 (2016)
125. M. de Souza Castilho, T. Laube, H. Yamanaka, S. Alegret, M.I. Pividori, Magneto immunoassays for plasmodium falciparum histidine-rich protein 2 related to malaria based on magnetic nanoparticles. Anal. Chem. **83**, 5570–5577 (2011)
126. A.T. Jones, M. Gumbleton, R. Duncan, Understanding endocytic pathways and intracellular trafficking: a prerequisite for effective design of advanced drug delivery systems. Adv. Drug Deliv. Reviews **55**, 1353–1357 (2003)
127. S. Marion, C. Wilhelm, H. Voigt, J.C. Bacri, N. Guillén, Overexpression of myosin IB in living *Entamoeba histolytica* enhances cytoplasm viscosity and reduces phagocytosis. J. Cell Sci. **117**(15), 3271–3279 (2004)
128. K. Montet-Abou, X. Montet, R. Weissleder, L. Josephson, Cell internalization of magnetic nanoparticles using transfection agents. Mol. Imaging **6**(1), 1 (2007)
129. I. Safarik, K. Pospiskova, E. Baldikova, Z. Maderova, M. Safarikova, in *Engineering of nanobiomaterials*, ed. by A. Grumezescu (Elsevier, Oxford, 2016), pp. 145–180
130. C. Wilhelm, F. Gazeau, J.C. Bacri, Magnetophoresis and ferromagnetic resonance of magnetically labeled cells. Eur. Biophys. J. **31**, 118 (2002)
131. K.E. McCloskey, J.J. Chalmers, M. Zborowski, Magnetic cell separation: characterization of magnetophoretic mobility. Anal. Chem. **75**, 6868–6874 (2003)
132. K.E. McCloskey, K. Comella, J.J. Chalmers, S. Margel, M. Zborowski, Mobility measurements of immunomagnetically labeled cells allow quantification of secondary antibody binding amplification. Biotechnol. Bioeng. **75**, 642–655 (2001)
133. B.N.G. Sajay, C.P. Chang, P.D. Puiu, Microfluidic immunomagnetic cell separation from whole blood. J. Chromatogr. B **1011**, 77–88 (2016)

134. D.W. Inglis, R. Riehn, R.H. Austin, J.C. Sturm, Continuous microfluidic immunomagnetic cell separation. Appl. Phys. Lett. **85**, 21 (2004)
135. L. Malic, X. Zhang, D. Brassard, L. Clime, J. Daoud, C. Luebbert, V. Barrere, A. Boutin, S. Bidawid, J. Farber, N. Corneau, T. Veres, Polymer-based microfluidic chip for rapid and efficient immunomagnetic capture and release of Listeria monocytogenes. Lab Chip **15**, 3994–4007 (2015)
136. J.H. Myung, S. Hong, Microfluidic devices to enrich and isolate circulating tumor cells. Lab Chip **15**, 4500 (2015)
137. Y.Y. Huang, K. Hoshino, P. Chen, C.H. Wu, N. Lane, M. Huebschman, H.Y. Liu, K. Sokolov, J.W. Uhr, E.P. Frenkel, J.X.J. Zhang, Immunomagnetic nanoscreening of circulating tumor cells with a motion controlled microfluidic system. Biomed. Microdev. **15**, 673 (2013)
138. E. Ozkumur, A.M. Shah, J.C. Ciciliano, B.L. Emmink, D.T. Miyamoto, E. Brachtel, M. Yu, P.I. Chen, B. Morgan, J. Trautwein, A. Kimura, S. Sengupta, S.L. Stott, N.M. Karabacak, T. A. Barber, J.R. Walsh, K. Smith, P.S. Spuhler, J.P. Sullivan, R.J. Lee, D.T. Ting, X. Luo, A. T. Shaw, A. Bardia, L.V. Sequist, D.N. Louis, S. Maheswaran, R. Kapur, D.A. Haber, M. Toner, Inertial focusing for tumor antigen–dependent and –independent sorting of rare circulating tumor cells. Sci. Transl. Med. **5**, 179ra147 (2013)
139. J.H. Kang, S. Krause, H. Tobin, A. Mammoto, M. Kanapathipillai, D.E. Ingber, A combined micromagnetic-microfluidic device for rapid capture and culture of rare circulating tumor cells. Lab Chip **12**, 2175–2181 (2012)
140. J. Kim, U. Steinfeld, H.H. Lee, H. Seidel, Development of a novel micro immune-magnetophoresis cell sorter, in *The 6th Annual IEEE Conference on SENSORS*, 28–31 October (Atlanta, 2007), p. 1081
141. M. Zborowski, J.J. Chalmers, Rare cell separation and analysis by magnetic sorting. Anal. Chem. **83**, 8050–8056 (2011)
142. Z. Wang, W. Wu, Z. Wang, Y. Tang, Y. Deng, L. Xu, J. Tian, Q. Shi, Ex vivo expansion of circulating lung tumor cells based on one-step microfluidics-based immunomagnetic isolation. Analyst (2016). https://doi.org/10.1039/c5an02554k
143. T.Y. Lee, K.A. Hyun, S.I. Kim, H.I. Jung, An integrated microfluidic chip for one-step isolation of circulating tumor cells. Sens. Actuat. B Chem. **238**, 1144–1150 (2017)
144. R.M. Mohamadi, J.D. Besant, A. Mepham, B. Green, L. Mahmoudian, T. Gibbs, I. Ivanov, A. Malvea, J. Stojcic, A.L. Allan, L.E. Lowes, E.H. Sargent, R.K. Nam, S.O. Kelley, Nanoparticle-mediated binning and profiling of heterogeneous circulating tumor cell subpopulations. Angew. Chem. Int. Ed. **53**, 1–6 (2014)
145. O. Osman, L.F. Zanini, M. Frénéa-Robin, F. Dumas-Bouchiat, N.M. Dempsey, G. Reyne, F. Buret, N. Haddour, Monitoring the endocytosis of magnetic nanoparticles by cells using permanent micro-flux sources. Biomed. Microdevices (2012). https://doi.org/10.1007/s10544-012-9673-4
146. H. Esmaeilsabzali, T.V. Beischlag, M.E. Cox, N. Dechev, A.M. Parameswaran, A.J. Park, An integrated microfluidic chip for immunomagnetic detection and isolation of rare prostate cancer cells from blood. Biomed. Microdevices **18**, 22 (2016)
147. T. Dong, Q. Su, Z. Yang, Y. Zhang, E.B. Egeland, D. D. GY, P. Calabrese, M.J. Kapiris, F. Karlsen, N.T. Minh, K. Wang, H. Jakobsen, A smart fully integrated micromachined separator with soft magnetic micro-pillar arrays for cell isolation. J. Micromech. Microeng. **20**, 115021 (2010)
148. A.-E. Saliba, L. Saias, E. Psychari, N. Minc, D. Simon, F.-C. Bidard, C. Mathiot, J.-Y. Pierga, V. Fraisier, J. Salamero, V. Saada, F. Farace, P. Vielh, L. Malaquin, J.-L. Viovy, Microfluidic sorting and multimodal typing of cancer cells in self-assembled magnetic arrays. P. Natl. Acad. Sci. USA **107**(33), 14524–14529 (2010)
149. J. Jung, S.-K. Seo, Y.-D. Joo, K.-H. Han, Label-free continuous lateral magneto-dielectrophoretic microseparators for highly efficient enrichment of circulating nucleated cells from peripheral blood. Sens. Actuat. B Chem. **157**, 314–320 (2011)

Chapter 8
Conclusion

Alain Bosseboeuf and Kukjin Chun

A lot of works on micro/nano biosystems has been already performed and they will deeply modify medecine practise and pharmaceutical research by providing low cost and fast biomedical analyses that could be done by non experts. To reach this goal, limitations of bioMEMS must not be underestimated and remaining challenges to overcome the commercialisation bottleneck must be well identified and overcome [1]. Indeed, beside legal and ethic issues, many labs-on-chip and μTAS proposed by research laboratories are still far from a possible introduction into the market. Despite the great opportunities for bioMEMS in healthcare and medical applications, it will take some time to reach its full potential. The same situation occurred in the past for MEMS and similar reasons can be found The challenges facing bioMEMS in the future are:

(i) a not suitable choice of materials and technology that does not warrant reliability or a real low cost of fabrication.

  – For example, PDMS (PolyMethylDiSiloxane) is a widely used polymer in research laboratories because it allows a rapid and easy prototyping but it is water permeable, allows adsorption of hydrophobic molecules, it has short term stability after surface treatment, it undergoes swelling in organic solvents, it requires very specific processes for electrodes integration and it is not suitable for fast processing as required for high volume production, So its uses in real applications is highly questionable. Similar or other

A. Bosseboeuf (✉)
Center for Nanoscience and Nanotechnology, CNRS, University Paris Sud-University Paris-Saclay, 10 Boulevard Thomas Gobert, 91120 Palaiseau, France
e-mail: alain.bosseboeuf@c2n.upsaclay.fr

K. Chun
Department of Electrical and Computer Engineering, Seoul National University, 1 Gwanak-ro, Gwanak-gu, Seoul 08826, South Korea
e-mail: kchun@snu.ac.kr

G. Barbillon et al. (eds.), *Engineering of Micro/Nano Biosystems*,
Microtechnology and MEMS, https://doi.org/10.1007/978-981-13-6549-2_8

issues may exist for other polymers or inorganic material, so their choice must be carefully done. It is also useful to consider other substrate materials than polymers, glass or silicon, like paper [2, 3], or a printed circuit board (PCB) [4].

- Many BioMEMS requires clean rooms and sophisticated equipements for their fabrication. So even if they use cheap materials they will be expensive to fabricate except if they have a low footprint and need to be fabricated in large quantities. This can nevertheless be acceptable if they offer a large added value by integrating a large set of functions and by minimizing the need of external equipements.

(ii) Tests or operation too far from real situations.

- Often, bioMEMs are tested with simple, well controlled or pre-processed samples containing the biological species to be detected. However, in real situations, it migh be necessary to handle and analyze more complex fluids. For example, for human diagnostics, this can be body fluids like blood, saliva, plasma, serum, urine,…So pretreatment or fractionation should be included.
- BioMEMs typically need external equipments for their operation for fluid injection, flow control, detection, power supply, etc. Ideally this should be also miniaturized, notably for portable point-of care systems.
- Miniaturization and low sample volumes often lead to a low signal/noise ratio and thus to the need of expensive, bulky and sophisticated electronic equipments or optical equipments for measurements. Improved techniques for sample concentration and hybridization signal amplification must then be developed. Likewise; integrated sensors, electronics and signal processing should be favoured when possible. CMOS lab-on-chip [5] or interfacing with a smartphone [6] are possible solutions. This was successfully demonstrated in the later case for cholesterol testing, label free bio detection, Elisa assays. and red and white blood cells counting.
- Packaging and resistance to environmental disturbances (temperature, humidity, chocs,…) is hardly considered although it might be a major issue to get robust micro/nano biosystems working in non-air conditioned rooms and stable at room temperature.
- Ease of use and ergonomy must be improved and quantitative and automated measurements must be realized.

Beside looking for solutions to overcome the commercialization barrier of bioMEMS, many investigations remain to be done. It is beyond the scope of this book to provide an exhaustive list of all current challenges in the broad field of micro/nano biosystems but some of them were yet identified in the different chapters. Concerning desease diagnosis by bioMEMS, many studies are focused on cancer, diabetes and aging diseases but many other diseases could also potentially benefit from an earlier and faster detection with micro/nanobiosystems to reduce their mortality such as malaria, HIV and Tuberculosis but also swine flu, Zika virus or Ebola virus which are a major concern nowadays.

Facing silver age, healthcare is more important than ever in in vitro diagnostics, pharmaceutical research, and drug delivery along with implantable medical devices. All of these applications are promising and require new innovative technologies such as integration with microelectronics and bio-compatible packaging with biodegradability. Neuroprosthetic devices are developed to replace lost sensory functions, e.g., hearing and sight and to control paralyzed and prosthetic limbs. The microelectrode array is one of the fundamental elements for interfacing with the neural tissue. For most neuroscience applications to investigate the neuronal circuits within the central nervous system, the array should have a significant number of stimulation and/or recording sites arranged in 3-D and an implantable signal processing electronics to process signals as well as signal transmitting and receiving while minimizing number of output leads by multiplexing. BioMEMS provides significant opportunities for improving drug delivery by miniaturazing passive and active components to reduce both the volume and power required and make them more portable or implantable with very small quantities at the point-of-care.

One of the emerging fields is the application to precision agriculture. In precision agriculture, monitoring and controlling the physiological conditions of plants is of critical importance for productivity and quality control. Technologies have been limited to indirect measurement of variables such as temperature, humidity, and solar intensity to figure out how the plants react to environment. However, the first chance is being unfolded to introduce bioMEMS for the direct sensing of in vivo plant states. Interrogating the in vivo plant state is not too different from taking the measurement in miniaturized environment and animal organs. Any small-scale technology developed for medical examinations can be readily modified and applied to monitoring internal reactions of plants such as sap flow rate, electrical conductivity, pH, and ionic composition. Such direct information will be useful for irrigation control and resistance against disease and environmental condition change. Precision agriculture is the area that requires high efficiency and minimum impact on environment.

We will conclude this book by giving a non exhaustive list of general guidelines useful for the design of state-of-the art micro-nano biosystems:

- As already mentionned, to improve the performances and added value of bioMEMS, the maximum amount of fluidic, biochemical operations and detections should be integrated on the same chip including pre-treatments steps like lysis, purification, pre-concentration, filtering, etc..... Whenever possible, hybrid integration of electronics or optics should also be considered.
- Highly selective bio markers or label free detection techniques are essential for cell-based diagnosis and sorting
- Biocompatibility and biofouling must be carefully considered to avoid channel clogging or sample contamination.
- Cell orientation and elongation might depend on fluidic microchannel geometry so this aspect must not be forgotten.

- Whenever possible, automatic fluidic flow like capillary flow, or measurements on stagnant samples, must be chosen respectively to eliminate on-chip or off-chip pumps and valves or the need of a microfluidic circuit.

References

1. M.I. Mohammed, S. Haswell, I. Gibson, Lab-on-a-chip or chip-in-a-lab: challenges of commercialization lostin translation. Procedia Technol. **20**, 54–59 (2015)
2. P. Lisowski, P.K. Zazycki, Microfluidic paper-based analytical devices (μPADS) and micro total analysis systems (μTAS): development, applications, and future trends. Chromatographia **76**(19–20), 101–1214 (2013)
3. Y. Xia, J. Si, Z. Li, Fabrication techniques for microfluidic paper-based analytical devices and theiur application for biological testing: a review. Biosens. Bioelectro. **77**, 774–789 (2016)
4. D. Moschou, T. Tserepi, The lab on PCB approach: tacking the μTAS commercialization bottleneck. Lab Chip **17**, 1388–1405 (2017)
5. Y. Ghallab, Y. Ismail, CMOS based lab-on-a-chip: applications, challenges and future trends. IEEE Circuits Syst. **14**(2), 27–47 (2014)
6. D. Erickson, D. O'Dell, L. Jiang, V. Oncescu, A. Gumus, S. Lee, M. Mancuso, S. Metha, Smartphone technology can be transformative to the deployment of lab-on-chip diagnostics. Lab Chip **14**(7), 3159–3164 (2014)

Index

G. Barbillon et al. (eds.), *Engineering of Micro/Nano Biosystems*, Microtechnology and MEMS, https://doi.org/10.1007/978-981-13-6549-2